# The Origin of Snakes
## Morphology and the Fossil Record

# The Origin of Snakes
## Morphology and the Fossil Record

Michael Wayne Caldwell
University of Alberta
Canada

**CRC Press**
Taylor & Francis Group
Boca Raton   London   New York

CRC Press is an imprint of the
Taylor & Francis Group, an **informa** business

CRC Press
Taylor & Francis Group
6000 Broken Sound Parkway NW, Suite 300
Boca Raton, FL 33487-2742

First issued in paperback 2021

© 2020 by Taylor & Francis Group, LLC
CRC Press is an imprint of Taylor & Francis Group, an Informa business

No claim to original U.S. Government works

ISBN-13: 978-1-4822-5134-0 (hbk)
ISBN-13: 978-1-03-217769-4 (pbk)
DOI: 10.1201/9781315118819

**Library of Congress Cataloging-in-Publication Data**

Names: Caldwell, Michael Wayne, author.
Title: The Origin of Snakes : morphology and the fossil record / Michael Wayne Caldwell.
Description: Boca Raton : Taylor & Francis, 2019. | Includes bibliographical references.
Identifiers: LCCN 2018061631| ISBN 9781482251340 (hardback : alk. paper) | ISBN 9781315118819 (ebook)
Subjects: LCSH: Pygopodidae. | Snakes--Morphology. | Snakes, Fossil.
Classification: LCC QL666.L27 C35 2019 | DDC 597.95/2--dc23
LC record available at https://lccn.loc.gov/2018061631

Visit the Taylor & Francis Web site at
http://www.taylorandfrancis.com

and the CRC Press Web site at
http://www.crcpress.com

To my wife, Amber Nicholson, to my mother, Geraldine Caldwell, and to the five greatest things I have ever done in my life—Garrett, Landon, Kylie, Devyn, and Liam—I dedicate this book. In memory, I also dedicate this book to my father, Delayne Caldwell, whose passion for literally everything lives on in me and in my five children and two grandchildren, Kayah and Tavish.

As the writing of this book was coming to an end, so too did the life of my friend and confidant, the world-renowned scholar of paleoherpetology, and in particular, "Serpentes," Professor Jean-Claude Rage (March 30, 2018). To his memory I dedicate the ideas expressed in this book, many of which he wrote about, wondered about, and expounded upon. In remembering the legacy that is his, I borrow from the walls of the Catacombs of his beloved Paris:

Chaque mortel paraît, disparaît sans retour; Mais, par d'illustres faits vivre dans la mémoire. Voilà la récompense et le droit de la gloire.

DELILLE, LIB. X

# CONTENTS

# INSTITUTIONAL ABBREVIATIONS

- AM–Albany Museum, Grahamstown, South Africa
- AMNH–American Museum of Natural History, New York, USA
- BMS–Murgermeister Museum Solnhofen, Solnhofen, Germany
- CMNH–Carnegie Museum of Natural History, Pittsburgh, Pennsylvania, USA
- DIP–Dexu Institute of Paleontology, Beijing
- FMNH–Field Museum of Natural History, Chicago, Illinois, USA
- HUJ-PAL–Hebrew University of Jerusalem, Paleontology Collections, Jerusalem, Israel
- GSI/GC–Geological Survery of India, Geological Collections, Jaipur, India
- IRScNB–Royal Belgian Institute of Natural Sciences, Brussels, Belgium
- MACN–Museo Argentino de Ciencias Naturales "Bernardino Rivadavia," Buenos Aires, Argentina
- MA RND–Musee Angouleme, Angouleme, France
- MCZ–Museum of Comparative Zoology, Cambridge, Massachusetts, USA
- MLP–Museo de La Plata, La Plata, Argentina
- MPCA–Museo Paleontologico "Carlos Ameghino," Cipoletti, Argentina
- MSNM–Museo di Storia Naturale di Milano, Milano, Italy
- NHM-BiH–Natural History Museum, Sarajevo, Bosnia-Hercegovina
- NHMUK–Natural History Museum, London, England
- NMW–Naturhistoriches Museum für Naturkunde, Vienna, Austria
- QMF–Queensland Museum, Brisbane, Australia
- Rh-E.F.–Natural History Museum of Gannat, Gannat, France
- SAMA–South Australian Museum, Adelaide, Australia
- SMF–Senckenberg Museum, Frankfurt, Germany
- SMNH–Slovenian Museum of Natural History, Ljubljana, Slovenia
- TMP–Royal Tyrrell Museum of Palaeontology, Drumheller, Canada
- UAMZ–University of Alberta Zoology Museum, Edmonton, Canada
- UF–University of Florida, Gainesville, Florida, USA
- USNM–National Museum of Natural History, Washington, DC, USA
- YPM–Yale Peabody Museum, New Haven, Connecticut, USA
- ZFMK–Zoologisches Forschungsmuseum Alexander Koenig, Bonn, Germany

# PREFACE

As a small boy, I was deeply fascinated by the natural world. I loved birds, insects, seashells, beaches, forests, leopard frogs, and ponds, glaciers, and mountains, but most of all, I loved hunting for fossils. In first grade, I discovered my school's gravel parking lot was a treasure trove of stony, petrified wood. My fossil-loving friend Chris Milner and I spent every recess kicking through the gravel looking for these leftover fragments of ancient life—always hoping to find a fossil dinosaur bone.

As a kid, I dreamed of nothing but being a paleontologist—scout's honor—I dreamed about it all the time, asleep and awake. For the adult me, the great surprise remains that after a few experiments with other degree programs and career pursuits, I actually ended up becoming not just a scientist, but a paleontologist. The second great surprise is that my career made an unexpected left turn, and I ended up devoting my time and energy to the study of fossil and living lizards, including, of course, snakes.

The "left turn" story begins with my doctoral dissertation project on limb evolution in extinct, aquatic reptiles. At the time, and for obvious reasons, such a project did not include snakes, but it did include several kinds of extinct marine lizards. After a graduate student tour of European museums in the spring of 1994, I had created a list of future projects, several of which included the small-bodied, elongate dolichosaurs, a group that had been of burning interest to about five people since Richard Owen first wrote about them in the late 1840s—I was very pleased to be number six.

At the same time, I also received a few tips from my soon-to-be postdoctoral supervisor, Olivier Rieppel, one of Bob Carroll's former students, that there were some "dolichosaurs" I should see in collections at Hebrew University in Jerusalem. The final player in my "left turn" story was added in the summer of 1995, at the annual meeting of the American Society of Ichthyologists and Herpetologists, where I met Michael S. Y. Lee.

Mike and I were at similar stages in our academic careers—just starting or just about to start postdoctoral positions. We also found that we had converged on similar hypotheses of squamate phylogeny. In my nascent data set, obscure aquatic lizards, that is, dolichosaurs, were "coming out" as the sistergroup to snakes in a clade including mosasaurs. Despite my lack of scholarly achievement as a snake guy, I prepared a manuscript announcing this momentous discovery, Mike prepared his as a separate paper, and we decided to submit our contributions together, as a diptych of sorts, to the journal *Nature*. Our papers were sent out for review (an achievement) and rejected (no surprise).

That rejection is important to this story because it inspired us to do more, but this time with actual data. In the spring of 1996, Mike and I went on a 2-month-long tour of Europe and the Middle East (using research funds from his Australian postdoctoral research grants; my Canadian postdoctoral fellowship had no research grant associated with it, so he paid the bill!) in order to examine the long-forgotten collections of small, elongate fossil marine lizards that were sprinkled around Europe and the Middle East. We met in Jerusalem, at Hebrew University, where Olivier Rieppel had assisted us with arrangements, to see two "dolichosaur" fossils, *Pachyrhachis problematicus* Haas, 1979, and *Ophiomorphus colberti* Haas, 1980.

I still remember that first day in collections, because it was right then and there that serendipity veered "left" and took over my career—Mike and I opened the drawer containing the specimens of the "dolichosaur" *Pachyrhachis*, to see, in perfect articulation, a fossil snake with back legs. In that spring of 1996, as I studied these fossils with Mike, I knew virtually nothing about living snakes, fossil snakes, snake origins, and the debate on snake phylogeny. Following what I have since dubbed the "Snake with Legs World Tour," Mike and I published the outcomes of our *Pachyrhachis* studies in *Nature* (Caldwell and Lee, 1997). On the Thursday afternoon in 1997 that *Nature* released the paper, I went to the grad student bar at the University of Alberta, the Power Plant, with my artist pal Tom Saunders, where bartender "Nemo" invented a tasty little shooter we called the "Snake with Legs" (see figure below); Tom painted and designed the T-shirt, and Mike Lee and I got deeper and deeper into the science of snakes in ways I could never have dreamed possible.

# SNAKE WITH LEGS

## World Tour 1996-97

| *The Blue Hole* (Jerusalem) | *Gästhaus Schnecken* (Frankfurt) | *many pubs* (London) | *The Power Plant* (Edmonton) |

"The Shooter"
*Stratigraphy of a Fine Drink!*

"The Fossil"

*"The Snake: Caught Between the Land and the Sea"*
(*Caldwell, Lee, and Saunders*)

To say the least, the years between 1996 and the present, 23 years in total, have been exhilarating, exasperating, inspirational, and infuriating. Still, despite being hammered by the painful hurricanes of dispute and debate, it has ultimately been deeply satisfying to be a part of this scientific arms race of ideas on everything snake. I have discovered some important things about myself as a person and as a scholar, and from the latter, I have learned much about the method, philosophy, and sociology of science. I have learned that each answer only rattles the question harder, and that not everyone always likes either your answers or your new questions, or even you through this process—but it does not matter. The resiliency of science is that it is ultimately immune to the passions and tribulations of its players, and this is good, because science is not an entity that can be owned; rather it is a vast and chaotic collection of tested and untested ideas, data and pseudodata, dogma and hypotheses, all of which are regularly under revisionist assault by a dizzying array of bullshit mixed with brilliance.

While I remain that small boy collecting petrified wood and dreaming about fossils, my adult dreamtime is now positively phantasmagorical and is populated by serpents great and small, with legs, with no legs, swimming in oceans, burrowing madly beneath the soil, dangling from the branches of primordial forests, and sometimes, yes, even flying! I have written this book to review what I think I know right now, and, make no mistake, without apology, this book is about how I see science, philosophy, method, snake lizards, paleontology, evolutionary biology, and scholarship. Again, I do not apologize for being critical, revisionist, and selective in what I have to say and what I have written about. But I also hope that what I have written is prescient with respect to whatever fantastical snake lizard beasts yet lie entombed in the rocky record that remains of this planet's fascinating and very ancient history, and that my thoughts on philosophy and method in paleontology and science have value in some way or another.

# ACKNOWLEDGMENTS

This book would never have become a reality were it not for the long-suffering persistence of Chuck Crumly. He asked me to write a book on "snakes" at the Society of Vertebrate Paleontology meeting in Los Angeles, late on the afternoon of October 30, 2013. When I signed the contract in the late winter of 2014, I thought a year would be long enough, but I could not see into the future, and so could not know that 10 months later, on October 4, 2014, my house would burn down. The stress of this disaster significantly delayed the timely and contractual writing of this book. For 2 years, the manuscript of this book languished in its digital world, untouched, before I could finally find the strength and creativity to really begin writing it again—I hope Chuck will one day forgive me.

Because this is a book about snake lizards, two people, formative in the timing and impact of their influence, require special mention. First, is Olivier Rieppel, mentor and postdoctoral supervisor— from tales of Paul Feyerabend and discourse on the philosophy of science and the meaning of pretty much everything (I still drink G & Ts and listen to Sheryl Crow), to a long and published series of fiery and polarized debates on squamate anatomy and systematics—his impact has been, and remains, profound. Second, is Michael S. Y. Lee—we were young together as we embarked on the "Snake with Legs World Tour," and when the road got rocky, our friendship and scholarly commitment to seeing our work completed turned into a lifelong collaboration that is now more than two decades old. I am proud of what we saw, thought, and wrote together, and what I learned from Mike and how it has influenced my own independent work as a scholar.

In 2005, my old friend Randall Nydam and I met up in Rio di Janeiro and began a collaboration that is now 14 years on and going strong. He is to be thanked beyond measure for his invaluable friendship, calming edits, skill as a comparative anatomist, and his any hour of the day call service as a sounding board for my impassioned rantings. I must also thank Allan Lindoe, the University of Alberta's long-standing paleontological technician (since 1966 or so)—best of the best of friends for more than 32 years, who has worked fossil beds with me around the planet, and has found and elegantly prepared numerous specimens of the ancient snake lizards Dinilysia and Najash. This book and its contents also suffer the gloried impact of Denis Lamoureux and his never-ending passion for philosophy, religion, science, and the evolution and creation debate. The same is true of Murray Gingras, King of the Haggis, mean in a kilt, and the best ichnologist and field geologist I know, and careful reader of Chapter 4. I would also be remiss were I to forget to tip a glass to the intellectual liberties created, given, and shared during my graduate student days in the Sid Vicious Sewing Sircle (SVSS) with Les Anthony, Tim Sharbel, Hinrich Kaiser, Hans Larsson, Rob Holmes, and Cliff Zeyl. I also want to thank Greg

Edgecombe and Zerina Johanson for three decades of friendship, inspiration, wine, and discourse, on all things science and evolution, all of which have deeply influenced what I have written in this book.

In these last 20 years, one thing has become clear above all else—I would be nothing without my brilliant graduate students and postdoctoral fellows. It is from working with them and their many talents and very bright minds that I am able to appear much smarter than I am, so thank you to—Braden Barr, Lisa Budney, Timon Bullard, Michelle Campbell, Jasmine Croghan, Cajus Diedrich, Alex Dutchak, Fernando Garberoglio, Paulina Jimenez-Huidobro, Takuya Konishi, Aaron LeBlanc, Laszlo Makadi, Erin Maxwell, Sydney Mohr, Alessandro Palci, Ilaria Paparella, Stephanie Pierce, Tiago Simões, Hallie Street, and Catie Strong. I must also highlight Alessandro Palci for his efforts and skills as a photographer and illustrator in creating a large number of the plates and figures in this book, not to mention his herculean effort in reading, correcting, and editing the entire volume between copyedits and galley proofs. And I also must thank Oksana Vernygora, Tiago Simoes, and Aaron LeBlanc for also reading major sections of the text and providing extremely helpful comments.

I also want to thank my friends, colleagues, and scholarly adversaries, several of whom I have never met, but all of whom have been invaluable contributors to my lifelong pursuit of knowledge on evolutionary biology, paleontology and of course, all things snake lizard and their kin—Sebastián Apesteguía, Jean-Claude Rage, Hans Larsson, Hans Dieter-Sues, Cristiano Dal Sasso, Jose Bonaparte, Jorge Calvo, Garth Underwood, Robert Carroll, Brian Chatterton, Mark Wilson, John Scanlon, Eitan Tchernov, George Haas, Sandra Chapman, Jakov Radovcic, Guillermo Rougier, Mark Hutchinson, Krister Smith, Agustin Scanferla, Laurie Vitt, Eric Pianka, Gavin de Beer, Adriana Albino, Robert Reisz, Robert Holmes, Alexandra Houssaye, Jason Head, Brent Breithaupt, Maureen Kearney, Jacques Gauthier, Johannes Müller, Hussam Zaher, Harry Greene, Susan Evans, Jack Conrad, Nick Longrich, and David Cundall.

# INTRODUCTION: SEEING AND KNOWING

Snake lizards are fascinating animals, with, in my opinion, an even more fascinating ancient past that can only be understood from their fossils, far and few between as they are. Without apology, this book is all about fossils and what we know, in particular, about ancient and modern snake lizards and their evolution from the evidence they left behind through more than 170 million years of earth history. A thesis thread that I will follow throughout this treatise is that no matter how information rich the modern fauna is, it tells no story of the kind to be found among the record of anatomy, morphology, diversity, disparity, paleoecology, and palebiogeography that we are now coming to understand from the expanding fossil record of snake lizards and their ancient kin. Through no fault of its own, the modern data set is simply mute on the real complexity of the ancient past and just how that dizzying array of biodiversity evolved to become the present world. A second thesis thread I will follow throughout this book is that snakes are lizards, and so I use the term "snake lizard" and will explain why in more detail in Chapter 1. And an important point that I must raise here in the beginning of this book is that it is "mine"—I am writing without apology about how I see science, philosophy, method, and, in particular, the history of snake lizard science as it has played out in the last few decades. I have taken the stance that scientific criticism is critical to what makes science different from other forms of inquiry and have written this book with that

position in mind. My criticisms are not personal, though they are criticisms of the scholarly contributions of people—I know how this feels and I know what it means when it is both received and given. Again, I make no apologies for what I write and fully expect to receive responses to my ideas and words shared under the same philosophy and approach to scholarly criticism.

## SOME THOUGHTS ON BEGINNINGS

Beginning some 390 million years ago, early in the evolution of vertebrate animals, a genetic groundplan arose that constrained the number of pectoral and pelvic fins, what would become limbs in tetrapods, to four. By comparison, body segments, or somites, in front of, between, and behind the limbs, were left to vary in number and size, and have continually done so, producing a great variety of forms throughout vertebrate evolution. These early "tetrapods" took ashore their fish-bodyplan of a bilaterally symmetrical set of front and rear limbs, each with its own girdle architecture, and never looked back, for a short while at least, until they reinvaded aquatic environments just about as quickly as they had left them. After all that evolutionary effort and innovation to build limbs and girdles and conquer new non-marine environments both watery and dry, numerous groups of tetrapodal "fishes," both terrestrial and secondarily aquatic, set about evolving toward body elongation and throwing away the limbs they had worked so hard to evolve in the first place; with seemingly little or no control on the number of body segments, limbless tetrapods have continually evolved elongated bodies as they also did away with their limbs. Snake lizards are but one of those groups, and they are the subject of this book. Like all other tetrapods, even though their fossil record has been poor for a very long time, we know what we know about the patterns and processes of their ancient origins and evolution only from fossils, and from nothing else.

Among modern groups of tetrapods, limblessness and limb reduction, coupled with body elongation, is extremely common (lizards and amphibians). For lizards, it appears that it is a morphological solution to adaptive experimentation that is common to almost every lineage except iguanians. For snakes among lizards, this major transformation from a four-limbed lizard bodyplan to a limb-reduced and eventually completely limbless lizard body plan is usually treated as a defining characteristic of the clade, yet it really is not. Snake lizards were not the first group of tetrapods to reduce or completely lose their limbs. From the fossil record, we know that that honor goes to several groups of ancient amphibian lineages known as aistopods, lysorophids, and their kin (Carroll, 1988), all of whom lived during the latter part of the Paleozoic Era. Even among living lizards today, limblessness is not the exclusive domain of snake lizards, as numerous anguids, skinks, gekkotans (pygopodids), teiids, dibamids, and amphisbaenians are elongate and show every imaginable combination of limb and girdle reduction and loss.

So if being elongate and limbless is not unique to snakes, no matter how well they have exploited that morphology, then what innovation, or suite of innovations, is/are responsible for their incredible success? The obvious answer is that there are many reasons for their success, but from a morphological point of view, almost all of them are to be found in their uniquely configured and elaborated heads. With some 3700 living species of snakes to examine, it is abundantly clear that there are numerous specializations that define the snake lizard skull. Cundall and Irish (2008) detailed the remarkable variety of snake lizard skulls in their treatise on the subject, making it clear that there are mobile and immobile components in a snake lizard skull and that the success of the various lineages revolves around their unique and novel solutions on how to modify the form and function

of those various skull units. Like all other groups of squamates, snake lizards are diagnosed in general, and in the particular, by the construction of their skulls—being long, skinny, and legless really has very little to do with it—and the same is true of other legless lizards and their limbed cousins (Evans, 2008).

As an example, if you comb through the osteology collections at any major museum and line up the skulls of three or four species of the Australian skink Lerista, you simply will not know which one of them, or all of them for that fact, are from species belonging to the completely legless and elongate forms and which are from the four-legged forms. You will know they are skinks, you might even be good enough with scincid cranial anatomy to recognize them as Lerista, but you won't know whether they had legs, and you surely won't confuse them with snake lizards because they do not have the skulls of any kind of snake lizard. The same holds true for the anguid and pygopodid lizards, and so on. They all retain traits that diagnose them to the major clades to which they belong despite the fact they are limbless or limb reduced. While only by analogy, because of course we do not have any four-limbed snake lizards alive today, the same connection must hold between limbless snake lizards and their ancient four-legged ancestors—what makes a snake a snake lizard is defined by skull anatomy the same way it is for a skink or an anguid lizard. Not only is this idea not rocket science, it is barely even "science," as it follows from nothing more than simple logic. So, there we have it, the head is key, not the body, to sorting out the origin and radiation of snake lizards from their four-legged ancestors. In a nutshell, that is another major thesis of this book—the head is the thing that matters most; it is the key puzzle piece to understanding the origins of snake lizards, not the body; and the fossil record is the only place we can go to find out anything about how this happened. I will argue throughout that limb reduction and loss, and postcranial evolution toward extreme elongation, is what happened to snake lizards after they became "snakes," not before. Twenty years of trying to find new answers to long-standing questions on what makes a snake lizard a snake lizard have resulted in this seemingly heretical, yet for me, utterly logical, thesis. In order to support that thesis, I have filled the pages of this book with my synthesis of the literature as I read it, blended with what I have come to understand about the actual data present in fossils, bones, rocks, molecules, soft tissues, and living snake and non-snake lizards, all served up alongside what is probably an unhealthy dose of my unpublished musings.

To do this correctly, that is, avoid complete chaos, requires no small amount of organization and an explanation of how I approach science in the way that I do. It is therefore imperative that I outline my philosophy of science as a student of snake lizard anatomy and evolution, and to follow in this introduction with a brief outline and explanation of the purpose of the various chapters.

## PHILOSOPHICAL APPROACH

Before I begin, let me apologize to all real philosophers who might read the next few paragraphs, and the rest of the book for that matter. Why? Because I am not a professional philosopher, and I have never studied as a philosopher such heady subjects as epistemology and ontology. However, I have read broadly on these subjects and thought long and hard about my philosophy as deeply as I think necessary, so that I am at least aware of the issues and their importance to my science and my scholarship. As a result, I can fairly state that I have epistemic and ontic perspectives that limit and circumscribe how I think about what I claim to "know" and what I think is "real." I try to adhere to the epistemic statement that all knowledge is relative, that all claims to knowledge, that is, hypotheses, regardless of how well they order the seeming chaos that are the data, are just "mostly good," and not "absolutely good," "right," or "true." I would not say, by comparison, that I am as much of an ontic relativist; I hold that "things" are real; that is, I accept as a first principle that objects have some measure of realness to them, with real connections to the past and a real opportunity to influence the future. Therefore, a sensible ontic statement for me is that an organism is real, and that I am an individual of a kind of organism. I also consider that objects have meaning and qualities only when a describer creates that meaning and or quality by presenting names and limits to observations concerning those objects. I also recognize the tensions between my epistemic relativism as a scholar (absolute truth is unattainable), my less relative ontic views (objects are real and have a past state even if I have not observed it, a current state I can "know," and a future state I can predict), and my psyche's passion for satisfyingly absolute answers to questions.

Therefore, in this relative world of my scholarly claims, as bounded by my empirical observations, I do hold that it is possible to "know" a great deal about, for example, snake lizard evolution. Though I consider it impossible to absolutely discover ancestors and truths, I will confidently hypothesize closest relatives (sistergroups) for species and clades of snake and non-snake lizards. I cannot pinpoint the time and place of the origins event that produced the first snake lizard, but I can hypothesize, based on my best relative knowledge, the parameters of anatomy and ancient ecology that provide insight and understanding of both that event and the anatomy of the earliest snake lizards. To fold these parameters into a well-articulated concept, I still use the words "ancestors" and "ancestral," so as to reference a concept of origins for a group of things, but I am not referencing a search or a hope of discovery of an actual ancestor represented by a fossil of an individual, species, or a kind—ancestors as I use them are purely conceptual because we cannot find them in any sense of my version of what is real. As a concept, ancestors are to be "found" in that fuzzy cloud of points that represent millions of individuals lumped together in populations and stacked on top of each other in time and space that exist somewhere below a clade and its closest sistergroup, where nothing is known—this is what I am referring to, as a concept, when I use the word "ancestor" or "ancestral." My ancestor in this sense can be constructed, like a collage of anatomies and habits and habitats, to predict the unknown form from which the clade we know something about was derived. I do not expect to find "real" ancestors because they do not exist (which sounds absolute and is not), but I do predict that in conceptualizing such a "kind," that one day a fossil, say a snake lizard with four legs, which would be an "ancestor" to snake lizards with two legs, or no legs at all, might well be found. If I were to summarize, I work to discover sistergroups in cladograms, but muse about ancestors and their conditions of form as probabilities I can reason down to third- or fourth-order hypotheses arising from the metadata upon which the cladogram was constructed.

As should be clear from the last paragraph or two, where I have used the language of cladistics, that I consider myself to be a cladist. I utilize the methods and language of cladistics in assessing and expressing the results of phylogenetic analyses of squamate interrelationships as inferred from cladograms built by clusters of synapomorphies. Combining my adherence to cladistics with my empirical approach to testing and falsification, I consider the testable statements in a phylogenetic analysis to reside exclusively within the empirical observations and experimenter decisions leading to character constructions, state descriptions, and the assignment of primary homologies for a taxon or group of taxa (see Simões et al., 2017). I do not consider either cladograms or phylogenies to be directly testable statements. Parsimony is not a test; it is a philosophical statement that justifies reliance on the "shortest trees" resulting from the Test of Congruence (*sensu* Patterson, 1982). Cladograms resulting from the Test of Congruence are not falsifiable independent of falsification of the character constructions used to construct them. Cladograms, and subsequently phylogenies, are constructs arising from character congruence, not the empiricism of individual character testing open to potential falsification via the Tests of Similarity (*sensu* Patterson, 1982). While it is true that if a character construction is falsified by superior empirical observation, and primary homology assignments modified in the character matrix, the resultant cladograms might change as the Test of Congruence finds new character/taxon arrangements, the falsification test still lies in the individual characters, not in the tree topology derived via the Test of Congruence. Therefore, my focus in this book, as it has been in my published research, is to continue to clarify the critical centrality of the empirically observable features of ancient and modern lizards of all kinds and, more specifically, snake lizards.

And, finally, I will include in this introduction to my way of seeing my science a brief review of my perspectives on evolution. For me, evolution as a process is simple—the summation of all processes that result in "change" at any biological level (e.g., molecular to morphological) in a self-organizing, self-replicating system over time. In cliché form, evolution is simply "change over time," nothing more—the specific tempo and mode of any one evolutionary event, though, are exceedingly complex and for the most part unknowable, but evolution is not. As a paleontologist, I am keenly aware of the centrality of the fourth dimension,

time, to my understanding of the pattern and process of evolution. It is no mere keeper of the Gate, but rather it is the "god" of this process we call evolution. It is within these unfathomable and vast tracts of time that biological variation accumulates, is sorted, selected for, selected against, ignored, promoted, and, in most cases, ultimately discarded—after all, 99.9999% of all things that have lived are now extinct, which means the vibrant energy that is the living world preserves but a fragment of the information and energy that has "been" these last 4 billion years and more. It is in that vastness of time that we find the hand that directs the end result of trillions upon trillions of microevolutionary flickerings around basepair substitutions, deletions, lateral gene transfers, and so on. Managed within vast tracts of time are the summations of small and large changes that contribute to the appearance of macroevolutionary transformations of phenotypes, that 100 million years later, or even 1 million years later, have seemingly produced, as if by magic, the substantive changes we see when we compare ancient fossil snakes to modern snake lizards or attempt to explain the great differences and divergence points between various groups of modern snake lizards. However, for this book, this is as far as I will go with process. As a primary research program in paleontology, process is of no real scholarly interest to me, as it is truly, in my view, an epistemic black hole—it is unknowable. I most certainly have, do, and will consider evolutionary processes in their broadest forms, but I only do so as speculations on higher-order hypotheses of adaptations, paleoecologies, and so on, that contribute explanatory power to phylogenies as evolutionary patterns of relationship. I remain content, epistemically and ontically, with deriving cladogenic patterns from the data of morphology mostly, and molecules increasingly, in my search for falsifiable character/primary homolog statements.

And, finally, stepping away from my philosophy of systematics to a broader philosophy of science, I recognize that my epistemic and ontic philosophies have directly affected my respect for existing paradigms describing snake evolution and, in some cases, actual hypotheses grounded in testable observations. Science survives because it was designed to revise itself by finding errors and mistakes of all kinds and fixing them—therefore, for me, there are no sacred cows in the science of snake lizard evolution and origins, no inerrant truths of morphology that dictate that the jugal is absent in snake lizards (see Palci and Caldwell, 2013), nor that scolecophidians are the most primitive snake lizards (see Caldwell, 2007a). These positivist statements using "is" and "are" are anathema to me, as are words such as "resolved" (Streicher and Wiens, 2017) and "solved"—the mere existence of these words in published works demands that professed absolute truth or a superior paradigm be tested and retested, rather than set aside to be revered as dogma for having forever solved the riddle. I find such epistemic claims of absolutes in science to be extremely problematic. I do not doubt that the great and needed debate only begins when ideas are questioned and then tested, and expresses one of two things through that debate—either a collision of data and interpretations, or a collision of relative versus absolute epistemic claims. This is what science is all about—the scholarly debate of data and hypotheses. It is exactly that approach to the many old and new questions about snake lizard anatomy, fossils, molecules, origins, and evolution that I take in the pages of this book.

# CHAPTER ONE

# Ancient Snakes, Modern Snakes

## "WHAT IS A SNAKE?"

---

In the absence of fossils we are ignorant of the real sequence of adaptations…
the fundamental modification shared by almost all snakes is the liberation
of the mandibular symphysis…This was the step that yielded sufficient
selective advantage to permit the extreme elongation of the body…

CARL GANS (1961:221)

I will begin this chapter with a thesis question, not a statement—Why do we refer to snake lizards as "snakes," rather than as snake lizards? We say "scincid lizard" in the literature or use the term "the lizard group Iguania," but no one ever says the "lizard group Serpentes" or ever, ever writes the proper adjective noun pair of snake lizard. I ask my question again: Why not? In this chapter, I explore this very important question. Here are the reasons as I see them, and some solutions to the problems that have been created.

In my experience, both the casual and scientific discourse on the subject of "snakes," "snakeness," or, "What is a 'snake'?" like the characterization of almost everything else in our physical world, has been organized in a very top–down manner. We naturally order everything around us by reference to ourselves first, and following that, we try to find order for all the other objects, things, and kinds around us. For example, I am me, you are like me, a snake lizard is not like us, but it is like other snake lizards, and for millennia was not seen as being overly similar to other kinds of lizards. My recognition of me, and how much you look like me, requires me to define what "like" or "likeness" means in terms of my essential characteristics of form and then to recognize the essential ones you share with me under that quality of "likeness"—those essential characters are used to define "human." Because "I" don't change periodically into a cube or a tomato, those human characteristics are deemed essential characteristics of my version of human, and likely yours, too.

When referencing an object that is not essentially human, we look for the essential characteristics of an object and define that object and objects like it around those features. Take, for example, a snake lizard—we recognize and define "snakeness," or ask the question, "What is a 'snake'?" by framing those recognition features, that is, defining characteristics, via reference to the modern snakes we see all around us. While this is perfectly sensible, it also creates limitations that can and do become problematic. For example, with respect to a fossil snake lizard, the usual approach is, "If the object possesses the essential characteristics that define all modern snakes, then the fossil is that of a snake; if it does not possess all of those essential characteristics, then it might be a stem snake (another term for ancestral snake); and if it possesses none of them, then it is a lizard." While this natural process of defining and grouping objects is sensible, it is also static because it is rationalized by essential characters linked much more deeply to thoughts of immutability, of archetypes, and of a permanence of form, and evolution is none of these things.

Seeing and thinking about the form of objects and how to categorize and group them is inherent in the organization of our brains. As far as we know, the philosophy and application of essential and immutable definitions of things was first written down by Plato (Allen, 2006) some two and a half millennia ago—that is, the ideal or essential characteristics of a kind of thing are necessary for it to be and function in the manner that it does, and that that form is essentially immutable and defines the reality of the thing (this ontologic of form has enormous epistemic implications, as I am sure you can imagine). Today, Plato's philosophy on form (Allen, 2006) is referred to as Platonic Essentialism, having arisen from his writings on Idealism. Aristotle, who followed Plato, built on essentialism with his own concept of homology, or *hómoios* in Greek, which means nothing more than "similar," in his book the *Historia Animalium*.

As an approach to what is real, I side with Plato and Aristotle—observed essential features are similar (*hómoios*) between individuals of the same species (these features are shared because of similar genetics, parents to offspring, between breeding members of the same species—Aristotle would have recognized that if a farmer breeds two black cows, they will likely produce a black calf that is similar to its parents). However, today, I also accept as an epistemic statement that goes well beyond the Greeks that the similarities between black cows, goats, and sheep are *hómoios* at the level of homology, a kind of similarity that finds much deeper common ancestry between different kinds of things beyond the mere genealogy of parents and offspring. I also differ from Plato, Aristotle, and just about everyone else leading up to the years 1858 to 1859, because I do not consider that those observed similarities reflect the presence of an ideal form with essential characters that are immutable to any natural process that might change them. I thus reject Owen's (1849) concept of the archetype (the fundamental body plan around which all organisms of kind were organized and modified, for example, the vertebrate archetype resembling the cephalochordate *Branchiostoma*). And further, I also reject Aristotle's last two stages of causality (see Rieppel, 1988a), "Causa Efficiens" and "Causa Formalis," which reflect a metaphysic of

design and a designer, respectively, in the creation of species, kinds, or ideal archetypes (i.e., there is no god at the root of my metaphysic).

Today, post-Darwin and Wallace (1858) and Darwin (1859), we work within another theory set that hopes to take us beyond the immutability of ideal or essential characteristics of form—it is called evolution. Working from this starting point, we step past the limitations of Plato (Allen, 2006) and Owen (1849) and conclude that such a philosophy satisfies our intellectual needs when considering limited subsets of a kind, such as an individual (I, for one, am pleased to immutably be me regardless of seeming mutability since conception), or even for temporally and spatially constrained populations of individuals of a kind. Platonic essentialism, and for that matter essentialism of any kind is not an intellectual monster as long as you recognize its presence (essential and immutable characteristics of an individual) and are able to move forward, backward, and around it when extrapolating beyond an individual of a kind to others of another kind (and to how they got those features), to very different kinds from the past and kinds that will likely exist and differ into the future.

For example, if the populations and species to which individuals of living things belong were indeed immutable objects, changing not at all from generation to generation, changing not at all through long periods of time, let alone over great tracts of geologic time, then they would have essential characteristics that define them, and such intellectual constructs would make sense. A modern snake would by virtue of its possession of the essence of its kind represent not only itself, but all snakes, from the first to the last. How easy would that be? The beauty of essential categories as they apply to an individual of a kind is that they help you "know" that a rattlesnake and its bite and venom will likely kill you, while the bite from a non-venomous, long, skinny, legless glass lizard will only hurt—knowing which one is which, based on their essential and recognizable properties, is terribly useful. But, if we need to consider alternatives to essentialism in order to consider time, and mutability over time in something like a snake lizard, that is, the evolution of new species and new kinds, then Plato's philosophy is wholly unsatisfactory and severely limiting in what it can predict. Were this the only limitation to essentialism, then it could easily be bypassed. The problem is that it is not so easy, as Plato's idealism and essentialism appear in fact to be fundamental constructions in the human psyche, and have long influenced scholarly thought on organisms, biology, paleontology, and evolutionary thinking, even though we have kept the best of Aristotle's *hómoios* and explained it with Darwin's (1859) mechanism of evolution.

Long before, during, and after Darwin (1859) published his treatise on the Origin of Species, his biological and paleontological colleagues were directing powerfully influential essentialist dialogues, oftentimes antievolutionary in their tone, but always essentialist in their vision of the fundamental immutability of organismal form (for reviews, see Hull [1988] and Rupke [1993]). For example, Georges Cuvier, the founder of comparative anatomy and vertebrate paleontology, considered species immutable, was a vocal antievolutionist (he had famously public arguments with Jean Baptiste Lamarck and Geoffrey St. Hilaire), and recognized four basic forms of organismal organization—vertebrates, arthropods, mollusks, and the "radiata," or such groups as jellyfish, anemones, and echinoderms (Cuvier, 1798). One of Darwin's most vocal and contemporary critics, Sir Richard Owen, also saw organisms as organized around immutable archetypes (Owen, 1849), both in terms of developmental processes (Owen was deeply influenced by von Baer's [1828] non-evolutionary approach to what are now referred to as von Baer's Four Laws of Development) and the immutability of the adult form. It would be easy to argue that Darwin's paradigm shift of evolution via natural selection should have reset the human predisposition toward essential features and immutability, but I am afraid it did not.

The problem for the modern student of evolution is twofold. First, they must move through the logical and necessary stages of observation and the subsequent listing of essential characteristics of an individual of a kind to create species diagnosis and taxonomic keys, recover autapomorphies of a terminal taxon, and so on. Second, and seemingly in contradiction to the first, they must ensure that they recognize these essential characters of an individual and of many individuals of a kind to be mutable and neither essentialist or idealized, both of which are different from essential characteristics, or taxic states (see Patterson, 1982), of individuals and kinds. With these caveats in mind, the empiricist observer must follow this empirical process with a listing of essential characters of the larger group to which a thing belongs to create higher-level diagnoses and clade definitions and to discover synapomorphies and symplesiomorphies of clades. Such a process and state of mind seem simple, and, as I argued earlier, there is no way around it—it is natural and needed. Where, then, lies the danger?

As actions and processes, observing and thinking are never either theory free nor assumption free. We all bring what we "know" to the observation of the next new thing and to the conclusions we draw about those observations. The problem can become gigantic, though, when the question being asked goes well beyond the mere empirical, that is, "What bone is that?" or, "What kind of snake is that?" to the larger questions we ask today, such as, "What is a snake?" or, "In an evolutionary sense, where did snakes come from?" How is this a problem? Well, in the few hundred years since Darwin (1859), these questions have traditionally and unwittingly been asked as questions deeply rooted in the historical burden of Platonic Idealism and Essentialism. How can that be? I am arguing here that the data set is everything, and you cannot ask a sensible non-Platonic mutability/evolution question if referencing the essential/immutable characteristics of only modern snake lizards and using those essential features (molecules or morphology) in isolation from the ancient essential features of fossil snake lizards to then attempt to construct hypotheses of the mutability of all snake lizards. As I have said and will state repeatedly through this discourse, the data of the fossil record are essential to any evolutionary perspective on the mutability of characteristics of any and all kinds.

As I read the literature on snake lizard evolution, I find it to be filled with this top–down approach to the concept of "snakeness," such that the modern assemblage serves as the singular probe of the evolutionary history of snake lizards, including the fossil snake lizard data set, with the intent to discover where "snakeness" most essentialistically (à la Plato and Aristotle) begins in the fossil record. As I hope to show, asking when the essential characteristics of modern "snakeness" appear in the fossil record of course has nothing to do with the origin of snake lizards from among their ancient lizard kin in the least! It is little wonder that we have been long in arriving to virtually nowhere in the 2500-some-odd years we have been poking this bear of a question (see Aristotle, *Historia Animalium*, Book I, Chapter VI). This approach, for many reasons, is remarkably similar to querying the icing for insight on the kind of cake that hides below that thin sugary coating. You will never arrive at a reasonable answer until you dig a hole right through to the bottom of the cake itself and ignore the characteristics of the icing.

The word "snake," in isolation of reference to the larger set to which it is merely an adjective, is used to describe a research program on snake lizards. What do I mean? I mean, in very real terms, that snakes are not snakes as though snakes are distinct from lizards and amphisbaenians, but are in truth a kind of lizard first, albeit a specialized kind of lizard, but a lizard nonetheless. To think of snake lizard evolution as distinct in some manner from lizard evolution—unless, of course, snakes are not lizards, but I think they are, as do most other researchers—is to bind the inquiry around the essential features that define "snakeness" (same problem, different nuance as explored above), and not to nest "snakeness" and the form inquiry we proceed with within the concept of lizard where it genuinely belongs.

Now, there will be eye-rolling at this seeming revelation of mine, as everyone is going to say something along the lines of, "Well, we all knew/know that, snakes are squamates and lizards are squamates, too," or, "Not talking about them as lizards is just common language, but everybody knows that they are lizards." In truth, I do not think I am making a point about something that is not commonly understood among scholars, but then, if absolutely true, why does no one call them snake lizards?

The essential characteristics of fossils, which are the soul of this treatise, did not even enter the

dialogue on "snakeness" until the description of a few vertebrae by Sauvage (1880); to say the least, their impact was zero, though in the last 20 years or so, that has finally changed (e.g., Caldwell and Lee, 1997). Therefore, the truth is, our notion of "snakeness" and the qualities of an ancient snake lizard and the evolution that such a concept has implied since Darwin (1859) has been guided by the leftover morphology, ecology, behavior of the living fauna of snake lizards and non-snake lizards (Plato's essential and/or ideal form). Clearly, such an approach is bound to be wrong if we are truly driven to ask questions that are framed under the theory of evolution, in which case, fossils are the only data set that can reconstruct the past and move the scholarly research program away from one of immutability and essentialist research cycles.

So, I would argue that we do not know that "snakes" are indeed lizards. In fact, it is not the least bit true to say that we do. A few researchers pay lip service to the idea, but no one, and I mean no one, really and truly thinks of snakes as lizards, because if you do, you make the kind of language you need for such an idea a practical reality and then use it. For Linnaeus (1740, 1756, 1758), lizards and snakes were not even closely related. He had classified "serpents," legless lizards (Table 1.1), and caecilians, along with amphisbaenians, in a single group. Today, people walking down the streets of London or Tokyo pretty much hold the same view as Linnaeus in that anything long, skinny, and legless is a serpent. Unfortunately, the problem leaks sideways into science as well, as I can assure you that lizard workers at the next herpetological symposium on lizards, wherever it is about to be held, are not going to talk about snakes during their platform presentations on lizard evolution, other than perhaps as a sidebar midtalk, as

### TABLE 1.1
*List of legless lizard families*

Amphisbaneia—190 species, all limbless (exception: *Bipes biporus* with robust forelimbs)

Anguidae—3 genera (*Ophisaurus, Anguis, Apodus*), 13 limbless species

Anniellidae—1 genus (*Anniella*), 6 species completely legless

Cordylidae—1 genus (*Chamaesaura*), 1 limb-reduced species (rear limbs only)

Dibamidae—*Dibamus* (22 species) and *Anelytropsis* (1 species), all completely limbless

Pygopodidae—7 genera (*Aprasia, Delma, Lialis, Ophidiocephalus, Paradelma, Pletholax, Pygopus*), 44 species, limbless with rear limbs represented only by a protruding scale

Gymnophthalmidae—1 genus (*Bachia*), 23 limbless species

Scincidae—8 major and globally distributed genera with hundreds of limb-reduced to completely limbless species

THE ORIGIN OF SNAKES

"derived anguimorphan lizards"; if I submitted a talk proposal to that lizard symposium they would bump me to the "snake" sessions. Snake lizards, or "snakes," likely because of the size of the clade (>3000 modern species) and the vast array of anatomical specializations that distinguish them from other lizards, are simply thought of as different enough to not be seriously contemplated as true lizards. The impact of this kind of categorical thinking is profound, and the rest of this chapter will examine how and why, and provide a solution, one that will run as a clear thread through the rest of the data and arguments in the rest of this book.

To further illustrate the problem, I will use as examples two excellent and very widely read books written independently of each other by three excellent herpetologists, *Snakes, the Origin of Mystery in Nature*, by Harry Greene (1997), which itself inspired *Lizards: Windows to the Evolution of Diversity*, by Eric Pianka and Laurie Vitt (2003). These three authors precisely state that snakes are squamates, and that snakes are lizards. From Greene (1997:17): "Actually, amphisbaenians and snakes are lizards in exactly the same sense that humans are primates, primates are mammals, and so forth," and from Pianka and Vitt (2003:1): "Snakes are merely one group of very specialized lizards."

So where is the problem? Well, Greene (1997), for the rest of his book, goes on to talk about snake lizards, snake lizard diversity and disparity, and finally snake lizard evolution, without ever referencing lizards again until his longish paragraph on Mesozoic lizards and early snake lizard evolution (Greene, 1997:271), where snake lizard origins are discussed not in terms of lizards and lizard evolution, but rather in the sense of the derivation of modern snake lizards from some kind of ancient snake lizard that was either a burrower or was terrestrial. If I were to paraphrase, and I will, Greene follows the received view that "snakeness" is understood by reference to modern snakes as snakes, not snakes as a kind of snake lizard derived from a kind of ancient snake lizard. The circularity of reasoning, common to all thought on the matter, is clear.

Pianka and Vitt (2003:273–277) give the question of snakes as a kind of lizard more air time in their Chapter 14 "Historical Perspective" section on Anguimorpha, where they state:

Snakes represent a putatively monophyletic adaptive radiation of exceedingly specialized but nevertheless highly successful anguimorphan

lizards. Like their ancestors, which were probably fossorial (burrowing lizards with reduced appendages or limb loss), snakes rely heavily on chemosensory cues to locate prey.

Importantly, though, Pianka and Vitt (2003) maintain the use of the term "lizard" as opposed to the term "snake" so that even the ancestor of snake lizards was a burrowing lizard, but not a snake lizard. When and how the burrowing lizard transitioned to "snakeness" is anyone's best guess, and so, like Greene (1997), Pianka and Vitt reinforce the perception that the ancient snake lizard was not a "snake" while still a lizard, but rather a kind of lizard from which snakes and snakeness evolved much later in the process of becoming a snake. There are at least two rather intractable problems with such a scenario and its accompanying characterization of snake lizards as something other than a kind of lizard. First, what distinguished the first snake lizard from its ancestral burrowing lizard ancestor? Second, why would the lizard ancestor of snake lizards have to be a legless or limb-reduced burrowing lizard in the first place, other than the presupposition that such an origins hypothesis is true even before there are data to support it?

Again, some of you are likely still reading this thinking—So what? I will answer that "So what?" question in several steps, because it is very important to get this right. The effect of hesitating to call a spade a spade, and the absence of the use of the proper language to describe a snake, that is, the noun "lizard" modified by the adjective "snake," is not trivial. This mistake, if you like, has an enormous effect on how we think about what a snake lizard is, where it came from, how it evolved, what it was once, what it is now, what an ancient snake lizard might look like, and whether "snakeness" was a characteristic of these ancient snake lizards while still a "lizard," or if it did indeed evolve later, that is, after legs were lost and the body became elongate (Figure 1.1a,b).

*Non-Trivial Mistake Number One:* What scientists think and write creates knowledge that when transmutated into non-technical writing and conversation becomes common knowledge; that is, snake lizards are members of a clade called Squamata that is composed of "lizards, amphisbaenians, and snakes." Because this language has been used for so long, and because people have had access to naïve personal experience with snakes for hundreds of thousands of years, these three squamates, as

**Figure 1.1.** Photographs of a native snake and iguanian lizards from the La Buitrera Paleontological Area, near Cerro Policia, Rio Negro Province, Patagonia; both lizards are sitting on the same rocks in which are found the fossils of the ancient snake lizard, *Najash rionegrina*. (a) A very tiny "Culebra ojo de gato," the cat's eye snake, or "False Tomodon," *Pseudotomodon trigonatus*; (b) "Lagartija verde," the rare Patagonian green lizard, *Liolaemus gununakuna*, which is very, very common to La Buitrera. (Photographs by the author.)

objects, seem to easily group into two different sets—legs and without legs. And so a knowledge definition of snake arises that in no way requires a reference or linkage to lizards in the least. This is wrong. And, it is even terribly wrong because it does not communicate what we know in the least.

First and foremost, what we do know (keeping in mind my epistemic and ontic boundaries) is that snakes are descended from an ancestor with four legs that was a very proper lizard (I doubt this not in the least; in fact, even Genesis I got this correct in the Old Testament of the Holy Bible, though the mechanism in that case was the Hand of God, not evolution), and second, that the loss of limbs is a condition of modern snakes, and many extinct snakes as well, but not the ancestral snake, which would at some point have had four perfectly fine legs while in possession of a very snakelike skull (Figure 1.1a,b). The correct common language should be: Squamata is a diverse clade of limbed and limbless lizards that includes varanid lizards, anguid lizards, scincid lizards, gekkotan lizards,

lacertid lizards, teiid lizards, iguanian lizards, dibamid lizards, amphisbaenian lizards, and yes, "snake lizards."

I continue on the logic of the point with an example of how misuse of words causes confusion. If snake lizards are a kind of anguimorphan lizard (e.g., Reeder et al., 2015), that is, if that is the clade in which they are hypothesized to nest, than they are more closely related to a wide variety of anguimorphans (glass lizards, monitor lizards, etc.) than any other anguimorphan lizard (including snake lizards) is to an iguanian lizard or to a teiid lizard. And yes, all three of the latter are referred to as lizards and therefore were discussed in detail in Pianka and Vitt's (2003) book on lizard evolution, while snake lizards were not. Greene (1997) and every other snake worker before them can be slightly forgiven for not referring to snakes as lizards, but only slightly. However, workers who have written extensively on lizards can find no grace in the exclusion of snake lizards from their text (the same holds true for another pet study group of mine, the mosasaurs and their kin, but that is a rant for another day). To continue, snakes are most definitely lizards, no matter how specialized, and their regretful absence from treatises on lizard evolution and paleontology (e.g., Camp, 1923) is no mere oversight, it is a misconception of what a lizard is versus a snake, and that misconception is disseminated as knowledge in exactly that manner—that snakes are different than lizards, in fact are not lizards—to both scientists and lay persons alike. Metaphorically, the "snake-lizard knowledge problem" is akin to an ornithologist writing a treatise on birds and excluding penguins from the book because no one calls them penguin birds, and they are so specialized that they do not fly and appear not to have done so for about 70 million years.

*Non-Trivial Mistake Number Two:* Logically, and by all the rules of modern systematics (taxonomy and phylogenetics), this also means that the group defined by the common term "lizards" (=Lacertilia) is *not* paraphyletic, and is the synonym of Squamata (and frankly has priority as it was a word born of the Latinization of Brogniart's [1800a,b] "sauriens," while Squamata was coined by Oppel [1811]). Talking about snakes as lizards would once again make Lacertilia monophyletic and communicate what we claim to know about their evolution—it would be epistemically and ontologically correct. This mistake also reflects the problem highlighted by Mistake Number One. Common parlance, no matter the source, refers incorrectly to the common word "lizard" as though it were a synonym of Lacertilia, and that lizards and Lacertilia are paraphyletic because the "group" does not include snakes and amphisbaenians—but it does. The problem is not phylogenetic; the problem, again, is language and how we think about what we just said. "Snakes" are lizards, anguimorph lizards of some kind by the look of it (e.g., Reeder et al., 2015), Lacertilia is the senior synonym of Squamata, and Lacertilia is not paraphyletic for the same reason that Squamata is not paraphyletic. I do not care that scholars and lay persons alike do not "think" of snakes as lizards, or that they confuse the use of the common term "lizard" for a synonym of Linnaeus' Order Lacertilia—not thinking about something the right way does not make the real truth false. An apple remains an apple even if you call it a banana, so snakes are a kind of lizard regardless of the language used to ignore this fact.

Non-Trivial Mistake Number Three: I am beginning this paragraph with a statement, "Snakes suffer from all of the same problems as do birds." Because snakes are imagined to be so different from their lizard kin as to warrant dropping the noun "lizard," thus forcing the adjective "snake" to become a noun, science follows suit and makes a mess out of bothering to discover the when, who, where, and how of the origin of snakes as lizards. Hands are tossed up in the air and any old ad hoc hypothesis is taken as a possibility, with the language of uniqueness applied to them as it has been long applied to birds (fortunately for birds, the language is being slowly changed so that we now understand dinosaurs are not extinct and that birds are indeed dinosaurs and not very unique at all, and have actually been around for a very, very long time). My point here, and it really is a thesis statement, is that like birds, the ancestry of snakes is not to be found among living groups of lizards, but is actually much older and rooted within long-extinct groups of lizards that were not and are not to be found among the extant crown of lizards even though we would likely systematize them within a group like Anguimorpha. This is why snake lizards resemble their "usual" sistergroups, for example, anguimorph lizards, amphisbaenian lizards, dibamid lizards, or recently iguanian lizards, about as much as an apple resembles a banana. Take for example the hypothesis

that snake lizards are placed within the crown of anguimorph lizards, with Pianka and Vitt (2003:16) using a dotted line to figure them as the sort of sistergroup to Lanthanotus borneensis within Varanoidea, or where Greene (1997:272) settles on a sistergroup relationship of all anguimorphs with his Serpentes. What do we learn here? Well, nothing, and these authors are guilty of nothing except regurgitating received wisdom. Like birds, the problem cannot be addressed without highlighting what we know from the fossil record: the likelihood that the snake lizard sistergroup is extinct, long extinct in fact, and currently not represented by fossil specimens we are aware of, is the most likely explanation for the problem we face in trying in vain to trace the ancestry and origin of snake lizards from among the groups remaining today that we call crown group lizards.

To illustrate with an example, if I hand you a live individual of the legless anguid lizard *Ophisaurus gracilis* and I ask you to identify what kind of lizard it is, if you have a passing notion that many limbed lizard clades have limbless forms that are not snake lizards, you will likely be able to sort out the problem of keying it out to the Anguidae as compared to other lizard groups and their limbed and limbless members. Why is this easy? Because you can use the skull, not the body (four legs and tail includes about 6000 living species of lizards), to link it up to a group of limbed lizards with similar skulls. If I give you a live individual of the crotalid snake lizard *Crotalus horridus* (timber rattlesnake), apart from being exceedingly careful in making any sudden movements, you will likely approach the question from a very different conceptual point of view and try and place it within Serpentes, paying no attention whatsoever to the 6000+ limbed lizards, because it is legless. What is important to note, though, is that when comparing it to other snake lizards, you will do what you would have done with *Ophisaurus*: you will take *Crotalus* and compare its head to the heads of other snake lizards to find big fanged snakes with beady eyes and triangular heads—in short, you skip the body and go straight to the head (as an example, compare the head of *Liolaemus* to that *Pseudotomodon* in Figure 1.1a,b).

Hmmm … why would you do this, and in doing this, what does it mean about the preconceptions you have brought to addressing the question I asked of you? Well, it means several things, the most important of which is that the architecture of the head holds the key to figuring out what something is, at least among this group of tetrapods. And second, at another level of comparison, it means one of two things: (1) either snakes are not lizards and the whole squamate/lizard thing is wrong, so I asked an impossible question, or (2) snake lizards are not descended from any living group of lizards and you need to be familiar with the fossil record to answer the question (my point above).

As should be clear, I believe the answer lies behind Door Number 2 as long as you remember that the head is the key. Recently, I coauthored a paper that said exactly that: the snake lizard head evolved before the snake lizard body, and snake lizard ancestry, clade origins, and their lizard sistergroup relationships, go back in time well beyond 170 million years ago (Caldwell et al., 2015). This is much, much older than was previously thought, involves a sistergroup relationship to an extinct group of lizards, not a member of an extant clade, though still likely nested within the crown of lizards (snakes are not stem squamates/lizards), and thus closely parallels the new story being developed for bird origins and phylogeny within dinosaurs. Likely the same problem exists for amphisbaenian lizards—ancient cladogenic events leading to the modern limbless forms with no obvious limbed clade within which they nest—though it is possible that amphisbaenians are stem lizards.

I have no idea why this language of "snake lizards" is so unpalatable, but it certainly seems to be. It is odd to me, considering that the term accurately reflects everything we have recently claimed to know about squamates/lacertilians, despite the differences between competing phylogenetic hypotheses (snakes in a clade with iguanians = Toxicofera [Vidal and Hedges, 2004], or snakes in a clade with Mosasauria nested within Anguimorpha high in the crown of Squamata [Reeder et al., 2015], or amphisbaenians as derived Lacertoidea [Müller et al., 2011]). But because its non-use causes so many misapprehensions and conceptual roadblocks, I highly recommend immediate implementation of the use of "snake lizard." For the remainder of this book, I am explicitly using "snake lizard" to make my point, though for brevity's sake, it would certainly be easier to use the word "snake."

I will end this section of Chapter 1 with a final example of how profound, pervasive, and misguided this problem is of the singular use of the term "snake," and, as you will see, also the use of the term "amphisbaenian" as distinct from "lizards," I am providing here several quotes to illustrate the common uses of the phrase, "Squamata includes lizards, amphisbaenians, and snakes." All of the authors of these quotes have simply followed convention, as did Greene (1997) and Pianka and Vitt (2003) in the use of the terms they have applied to groups and groups within groups, even when their phylogenetic hypotheses found some of those named groups to be literally nested as highly derived clades within the crown of their cladograms and phylogenies.

From the important Tree of Life treatise on squamate phylogeny by Gauthier et al. (2012:3), the first sentence of their abstract reads as follows:

> We assembled a dataset of 192 carefully selected species—51 extinct and 141 extant—and 976 apomorphies distributed among 610 phenotypic characters to investigate the phylogeny of Squamata ("lizards," including snakes and amphisbaenians).

And from Palci and Caldwell (2013:973):

> Snakes are a very successful and diverse group of squamate reptiles, numbering about 3,000 living species (Greene, 1997; Caldwell, 2007a,b).

And from Reeder et al. (2015:2):

> Squamate reptiles (lizards and snakes) are an important and diverse group of terrestrial vertebrates, with >9,000 species.

Bearing in mind the key message of Chapter 1 to this point, I will follow up with a metaphor that I hope hammers home my point, and it bears no citation, as I have created it here:

> The clade Aves is composed of birds, penguins, and hummingbirds.

Your response is likely, "Huh? But they are all birds, why highlight them as different?" My response is, "You are correct. It makes no sense."

## ANCIENT AND ANCESTRAL SNAKE LIZARDS

There is yet another major concept I need to explore here, and that is "Genealogy versus Phylogeny"—I will begin by using me as an example. As an individual organism from a sexually reproducing hominid species, I have two "Caldwell" ancestors, male and female, that link backward in time to increasingly ancient pairs and multiples of pairs of males and females, to create a pyramid-scale complex of genetically linked human beings in a very ancient genealogical nexus with me at the top. All living humans have a similar pyramid shape to their ancient cloud of genetic linkages, as every individual is himself or herself a temporal endpoint of their own lineage of origination. While each individual's story appears to be a pyramid, this is not true for the species—that story is an expanding and contracting cloud of all linkages of individuals within populations and populations to the "species" to which the individual belongs, through space (three dimensions) and time (fourth dimension). The expansions and contractions represent population-level and thus genetic-level fluctuations (i.e., flow, drift, etc.), linked to numerous population-level adaptive radiations and population-level near-extinction events. What is abundantly clear is that a "species" is an exceedingly dynamic entity in both time and space and that no single individual or pair of individuals can actually be claimed to sit at the origin point of species (unless divinely created). The origins of a species is likely not much of a remarkable or notable event because it would be lost in the poorly defined space and time of gene-level fluctuations within a set or subset of individuals within a population. Therefore, "ancestors," at best, are an unidentifiable "ancestral population" of indeterminate and varying size, with very fuzzy and very dynamic boundaries in space and time. The species arising out this ancestral population—which could be hundreds of thousands of individuals, or more—may or may not be immediately recognizable as phenotypically different from the "ancestor." It is likely that genetic isolation and vicariance play a strong role in terms of fixing phenotype in a population or incipient species. I suspect there is no constant tempo and mode to speciation because there is no constant tempo and mode of variation within and between ancestral populations.

What this means for ancient and modern snake lizards is that there is no such thing as an empirically delimitable "ancestor," and by extension to the topic at hand, no such thing as an "ancestral snake lizard." However, while empirically one can neither find nor observe the ancestral anything, it does not mean conceptually, in terms of hypothesis and theory, that we cannot develop, no matter how fluid, an epistemically sound concept of an ancestral Caldwell primate, or in the case of this discussion, an ancestral snake lizard.

As a cladist, I search for sistergroups linked together by my understanding of character states and how those states when tested via congruence reconstruct synapomorphies—these are easy concepts. As an interpreter of phylogenies, I am trying to discover qualities that can be read from those patterns of relationship that reflect evolutionary pattern and process, the inheritance of ecologies or functional morphologies, the evolution of homologs within and between groups, paleobiogeographies and vicariant distributions, and so on. As a paleontologist, I am keenly aware that somewhere in the ancient past, hopefully locked away in the rocks, is a fossil of an organism that will reflect some aspect of my "ancestral snake" paradigm and further inform the empirical data set and make corrections to cladograms, phylogenies, and third- and fourth-order hypotheses. Such concepts, though they reflect nothing that ever was "real," are predictive, as they create models of an ancient thing that we can use to probe the data set of the fossil record for possible anatomical combinations that we might expect to find hidden away in the rocks. As such, they have great value.

## "SNAKENESS": A HISTORY OF NAMES AND AFFINITIES

The use of the term "modern snake" is not modern at all, which means its current usage comes laden with a cloudy burden of previous thought and use (see the Non-Trivial Mistakes above). There are no references at all, pre-Darwin (1859), where scholars used the word "modern" as an adjective of "snake," nor to any other group of animals or plants other than to humans, usually in an archaeological or anthropological sense of "ancient man" as opposed to "modern man." Importantly, however, this temporal reference of ancient versus modern is nearly immediately apprehended in the scholarly evolutionary literature, post-Darwin (1859), to make reference to living groups of organisms versus ancient or extinct groups of organisms. For example, Owen (1863) and Huxley (1868a,b) began referring to "modern birds" when they were writing on their comparisons of birds to the extinct *Archaeopteryx*; Cope (1869) refers to "modern iguanas" when discussing aquatic evolution in the extinct mosasaurs and plesiosaurs; the earliest distinct use of "modern snake" appears to have been Marsh (1871:323) when he compared the vertebrae of a new North American fossil snake, *Boavus occidentalis*, to the vertebrae of two European fossil snake lizards, as opposed to those of "modern serpents," which he noted were different from the fossil forms. The use of the term "modern snake" has continued to the present in the scholarly literature and is extremely widespread (see Caldwell et al., 2015; Hsiang et al., 2015).

So what is the problem? Well, to understand the problem, it is reasonable to ask the question, "What do you mean when you use the term 'modern snake'?" The answer, as I will illustrate, is that "modern snake" and "snake" are virtually synonyms, and if you exclude the concept of snakes as lizards, as discussed earlier, it becomes nearly impossible to characterize a fossil snake lizard as a snake lizard because such animals, understandably, are not modern snake lizards and display an uncomfortable level of their ancestral lizardness.

## In the Beginning

A quick search for the word "snake" on the internet turned up 362,000,000 million hits, not bad considering that "Jesus" resulted in 904,000,000. A slightly deeper search revealed that nestled neatly within the 362 million snake hits were an endless number of mistaken identity postings—worms identified as snakes, legless lizards (Table 1.1) confused with snakes (this one is understandable at a certain level), spittlebug spit mythologized as snake spit, and of course the iconic story of Lucifer, the four-legged snake of the Garden of Eden, who for his sin of temptation was made legless:

> 14 - And the Lord God said unto the serpent, Because thou hast done this, thou art cursed above all cattle, and above every beast of the field; upon thy belly shalt thou go, and dust shalt thou eat all the days of thy life:

> 15 - And I will put enmity between thee and the woman, and between thy seed and her seed; it shall bruise thy head, and thou shalt bruise his heel.

KING JAMES VERSION, OLD TESTAMENT<br>GENESIS III: 14–15

The deep, deep role of snakes in human mythologies and cultures, perhaps more so than any other living or mythical animal (a topic well beyond the intention of this treatise), is an incredibly fascinating subject that is worth all the scholarly attention it receives. But for me, as a student of snake evolution, the single most important aspect of this cross-cultural, cross-mythology, cross-prehistoric and historic time iconization of snakes is that almost every living human being, for millennia upon millennia, has had a set of rules to diagnose and define what a snake is. The only difference for "science" in the last 200 years is that we have tried to codify a suite of morphological features that make a common "snake" into a biological entity as a snake lizard that is changing through time. Science also tries to explain the linkage of biology and time within the context of a suite of non-supernatural phenomena such as genetic inheritance, evolution, and deep geological time. Truthfully, I personally like the whole "duality of good and evil" thing that "snakes" represent across many mythologies, alongside the biblical origin myth of leglessness as an angry god's justice for a great sin, but, sadly, empirical observation and hypothesis testing come up zero on those two points while returning a great deal more information on the evolution and relationships of snake lizards to each other and to their closest lizard relatives.

## Somewhat after the Beginning

Divine creation aside, the biological question, which of course is an evolutionary question in the end, is: What is a snake lizard? Like many revelations from the Renaissance, the Greeks seem to have been the first to record such a diagnosis, and, again, not at all surprisingly, it is the philosopher and great typologist Aristotle, who nearly 2500 years ago, writes it out in plain language:

> An animal that is blooded and capable of movement on dry land, but is naturally unprovided with feet, belongs to the serpent genus; and animals of this genus are coated with the tessellated horny substance. Serpents in general are oviparous; the adder, an exceptional case, is viviparous.

ARISTOTLE, HISTORY OF THE ANIMALS<br>BOOK I, CHAPTER VI

Quite importantly, Aristotle did not stop here, but went on to differentiate "snakes" from lizards by noting that "snakes" were elongate and limbless, whereas in general, lizards were not:

> The serpent genus is similar…to the saurians among land animals, if one could only imagine these saurians to be increased in length and to be devoid of legs. That is to say, the serpent is coated with tessellated scutes, and resembles the saurian in its back and belly…
>
> ARISTOTLE, HISTORY OF THE ANIMALS
> BOOK II, CHAPTER XII

Aristotle then goes on to list a large number of anatomical features shared between snake lizards and non-snake lizards. Frankly, this is a rather remarkable typological diagnosis, conceived of in very modern terms from specific observations. I am struck, in particular, by the exceptional perceptiveness, almost a presage of evolutionary thinking, of the statement, "if one could only imagine these saurians to be increased in length and to be devoid of legs," to which I would add as clarity, "then lizards would be snakes and snakes would be but a form of lizard." Aristotle's attention to anatomy, both external and internal, as the key elements of his diagnosis are astute and strongly indicate he was creating groupings based not on generalizations of form, but on specifics. There is nothing to suggest he was thinking in transformative, or evolutionary, terms, but it is clear he was developing his criteria based on typology, or what today we refer to as taxic criteria, that is, taxic homology.

As I hope I am beginning to communicate, the idea of a "something" is slow to build, but build it does, and as we come to the present, oftentimes we find that the "something," with its definitions and boundaries, does not really describe what we wanted it to, nor what we thought it did. But I am getting ahead of myself here; there is a great deal more history to explore in developing what seems to be so simple—our idea of "snakeness."

## Three Hundred Years Ago

Following Aristotle, which spans more than a 2000-year gap, we come to the less-than-definitive thoughts of Linnaeus (1735, 1740, 1756) where he first applies the term "Serpentia" to a group inclusive of amphibians and reptiles, followed by the "Ordo Serpentia" (*Systema Naturae*, 2nd edition) that included "serpents," legless lizards, and caecilians, and, finally, in the 1756, where he added amphisbaenians to the group (Table 1.1). In the tenth edition of the *Systema Naturae*, Linneaus modified "Serpentia" to "Serpentes," the term still in use today. The odd thing in today's parlance, and in particular by comparison to Aristotle (see above), was that Linnaeus considered Serpentes amphibians and diagnosed them as sharing the following features, "*Corpus nudum aut squamosum, dentes molares nulli, omnes acuti, pinnae nullae radiate*," which, translated from the Latin, means, "Body either naked or covered in scales, lack of molar teeth, all teeth are pointed, absence of ray fins." Most importantly, because it links to the central thesis of this treatise, he considered the diagnostic features of his Serpentes among his Amphibia to be the absence of limbs or feet, "*pedes nulli*" (Linnaeus 1740, 1756). By the 10th edition (Linnaeus, 1758), he had modified that anatomical diagnosis to read, "*Os respirans, pes pinnaeue nullae*," and translated, "Organisms that breathe from their mouth and lack feet or fins." Thus, for Linnaeus, being limbless was a "snake" feature that trumped all other features of external or internal anatomy, including morphologies that today we confidently use to diagnose very different groups of vertebrates such as snake lizards, other legless lizards, and caecilians. Without doubt, Linnaeus's perspectives on "snakeness" strongly influenced successive generations of scholars (e.g., Brongniart, 1800a,b) even as they labored to separate amphibians from reptiles, and legless lizards and caecilians from snakes.

The first suggestion to remove caecilians from Serpentes (Linnaeus, 1758) was made by Oppel (1811), who placed them in his Order Nuda and created Saurii for legless lizards along with all other lizards and crocodilians. Oppel (1811) still placed amphisbaenians and snakes in his Ophidii, but moved Ophidii and Saurii into his Squamata, which included, as noted, crocodilians. Oppel's (1811:53) diagnosis of Ophidii was, "*Corpus elongatum, cylindricum, pedibus, sterno, pelvique carens, squamis obtectum. Penis duplex*," or "Body elongate and cylindrical, lacking feet, sternum and pelvis, covered with scales. Paired penis." Slowly but surely, improvements were being made as successive scholarly works considered, revised, and improved upon the data, concepts, and conclusions of their predecessors (e.g., Merrem, 1820; Gray, 1825).

## The Golden Age

The modern composition of Linnaeus's Serpentes was first proposed by Wagler (1830), who recognized "snakeness" much more restrictively, limiting it to blind burrowing snake lizards, the scolecophidians, and all other modern snake lizards now recognized in the clade Henophidia (Hoffstetter, 1939). Wagler's (1830) diagnosis of Serpentes was simple and straightforward—"*tomia mandibulae in apice ligamento connexa*," or, "mandibles connected by a ligament at their apex." For the early nineteenth century, this was a simple and straightforward diagnosis and delimitation of the group based on observable anatomy, a bit brief, but clear and concise. From this point until the present, the majority of the contributions to understanding snake lizards focused on organizing the species into some kind of sensible ranking of higher taxa nested within "Serpentes." "Snakeness," at least in terms of the modern fauna, had been solved to everyone's agreement, and the problem at hand was to organize the ingroup relationships, so to speak.

It is during this settled but yet Golden Age of herpetology, some of it pre-Darwinian in terms of both timing and approach (e.g., Müller, 1831; Duméril and Bibron, 1844; Duméril, 1853), and some of it much later and definitively evolutionary in intent and hypothesis construction (e.g., Cope, 1864, 1900; Boulenger, 1893, 1894, 1896; Nopcsa, 1923; Hoffstetter, 1939), that the fully modern concept of "snake" and "modern snake" becomes entrenched in both observation and hypothesis.

For example, the important but constraining term, along with anatomical diagnoses, Macrostomata, was proposed by Müller (1831) to group snakes with a large gape from those with a small or limited gape, the Microstomata. Among modern or extant snake lizards, the Microstomata included the blind burrowing snake lizards, or Scolecophidia (i.e., Typhlopidae, Leptotyphlopidae, and the Anomalepididae) and the Anilioidea (i.e., *Anilius, Cylindrophis,* and the Uropeltidae). Oddly, and in contrast to Gray (1825), Müller (1831) revived the placement of amphisbaenians within his Microstomata. While the placement of amphisbaenians back into Serpentes is paradoxical, what is important to consider from Müller's (1831) work is the nature of the transformation he is suggesting, pre-Darwinian processes of evolution, on small gape being the logical prelude to big-gaped snakes. The historical burden of such suggestions is profound, particularly as concerns the implications for the mechanisms, tempo, and mode of evolution, and for phylogenetic relationships, as a feature, such as small gape or large gape, can become "knowledge," both in terms of a taxic diagnosis and a transformational sequence. In the case of snake lizards, it does not mean that it is not, but it does not mean that it is, either.

## Gilded Age of the Fossil

A member of the Golden Age of Herpetology, E.D. Cope, a herpetologist and paleontologist who swiftly absorbed Darwin's arguments on evolution, but was always interested in anatomy as a means of organizing taxa around evolutionary relationships, began to include fossil taxa and their anatomical features into his classifications inclusive of modern lizards, including snake lizards. Cope thus became the "Father of the Gilded Age of the Fossil," and I say "gilded" because in fact Cope was the only paleontologist of his era to also work intensively and with great expertise on the modern herpetofaunal assemblage and to blend/gild that data set with the fossil data set in order to understand the big picture of their evolutionary history. His use of fossils to generate a hypothesis of snake lizard phylogeny (Cope, 1869) stands as nearly the oldest hypothesis of its kind in vertebrate paleontology and evolution, second only to the hypothesis on bird phylogeny and origins made by Huxley (1868a,b) only one year before.

Cope (1864) began this effort as a herpetologist, building upon Wagler (1830) and Duméril and Bibron (1844) by refining the concept of the latter's "Scolecophides" to the Latinized Scolecophidia, removing amphisbaenians from the Scolecophidia, and limiting the Scolecophidia to the modern concept of blind burrowing snake lizards sharing distinctive scolecophidian anatomical features. But it is his refinement of Wagler's (1830) diagnosis of Ophidia where he notes that the ligamentous connection between the dentaries is also found in fossil marine lizards, specifically the mosasaurs, where he went beyond what had been done before.

Cope (1869) unleashed a firestorm of controversy over his hypothesis of relationships that held that snake lizards were more closely related to mosasaurs than to any other kind of lizard, including varanids. Unlike all of his predecessors, and likely not intentionally, but certainly by intuition, Cope had stepped away from considering "snakeness" to be an essentialistic character of "snakes"; in other words, he left Plato behind, and moved instead to define snake lizards by their anatomical features as the evolutionary inheritance of their lizard ancestry—he went Aristotelian on snakes. Not too surprisingly, and almost without exception, Cope's critics and supporters considered that he was arguing that snakes evolved from mosasaurs (Owen, 1877; Marsh, 1880; Baur, 1892, 1895, 1896; Boulenger, 1893; Williston, 1898, 1904; Nopcsa, 1903, 1908, 1923, 1925; Janensch, 1906; Bolkay, 1925).

Cope (1869:258) wrote: "On account of the ophidian part of their affinities, I have called this order the Pythonomorpha." He was clearly making direct reference to mosasaurs and hypothesizing their relationships based on shared anatomical features—forward thinking to modern paradigms where we constrain relationship hypotheses, phylogeny, to data. Cope (1895a,b, 1896a,b) continued to argue for a closer relationship between snakes and mosasaurs within lizards, and not a relationship of ancestor to descendent where snakes evolved from mosasaurs, until he died (Figure 1.2). Unfortunately, the truths about Cope's position were far less interesting to his critics then their retelling a "truth" that allowed them to push over their straw man, a problem that has continued to the present (e.g., Rieppel, 2012)—but that is a story for later in this book (Chapter 5, to be precise). Framed in this manner, so began the grand debate among paleontologists and herpetologists that would last for more than 50 years, with lines drawn in the sand stretching

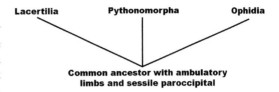

**Figure 1.2.** Phylogeny of Lacertilia, Ophidia, and Pythonomorpha from Cope (1895a:858). (Note: He indicates at the base of his phylogeny that all three share a common ancestor but that one is not derived from the other, only more closely related.)

across the Atlantic Ocean, from London to Brussels to Budapest, and from Princeton to New York and New Haven. Fossils, mostly mosasaurs and their kin, were now at the center of the debate; what is unfortunate, in that there were so few fossil snakes available at the time, is that until Nopcsa (1908, 1923), these specimens were hardly mentioned in the discussion of what constituted "snakeness." Nopcsa (1908, 1923) managed to make lasting contributions to "snake" nomenclature, evolution, and phylogeny (Figures 1.3 and 1.4); he recognized and named

**Figure 1.3.** Phylogeny of Mesozoic Platynota from Nopcsa (1908).

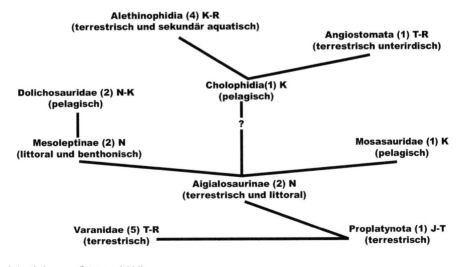

**Figure 1.4.** Phylogeny of Nopcsa (1923).

the "Alethinophidia" as distinct from the Angiostomata (junior synonym of Scolecophidia), and considered both to have arisen from within the fossil snakes he knew about, his "Cholophidia" (*Archaeophis provaus*, *Pachyophis woodwardi*, *Palaeophis toliapicus*, and *Simoliophis rochebrunei*). The only odd thing about Nopcsa's (1923) contribution is that he did not discuss the excellent skull of *Dinilysia patagonica* Smith-Woodward, 1901, even though he named *Pachyophis woodwardi* after Smith-Woodward.

The problem for Cope and everyone else in the mid- to late nineteenth century was simple: there were very few fossil specimens of anything vaguely snake lizard-like. In the middle part of the nineteenth century, there were two snake lizard species known that included more than one or two isolated vertebrae: *Archaeophis proavus* Massalongo, 1859 (Mt. Bolca, Italy) and *Palaeophis toliapicus* Owen, 1841 (London Clay, United Kingdom). Both of these snake lizard fossils were from Eocene-aged rocks, so they were geologically very young, and neither was recognized from good skull remains. By the late nineteenth century, two new taxa had been added to the data set. *Simoliophis rochebrunei* Sauvage, 1880, from southwest France, was known from a large number of isolated vertebrae with an odd boxlike morphology to the centrum, and Marsh (1892) had described an Upper Cretaceous snake from the Western Interior Basin of North America, *Coniophis precedens*, but the type and subsequent materials, like *Simoliophis*, were little more than isolated vertebrae. A few years later, Smith-Woodward (1901) would describe an excellent skull of a snake from the Late Cretaceous of Argentina, *Dinilysia patagonica*, but Cope was already deceased by this time, and the debate he had started was dying with him. This beautiful Mesozoic snake skull was far from the prying eyes and minds of herpetologists and paleontologists alike, tucked safely away in the collections of the Museo de La Plata in La Plata, Argentina—for reasons that remain mysterious, the importance of *D. patagonica* remained completely overlooked for more than a century (Caldwell and Albino, 2002; Zaher and Scanferla, 2012), despite Estes et al.'s (1970) attempts to reintroduce it to paleontologists and herpetologists alike. So, with little attention paid to *Dinilysia*, by the early part of the twentieth century, the snake lizard fossil record had increased by two more specimens, both of which were found in marine-deposited limestones with vertebral anatomies similar to *Simoliophis*, though,

frustratingly, neither possessed a complete skull: *Pachyophis woodwardi* Nopcsa, 1923, *Mesophis nopcsai* Bolkay, 1925.

## The Phoenix of Neontology

With barely a whimper, as the roaring 20s came to a close, and with the world poised on the event horizon of the Great Depression, paleontologists and the data set of the fossil record virtually disappeared from the debate on "snakeness" shortly after Bolkay (1925) described *Mesophis nopcsai*, a Cenomanian-aged snake lizard from the same locality as *Pachyophis woodwardi*. Why? Perhaps the paucity of the fossil record and its lack of insight on "snakeness" had banned them from the debate; it is also true that as the neontological data set of modern snake lizard diversity was expanding, the excitement of trying to answer these questions from these new data points overwhelmed the rather smaller data set from fossils. The final nail in the coffin of paleontological contributions was hammered home by the great North American paleontologist and herpetologist, Charles Camp (1923:313), in his "Classification of the Lizards." Though an admirer of Cope's use of specific anatomical data and the legacy of that work, he could not abide mosasaurs and their kin as some kind of snake lizard relative, to the exclusion of nesting them, without "snakes," inside some kind of platynotan superfamily. Camp (1923) provided his own hypothesis of the descent of snakes, and thus the condition of "snakeness," but oddly elevated them to Suborder status within his Order Squamata, despite recognizing them to be a kind of derived lizard:

> The snakes appear to have arisen from anguimorphid, grass-living lizards... burrowing snakes (Typhlopida and allied families) parallel the burrowing lizards in many profound ways and would seem to be derived from autarchoglossid stock for that reason and because of the paleotelic characters they hold in common with the Anguimorpha.

So, with fossils and a fossil lizard as their closest relative now gone, the paleontological interest in snake lizards disappeared. Arising from their ashes, the neontological herpetologists returned with great vigor and resumed control of the data set driving the discussion on "snakes." Mahendra (1938) assessed the anatomical features of leptotyphlopids and typhlopids and proposed the first ingroup phylogeny of snake lizards with scolecophidians as basal and a burrowing origin for the group as a possibility. Hoffstetter (1939) subdivided Nopca's (1923) Alethinophidia into the Henophidia (anilioids, boas, and pythons) and Caenophidia (*Acrochordus* and colubroids). Walls's (1940, 1942) influential work on ophidian eyes and visual cell systems allied degenerate and simple with primitive, and added vigor with a prosaic myth on burrowing as primitive to the growing burrowing origins hypothesis. Subsequent work followed the same approach of refining "snakeness" within snake lizards and not the concept of "snake" itself by reference to concepts of snake ancestors, nor to fossils (e.g., Haas, 1930; Bellairs and Boyd, 1947, 1950; Bellairs and Underwood, 1951; Underwood, 1967, 1970; McDowell, 1974, 1975, 1979; Rieppel, 1979a,b).

From Mahendra (1938) or Hoffstetter (1939) to the present, the principle focus of neonotological herpetologists has been a continued refining of snake lizard ingroup relationships, along with, in some cases, efforts to hypothesize the closest lizard sistergroup to snake lizards (e.g., Cundall et al., 1993; Heise et al., 1995; Slowinski and Lawson 2002; Vidal and Hedges 2002, 2004; Wilcox et al., 2002; Lawson et al., 2004; Townsend et al., 2004; Vidal et al., 2007a,b; Wiens et al., 2008, 2010, 2012; Douglas and Gower 2010; Pyron et al., 2011, 2013, 2014). But it was not until Conrad (2008) that a burrowing lizard sistergroup was finally recovered (see also Gauthier et al., 2012).

## The Return of the Fossil as King

The most important change to our concept of "snakeness" has been the return of paleontological data to the debate, but it has been slow because of the small number of informative fossil snake lizards, as has been realized for some time (e.g., Bellairs and Underwood, 1951; Bellairs, 1972; Cundall and Irish, 2008). Estes et al. (1970), to surprisingly little fanfare, redescribed the type, and at that point, only known skull of *Dinilysia patagonica*. It was not until several new legged snake lizards were recharacterized (Caldwell and Lee, 1997) or newly described (Rage and Escuillé, 2000, 2003; Tchernov et al., 2000) that fossil snake lizards were reintroduced as central to the enlightenment of our understanding of snakeness, origins, phylogeny and evolution (in spite of the fact that Cundall and Irish [2008:358] accurately and hilariously referred to the resulting scholarly exchange as a "a blizzard

of literature on a few controversial fossils"). In truth, the blizzard over not much data reignited interest in the fossil record and has resulted now, some 20 years later, in a wealth of new data. It is also true that data from fossils have re-entered both scholarly and lay conversations and are now broadly recognized as critical to addressing evolutionary questions of any kind (e.g., Gauthier, Estes et al., 1988; Gauthier, Kluge et al., 1988).

And, much to my pleasure, *Dinilysia patagonica*, and the pioneering efforts of Estes et al. (1970), are finally at play, where they belong, in the complex arena of snake lizard evolution (Caldwell and Albino, 2002; Caldwell and Calvo, 2008; Zaher and Scanferla, 2012). As of the writing of this book, snake lizards are now known from fossils with legs, found very clearly on animals that lived in both the oceans and in xeric-mesic (arid-semi-arid) terrestrial environments (Caldwell and Lee, 1997; Rage and Escuillié, 2000; Tchernov et al., 2000; Apesteguía et al., 2001, 2016); the race to identify and describe the first four-legged snake is on, though mistakes in identification are being made (e.g., *Tetrapodophis amplectus* Martill et al., 2015). Skulls preserved in three dimensions, and assessable using computed tomography scanning technologies, are providing unexpected insights on the phylogeny and function of the inner ear complex and braincase construction (Caldwell et al., 2015; Yi and Norell, 2015; Palci et al., 2017; Palci et al., 2018). Fossil snake lizards are now found in the Middle to Upper Jurassic of ancient Laurasia, in a very complex suite of environments, and appear, based on known postcranial remains, to have had non-snake-like bodies (Caldwell et al., 2015), but snakelike skulls, in direct contrast to the hypothesis made by Longrich et al. (2012) from their *Coniophis precedens* chimera—and now the thesis argument of this chapter comes full circle and begins to reveal why the term "snake lizard" is so critical. Scolecophidians still have a very poor fossil record, but in the context of their anatomical features, revealed in the context of fossil snake lizards, appear extremely derived, with numerous features contradicting their basal position at a clade point with Alethinophidia and instead strongly indicating the higher snake, if not colubroid, affinities of their anatomical features. Mosasauroids and their kin are now robustly hypothesized as the sistergroup to snake lizards nested near but within the base of a radiation of fossil and modern anguimorph lizards (Reeder et al., 2015). In short, if Edward Drinker Cope has a ghost, and could possibly be sharing a fine cognac and cigar with Franz Nopcsa's spirit self, then they are likely smiling, because the fossil has returned as King, and there are enough of them now to predict a much older version of the ancient snake lizard than we have ever been able to do before.

## "SNAKENESS," "MODERN SNAKES," AND "SNAKE LIZARDS"

As of January 2018, the Reptile Database (Uetz et al., 2018) listed 3645 species of living snake lizards conventionally systematized into two major groups, the Scolecophidia (442 species) and Alethinophidia (3203 species). Beginning with Aristotle and racing forward to the year 2018, nearly 2500 years of scholarly investigation have revealed an amazing diversity and disparity of modern forms; fossils have played a role, but have had little influence on the concept of "snakeness" or just, "What is a snake?" I hope I have made it clear that it is from this long history of discourse, blended with this vast array of living snake lizards, displaying a bedazzling array of differences and similarities (i.e., long, skinny, legless, blind, toothy, nearly toothless, harmless, earless, eyelidless,

and venomous), that the concept of "modern snake" has arisen in common and scholarly usage (Figure 1.5). As I think I have both demonstrated and argued, the cherished terms of "snake," "snakeness," and "modern snake" are Platonic categories that by their very construction reflect the essential characteristics of modern snakes to define all snakes. The problem is as intractable as it is circular—we are left to define all "snakes," modern and ancient, by "snakeness" as referenced from "modern snakes" themselves—3645 species of long, skinny, legless snakes.

My goal is to step past limiting concepts of ancestral states where long and skinny defined the snake progenitor (e.g., Pianka and Vitt, 2003; Longrich et al., 2012) and conceive of an ancient snake lizard as a short-bodied, four-limbed, long-tailed lizard with a particular configuration of the head sharing characteristics still possessed by all modern snakes (Caldwell et al., 2015). This would mean, like all groups of "lizards" known today, there are limbless to limb-reduced forms that look like "snakes," but yet are recognized as derived members of the Scincidae (e.g., *Lerista*) or Anguidae (*Ophisaurus*) or Gekkota (*Pygopodus*) because of their skull construction. In other words, modern snake lizards are members of an ancient clade of lizards,

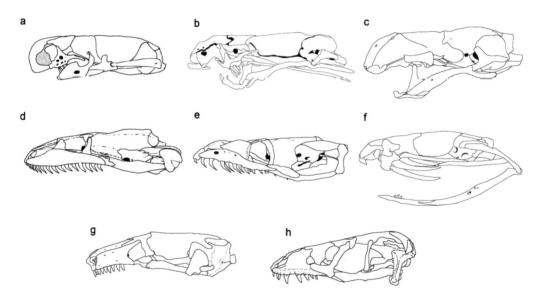

**Figure 1.5.** A comparison of osteological features of scolecophidian cranial osteology with that of fossil and extant snakes and lizards and of the vertebral elements of a Miocene scolecophidian with that of an extant taxon. (a) Skull of *Leptotypholops*; (b) skull of *Anomalepsis* (Redrawn from Haas, 1968); (c) skull of *Typhlops* (Redrawn from Mattison, 1995); (d) reconstruction of skull of *Dinilysia*; (e) skull of *Python* (Redrawn from Rage, 1984); (f) skull of *Atractaspis* (Redrawn from Mattison, 1995); (g) skull of *Dibamus* (Redrawn from Rieppel, 1984); (h) skull of *Heloderma* (Redrawn from Rieppel, 1980a).

whose ancient snake lizard ancestor was a "snake" in every sense because of its skull, not because of its body, axial and appendicular skeleton, and musculature—it was a four-legged lizard with a snake skull architecture. What happened over snake lizard evolution? The four-limbed members of the ancient snake lizard clade have obviously all gone extinct. It is even likely that snake lizards were the first lizard clade, shortly after the early-mid Triassic divergence event that led to lizards among lepidosaurian diapsids (Simões et al., 2018a) experimenting with limb reduction, limb loss, and axial elongation. Snake lizards are an ancient clade—imagine, 100 million years into the future, if all modern legged scincids have gone extinct, and the clade has radiated around the limbless forms surviving among only the Lerista–complex.

By using a new category, the term "snake lizard" helps to constrain us to terms and definitions that permit ontologic and epistemic consideration of "ancient snake lizard" in a sensible fashion. To do otherwise is simply to continually push a circular concept of modern snake further and further into the past. Snakes with legs are real entities defined by fossil forms, not extant forms (e.g., Caldwell and Lee, 1997) because snakes are snake lizards. Four-legged forms with short bodies also existed, whether their fossils were ever formed or will ever be found. So, what is a snake? It is a "snake lizard," which is certainly a synonym of "snakeness," but one that is a more effective communicator of what is of epistemic concern, is ontologically delimited, and generates new questions in a manner consistent with what we claim we know—the terms "snake" and "snakeness" are thus abandoned for the rest of this treatise, though we can certainly now discuss "ancient snake lizards" and "modern snake lizards" and have these terms make perfect sense. And, to be clear, "ancient snake lizards" are most certainly members of the ingroup Serpentes and thus can also be thought of as "stem group snake lizards."

## "What Is a Snake Lizard?"

Like all groups of lizards, the earliest snake lizards would have been four-legged animals with relatively short bodies, long tails, and heads that had already evolved a number of cranial specializations that still characterize modern snakes today. Modern snakes are uniform in their shared absence of limbs, though many non-colubroid forms still possess pelvic girdle remnants; the key point to remember, though, is that axial elongation and limblessness are not unique to modern snake lizards among lizards, nor did they characterize the earliest snake lizards; this state of limblessness and elongation is a specialization of highly derived snake lizards that shows up in the fossil record at least 100 million years ago. Among modern snake lizards, these features have aided them in the relatively recent adaptive radiation around the globe into a myriad of habitats and niches. But, it is not and was not the key morphological specialization that characterized the origin and evolution of snake lizards hundreds of millions of years ago. Were it the other way around, and limblessness and elongation evolved first and the snake skull was that of a lizard of a non-snake kind, then snake lizards would be the only group of lizards to have evolved in a non-specialized head manner first. Such a pattern of body before head cannot be found in any group of living or fossil lizards. So if anguids, skinks (and dibamids), pygopodids, and teiids all have legless and elongate members within many subfamilial lineages, all of whom retain the cranial architecture of their clade but have lost their limbs and evolved axial elongation, why would we expect snake lizards to show the reverse and be derived instead from some generic legless lizard (Table 1.1) of another non-snake kind? This makes no sense except in the light of essentialistic expectations of discovering snakeness by reference to snakeness. I personally reject the non-logic of this approach and the historical and intellectual burden it represents.

From a paleontological and evolutionary perspective, the next question must be, "When did this happen?" In other words, when in the rock record, and where as well, should paleontologists go hunting for new fossil snake lizard remains? My ballpark guess would be some time prior to the appearance of Eophis, Diablophis, and Portugalophis in the Middle to Late Jurassic (>170 million years ago) (Caldwell et al., 2015). If the Middle Triassic datum of the stem lizard Megachirella is any indication, I would suggest beginning the search from the Middle Triassic forward to Middle Jurassic (Simões et al., 2018a). The entire Jurassic remains largely unexplored for snake lizards and other lizards generally, and so will no doubt produce unexpected twists and turns in the story of snake lizard evolution, both in terms of diversity and morphological disparity (and thus unexpected sistergroup relationships and snake lizard clade origins). But, at this point, it seems abundantly clear that the Triassic Jurassic needs the attention of those researchers deeply interested in the next major discovery in the story of snake lizard evolution, for it is here that the fossil record data point to the deep origins of lepidosaurians within this seemingly explosive radiation of Pangaean sauropsid reptiles. If this is the case, then why has there been so much interest and seemingly so much data to support an origin of snake lizards in the later part of the Early Cretaceous (e.g., Longrich et al., 2012; Hsiang et al., 2015)? As I will argue throughout this book, we have suffered for a long, long time the intellectual burden of a top–down view of the organization of life, snake lizards included, and I suspect the Early Cretaceous represents just that—the origin of the modern snake lizard "archetype," but certainly not the ancient snake lizard from the very blurry concept of a Middle Triassic–Middle Jurassic ancestral snake lizard.

Despite this top–down problem, the modern assemblage is a critical starting point for any student of snake lizard evolution, both in terms of species and clade diversity and in terms of morphological disparity and anatomy. It is therefore necessary to understand modern snake lizard anatomy and morphology in order to look backward into the fossil record and begin to understand the fragments we find preserved there. The key step after that, though, is to remember you are looking top–down, that you have formed your assumptions and preconceptions of what you saw first, but now must reverse that view and inform the present using the data of the past.

## THE SKULL OF A MODERN SNAKE LIZARD

To be clear, there is no such thing as a typical modern snake lizard, which of course is one of the most profound problems systematists and comparative anatomists alike have faced in trying to understand the evolution and phylogenetic relationships of modern snake lizards from the "within-group study" of the modern assemblage. I have been straightforward in my certainty that the fundamental key to the problem lies in the fossil record, as we are indeed trying to unravel a seemingly infinite number of events that began hundreds of millions of years ago and have not

stopped creating the modern assemblage and its morphological disparity ever since. The problem, of course, is that the modern assemblage of snake lizards are all currently alive as individual entities and preserve no direct information on what their ancient progenitors looked like in the past. While we can link modern snake lizards to each other with reasonable certainty in sensible clade structures; that is, the various pythons look like each other and share numerous similarities with boas, the task is made much more difficult when linkages are sought between pythons and scolecophidians, or anilioids and viperids. Not only are there little anatomical data that overlap between these forms (scolecophidians and any other snake, anilioids and vipers, etc.) and the clades we have constructed to contain them, they also do not come with a date stamp on the geologic time period of origin of each clade (and it is likely very recent for the balance of modern species, as they are all colubroids anyhow!), and the clades are so derived that they provide virtually no information on their presumably common ancient snake lizard ancestor. Cundall and Irish (2008:593) phrased the problem rather nicely in their summary paragraph, though at the time they did not hold out much hope for deliverance via the fossil record:

> Like most diverse clades, snakes present a fascinating evolutionary puzzle. Despite the wealth of information on their skulls, links between major clades are missing, and hence, the origins of the major skull design patterns remain obscure. Continued analysis of skulls alone may not solve these problems.

To put it another way, the problem has been like trying to understand apples by only studying apples—you can come to understand what they are, but not what they were before they became what they are. And, should the problem of origins be so complex that modern kinds of apples do not include representatives of ancient apple kinds (i.e., some of these lineages are now extinct), then the apple data you need to sort out this problem of relationships and origins is unavailable. The same is true of modern snake lizards—despite the diversity and disparity of the modern assemblage and its numerous living lineages, there are far more extinct ones than living ones and thus the necessary key data are absent if the only data being examined come from the modern assemblage.

THE ORIGIN OF SNAKES

In trying to solve this problem, as noted here, and as quoted above from Cundall and Irish (2008), numerous authors, for several hundred years, have studied, pondered, and debated, hypothesized and then lucubrated to the annals of scholarly history their legacies of thought on the development, morphology, evolution, anatomy, and function of the vertebrate skull and skeleton, and, more specific to my goals, on the skulls and skeletons of lepidosaurs, including sphenodontians, typical lizards, and, of course, snake lizards. The list of smaller single-paper studies is virtually endless and can be found populating the pages of the "References" chapter of this book. On the other hand, there is also a long list of true monographic studies on vertebrate skulls, and more specifically the skulls of lizards, some specific to snake lizards, that are truly invaluable in their length, comparative intent, and content. Most notable and of impact are the works of Cuvier (1805, 1824), Huxley (1858, 1864), Parker, 1878, Gaupp (1900), De Beer (1937), Romer (1956), Bellairs and Kamal (1981), Hanken and Hall (1993a,b,c), Evans (2008), McDowell (2008), and Cundall and Irish (2008). In fact, from my perspective, the last one in that list, Cundall and Irish (2008), is perhaps the most important contribution on snake lizard skull anatomy and evolution to date, both in terms of thoroughness and quality and quantity of illustration, and consideration of the fossil record, even though it is only 10 years old now.

As I have argued in this chapter, will argue in the rest of this book, and have argued in the literature (e.g., Caldwell et al., 2015), the head is the thing that has driven the innovation and evolution of snakes, not the body. To characterize the skull of modern snake lizards, I have played a little game, Sesame Street style, called "one of these things is not like the other," which at the end leaves me with four snake lizard skull groupings. What I end up with in this arrangement of morpho-groupings around skull anatomy, which is difficult considering the bewildering disparity of modern phenotypes (Cundall and Irish, 2008; McDowell, 2008), are the following four categories: (1) Type I–Anilioid; (2) Type II–Booid-Pythonoid; (3) Type III–Colubroid (excluding viperids and elapids); and (4) Type IV–Viperid-Elapid-Scolecophidian.

Considering there are nearly 4000 living species of snakes, this seems like a small number of skull types. I am fully cognizant of the accusations of "glossing over disparity" that the apparent limitations of just four categories seems to bring. However, doing so has a purpose, and that is to build from the general to the specific condition of form in the organization of the skulls of modern snake lizards, recognizing, of course, that the general condition of form is problematic as a first instantiation of a category. Within this metaphor of form, I am applying a von Baerian–style approach to categorizing adult modern snake lizard skull morphology around a small set of generalized morphotypes (von Baer, 1828); that is, I am trying to avoid getting lost in the forest for the very reason that it is full of individual trees.

Apart from identifying my reasons for trying to identify the most general condition of form (an essentialist category in the extreme), a second challenge is to reinforce the understanding that these four categories are by no means tightly defined skull groupings, quite the opposite, in fact. At the base of each "cloud" of skulls (not at its centroid), making each cloud effectively a cone in space, is an essential type for the category, and all of the "like" skulls hover within that spatial cone like bees around a cone-shaped hive; one of the categories has two seemingly different skull types in it, sitting at its base together, which, as you will see based on my criteria, are not so different as they might seem on first inspection (Type IV–Viperid-Elapid-Scolecophidian). There are points in these cone-clouds with extreme morphologies compared to the base skull(s), and they define the boundaries of these very fuzzy cone-cloud sets with little or no overlap with other sets. Some, like the Type I–Anilioid, are easier to define because the numbers of species displaying that skull type are small (thus, small cone-cloud); others, such as Type III–Colubroid (excluding viperids), include thousands of skulls and are potentially divisible into numerous smaller sets of cone clouds, an argument I could accept, but the detail of which is not relevant to my core argument. In any case, I have kept my criteria for recognizing skull types extremely simple, in part to downplay unique morphologies of a kind or of a particular bone and to focus instead on broad, shared similarities. In constructing exemplars of the base of my cone-clouds, I have tried to avoid functional or evolutionary transformation arguments in the delimitations, as much as I have avoided deferring to the received wisdom of sistergroup relationships among groups. I do recognize, though, that theory-free observation is impossible and will admit it seeps into my categories regardless of the efforts to expunge it. The phylogenetic implications of my categories are obvious, and are also most certainly linked to my understanding of the skull categories present among the fossil snake lizards profiled in this book. Despite this, my goal remains to present broader skull groupings that can be used to create general conditions of form that serve as predictive tools for modeling for the skull and skeleton of ancient snake lizards.

I will build toward the characterization of my four skull categories with a review of snake lizard skull anatomy, development, and embryology, framed around the basics of the vertebrate skull, and finish with the descriptions of the categories. It is essential to know the modern snake lizard skull inside and out if the ancient snake lizard skull is to be conceptualized in any sensible manner. I also hasten to point out that this cannot be a comprehensive review of snake lizard skull anatomy on the scale of Cundall and Irish (2008), a study which I think I have made clear I admire greatly for its thoroughness and approach, and refer the student of snake lizard skull anatomy to that treatise for details that cannot be presented here.

## The Basics of Skull Basics

*Development Units*—The vertebrate skull is composed of three developmentally distinct components— (1) the chondrocranium, which is preformed in cartilage, and in lower vertebrates such as sharks, is composed only of cartilage, while in higher vertebrates, such as snake lizards, it is formed of cartilage and endochondral bone (see de Beer, 1937); (2) the splanchnocranium or visceral arch skeleton (similar phylogenetic distribution of cartilage and bone as the chondrocranium; and (3) the dermatocranium, which is intramembranous bone that does not have a cartilage precursor stage and forms by osteocyte differentiation and accretion directly within the mesoderm. In snake lizards, the chondrocranium forms the floor and walls of the bony and cartilaginous braincase inclusive of the nasal capsule, orbits, and braincase *sensu stricto* plus the otic capsules (basioccipital, basisphenoid, prootics, exoccipital-opisthotics/otoccipitals, supraoccipital). The splanchnocranium forms the core of the upper and lower jaws (the Palatoquadrate and Meckel's cartilages, respectively, though these cartilages are sheathed by dermatocranial bones),

and also forms the cartilage precursors of the quadrate, articular, and stapes/columella. The dermatocranium forms the roof of the braincase, the bony palate below the chondrocranium; sheaths the upper and lower jaws with a variety of elements; and supports and covers the remainder of the snout, face, and temporal regions (i.e., premaxilla[e], and the paired maxilla, nasal, frontal, prefrontal, parietal, supratemporal, squamosal, postorbital, postfrontal, jugal, lacrimal, dentary, splenial, coronoid, prearticular, surangular, parabasisphenoid, vomer, septomaxilla, palatine, pterygoid, and ectopterygoids).

Embryologically, the skull is an exceedingly complex region of the body; our understanding today is well beyond the pattern observations of de Beer (1926, 1937), and it is now understood, though very incompletely, at the cell proliferation and germ cell level (Hall, 2005). The cartilaginous and bony tissues of the three developmental units develop around incredibly complex interactions between all three primary germ cell lineages (ectoderm, mesoderm, and endoderm). However, the most profound impact on skull bone and cartilage development, which is now only beginning to be understood, begins early in embryogenesis as neurulation begins and the neurectodermally derived neural crest cells proliferate on the left and right crests of the developing neural tube. As these cells proliferate and migrate around the head and body, they interact with the undifferentiated and differentiated mesenchyme that will become the chondrocranium, splanchnocranium, and dermatocranium in the formation and specification of position, form, and complexity of those cartilages and bones. The details of these processes, and the patterns they create, are well beyond the scope of this book (see Hall, 2015), but they are critical to the evolution and development of the snake lizard skull, for here we find the embryological and developmental processes that control many aspects of morphology—perturb one or two steps early on in development, perhaps by just minute variations of even the timing of expression for one or two genes, and the end result is altered morphology. Add in more complex variation driven by mutations, plus the spectrum of poorly understood epigenetic effects on phenotype, and the complexity becomes beyond conceptualization.

*Kinetic Units*—The kinesis or mobility of skull elements, particularly those that can be functionally linked to feeding and the ingestion of prey, has rendered the snake lizard skull unique among amniote tetrapods; even by comparison to typical lizards, which do show some degree of skull element kinesis, snake lizards have taken kinesis to its acme. In fact, skull element kinesis is so pronounced that the highly derived skulls and skull elements of many viperids, as an example, are nearly as kinetic as the skulls of many modern teleost fishes.

Considering that skull kinesis is a defining feature of the snake lizard skull, it requires some definition, as the presence and degree of this kinesis will require characterization for understanding and modeling the ancient snake lizard skull. In the sense I am using the word kinesis, it reflects two possibilities for movement: (1) flexion, extension, protraction, retraction, levation, rotation, and so on, at an actual joint surface, whether it is syndesmotic, synarthrotic, or synovial; (2) the passive or active extension–contraction of a ligament or tendon connecting elements that do not share a bony joint surface. It is critical to understand kinesis, as the snake lizard skull is a complex construction of bony elements, some of which are firmly sutured, if not fused, to each other (e.g., the bony chondrocranium), around which are suspended by ligaments and tendons, and a couple of true synovial joints, a large number of bony dermatocranial and splanchnocranial elements.

The cranial skeleton of a snake lizard can be considered in three parts, the snout or nasal capsule, the orbital region, and the braincase + otic capsule. In the typical vertebrate condition, the cranial region demonstrates kinesis at three joints, the prokinetic, mesokinetic, and metakinetic joints, all of which define the boundaries between the three cranial compartments (nasal, orbital, braincase/otic). In snake lizards, the general condition is enhanced prokinesis at the naso-frontal joint as compared to typical lizards (with the addition of a rhinokinetic joint between the premaxilla and nasals), while the meso- and metakinetic joints essentially become rigid and fixed (in typical lizards, the mesokinetic joint is between the frontals and parietals, while the metakinetic joint is located between the supraoccipital and parietal[s]). As an exception to this general statement of cranial kinesis, the reverse is observed in the blind, burrowing scolecophidians, where the prokinetic joint is effectively reduced to zero (though other elements

of the joint characterization remain, that is, contact of the posterior process of the septomaxilla with the base of the paired frontals), but because of reduced ossification, the cranial skeleton is now quite flexible at the mesokinetic joint, and to a lesser degree at the metakinetic joint boundary.

While it is not conventional to do so, I have separated the maxillary and pterygopalatal skeleton from the cranial and snout skeletons in snake lizards and linked them together in my consideration of the varied skull units. I have done so in violation of the convention of considering them with the cranial and/or snout skeleton because these bones are not integrated or fixed to any one unit of the cranial or snout skeleton, but rather share a varying number of points of contact linked to great or lesser degrees of maxillary and integrated pterygopalatal mobility that functionally correspond to the degree of kinesis required for snake feeding (as widely varied as it is). The maxillary skeleton does share a variably kinetic series of joint surfaces in most snake lizards with the prefrontal and palatine anteriorly and the ectopterygoid-pterygoid complex posteriorly, making it a functional unit in all modern snake lizards, even if it is not mobile in some groups.

Bony Skull Units—I am recognizing here six skeletal/functional/evolutionary units in the snake lizard skull that are independent of each other on the one hand (position relative to each other) but are also clearly linked to each other at differing levels of organization, for example, their unique function (feeding mechanics), ontogeny and development (chrondrocranial, dermatocranial, or splanchnocranial origins), or anatomy (joint, ligament, and tendon systems). These six units are: (1) cranial, (2) circumorbital, (3) snout, (4) maxillary–pterygopalatal, (5) suspensorial–mandibular, (6) hyobranchial.

## Skull Units: Details

*Cranial Unit, Nonscolecophidians*—For the purpose of highlighting significant variation within and between modern snake lizards, I will describe scolecophidian snake lizard cranial anatomy separately from the characterization of all other modern snake lizards. The cranial unit, or cranial skeleton, includes the chondrocranial and dermatocranial elements forming the floor, walls, and roof of the brain and sensory organs, from the nasofrontal joint moving posteriorly to the exit of the spinal cord from the foramen magnum. The chondrocranial and dermatocranial bones of the typical nonscolecophidian snake lizard cranial skeleton include the frontals, unpaired parietal, supraoccipital, prootics, exoccipitals-opisthotics [otoccipitals], basioccipital, basisphenoid, and parasphenoid rostrum. The cranium is characterized by the virtually complete bony enclosure of the brain and optical and olfactory tracts to form a solid tube of bone, and forms in three parts: (1) by the development of descending flanges of the paired frontals that posteriorly contact the anterior descending processes of the unpaired parietal, enclose the olfactory tracts (often with paired median frontal pillars in higher snake lizards), and meet at the ventral midline around the elongate basisphenoid rostrum where that process extends anterodorsally between the orbits—the effect is the formation of a bony medial wall of the orbit with an optic foramen usually shared with the parietal; (2) by the development of descending flanges of the parietals around the fore- and midbrain that anteriorly meet the descending processes of the frontals, posteriorly contact the expanded alar process of the prootics, and ventrally contact the dorsal edges of the right and left margins of the basisphenoid and parabasisphenoid; and (3) by the expansion of the prootic alar process to form a continuous lateral contact with the parietal as well as an elongate contact dorsally and posteriorly with the parietal, to the contact with a significantly expanded supraoccipital that also broadly underlaps the parietal. Meso- and metakinesis are thus lost in typical modern snake lizards due to significant bony overlap/underlap at the frontoparietal contact as well as posteriorly with the expansion of the supraoccipital. Importantly, the supraoccipital also forms a significant portion of the roof of the bony labyrinth of the otic capsules (and contains portions of the semicircular canals) and, sagittally, the braincase. The bony labyrinth and chamber of otic capsules are floored by the basioccipital, the lateral and medial walls are shared by the prootic and exoccipital-opisthotic and the roof by the supraoccipital; the labyrinth for the semicircular canals weaves around and through the walls of the prootic and exoccipital-opisthotic.

In very general terms, the typical non-scolecophidian modern snake lizard braincase demonstrates a general condition of proportions where the snout and orbital regions, as defined by premaxilla and nasals and the paired frontal bones, respectively, are short in comparison to the parietal parasphenoid rostrum portion, which is generally elongate and expanded (in typical lizards, the opposite condition holds, where the snout and orbital region are elongate, while the cranial region is quite short). In modern snake lizards, the otic region varies in mediolateral dimensions (i.e., narrower than the width of the braincase at the prootic contact with the parietal, to bulbous and inflated posterior to that same sutural contact), but is generally shortened in relationship to the parietal contribution of the braincase.

Other notable characters of the typical modern non-scolecophidian snake lizard cranium include the presence of a "laterosphenoid" ossification dividing the lateral opening of the trigeminal foramen within the prootic into two large superficial foramina ($V_2$ Maxillary Branch exits anteriorly and $V_3$ Mandibular Branch posteriorly), and often even a third much smaller and more ventral foramen is present ($V_4$ Dorsal Constrictor Branch), along with a fourth for a branch of V4 for the protractor pterygoidus nerve. While much has been written on the embryological and developmental origins of this pillar or plate of bone and thus its amniote homologies (e.g., from Parker [1879] to Romer [1956] to Rieppel [1976], Haluska and Alberch [1989], Presley [1993], Cundall and Irish [2008], and McDowell [2008]), these homologies remain unclear. Classical embryologists (e.g., Parker, 1879; Peyer, 1912) referred to the "bony strut" as the alisphenoid, the term applied to a similar strut/plate in mammals. As noted by Rieppel (1976) and Gregory and Noble (1924), seeking to differentiate this bony strut in snake lizards from that of mammals suggested the name "laterosphenoid." De Beer (1926:312) refers to and describes his "laterosphenoid" in his short paragraph on "Lacertilia" distinct from snake lizards in this case, as:

In some forms the pila metoptica may ossify. To this bone, a post-optic ossification of the original skull-wall, the name laterosphenoid must be applied.

Reading de Beer (1926) must be done carefully because the nomenclature is not the same as current usage and it is possible to mix up the terms and thus nerve trunks and homologies (e.g., he refers the trigeminal nerve in three parts, "profundus" [which in snake lizards runs deep/medial to the prootic], V1 and V2, which in modern usage are the $V_1$ Opthalmic Branch, $V_2$ Maxillary Branch [which has two large sub-branches], and $V_3$ Mandibular Branch; de Beer [1926] makes no mention in his text of the $V_4$ Dorsal Constrictor Branch). The laterosphenoid term that de Beer (1926) applies to lizards is not the same structure he refers to as his "pleurosphenoid" in snake lizards, which arises from the basal plate of the chondrocranium to meet or almost meet the anterior tip of the auditory capsule of the incipient prootic, thus forming a prootic cleft (incisura prootica) for the exit $V_2$ (Maxillary Branch) and $V_3$ (Mandibular Branch) from the braincase. De Beer (1926:331) qualifies his laterosphenoid bone as:

In snakes the cartilaginous primitive skull-wall is much reduced. The foramen prooticum appears not to be bounded by a true pila prootica, but a cartilage appears which may ossify into a laterosphenoid.

The pleurosphenoid process, which is neither a pila antotica nor a pila metoptica, arises from the basal plate to meet what de Beer (1926) recognized as a cartilage element that appears near the alar process of the prootic (which is not a remnant pila prootica) to co-ossify with the prootic first, then to co-ossify with his pleurosphenoid process. As described by Rieppel and Zaher (2001) for Achrochordus, the laterosphenoid clearly has two components: de Beer's (1926) "pleurosphenoid" is the more basal and chondrocranial portion, while the dorsal portion appears to be derived (contra de Beer [1926]) from a presumably dermatocranial ossification (see Bellairs and Kamal, 1981) associated with the prootic. To provide important phylogenetic context, de Beer (1926) was studying embryos of Tropidonontus, or, in modern parlance, the colubrid snake lizard Natrix natrix (the grass snake) and the elapid snake lizard Pseudechis porphyriacus (the red-bellied black snake), both of which belong to higher colubroids, while Achrochordus is generally in the sistergroup position to that group. A laterosphenoid is certainly present in modern alethinophidian snake lizards (see Cundall and Irish, 2008) but is absent in scolecophidian snake lizards, even though embryological studies have not sampled other alethinophidians as heavily as they have higher snake lizards.

The question is the utility of the "unique" laterosphenoid as a defining feature of all snake lizards, with the exception of course of scolecophidians where it is absent (for a concise review, see Cundall and Irish, 2008), particularly when fossil snake lizards are considered (see later chapters in this book) where the laterosphenoid is absent (e.g., Dinilysia patagonica [Estes et al., 1970]). Does this mean they are not snakes lizards, as, yet again, a feature seemingly defining all modern snake lizards except for the obviously primitive scolecophidians is absent?

Following on a subtheme of this book, which is that scholars and their science naturally adopt a top–down approach to understanding the world in which we live, I continue with an examination of the snake lizard autapomorphy of the laterosphenoid (note: I am considering the more dorsal component of de Beer [1926] and the same of Rieppel and Zaher [2001] when I speak of the laterosphenoid here). Presley (1993) wrote a very interesting paper about the problematic use of adult osteology to look for homologs at earlier stages in an organism's ontogeny when all that can be observed are non-osteological features. In this light, Presley (1993) assessed de Beer's (1926, 1937) conclusions on the laterosphenoid in snake lizards and presented a consequential counterargument to the long-held position expressed by de Beer (1926, 1937) that the laterosphenoid could not be derived from the primitive vertebrate pila lateralis. De Beer (1926) reached this conclusion based on his observations of the position of the lateral head vein in Natrix (common grass snake) as compared to the lateral head vein in the chondrostean fish Amia. As Presley (1993) noted, de Beer's (1926, 1937) conclusions on the laterosphenoid are completely dependent on the identification and thus primary homology of the lateral head vein of Natrix, and thus all snake lizards, as being the same vein as he illustrated in Amia. In other words, de Beer's (1926, 1937) topological criteria may well have been in error regarding his certainty that the snake lizard "cartilage ossification" he observes ossifying to the prootic is not a portion of the pila lateralis (if Bellairs and Kamal [1981] are correct and this is dermatocranial ossification, then both de Beer [1926] and Presley [1993] are in error, and Rieppel [1976] is correct). Presley's (1993) review of de Beer (1926, 1937) strongly indicates that the latter may have been incorrect in the identification of the lateral head vein, as his illustrations of both species (Fig 40 [de Beer 1926:291] vs. Fig. 85 [de Beer 1926:314]) do not identify the same venous structures in relationship to the pila lateralis and the trunk and roots of the trigeminal nerve in *Amia* as compared to *Tropidonotus*. If the venous homologies are inaccurate, argues Presley (1993), and these landmarks are de Beer's (1926, 1937) "topological relations" (sensu Patterson, 1982), then as Presley (1993) argues, the embryonic chondrocranial structure of the pila lateralis can be hypothesized to be the homolog of the similar cartilage leading to the laterosphenoid in snake lizards. Importantly, this means that contra Rieppel (1976), the laterosphenoid is potentially not a de novo ossification in snake lizards and thus yet another required synapomorphy for membership in the group. Also, if the laterosphenoid simply reflects ossification of the chondrocranial pillar of the pila lateralis to link with the prootic, then it is, in fact, a much more basal character that is not lost in snake lizards generally and reacquired in some special manner by a de novo evolution of a basal plate feature linked to a new condition of the prootic. Scolecophidians might very well represent nothing more than a pedomorphic condition where basal plate development remains at a much earlier stage of ontogeny and never produces the pila lateralis in the first place, or perhaps it fails to ossify; unfortunately, because of accessibility, the embryology of scolecophidians remains extremely understudied.

As pure speculation, I would also suggest a fourth possibility, in consideration of the unusual development of the pterygoquadarate cartilage in snake lizards. The pterygoquadrate cartilage is splanchnocranial in origin (which means, like all cranial bones, it is significantly influenced by neural crest cell activity) and in typical lizards during early craniogenesis is a three-pronged element composed of quadrate and ascending pterygoid processes; as craniogenesis proceeds, five cartilage elements remain, with the ascending process later ossifying to become the epipterygoid and the posterior ramus forming the quadrate (there are three other more basal remnants as well). In snake lizards, the only splanchnocranial element formed in cartilage from ptergyoquadrate condensation is the quadrate proper. As purely speculative as this may be, in consideration of the loss of the cavum epitericum during snake lizard embryogenesis and the relative position of the epiptergyoid condensation to the pila lateralis,

basal plate (pleurosphenoid eminence of de Beer [1926]), and the prootic with the "laterosphenoid" ossification of uncertain dermatocranial (Bellairs and Kamal, 1981) or cartilage origins (de Beer, 1926), it is possible that this remnant ossification is derived from the remains of a splanchnocranial epipterygoid condensation.

With that said about the snake lizard laterosphenoid, and even though this book is certainly about snake lizards, it is worth briefly comparing the braincase/cranial skeleton to a typical non-snake lizard, even though very little of the snake lizard braincase/cranial skeleton is similar to that of a lizard (see above as an example). The typical lizard braincase (amphisbaenians and dibamids excluded here—see Evans [2008]) is not a complete cranial tube or cylinder as in snake lizards. The roof (frontal[s], parietal[s], and supraoccipital) and floor (basioccipital and basisphenoid) are relatively complete, as are the walls of the braincase at their most posterior (opisthoic, exoccipital, prootic), but anterior to the prootic contribution to the trigeminal nerve trunk exit below the prootic alar process, the braincase receives no bony contribution from either the parietal or frontal (though some lizards have a small section of the olfactory tracts surrounded by decensus frontali that wrap around the tract and meet at the midline). The forebrain and parts of the midbrain are instead surrounded by a dense sheet of fibrous tissue—a membranous braincase (see Bahl, 1937; Evans, 2008)—that is supported on its leading edges by the right and left paired orbitosphenoids with the parasphenoid rostrum supporting the base on the midline. The optic nerves (Cranial Nerve II) in typical lizards exit the membranous braincase via the optic septum created along the leading edges of the paired, hatchet-shaped orbitosphenoids. In contrast to snake lizards, the membranous braincase and the chondrocranial braincase (ossified and cartilaginous elements) are perforated by nerves and vasculature and are supported in the midline by the parasphenoid rostrum as it rises to the frontals between the eyeballs; by the cartilaginous cranial skeleton; and by a number of small bony ossicles, for example, the orbitosphenoid and epipterygoid, variously articulating with the cartilaginous cranial skeleton and the chondro and dermatocranial elements and connected as a whole by the membranous braincase. It is, indeed, quite different in appearance, though not so much

THE ORIGIN OF SNAKES

as it would appear—the remarkable differences really focus on the degree of development of the decensus frontali and parietali and the fate of the few bony elements associated with the membranous braincase anterior to the prootic. Other than that, nerve exits remain at similar junction points in the braincase, the supraoccipital forms the roof of the bony otic capsule, and the metotic fissure is common to both in position and effect on the anatomy of the otic capsule and vagus/jugular foramen.

*Cranial Unit Scolecophidians*—The cranial unit of many scolecophidians varies in important ways from that of other modern snake lizards (Cundall and Irish, 2008), while some scolecophidians are surprisingly similar (Pinto et al., 2015). Though the details of these differences are often ignored, the reasons for these differences are important, but are, of course influential on how scolecophidians relate to all other modern snake lizards. There are two principal and somewhat contrasting hypotheses that drive the polarity of primitive versus derived for all modern snake lizards, and, of course, because they are essential characters of snakeness, for all ancient snake lizards as well: *Hypothesis One*—these morphologies are primitive/plesiomorphic for all snake lizards and represent retained lizard characteristics in what is clearly the primitive, blind, fossorial stage of snake lizard evolution (Rieppel, 2012). *Hypothesis Two*—these are highly derived and extremely variable anatomies representing, among many things, heterochronic developmental processes (pedomorphosis, neoteny, etc.) and are not primitive in the least, suggesting scolecophidian snake lizards are highly derived snake lizards; not at the base of the modern snake lizard radiation; and certainly not representative of all snake lizards, including ancient snake lizards (Caldwell, 2007a).

Beginning with the frontals, all modern scolecophidians, and there are no exceptions I am aware of, share with all other modern snake lizards the possession of paired frontals. However, though the frontals are paired, List (1966) illustrated a specimen of *Leptotyphlops bakewelli* that also displayed a small sagittal fragment of bone toward the anterior portion of the suture separating the right and left frontals. And posteriorly, the second morphology that appears to be common to scolecophidians and other modern snake lizards is that the exoccipital and opisthotic are always fused to form the otoccipital (even though, as I

will discuss, the prootics and supraoccipital[s] can variably fuse to the exoccipital-opisthotic). Apart from these two morphologies, there appear to be few commonalities between scolecophidians and between scolecophidians and other modern or ancient fossil snake lizards (to date, ancient fossil snake lizards possess similar cranial skeleton anatomies with all other modern snake lizards except scolecophidians).

Cundall and Irish (2008) summarized in detail the research on the wide variety of scolecophidian cranial skeletons, and there are numerous more recent studies that have added to that work (e.g., Rieppel et al., 2009; Pinto et al., 2015). I will summarize it here. The parietals can be paired, unpaired, and semifused (i.e., an unfinished suture is apparent between the right and left sides). The supraoccipital is sometimes present as a single element; in some taxa as a broad, more typical snake lizard element; paired in some scolecophidians; or fused into a single element with the prootic and exoccipital-opisthotic. As already mentioned, the prootic often fuses with the exoccipital-opisthotic and in some taxa includes the supraoccipital so that the entire otoccipital region is a single mass of bone. In some anomalepidids, the supraoccipital is a single median element that does not participate in forming the lateral roof of the bony otic capsule—this is a radical departure from the usual lizard condition, including all other snake lizards.

Other features of consequence, as noted above for non-scolecophidian snakes, include the laterosphenoid ossification and the exits for Cranial Nerves $V_2$ and $V_3$. In scolecophidians, the laterosphenoid is generally considered absent, though there are exceptions (Abdeen, 1991a,b,c; Cundall and Irish, 2008). However, my remark on this character, though it might be considered plesiomorphic for all modern snake lizards, is to point out that there is seldom a defined trigeminal foramen at all (there are some scolecophidians, mind you, where the foramen is well developed and contained within the prootic as in most modern snake lizards), but rather just a gap between the prootic and parietal. This appears to be the result of significantly reduced ossification at the suture between the dermatocranial descending flange of the parietal, coupled with reduced chondrification and ossification of the chondrocranial prootic such that the bones do not meet, and, more importantly, a trigeminal foramen does not form within the prootic. It is also clear in some scolecophidians that the basisphenoid might also contribute to this "foramen," as the trunk of the trigmeninal nerve is simply exiting the braincase at the three-point gap created by the basisphenoid, prootic, and parietal. The absence of a laterosphenoid ossification is not surprising considering that the foramen itself is generally absent to poorly defined. In any case, none of these conditions of form for the foramen resemble at all the condition of a typical lizard (trigeminal foramen exits below the alar process of the prootic in the prootic notch, and there is no descending flange of the parietal, and so the anterior margin of the foramen is contained within the tectal membrane, not within the prootic, and not within the parietal).

*The Circumorbital Unit*—The circumorbital unit of modern snake lizards includes the bones framing the orbit anteriorly, and in part dorsally and posteriorly, but does not include the frontal where it frames the superior portion of the orbit, nor the maxilla where it underlies the orbit; these two elements are grouped here within the cranial and maxillary-pterygopalatal-mandibular units. The circumorbital unit includes the jugals (past convention referred to it as the postorbital [see Chapters 2 and 3], but I follow Palci and Caldwell [2013]), the postfrontals, prefrontals, and supraorbital/palpebrals (when present). In modern snake lizards, the circumorbital series is variably reduced from the complete series listed here to only the prefrontal in the most extremely reduced variation of form, for example, most scolecophidians (though some retain a free-floating jugal/postorbital [Cundall and Irish, 2008; Rieppel et al., 2009), anomochilids, uropeltids, and atractaspids. When the series is complete, the jugal of modern snake lizards is very variable in form, but is always in contact with the posterior border of the postfrontal posterior to the frontoparietal suture when a postfrontal is present. The jugal seldom extends ventrally to contact the posterior ramus of the maxilla, but can often be close to the maxilla and the lateral process of the ectopterygoid (together, these three, if the jugal is present, share the insertion of the quadratomaxillary ligament [see Palci and Caldwell, 2013]). The posterior process of the jugal, that is, the quadratojugal process, is lost in modern snake lizards, as is the anterior or suborbital ramus of the jugal. The supraorbitals, or palpebrals, are variably present in modern snake lizards, lying along the dorsal orbital margin when present, in a position just

posterior to the frontal-prefrontal articulation. The postfrontal is a small, generally shield-shaped element that is topologically located at the fronto-parietal suture and may extend a variable distance above the orbit. In all modern snake lizards, the prefrontal is the only circumorbital element that is never lost and remains in its plesiomorphic position, lateral to the prokinetic joint at the naso-frontal contact, forming a relatively fixed joint with the frontal, and in many modern snake lizards, usually quite firmly with the nasals as well. Likely the most important contact, at least in terms of function, is with the maxillary, where it forms a rather rigid, though possibly flexible, joint in non-macrostomatans and basal macrostomatans and is extremely mobile in viperids and most scolecophidians. The prefrontal also maintains a varying degree of contact with the palatine and vomer.

*The Snout Unit*—The snout unit of modern snake lizards includes the dermatocranial bones of the nasals, unpaired premaxilla, vomers, and septomaxillae, as well as the chondrocranial skeleton inclusive of the nasal/sphenethmoidal cartilages. The complete suite of snout unit elements is present in all modern snake lizards, though there are several recognizable forms emphasizing more or less kinesis within the snout unit and between the snout and cranial units at the naso-frontal prokinetic joint. In all modern snake lizards, the premaxilla is relatively small, bears a short ascending or nasal process, seldom bears teeth (though teeth are present on the premaxilla in pythons and their kin), and is not fused to the maxillae, but may just slightly come into contact with the maxilla or be widely separated from the maxilla. The vomers generally form the floor of the snout unit, providing protection in the roof of the mouth for the sensory system of the Jacobson's Organ, extending from their contact with the premaxilla posteriorly to or close to the palatines and form a parasagittal partition of bone below the septomaxillae. In turn, the septomaxillae overlie the vomers, contact the premaxilla dorsal to the vomerine processes, underlie the nasals, and form a roof over the Jacobson's Organ as well as forming walls and protection for the nasal capsules and ethmoidal tissues (the sagittally placed nasal/sphenethmoid cartilage framework is medial to the septomaxillae, resting on or alongside the vomers). The septomaxilla is a complex and very thin scroll-like bone with multiple processes, the postero-median of which, in many snakes, extends to the naso-frontal joint and, along with the vomer and nasals, can create a highly specialized hingelike joint with the frontals (see Rieppel, 1978; Cundall and Irish, 2008). The nasals form the dorsal surface of the nasal capsule/snout unit, varying in form by either being severely reduced towards the naso-frontal or prokinetic joint, or being much larger and more platelike and partially to completely hiding the underlying septomaxillae in dorsal view.

The Maxillary-Pterygopalatal Unit—The maxillary-pterygopalatal unit of modern snake lizards includes the dermatocranial bones of the maxillae, ectopterygoids, palatines, and pterygoids. This suite of four elements, on any one side of the skull, functions as an articulating kinetic platform for the marginal and palatal attachment of teeth; only the ectopterygoid remains toothless in modern snake lizards, though the other three elements are variably toothless across the diversity of modern snake lizards. In most modern snake lizards, the maxilla articulates medially with the palatine, usually via a significant overlapping joint where a substantive lappet of the maxilla (palatine process of the maxilla) underlaps a usually larger maxillary process of the palatine. At its posteriormost extent, the suborbital process of the maxilla underlaps a clasping and/or overlapping process of the ectopterygoid, which has a ramus articulating with the ectopterygoid process of the pterygoid. The pterygoid is an extremely elongate element in all modern snake lizards that articulates anteriorly with the posterior ramus of the palatine, laterally with the ectopterygoid, dorsomedially with the pterygoid process of the basisphenoid (in modern snake lizards, varying from a stout process to little more than an eminence on the basisphenoid), and posteriorly, via its longest ramus, with, or close to, the medial face or facet of the quadrate. Anatomical variation between various clades of modern snake lizards around just these four bones is immense. In non-macrostomate anilioids (anomochilids are recognized here as derived, but likely derived from the anilioid condition [Cundall and Rossman, 1993; Cundall 1995; Cundall and Irish, 2008]), the palatine is more rigidly articulated with the maxillae in a clasping joint formed by the maxilla with the single process of the palatine; in addition, the maxilla is quite rigidly fixed to the prefrontal, thus reducing maxillary and, overall, maxillary-pterygopalatal kinesis with the palatine. As was noted by Cundall and Irish (2008), there is no notable mobility of the left or right tooth-bearing complexes, though the palatine-pterygoid joint is not tight, and the palatine can flex forward in a limited ratcheting maneuver.

In macrostomates such as pythons, boas, and so on, the degree of kinesis increases for each right left maxillary-pterygopalatal unit. This is accomplished anatomically, by a smoothing out of the prefrontal joint surface with the maxilla, overlapping joints at all contact points between the palatine and ectopterygoid with the maxilla, and a tightening of the palatine pterygoid joint so those two elements function like a single joint. In addition, and as noted above, the pterygoid processes of the basisphenoid are reduced to modest-sized processes to mere eminences if they are visible at all (e.g., tropidophiines).

In colubroids, not too surprisingly, the anatomy is extremely variable, with a number of typical colubroids resembling highly kinetic versions of more basal macrostomatans. However, among the more highly derived and diverse clades of colubroids, and not too surprisingly including scolecophidans, the dominant trend is a bizarre mix of element reductions coupled with elongations; a reduction of teeth and tooth attachment sites on the palatine and pterygoid, sometimes with loss of the maxillary dentition; and a shift of kinesis toward extreme maxillary mobility. And, finally, the pterygoids lose their articulations with the basisphenoid and the quadrate, as the quadrate ramus of the pterygoid becomes extremely posteriorly elongate. A unifying trend (not a character state or primary homology) is that as the maxilla shortens to the degree where it can support only one or two functional teeth and becomes highly mobile, articulating in a socket on the prefrontal (viperids and scolecophidians), the palatine becomes severely abbreviated, while the ectopterygoid and pterygoid become extremely elongated. Reduction is so advanced in scolecophidians that in some groups the ectoptergyoid is lost, or perhaps fuses to the posterior ramus of the maxilla (List, 1966; Pinto et al., 2015).

The Suspensorial–Mandibular Unit—The suspensorial–mandibular unit is composed of dermatocranial and splanchnocranial elements, including the supratemporals, quadrates, and the mandibular elements inclusive of the Meckel's cartilage, and the dermatocranial sourced articulars, prearticulars, surangulars, coronoids, dentaries, and splenials.

Not too surprisingly, even though I have separated the suspensorial-mandibular and maxillary-pterygopalatal units into two systems, the trends for the mandibular component of the former more closely parallel the anatomical trends in the latter (i.e., elongations, reductions, toothlessness, etc.), as do the quadrate and supratemporal.

In brief, within non-macrostamatans such as anilioids (again, recognizing anomochilids as "special"), the dentary and compound bones (fusion of the articular, surangular, and prearticular bones) are roughly equivalent length, and the quadrate is robust, short, oriented vertically in its articulation with the glenoid fossa of the compound bone; has a large suprastapedial process and thus describes an inverted "j"-shape; and articulates on its medial face with a short, robust, and firmly sutured supratemporal. The latter element sits in an "L"-shaped facet formed by both the prootic and exoccipital-opisthotic, with the open face of the "L" facing laterally. Though not a component of the suspensorial-mandibular unit (though it does share a splanchnocranial origin with the quadrate), the columella and its very characteristically broad foot-plate in non-macrostomatans such as *Anilius*, *Cylindrophis*, *Anomochilus*, and, yes, the purported macrostomatans *Xenopeltis* and *Loxocemus* as well (see Chapter 3 for more detail on this matter) bears some discussion and highlighting here. As noted, the footplate is extremely enlarged, such that its circumference is nearly as large as the inflation of the bony otic capsule itself. The columella is short and robust and sharply angled posterodorsally toward a pronounced if not large suprastapedial process of the quadrate; articulating with the distal tip of the columella is a small intercalary/extracolumellar element, which fills the gap between the columella and the medial surface of the suprastapedial process.

In basal macrostomatans, the general condition of equivalent length of the dentary to the compound bone remains the same, as the general vertical orientation of the tall, less robust quadrate (with virtually now suprastapedial process). However, the supratemporal is no longer suturally attached to the prootic and exoccipital-opisthotic, and it is also not short, but quite elongate. The supratemporal now overlaps the exoccipital-opisthotic, the supraoccipital, the prootic, and the parietal and is anteriorly elongated, approaching the anterior border of the prootic above the anteriormost trigeminal nerve foramina ($V_2$). The supratemporal

is also posteriorly elongate, extending well past the occipital condyle of the basioccipital, thus placing the suspensorial articulation between the supratemporal, quadrate, and compound bone distal or caudal to the braincase. In contrast to the columellar anatomy noted above for non-macrostomatans such as *Anilius*, and so on, the macrostomatan anatomical plan requires the presence of an elongate and very thin columellar shaft that is angled posteriorly, but not dorsally, and contacts the shaft of the quadrate at midheight, not the medial face of the suprastapedial process (which is absent). The columellar shaft often firmly anchored via cartilage or a ligament to a small process on the quadrate.

The colubroid condition trends again toward extreme elongation and reduction, and is observed in its extremes among atractaspids and scolecophidians. The dentary changes from being equivalent in length to the compound bone to becoming extremely small (from 1/3 the length of the compound bone to less than about 1/6 of the length of the compound bone). The quadrate and supratemporal follow this same trend, though it is highly variable among higher modern snake lizards. The supratemporal becomes even more elongate anteriorly, and in bizarre cases extends onto the parietal to the posterior margin of the orbit next to the jugal (*Heterodon* sp., Cundall and Irish, 2008); it also extends posteriorly and again in some snake lizards, nearly an equivalent distance to the length of the skull (*Heterodon* sp., Cundall and Irish, 2008). And, most confusingly, the supratemporal can also simply be lost or reduced to a mere thread of bone, leaving the quadrate to articulate with the lateral wall of the exoccipital-opisthotic. Such extreme reductions of the supratemoral are observed in scolecophidians. To make matters more complex, the quadrate bone also becomes exceedingly elongate and posteriorly directed in order to match the elongation of the supratemporal and the extreme posterior placement of the quadrate articulation with the compound bone. In highly derived colubroid anatomies, such as those observed among viperids, an elongate posteriorly directed quadrate makes perfect sense. In scolecophidians, which lose the supratemporal, the quadrate is still elongated, but this time the distal condyle is directed anteriorly, if indeed the entire element is not horizontal.

*The Hyobranichal Unit*—The hyobranchial unit, or hyobranchial skeleton, includes the splanchnocranial elements of the hyoid apparatus (remnants of the ancient visceral arch/gill arch system) that support the larynx and tongue musculature (Romer, 1956; Rieppel, 1981). In most vertebrates, the hyobranchial skeleton is a key component of the specialized mechanics of pharyngeal function associated with feeding (Romer, 1956; Carroll, 1988). The functional linkage of the more posterior cartilages of the splanchnocranial hyobranchial skeleton, with the more anterior bony and cartilaginous elements of the same skeleton (Meckel's and palatoquadrate cartilages, quadrate and columellar bones) is unsurprising. That this splanchnocranial skeleton is also linked functionally and developmentally to the tooth-bearing dermatocranial skeleton is a key innovation of the gnathostome heritage of snake lizards (Carroll, 1988).

The typical lizard hyobranchial skeleton is composed of two unpaired median elements, the anteriorly elongate entoglossal process (or processus lingualis, which extends forward into the tongue to support the tongue musculature) and the more posterior basihyal (or corpus hyoideum) cartilage. Arising from an anterior facet on the basihyal are the right and left paired hypohyals (or hyal cornua), which extend anterolaterally and on their anterior tips articulate with the posterolaterally directed ceratohyals. From a pair of posterior processes on the basihyal arise the right and left paired first ceratobranchials (Ceratobranchials I); many lizards have an additional pair of ceratobranchials (Ceratobranchials II), and these are placed medial to the first pair, articulating to a paired facet on the basihyal. In many fossil and modern lizards, the hyobranchial skeleton ossifies and is relatively well known in fossil taxa (e.g., Rieppel and Grande, 2007). In terms of position in pharynx, the basihyal element usually is placed above the arytenoid and cricoid cartilages of the larynx; ceratobranchials I and II extend posteriorly past the laryngeal cartilages supporting the musculature of the pharynx.

Among snake lizards, the hyoid apparatus is extremely reduced by comparison to the typical lizard anatomy described above. The most common condition is a simple tuning fork–shaped apparatus, which presumably includes the first ceratobranchials fused to some remnant of the basihyal (colubroids, boids, etc.). A further state of reduction is observed where the basihyal element is lost and all that remains are the right and left ceratohyals. McDowell (1974, 2008) held

a contrasting opinion—the ceratobranchials are lost, and the posterior processes are basihyal extensions, not ceratobranchials; it would follow that the more reduced condition would also be remnants of the basihyal extensions, not ceratobranchials.

## The Bare-Bones Summary: Four Skull Categories

Everyone likes to create meaningful categories as the explanans for their interpretations of data and hypotheses—I am no different. I therefore present here, as my concluding remarks on "What is a snake?" and my assessment of what a modern snake lizard is based on this brief review of the osteology of their skulls, blended with my long-term consideration of the history, anatomy, and evolution of lizards, and in particular snake lizards (both ancient and modern), the four skull categories identified earlier: (1) Type I–Anilioid, (2) Type II–Booid-Pythonoid, (3) Type III–Colubroid (excluding viperids and elapids), and (4) Type IV–Viperid-Elapid-Scolecophidian. As a caveat, I do not consider these four skull types to represent evolutionary groupings, though they could, and I certainly do not consider the features I have employed in these broad groupings to be of any utility in the search for primary homologs, let alone be subjected to a test of congruence. My types or categories are a blend of grades, overall similarity, thumb-scale morphometrics, and qualities that are intended to represent generalities, not specifics. Such generalities provide fertile inductive statements that can be used to probe other data and ideas, and, in that process, provide potentially more empirically stringent observations and deductive hypotheses arising as a result. No modern snake is an archetype throwback to the ancient snake lizard at the time of the origins of snake lizards, though it is clear to me that some modern snake lizards are very similar to some ancient snake lizards at the time that essential type was evolving in the Late Mesozoic. I begin with Type I—Anilioid, the first of four snake lizard skull types.

*Type I—Anilioid* (Figure 1.6)—The Anilioid type is typified by the skull of *Cylindrophis ruffus* (USNM 297456). Defining anatomies and morphologies of Type I are: (1) *Key feature*: the unique anatomy of the incredibly large footplate of the columella; (2) *Key feature*: strongly inflated otic capsule to accommodate the massive size of the bony

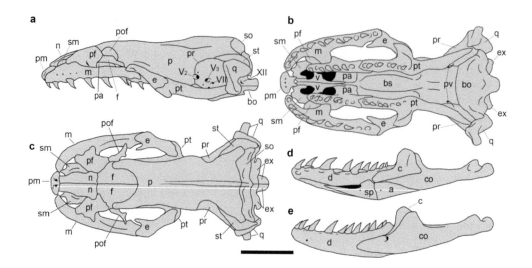

Figure 1.6. Skull Category I—Anilioid Type: Skull of *Cylindrophis rufus* (USNM 297456). (a) Left lateral view; (b) ventral view; (c) dorsal view; (d) lower jaw in medial view; (e) lower jaw in lateral view. (Abbreviations: V₂, foramen for maxillary branch of trigeminal nerve; V₃, foramen for mandibular branch of trigeminal nerve; VII, foramen for facial nerve; XII, foramen for hypoglossal nerve; a, angular; bo, basioccipital; bs, basisphenoid; c, coronoid; co, compound bone; d, dentary; e, ectopterygoid; ex, exoccipital-opisthotic; f, frontal; fe, femur; il, ilium; is, ischium; j, jugal; m, maxilla; n, nasal; o, optic fenestra; p, parietal; pa, palatine; pf, prefrontal; pm, premaxilla; pof, postfrontal; pr, prootic; pt, pterygoid; pu, pubis; pv, posterior opening of vidian canal; q, quadrate; sm, septomaxilla; so, supraoccipital; sp, splenial; st, supratemporal; v, vomer.) All images are drawn to the same scale. Scale bar equals 5 mm. (Images courtesy of copyright Alessandro Palci, Flinders University, Adelaide, Australia.)

labyrinth for the sacculus, utriculus, and lagena/cochlea (the otolith-statolith is also enormous, nearly filling the bony otic space); (3) *Key feature*: orientation of the shaft of the columella toward its articulation with an intercalary element and the suprastapedial process of the quadrate; (4) crista circumfenestralis absent to incomplete (Type 1 and 2 of Palci and Caldwell [2014]); (5) "j"-shaped to "C"-shaped quadrate with large to enormous suprastapedial process; (6) supraoccipital, narrow exposure across posterior width of skull roof, with significant underlap of parietal, forming facet with prootic and parietal for supratemporal; (7) *Key feature*: relatively rigid maxillary-pterygopalatal unit (see above); (8) prokinesis present, but limited; (9) coronoid present, dentary equivalent length to compound bone. Modern snake lizards included in this rather small, but elite, Type I include species included in the genera *Cylindrophis, Anilius, Xenopeltis, Loxocemus,* and *Anomochilus* and species assigned to the various genera in the Uropeltidae. While I have certainly depended on my own observations of cranial construction to arrive at this grouping, I note that I am not the first to make such an argument, as I was preceded by many years by McDowell (1987) and with respect to similarities

in jaw muscle anatomy by the anatomical works of Haas (1955:5), in particular on *Loxocemus bicolor* and its then phyletic position as a pythonine, with which he disagreed:

> Many primitive snakes exhibit an approximation to this bipinnate arrangement of the two muscles; examples are Xenopeltis, Calamaria, Atractus, the Ilysiidae, and the Uropeltidae.

And from Haas (1955:7):

> Some importance must be given to the presence of an occipital portion of the depressor mandibulae in both Xenopeltis and Loxocemus... This peculiarity is known in Ilysia and Cylindrophis (perhaps the most primitive snakes in existence), in one boid genus, Eryx, and in the colubrid genus Calamaria.

*Type II—Booid-Pythonoid* (Figure 1.7)—The Booid-Pythonoid Type is typified by the skull of *Python regius* (UAMZ 3818), but by comparison to Type I, this is a much more generically rich group, though disparity or variation is surprisingly low. Defining anatomies and morphologies of Type II

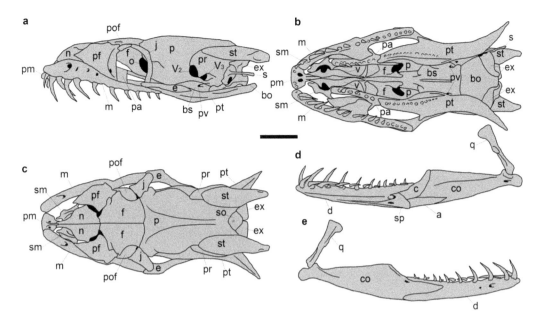

Figure 1.7. Skull Category II—Booid-Pythonoid Type. Skull of *Python regius* (UAMZ 3818). (a) Left lateral view; (b) ventral view; (c) dorsal view; (d) right lower jaw in medial view; (e) right lower jaw in lateral view. (Abbreviations: V₂, foramen for maxillary branch of trigeminal nerve; V₃, foramen for mandibular branch of trigeminal nerve; VII, foramen for facial nerve; a, angular; bo, basioccipital; bs, basisphenoid; c, coronoid; co, compound bone; d, dentary; e, ectopterygoid; ex, exoccipital-opisthotic; f, frontal; fe, femur; j, jugal; m, maxilla; n, nasal; o, optic fenestra; p, parietal; pa, palatine; pf, prefrontal; pm, premaxilla; pof, postfrontal; pr, prootic; pt, pterygoid; pu, pubis; pv, posterior opening of vidian canal; q, quadrate; s, stapes; sm, septomaxilla; so, supraoccipital; sp, splenial; st, supratemporal; v, vomer.) All images are drawn to the same scale. Scale bar equals 5 mm. (Images courtesy of copyright Alessandro Palci, Flinders University, Adelaide, Australia.)

are: (1) *Key feature*: tall slim quadrate that does not bear a posteriorly directed suprastapedial process; (2) quadrate is vertical or held on a incline with glenoid condyle directed posteriorly; (3) *Key feature*: shaft of columella is thin and elongate and attaches/articulates to the shaft of the quadrate on distinct pedicle; (4) *Key feature*: supratemporal elongate and extends posterior to occipital condyle; (5) *Key feature*: supratemporal articulation is not sutural, but overlapping on parietal and prootic; (6) quadrate rami of pterygoids extend posteriorly past the occipital condyle; (7) low or no facial process of maxilla; (8) maxilla very kinetic with respect to prefrontal and palatine articulations; (9) *Key feature*: postfrontal and jugal usually both present; (10) high degree of prokinesis; (11) Type 3 crista circumfenestralis (Palci and Caldwell, 2014). Modern snake lizards included in this type include genera within the Pythonidae, Booidae, and Ungaliophiidae, and species in the genera *Tropidophis, Casarea, Sanzinia, Calabaria, Eryx*, and *Charina*.

Type III–Colubroid (*excluding viperids and elapids*) (Figure 1.8)—The Colubroid type is typified by

the skull of *Natrix natrix* (SAMA R36836); finding an exemplar for the colubroid radiation (~3025 species [Uetz et al., 2018]) of skull types was not easy, is likely to be contentious, and thus bears some discussion as to the choice of exemplar of this kind of snake skull. To begin with, based on unique aspects of skull morphology and function, I excluded viperids (348 species) and elapids (369 species), leaving a mere ~2308 species of modern colubroid snake lizards. Consideration of the form of the skull and which taxon to highlight was multifaceted and considered availability of specimens; common usage by a variety of researchers; and existing osteological, soft tissue, and embryological studies. As it happens, the cosmopolitan distribution of *Natrix natrix*, or closely related forms now removed from the genus, has made it a snake lizard of choice for biological research for decades. Availability and ease of access made it a perfect choice for embryological studies long before de Beer (1926, 1937), and it has continued to be used for the same (e.g., Winchester and d'Bellairs, 1977). There are also numerous soft tissue, molecular, and osteological studies

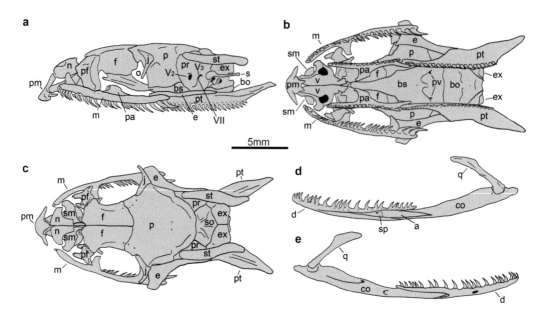

Figure 1.8. Skull Category III—Colubroid Type. Skull of *Natrix natrix* (SAMA R36836). (a) Left lateral view; (b) ventral view; (c) dorsal view; (d) right lower jaw in medial view; (e) right lower jaw in lateral view. (Abbreviations: V$_2$, foramen for maxillary branch of trigeminal nerve; V$_3$, foramen for mandibular branch of trigeminal nerve; VII, foramen for facial nerve; XII, foramen for hypoglossal nerve; a, angular; bo, basioccipital; bs, basisphenoid; c, coronoid; co, compound bone; d, dentary; e, ectopterygoid; ex, exoccipital-opisthotic; f, frontal; fe, femur; il, ilium; is, ischium; j, jugal; m, maxilla; n, nasal; o, optic fenestra; p, parietal; pa, palatine; pf, prefrontal; pm, premaxilla; pof, postfrontal; pr, prootic; pt, pterygoid; pu, pubis; pv, posterior opening of vidian canal; q, quadrate; sm, septomaxilla; so, supraoccipital; sp, splenial; st, supratemporal; v, vomer.) All images are drawn to the same scale. Scale bar equals 5 mm. (Images courtesy of copyright Alessandro Palci, Flinders University, Adelaide, Australia.)

focused on *Natrix* (e.g., Malnate, 1960; McDowell, 1961; Rossman and Eberle, 1977; Rupik, 2002; Kindler et al., 2017). The availability of this snake lizard, or related forms, made it a sensible choice as the exemplar of Type III—Colubroid (excluding viperids and elapids).

After the removal of viperids (348 species) and elapids (369 species), the disparity of skull morphologies observed among members of the Clade Colubroidea is significantly reduced, and in fact results in a surprisingly uniform skull architecture. Defining anatomies and morphologies of Type III are: (1) *Key feature*: elongate (sometimes extremely elongate, e.g., *Heterodon platirhinos*) quadrate (slim to robust) with a usually posteriorly directed distal articular condyle, no suprastapedial process; (2) shaft of columella thin and elongate and attaches/articulates to the shaft of the quadrate on distinct pedicle (note: distal tip is often unossified, but is not an extracolumellar cartilage); (3) *Key feature*: supratemporal extremely elongated, extending up to half of its length past occipital condyle; (4) supratemporal articulation is not sutural, but overlapping on parietal and

prootic; (5) *Key feature*: quadrate ramus of pterygoid extremely elongate, extends posteriorly past the occipital condyle, often with teeth extending past occipital condyle; (6) *Key feature*: maxilla generally very thin (dorsoventrally) throughout its length, but very long; (7) *Key feature*: maxillae extends well past the posterior margin of the orbit as defined by the position of the jugal; (8) maxilla very kinetic with respect to prefrontal and palatine articulations; (9) jugal present, postfrontal lost; (10) *Key feature*: high degree of maxillopterygopalatal kinesis; (11) high degree of prokinesis, with well-developed joint between septomaxilla, nasals, vomers, and frontals; (12) *Key feature*: nasals small and only retain median process contact with frontals, no prefrontal contact; (13) compound bone slim and elongate, subequal in length to dentary; (14) *Key feature*: splenial and angular highly reduced, no coronoid nor even a coronoid eminence; (15) Type 3 and Type 4 crista circumfenestralis (Palci and Caldwell, 2014); (16) *Key feature*: generally a very high maxillary, dentary, palatine, and pterygoid tooth count (~200 all tooth-bearing elements).

THE ORIGIN OF SNAKES

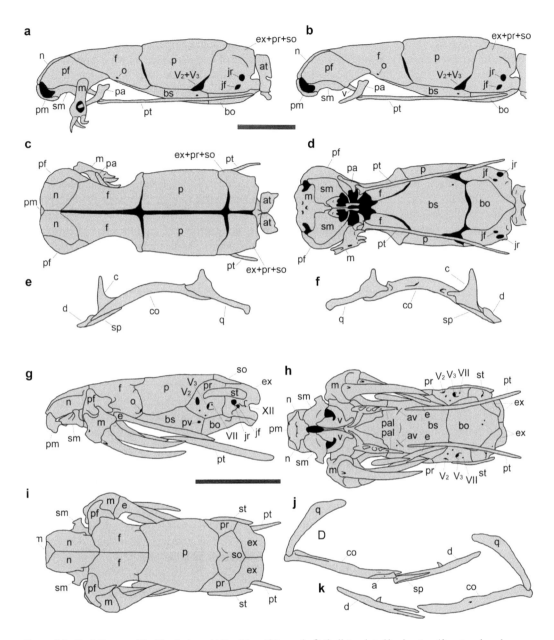

**Figure 1.9.** Skull Category IV—The Scolecophidian-Viperid Type: (a–f) Skull *Ramphotyphlops braminus* (drawings based on computer tomography of UAMZ 553), (g–k) skull of *Atractaspis aterrima* (drawings based on computer tomography of AMNH R-12352). (a) Left lateral view; (b) left lateral view, maxilla removed; (c) dorsal view; (d) ventral view; (e) right lower jaw in lateral view; (f) right lower jaw in medial view; (g) left lateral view; (h) ventral view; (i) dorsal view; (j) right lower jaw in medial view; (k) right lower jaw in lateral view. (Abbreviations: $V_2+V_3$, common foramen for maxillary and mandibular branches of trigeminal nerve; $V_2$, foramen for maxillary branch of trigeminal nerve; $V_3$, foramen for mandibular branch of trigeminal nerve; VII, foramen for facial nerve; XII, foramen for hypoglossal nerve; a, angular; at, atlas; av, anterior opening of vidian canal; bo, basioccipital; bs, basisphenoid; c, coronoid; co, compound bone; d, dentary; e, ectopterygoid; ex, exoccipital-opisthotic; f, frontal; is, ischium; jf, jugular foramen; jr, foramen for the juxtastapedial recess; m, maxilla; n, nasal; o, optic foramen; p, parietal; pa, palatine; pf, prefrontal; pm, premaxilla; pr, prootic; pt, pterygoid; q, quadrate; sm, septomaxilla; so, supraoccipital; sp, splenial; st, supratemporal; v, vomer.) Scale bar equals 1 mm. (Images courtesy of copyright Alessandro Palci, Flinders University, Adelaide, Australia.)

*Type IV–Viperid-Elapid-Scolecophidian* (Figure 1.9)—
This type, Type IV, is the most unusual and, I
believe, unexpected category, as it revolves around
similarities, morphologies, functions, and so on,
of the maxilla, pterygo-palatal complex, "crista
circumfenestralis," and dentary and compound
bone between seeming extremes—viperids/
elapids, atractaspids, and scolecophidians. As with
all of the "types" presented so far, the anatomies,
functions, and architectures are broken down
around trends arising from the general condition
of form to the more specific or extreme condition of
form. To communicate these trends and similarities
in a manner that makes sense, I have chosen as
exemplars the typhlophid snake lizard *Ramphotyphlops
braminus* (UAMZ 553), and the lamprophiid elapid
*Atractaspis aterrima*. I recognize that prior to Pyron
et al. (2010), the Lamprophiidae was considered
a kind of colubrid, but will add to Pyron et al.'s
(2010) findings that the cranial architecture in
no way supports such an affinity at the level of
simple groupings (Type I–IV), let alone in terms
of a phylogenetic analysis and clade relationships.
In truth, though, while *Atractaspis* shows a striking
number of similarities to *Rhamphotyphlops* in numerous
characters, other Type IV snake lizards demonstrate
the same trends in their extremely modified form
that are observed in scolecophidians.

The anatomies and morphologies of Type IV
recognize the extremes of structure and function
not observed in any other kinds of modern snake
lizards: (1) *Key feature*: elongate quadrate and absence
of a suprastapedial process); (2) shaft of columella
thin and elongate and attaches/articulates to the
shaft of the quadrate often on distinct pedicle; (3)
*Key feature*: Some Type 3, but mostly Type 4 crista
circumfenestralis (Palci and Caldwell, 2014); (4)
*Key feature*: supratemporal short, not extending past
margin of foramen magnum when present (usually
absent in scolecophidians) elongated, extending
up to half of its length past occipital condyle;
(5) *Key feature*: extremely tiny contact to no contact
of supratemporal with parietal (usually restricted
to prootic and supraoocipital); (6) *Key feature*:
quadrate ramus of pterygoid extremely elongate,
may extend significantly past occipital condyle,
not toothed; (7) *Key feature*: pterygoid usually does
not bear teeth, though there are a few exceptions;
(8) *Key feature*: maxillae extremely foreshortened,
does not extend past anterior edge of prefrontal;
(9) *Key feature*: maxilla mobile, articulating in saddle
joint with prefrontal or on lateral face of prefrontal
(when retracted, folded posteriorly; when rotated,
rotates forward and extends fangs); (10) *Key feature*:
extremely reduced maxillary dentition, usually
from 1 to 4; (11) *Key feature*: palatine shortened
(12) jugal present (lost in burrowing elapids,
typhlopids and leptotyphlopids), postfrontal
often lost; (13) *Key feature*: maxillopterygopalatal
kinesis; (14) *Key feature*: high degree of prokinesis
(except scolecophidians), with well-developed
joint between septomaxilla, nasals, vomers, and
frontals; (15) nasals small (except scolecophidians)
and only retain median process contact with
frontals, no prefrontal contact; (16) *Key feature*:
compound bone slim and extremely elongate; (17)
*Key feature*: dentary highly reduced in size, less than
1/4 length of complete mandible; (18) *Key feature*:
splenial very small, angular highly reduced to
lost; (19) highly reduced dentition, development
of oversized maxillary teeth.

THE ORIGIN OF SNAKES

# CHAPTER TWO

# Ancient Snake Lizards

## THE FOSSIL RECORD

---

The debate over ancestry is fueled by the absence of fossils clearly intermediate between snakes and some other squamate taxon and the tantalizing clues provided by a few Cretaceous fossils with partial skulls...

CUNDALL AND IRISH (2008:352)

Modern snake lizards, distributed as they are into their extant crown clades, can be traced backward in time through a patchy fossil record that begins ~60 million years ago in the early part of the Cenozoic in the Paleocene and Eocene epochs (Rage, 1984; Holman, 2000). Ancient macrostomatan snake lizards of the "booid" kind (used very loosely here to characterize early macrostomatan snake lizards that are likely nested within the crown clades) are well preserved and represented, in particular, in Eocene-aged rocks (Rage, 1984, Holman, 2000; Scanferla et al., 2016); the best example of a Paleocene snake lizard of the "booid-type," or Type II—Booid-Pythonoid, *Kataria anisodonta*, was recently described from Lower Paleocene sediments in Bolivia (Scanferla et al., 2013). The earliest evidence for the bulk of modern snake lizard species nested within the crown Colubroidea (>3000 species) begins approximately 25 million years ago or so (see Rage, 1984; Holman, 2000). When the colubroids appear in the fossil record their radiation is swift, diverse, and global. The Cenozoic is quite clearly the age in geologic time during which the crown clades of modern snake lizards

originate, but, more importantly, radiate and diversify to produce the modern fauna—some call it the Age of Mammals, but in truth, it is the Age of Squamates/Lizards. Today, there are approximately 5500 species of modern mammals, give or take a few from ongoing extinctions, while the number of squamates/lizards is now moving beyond 10,000 regardless of extinctions. While it is true that mammals represent the modern megafauna, if an "Age" means something along the lines of "ruling," then megafauna win...but if it means evolutionary success and radiations, then squamates and birds are the winners by far.

The focus of this chapter, though, and the purpose of this book, is not to argue about "Age of...," but rather to illuminate the roots of the great tree of snake lizard evolution and interrelationships, because it looks nothing at all like the modern diversity of macrostomatan snake lizard "leaves" and "flowers" on the tips of the living clade branches. Long before the ancestral populations of modern snake lizards appeared, there were ancient snake lizards, which from their temporal and spatial fossil record, were spread around the globe, during the "Age of Dinosaurs," and had occupied a wide variety of ancient environments and available niches. Modern snake lizards are representative of little more than a fresh start driven by a relatively recent and clearly explosive radiation and evolution of colubroids (Greene, 1997).

What this means is that the fossil record is the only data set we have that provides insight on the morphology and evolution of ancient snake lizards (Rage, 1984). That is, if we wish to ask questions about snake lizard evolution beyond the last 25 million years or so, then the only data set that provides that information is hidden with the sedimentary rock record of this planet. Only in the fossil record do we have information on the patterns of morphological change through time that have led to the wildly beautiful diversity of the organisms that populate the modern world—I repeat, it is the only source of information we have. The available data of any kind, molecules, morphology, behavior, and so on, from modern organisms are nothing more than glimmering biological reflections of an exceedingly ancient past, not a record of it. Neither the genotype nor phenotype is in any sense even inherited from a 100 million-year-old ancient ancestor, except in our imaginations. Genes are constantly emended facsimiles of the parent gene they were copied from, while phenotype reflects genetic and epigenetic processes in the transformation of gene-prescribed proteins to a whole organism built of carbohydrates and fats, cells and tissues, and organs. The linkage of a modern *Thamnophis sirtalis sirtalis* to the 96 million-year-old fossil snake lizard *Pachyrhachis problematicus* is at best tenuous, as we are limited in making our logical and empirical connections through application of the concept of homology; homology in this sense is heuristically driven by the recovery of synapomorphies, and not by a "real" chain of molecules or soft tissues that connects a garter snake to *Pachyrhachis*. I accept that the concept of these connections, that is, homologies, is a reflection of reality (my ontologic in this case), because I reject the metaphysic of divine intervention. I also accept the concept of these connections based on a single data set, and that is the morphological character data preserved in the fossil record that link *Thamnophis* and *Pachyrhachis* to each other, to the exclusion of other forms of tetrapods, over deep time. The fossil record is, therefore, the only, and I mean *only*, insight on ancient organisms and the evolutionary experiments and innovations of

the deep past that created the present as we know it. We must have data from the ancient past as a necessary explanans for the present; while the present can be a key to understanding processes in the past (e.g., gravity worked 1 billion years ago as it does today), it does not permit us to understand how the present came to be in the absence of pattern data from the past (i.e., snake lizard phylogeny cannot be accurately reconstructed using only the molecules and morphology of modern snake lizards). The data set from the modern world—biological, geological, chemical, and physical—provides insight on processes at all levels or organization that cannot be obtained from the paleontological data set. However, that modern data set can only be used in a very limited sense, to extrapolate backward along the vector of the fourth dimension to points in time that are not very deep at all. I realize such claims are made all the time, that is, rates of evolution of genes in nuclear versus mitochondrial DNA, but in truth, such reasoning is deeply circular as you are modeling evolutionary process by rate claims that are then used as justification for heuristic recovery of evolutionary pattern—not good—in order to claim discovery of exceedingly ancient phylogenetic relationships. Fossils are better— they are real, ancient, and do not require circular reasoning from process to pattern; they are pure pattern. Now, do not get me wrong, scientific probing and hypothesis construction from the data set of the modern is critical to our understanding of everything—I use it all day long—but it has its limits and these must be understood when we want to probe ancient time.

Therefore, it is not a cliché to state that 99.99999% of everything that has ever lived during this 4.6-billion-year journey through time and space is extinct, because it is true. The teeming biodiversity we see out of our windows today, all 10 million or so species of microbes, plants, and animals, is an infinitesimal fraction of the cumulative biodiversity and biodisparity that have come and gone on this planet. Only the fossil record, and nothing else, tells us something about that past biodiversity, even if that data set is frustratingly fragmentary and incomplete. And make no mistake, as a paleontologist I can assure you that the fossil record is not complete, and is in fact deeply biased by time, taphonomy, weathering, erosion, collection biases, and the fact that many organisms simply left behind no

THE ORIGIN OF SNAKES

preserved record. Not surprisingly, even Darwin (1859:310–311) wistfully opined for a more complete fossil record because he understood the insight those data would provide, when he wrote:

> For my part, following out Lyell's metaphor, I look at the natural geological record, as a history of the world imperfectly kept, and written in a changing dialect; of this history we possess the last volume alone, relating only to two or three countries. Of this volume, only here and there a short chapter has been preserved; and of each page, only here and there a few lines. Each word of the slowly-changing language, in which the history is supposed to be written, being more or less different in the interrupted succession of chapters, may represent the apparently abruptly changed forms of life, entombed in our consecutive, but widely separated formations.

It has been recognized from the very beginnings of the sciences of geology and paleontology that the fossil record is an incomplete data set of past organisms, events, patterns, and processes. The real problem with fossil data resides in what kind of information you think it should provide. If you think the fossil record should provide data revealing a continuous progression of forms, gradually evolving over millions of years, so perfectly preserved that you would find perfect missing links to the world of the present, you will be disappointed. In other words, if you suffer from a twenty-first century version of logical positivism, if you are looking for absolute answers to questions of evolutionary pattern and process, the fossil record is not for you (e.g., Streicher and Wiens, 2017). Frankly, neither is the neontological record (i.e., as a source of absolute answers), but that is another problem of a different kind. If however, your epistemic and ontic goals are rather more relative, then the fossil record in all of its incompleteness is the only data set that can reveal the complexities of the past that have led to the diversity of the present. The fossil record of snake lizards is becoming beautifully informative on many questions surrounding the evolution of snake lizards, though, again, and be reminded, it is also frustratingly mute on a great number of other very important questions. Cundall and Irish (2008) correctly stated as such in their treatise on modern snake lizard skulls, as noted in the quote chosen as the header for this chapter. However, in a decade, I am pleased to say, that all this had changed.

The geologically oldest snake lizards, from the Middle Jurassic to Lower Cretaceous, are now known from a number of jaws, skull elements, an isolated braincase, vertebrae, and ribs from both Laurasian and Gondwanan localities (Caldwell et al., 2015). The previous record of "geologically oldest snakes" was held by five isolated fossil vertebrae found in both Gondwanan (three vertebrae) and Laurasian (two vertebrae) localities. Two of the isolated vertebrae from Gondwanan Africa were described by Cuny et al. (1990) and considered to be from Upper Albian–aged rocks. One specimen was left in open nomenclature, while the other was assigned to Hofstetter's (1968) Lapparentophiidae; Hofstetter's (1959) *Lapparentophis defrennei*, described from a single vertebra, was recognized as coming from younger rocks (Lower Cenomanian) by Cuny et al. (1990). Gardner and Cifelli (1999) described the two North American vertebrae (Lower Cenomanian) and assigned them to *Coniophis* sp. but not to *Coniophis precedens* Marsh, 1892, a Maastrichtian-aged snake lizard, the type species for the genus. Rage and Escuillé (2003) identified as a lizard an isolated vertebra from the Barremian of Spain that had originally been identified as a "snake" by Rage and Richter (1994), but not named; in light of the newly described middle to Upper Jurassic snake lizards (Caldwell et al., 2015); it may well be worth revisiting that specimen and reassessing the morphology against the Jurassic vertebrae. Unfortunately, all of these previous "oldest snakes," are only known as vertebral form-taxa and present very little useful information. The newly described Jurassic specimens are the first known snake lizards to be recognized from non-vertebral elements.

The quality of the Mesozoic fossil record is profoundly improved by the discovery of articulated Upper Cretaceous snake lizards from marine deposited rocks from the Lower to Middle Cenomanian: *Pachyophis woodwardi* Nopcsa, 1923; *Mesophis nopcsai* Bolkay, 1925; *Pachyrhachis problematicus* Haas, 1979, 1980a,b; *Haasiophis terrasanctus* Tchernov, Rieppel, Zaher, Polcyn and Jacobs, 2000; *Eupodophis descouensi* Rage and Escuillié, 2000. Three of these five snake lizards retain small but well-developed hindlimbs, all of them show aquatic adaptations, and three possess well-preserved skulls that show very important and unexpected patterns of cranial element homologies (Nopcsa, 1923; Bolkay, 1925; Caldwell and Lee, 1997; Lee and Caldwell, 1998; Scanlon et al., 1999; Rage and Escuillié, 2000; Tchernov et al., 2000; Rieppel and Head, 2004; Palci et al., 2013; Palci and Caldwell, 2013, 2014).

From sediments deposited in freshwater environments, with possibly some aeolian influence, the record from Gondwanan South America has moved from a single specimen reported on in 1901 (Smith-Woodward, 1901) to the extraordinarily rich record, both in terms of quality of preservation and in terms of insight on the evolution of snake lizard anatomy. *Dinilysia patagonica* Smith-Woodward, 1901, from the Santonian of Argentina, known for more than 100 years from only a single specimen and some fossil scraps, is now known from literally dozens of skulls and one well-articulated specimen (Estes et al., 1970; Caldwell and Albino, 2002; Caldwell and Calvo, 2008; Zaher and Scanferla, 2012). Though controversy over the interpretation of morphologies continues, the new materials of *D. patagonica* have provided extremely valuable insights on the evolution of snake lizard anatomy. The newest addition to the snake lizard fossil record from Gondwanan South America is a hindlimbed animal, *Najash rionegrina* (Apesteguía and Zaher, 2006), found in Upper Cenomanian–aged sandstones from the Candeleros Formation in the Neuquén Basin of Argentina (Apesteguía et al., 2001; Leanza et al., 2004); other snake lizard fossils of a similar if not older age are also known from Brazil, though these are only isolated vertebrae (Hsiou et al., 2014). Another well-preserved snake lizard from the Maastrichtian of Gondwanan India, *Sanajeh indicus* Wilson, Mohabey, Peters, and Head, 2010, appears to be morphologically very similar to both *N. rionegrina* and *D. patagonica*, though it is substantially younger and was found intermingled with sauropod dinosaur eggs and embryos in a nesting site. Fragmentary remains of another morphologically similar snake lizard from the Maastrichtian of Gondwanan Madagascar, *Menarana nosymena* Laduke, Krause, Scanlon, and Kley, 2010, include both vertebrae and intriguing basicranial remains, thus broadening the Gondwanan radiation of snake lizards, as well as their subsequent adaptive radiations on their respective Gondwanan relictual landmasses (Hoffstetter [1961] also described vertebrae of a madtsoiid from the Upper Cretaceous of Madagascar, *Madtsoia madagascarensis*). All of these rather well-preserved

Upper Cretaceous Gondwanan snake lizards are big-bodied, big-headed animals found in sediments deposited in a variety of freshwater environments.

The fossil record of snake lizards in the later part of the Cretaceous, and into the early Cenozoic, particularly from Laurasia, is largely restricted to isolated vertebral remains, most of which have been assigned to the vertebral form taxon *Coniophis*. Unfortunately, form taxa, named or not, when known from such uninformative elements, provide little information beyond the obvious—that there were snake lizards in late Mesozoic and early Cenozoic that we know little about. The problem is, however, that form taxa are not sufficient for defining ancient snake lizard taxon concepts, but rather are concept fragments of incomplete snake lizard fossils, and in the case of vertebral form taxa, are exceptionally uninformative. Far too often, as in the case of *Coniphis*, they ascend to the status of snake lizard taxon concept in the absence of anything but chimeric assemblages of bits and pieces and are used as terminal taxa in phylogenetic analyses (Longrich et al., 2012).

A recent exception to the early Cenozoic record of snake lizards is the Paleocene-aged specimen recently described as *Kataria anisodonta* Scanferla, Zaher, Novas, Muizon and Céspedes, 2013. This specimen includes only a beautifully preserved skull, which is unfortunate, as some portion of the postcranium would have been invaluable in assessing the isolated vertebral remains common in deposits around the world from this critical time period. This is because, by the Eocene, well-preserved snakes, specifically a diverse and well-distributed assemblage of booidlike animals, are known from deposits in North America (Marsh, 1871; Rage, 1984; Holman, 2000) and western Europe (Schaal and Ziegler, 1992; Baszio, 2004; Schaal and Baszio, 2004). In addition, there are an enormous number of well-preserved vertebrae, and in many cases, articulated specimens, known from Paleocene and Eocene rocks that preserve information on obscure groups of small to giant-sized aquatic snake lizards (marine to esturiane environments), often assigned to the booids at some family or subfamily level (Owen, 1841; Massolongo, 1859; Lucas, 1896; Andrews, 1901; Rage, 1984; Holman, 2000; Head et al., 2009). The Paleocene-Eocene record of snake lizards from Europe, and now parts of North Africa, also includes the earliest scolecophidian vertebrae (Auge and Rage, 2006).

By the end of the Oligocene, that is, at the P/N boundary, the modern snake lizard fauna is represented in the fossil record by fragmentary specimens assignable to scolecophidians and most of the modern groups of alethinophidians (Rage, 1984; Holman, 2000; Mead, 2013). It is therefore the Mesozoic record that the remainder of this chapter looks at in detail, for both the general aspects of their anatomical organization and insights on specific and very problematic anatomies seemingly without homology in modern snake lizards.

The Mesozoic record spans approximately 100 million years of earth history, from the upper Middle Jurassic (~167 mya) to the end of the Mesozoic at the Cretaceous–Paleocene Boundary (~65 mya). This is a significant amount of time when you consider that the Cenozoic is only 65 million years old, a time during which the snake lizards arising during and surviving through to the end of Mesozoic evolved toward the forms and diversity defining the modern assemblages around the world.

What this means, is that the 100 million years of the Mesozoic are the temporal keys to the present in a very real sense; the problem of course is that there are very few preserved moments (fossils) from this period of time, and the stories that those keys unlock are therefore far from being interconnected and thus far from straightforward. By logic, the origin of snake lizards must have occurred long before 167 million years ago, as those earliest snakes would have evolved sometime before we first find them in the fossil record. It is also logical to predict that early experiments in snake lizard adaptive radiations and evolution produced ancient snake lizard lineages not alive today, and that they likely possessed different character suites than modern groups. In short, the ancient Mesozoic record should, predictably, provide evidence of evolutionary experimentation, unexpected clade structures, odd combinations of morphologies, and everything but a simple, straightforward linear trajectory of snake lizard evolution bringing the group to the present. From my perspective, this is exactly the case. The story is not a tidy, linear one, but a chaotic and unexpected one, making it far more interesting than has ever been imagined.

It is also revealing that I can link snake lizard skull evolution from the Jurassic forward to the categories of form developed towards the end of Chapter 1: (1) Type I–Anilioid, (2) Type II–Booid-Pythonoid, (3) Type III–Colubroid (excluding viperids and elapids), and (4) Type IV–Viperid-Elapid-Scolecophidian. At this point, the Mesozoic fossil record of snake lizards skulls, I suppose not too surprisingly, has presented only two skull types: Type I–Anilioid and Type II–Booid-Pythonoid. Where possible, based on quality of preservation, I have identified for the Mesozoic fossil snake lizards highlighted the "Type" to which they correspond. I need to be very clear on one point, though, as the tendency will be to then think of ancient snake lizard skulls within a conceptual framework bounded by reference to the modern—such top–down essentialism is not the point. The key point to remember here is that the osteological types will eventually be defined against the fossil snake lizard types, as these are indeed the progenitors of the modern. For the time being, therefore, in order to link the ancient to the characterization of the modern snake lizard given in Chapter 1, I will apply the categories developed for the modern to the descriptions of ancient snake lizards given here.

The descriptions and data presented in this chapter provide anatomical portraits of fifteen moderate to extremely well-known Mesozoic snake lizards. Some of them are beautifully preserved, while others are tantalizing fragments that, surprisingly, communicate important data on key aspects of snake lizard evolution. Form taxa represented by nothing more than isolated vertebrae are largely ignored, with the exception of the rather famous *Coniophis precedens* Marsh, 1892, and the first described Mesozoic snake lizard, *Simoliophis rochebrunnei* Sauvage, 1880—this brings the total list of taxa detailed in this chapter to seventeen. Even though I do not consider it a snake lizard (Caldwell et al., 2016; Lee et al., 2016), I also devote the end of the chapter to an overview of published and in-prep data rebutting the consideration of *Tetrapodophis amplectus* Martill et al., 2015, as a snake lizard, regardless of its possession of four legs. The anatomical portraits of these ancient snake lizards are organized temporally by geologic period and age, and geographically by reference to ancient paleogeography (e.g., Laurasia or Gondwana), and then by the name of the modern "place" in which their fossils are found (e.g., England, Argentina, etc.).

## MID TO LATE JURASSIC—LAURASIA

### Bathonian—England

**Eophis underwoodi** Caldwell et al., 2015 (Figure 2.1a)

The currently oldest known snake is represented by three dentary elements (NHMUK R12355, R12354, and R12370 [NHMUK—Natural History Museum, London, England]), with a single maxillary fragment referred to the taxon (NHMUK R12532). All of the specimens come from the Forest Marble Formation, Washford Quarry, as exposed at the Kirtlington Cement Works Quarry, Oxfordshire, England (Middle Jurassic; Bathonian). While the first description of the materials by Evans (1994) included a great deal more squamate and possibly sphenodontian specimens in the referral to *Parviraptor* cf. *P. estesi* Evans, 1994, Caldwell et al. (2015) recognized that chimeric assemblage and identified the dental elements as those of a small snake lizard.

Anatomical features common to all recognized fossil and modern snake lizards are present in these jaw fragments. The dentary fragments possess three-sided alveoli forming distinct interdental ridges, presumably composed of alveolar bone (Budney et al., 2006), with a nutrient notch at the base of the alveolus and with an interdental ridge canal piercing the alveolar bone between alveoli. The one element preserving the mandibular symphysis shows that that face was smooth and broad and did not form a tight joint with the opposite dentary. The lateral surface of the dentary is decidedly convex and possesses 4–5 large mental foramina. The fragment of the left maxilla shows similar construction and features of the three preserved alveoli to those of the dentary fragments. Because of the incompleteness of the skull (dentary fragments and one maxillary fragment), *Eophis* cannot be assigned to any of the skull types as defined in Chapter 1.

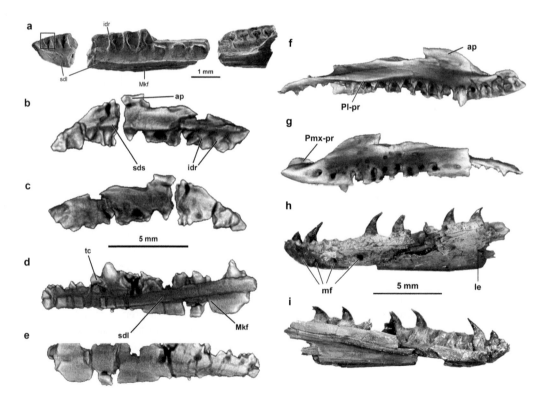

**Figure 2.1.** Fossil remains of Middle to Late Jurassic snake lizards (Middle and Upper Jurassic). (a) *Eophis underwoodi*, fragments of three right dentaries; (b) *Diablophis gilmorei*, right maxilla, medial view; (c) *Diablophis gilmorei*, right maxilla, lateral view; (d) *Diablophis gilmorei*, right dentary, medial view; (e) *Diablophis gilmorei*, right dentary, lateral view; (f) *Portugalophis lignites*, left maxilla, medial view; (g) *Portugalophis lignites*, left maxilla, lateral view; (h) *Portugalophis lignites*, left dentary, lateral view; (i) *Portugalophis lignites*, left dentary, medial view.

**Portugalophis lignites** Caldwell et al., 2015 (Figure 2.1f–i)

*Portugalophis lignites*, Caldwell et al., 2015, had not been formally described, though it has been informally referred to as *Parviraptor* Evans, 1994, by Brochinski (2000) in her overview of *Lizards from the Guimarota Coal Mine*. The holotype specimen designated by Caldwell et al. (2015) is a left maxilla (holotype), MG-LNEG 28091 (MG-LNEG—Museu Geologico, Lisboa, Portugal); the paratype is a left dentary with a good number of well-preserved teeth (MG-LNEG 28094), and a partial left maxilla without teeth was also referred to the taxon (MG-LNEG 28100). *Portugalophis* differs from *Eophis underwoodi*, *Diablophis gilmorei*, and *Parviraptor estesi*, in a number of ways, quite apart from being the largest-bodied snake lizard known from the Upper Jurassic. For example, the dentary tooth bases clearly extend into the subdental gutter; on the maxilla, the premaxillary process is quite wide, and the medial surface of the ascending process of the maxilla is quite concave. Because of the incompleteness of the skull, *Portugalophis* cannot be assigned to any of skull types as defined in Chapter 1.

Kimmeridgian—Colorado

**Diablophis gilmorei** Evans, 1996 (Figure 2.1b–e)

Originally described as *Parviraptor gilmorei* Evans, 1996, this small to medium-sized Upper Jurassic snake lizard is recognized from numerous skull and axial skeletal remains, all found in several localities within the Fruita Locality in the Brushy Basin Member, Morrison Formation, Mesa County, Colorado, US. Evans (1996) only assigned the maxilla and dentary to *P. gilmorei*, but Caldwell et al. (2015) referred a substantially larger number of elements to the species and, noting differences with the type species *P. estesi*, named a new genus. *Diablophis gilmorei*, includes the holotype material (LACM 4684/140572 [LACM—Los Angeles County Museum, Los Angeles, USA]) inclusive of a broken right maxilla, broken right mandible, and broken axis vertebra. The referred materials include: numerous precloacal vertebrae and one possible sacral vertebra (LACM 4684/140572); a fragmentary right dentary (LACM 5572/120732); and four precloacal vertebrae and one caudal vertebra with quite large transverse processes

(LACM 4684/120472). *Diablophis gilmorei* is a small-bodied snake lizard that differs from *Parviraptor estesi* and *Portugalophis lignites* in having a smaller palatine process and lacking the strong medial deviation of the anterior end of maxilla; it also differs from *P. lignites* and from *Eophis underwoodi* in having a more fully developed subdental lamina relative to the size of the dentary. Like all snakes, the dentary teeth are attached to the margins of an alveolus. Similarly to *Eophis*, there are three-sided alveoli and the teeth are conical, and strongly recurved, though there are no tooth tips preserved. The neural spines are tall and the synapophyses are massive and vertical; importantly, the neural canals show the "trefoil" organization often present in fossil and modern snakes, and there are small zygosphene-zygantrum articulations. Unfortunately, and consistent with *Eophis* and *Portugalophis*, the incompleteness of the remains does not allow assignment of *Diablophis* to any of the skull types as defined in Chapter 1.

## EARLY CRETACEOUS—LAURASIA

Berriasian—England

**Parviraptor estesi** Evans, 1994 (Figure 2.2)

The story of *Parviraptor estesi* Evans, 1994, is a complex one, having most recently been re-evaluated by Caldwell et al. (2015) who identified the type maxilla as that of a snake lizard, thus drastically revising the original concept that it was a platynotan lizard. Evans (1994, 1996) had developed a generic concept for *Parviraptor* from a temporally and spatially distributed complex of disassociated remains from the Bathonian of Kirtlington, England (see *Eophis underwoodi*); isolated remains from the Kimmeridgian of Portugal (see *Portugalophis lignites*); two blocks of unassociated skeletal elements from the Berriasian of Swanage, England (*Parviraptor estesi* and *Parviraptor* aff. *estesi*); and the remains from the Kimmeridgian of Colorado (see *Diablophis gilmorei*). That generic concept was challenged by Caldwell et al. (2015), who recognized instead four separate monotypic genera and left one set of fossils in open nomenclature (NHMUK R48388 [NHMUK—Natural History Museum, London, England]). The type specimen for *Parviraptor estesi* Evans, 1994, was identified as the maxilla-bearing block, NHMUK R48388; the second slab, not associated with the first in the least, NHMUK R8551, preserves a right frontal,

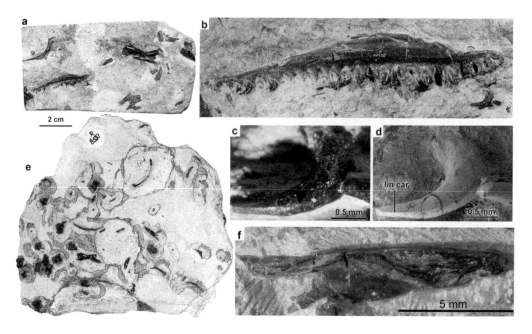

Figure 2.2. *Parviraptor estesi* and aff. *Parviraptor estesi*. (a) NHMUK 48388—Type block from the Purbeck Limestone Formation, Durlston Bay, Swanage, Dorset, England (Upper Jurassic; Tithonian/Lower Cretaceous; Berriasian); (b) medial view of *P. estesi* left maxilla on type block, note anterior maxillary tooth to far right; (c) enlargement of isolated maxillary tooth; (d) SEM of isolated maxillary tooth, note: lingual carina; (e) block with frontal and vertebrae referred to aff. *Parviraptor estesi* (NHMUK R8551); (f) right lateral view right frontal, from NHMUK R8551. (Abbreviations: lin car, lingual carina.)

atlas, and associated precloacal vertebrae of a snake lizard, but one that cannot be obviously referred to *P. estesi*, as there are no comparable elements.

The type specimen of *Parviraptor estesi* is a nearly complete right maxilla with twenty-four tooth positions preserved, that is, exposed in medial view. It possesses a very snakelike, long low, ascending process extending from tooth position 4 to tooth position 16/17. The anterior superior alveolar foramen is large, is positioned at the front of the ascending process, and is exposed dorsally. The premaxillary process is turned medially but lacks any apparent sutural faces as in snakes. The supradental shelf is narrow and thin and bears a prominent, though medially damaged, palatine process at the posterior end of ascending process. The posterior portion or suborbital ramus of the maxilla is long, and bears at least ten tooth positions posterior to the prefrontal facet and palatine process; the morphology of this condition is very snakelike (tooth implantation, tooth morphology, tooth size), though not unique to snakes, as many extant geckos also possess a toothed suborbital ramus (though implantation is different and tooth type differs, as does tooth size and the morphology of the ramus). The maxillary tooth positions of *P. estesi*

are defined by three-sided alveoli with no medial border and the presence of a small posterolingual notch forming basal nutrient foramen. The preserved teeth are attached to the rims of alveoli, appear to be conical with circular cross-sections, are recurved, and bear labial and lingual carinae; all conditions are common to snake teeth and their mode of attachment. The maxilla of *Parviraptor estesi* differs from the maxilla of *Coniophis precedens* in lacking a medial process at the anterior end and in possessing a greater degree of recurvature of the teeth. By comparison to *Portugalophis lignites*, it has a much narrower premaxillary process. Compared to *Diablophis gilmorei*, *P. estesi* possesses a larger palatine process and lacks the medial curvature of the anterior end of the maxilla.

While Caldwell et al. (2015) could not find comparable features on the second slab (NHMUK R8551) that permitted assignation to *Parvirpator*, let alone *P. estesi*, they did discuss the fossil remains on the slab in the context of them being snake lizards. While the vertebrae are relatively non-descript, they do show snake lizard features (Table 2.1). In particular, though, the frontal preserved on that slab shows a number of important features shared only with snake lizards (orbitosphenoid

THE ORIGIN OF SNAKES

TABLE 2.1

**TABLE 2.1**

*Diagnostic vertebral features of snakes present in* Diablophis *and* aff. Parviraptor

| Vertebral anatomical traits (observed in *Najash* and *Coniophis*) | *Diablophis gilmorei* | *aff. Parviraptor* |
|---|---|---|
| Well-developed synapophyses continuous with prezygopophysis | Present | Present |
| Synapophyses divided in dorsal & ventral surfaces | Present | Present |
| Shallow angle of prezygapophyses | Present | Present |
| Accessory processes absent; rib articulations lateral, not ventral as in modern snakes | Present | Present |
| Synapophyses flare laterally from cotylar margin; posteriorly centrum expanded (except *C. precedens*) | Present | Present |
| Small zygosphenes & zygantra; zygosphenes not well developed | Present | Present |
| Possession of "trefoil" organization of neural canal | Present | Present |
| Round condyle & cotyle | Present | Present |
| Constriction at base of condyle | Present | Present |

process/suboptic shelf of the descensus frontalis, complete decensus frontalis, as well as several morphologies that are more generalized "lizard"). Whether the vertebrae and frontal preserved on NHMUK R8551 belong to *P. estesi* is really of no consequence, as the material shows remarkable snake lizard features that one would predict should be present in an ancient snake lizard. They are similar to, but not identical to, the vertebrae and frontals of modern snake lizards, again, as would be predicted in a snake lizard 170 million years older than anything alive today.

Parviraptor, along with *Diablophis*, *Eophis*, and *Portugalophis*, like all data points along the continuum of information, raise more questions than they answer about snake lizard evolution. However, what is clear is that the snake lizard "head" defined the group at least 170–145 million years ago, like it does today for all groups of modern lizards, limbed or limbless. Whether the mid-Mesozoic group of snake lizards were already radiating around a limbless to limb-reduced body plan remains to be established. The relatively "primitive" configuration of the vertebrae, and the presence of a stout transverse/sacral rib, would suggest a more generalized four-legged lizard body plan as compared to the elongate, limb-reduced body plan of snake lizards such as *Pachyrhachis problematicus* or *Haasiophis terrasanctus* some 65–55 million years later. And herein lies an important and often overlooked fact: there is an equivalent amount of time separating *Eophis* and *Parviraptor* from *Pachyrhachis* and *Haasiophis* as there is between you and me and the Cretaceous Extinction Event. As for *Eophis*, *Portugalophis*, and *Diablophis*, the lack

of referable cranial material to *Parviraptor* and the incompleteness of the specimen prohibits referral to any of the skull types as defined in Chapter 1. However, the right frontal on NHMUK R8551 is suggestive of a snake lizard cranial anatomy/skull type that is not observable amongst modern snake lizards and could be considered "Type 0"—"Parviraptorine," where the frontal is long, not short, and facets for the prefrontal and jugal/postfrontal are deep and long, but where the jaws are more similar, as are the teeth, to the condition of Type I—Anilioid—intriguing, to say the least.

We should expect, let alone predict, that ancient snake lizards would be different than modern snake lizards. There is absolutely no fossil evidence of a single caenophidian-grade (Skull Types III and IV) (achrochordids + colubroids = ~3000 living species) snake older than the beginning of the Cenozoic (Rage, 1984; Holman, 2000), with most of that positively identifiable fossil record starting in the Late Oligocene/Miocene through to the Pleistocene. The fossil record of scolecophidians is similar—three possible vertebrae from the Paleocene/Eocene and a reasonable fossil record from the Miocene to the present (Augé and Rage, 2006; Rage, 1984; Holman, 2000). At some point, we must be fair to the data—the presence of fossils means something, and in many cases, like the one I have highlighted here, their absence through long tracts of geological time also means something, that is, that the groups/clades they belong to had not yet evolved, "ghost lineages" aside (Norell, 1992). I will explore this question a little later in this volume (Ch. 6), but suffice it to say here, the fossil record of ancient snake lizards from the mid-Mesozoic is clear on the kind of body plan, in particular the skull, that was at the center of early snake lizard evolution—big heads, big teeth, big bodies, and likely limbed-lizardlike bodies on globally distributed snake lizards living in a wide variety of environments.

## EARLY CRETACEOUS—GONDWANAN AFRICA

### Valanginian—South Africa

**Unnamed Snake Lizard Cranium** (Ross et al., 1999) Type I—Anilioid (Figure 2.3)

This unique specimen (AM 6001 [AM—Albany Museum, Grahamstown, South Africa]), an isolated and three-dimensionally preserved braincase and a partial basisphenoid, was originally described in great detail by Ross et al. (1999) and assigned, unnamed, to the "Squamata." However, the specimen, currently under redescription and rediagnosis, is now recognized to be the partial skull remains of an ancient snake lizard, in this case found in rocks from right near the Upper Jurassic–Lower Cretaceous boundary near Kirkwood, South Africa. The original description by Ross et al. (1999) described the otic capsule, noting a squamate style fissura metotica, the presence of the recessus scala tympani, a portion of the cavity for the saccule, lagenar basin, perilymphatic foramen and shelf, evidence of the internal and external semicircular canals, and in general, foramina for cranial nerves VI, VIII, IX, X, and XII; I would add here, most importantly, that there is no crista circumfenestralis (Palci and Caldwell, 2014) and that the specific

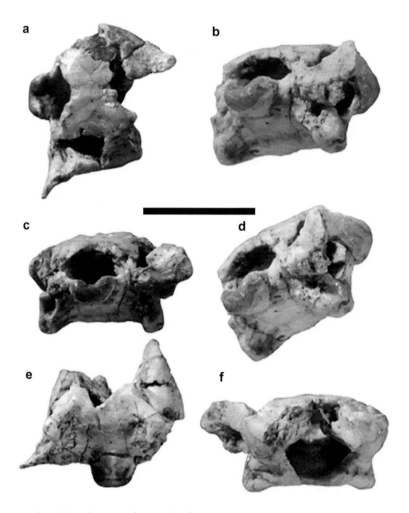

Figure 2.3. Unnamed snake lizard cranium from South Africa, Early Cretaceous (Valanginian). (a) Dorsal view; (b) right posteroventral view; (c) posterior view; (d) ventral oblique view; (e) ventral view; (f) posterior view specimen.

THE ORIGIN OF SNAKES

features of the otooccipital region are most similar to those of *Najash, Dinilysia, Sanajeh* (and in part, *Menarana,* as the braincase is very incomplete), and living anilioids such as *Cylindrophis* and *Anilius.* This unnamed South African snake lizard is an extremely important data point in the mid-Mesozoic evolution of snake lizards, as it is both Gondwanan in time and space, nearly as old as the oldest known snake lizards (Caldwell et al., 2015), and is likely the oldest known madtsoiid snake lizard, a group currently recognized as terrestrial in habits.

## LATE CRETACEOUS—AFRO-ARABIAN PLATFORM

There are currently three species of marine snakes with rear limbs known from the Lower Cenomanian (Upper Cretacous) of modern Lebanon and the West Bank—*Pachyrhachis problematicus* Haas, 1980a,b, *Haasiophis terrasanctus* Tchernov et al., 2000, and *Eupodophis descouensi.* One hundred million years ago, these three species were living far from the continental shorelines of either ancient Gondwanan Africa or the island chains and landmasses of the mostly submerged central and western Europe. This is not to say that these three-legged snake lizards, found in carbonate rocks on the Afro-Arabian Platform, only lived out to sea on this shallow basin-reef system, except that we do not have fossil examples of them from closer to any recognized continental shorelines. The anatomy of these three snake lizards is reviewed separately, but the examination of the paleoenvironment blends all three together, as two of the snake lizards come from the same quarries, and the third, *Eupodophis,* is from a virtually identical set of lithofacies and thus depositional environments.

### Cenomanian—West Bank

**Pachyrhachis problematicus** Haas, 1979—Type II Booid-Pythonoid (Figure 2.4)

*Pachyrhachis problematicus* Haas, 1979 is known from two specimens, originally described as two separate taxa: *Pachyrhachis problematicus* Haas, 1979 (HUJ-PAL 3659 [Hebrew University, Jerusalem—Paleontology]) and *Ophiomorphus colberti* Haas, 1980a,b (HUJ-PAL 3775) (*Ophiomorphus* was renamed to *Estesius* by Wallach [1984]). Caldwell and Lee (1997) and Lee and Caldwell (1998) argued for a synonymy of the two specimens under the

senior synonym, *P. problematicus* Haas, 1979 (Haas, 1979, 1980a,b); to date, no arguments have been presented that counter the data used by Caldwell and Lee (1997) to support their synonomy. Both specimens indicate that *P. problematicus* was a large-bodied snake lizard with a relatively small and narrow head, a long body composed of numerous pachyostotic vertebrae and ribs, a distinct cervical region where pachyostosis is reduced to absent, a small pelvic assemblage composed of distinct girdle elements, and small hindlimbs, though phalangeal elements are not known.

Aspects of the skull and postcranial skeleton have assessed and reassessed since Caldwell and Lee (1997) and Lee and Caldwell (1998), with the most recent works focusing on positively identifying the presence of jugal in *Pachyrhachis,* at least four mental foramina along the dentary, a large number of unfused intercentra on anterior vertebrae in the precloacal column, and the presence of at least one sacral vertebra with an unfused sacral rib (Palci et al., 2013). The confirmation of these features is an important addition to the data presented and hypothesized by Caldwell and Lee (1997) and is particularly important, as they are features not found in any modern snake lizard. However, 21 years after the publication of Caldwell and Lee's (1997) study of *Pachyrhachis,* no one would now argue it is not a fossil snake lizard, yet the data it presents would suggest that it does not share these putative key synapomorphies with modern snake lizards. *Pachyrhachis* is a superbly well-preserved Cenomanian-aged snake lizard, unlike most of the other snake lizards described so far; what its anatomical features suggest most strongly is that understanding the anatomy of ancient snake lizards, and thus the features that define these animals, can only be understood by reference to what is preserved in the fossil record. The anatomy of modern snake lizards is extremely highly derived, both within the clades to which these lizards belong and as compared among lizards of all kinds more generally.

### Cenomanian—West Bank

**Haasiophis terrasanctus** Tchernov et al., 2000— Type II Booid-Pythonoid (Figure 2.5)

*Haasiophis terrasanctus* Tchernov, Rieppel, Zaher, Polcyn and Jacobs, 2000 (HUJ-PAL 695) is known from a single, beautifully articulated specimen (see

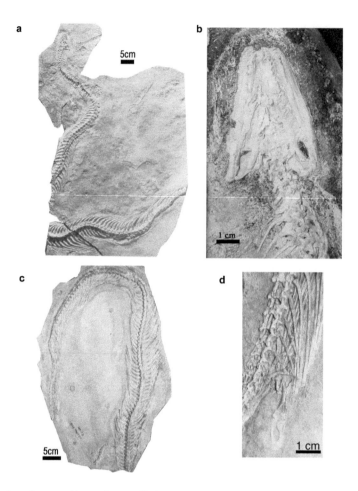

Figure 2.4. *Pachyrhachis problematicus*. (a) Dorsal view of holotype (HUJ-PAL 3659), as photographed during the late 1970s (From file photo provided courtesy of E. Tchernov); (b) Ventral view of the skull of holotype (HUJ-PAL 3659) (Photo by author); (c) dorsal view of referred specimen (HUJ-PAL 3775) (Photo by author); (d) pelvis and hindlimb of referred specimen (HUJ-PAL 3775) (Photo by author).

also Rieppel et al., 2003a) of a much smaller-bodied individual than the two specimens of *Pachyrhachis problematicus*. Like *Pachyrhachis*, the specimen of *Haasiophis* preserves a beautifully exposed rear limb with two proximal metatarsals preserved—again, there are no phalangeal elements preserved. Importantly, the skull of *Haasiophis* is very different from that of *Pachyrhachis* in its general architecture and anatomy. As such, it is clear that there were very diverse clades of snakes, all retaining rear limbs, present by the Cenomanian.

Characteristically, the skull of *Haasiophis* presents a small, narrow, edentulous premaxilla; there are twenty-four tooth positions on the maxilla, eight on the palatine, fifteen to seventeen on the pterygoid, and twenty-six on the dentary (visible teeth are clearly striated where those of *Pachyrhachis*

are not, and there are a great many more teeth in *Haasiophis* than in *Pachyrhachis*). All total, there are 154 precloacal vertebrae and five ribs with lymphapophyses. Importantly, since the original description by Tchernov et al. (2000) and the follow-up study by Rieppel et al. (2003a), Palci et al. (2013) have modified and added to those previous studies by finding new and unexpected anatomies for *Haasiophis*: counter to Tchernov et al. (2000) and Rieppel et al. (2003a), there is no evidence of a laterosphenoid, there are distinct and identifiable chevron bones in the caudal vertebrae and a large number of unfused intercentra on anterior vertebrae in the precloacal column (similar to *Pachyrhachis*). Again, these features are not considered to be synapomorphies of Serpentes, or better yet, modern snake lizards.

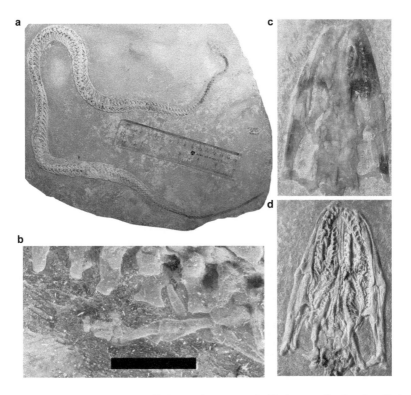

Figure 2.5. *Haasiophis terrasanctus*. (a) Dorsal view of holotype (HUJ-PAL 659); (b) close-up of limb and girdle (HUJ-PAL 659); (c) dorsal view of skull, photographed through acrylic embedding medium; (d) ventral view of skull.

## Cenomanian—Lebanon

**Eupodophis descouensi** Rage and Escuillié, 2000—
Type II Booid-Pythonoid (Figure 2.6)

The holotype specimen of *Eupodophis descouensi* Rage and Escuillié, 2000 (Rh-E.f. 9001, 9002, 9003 [Rh-E.f.—Natural History Museum of Gannat, Gannat, France]) and referred specimens as assigned by Rieppel and Head (2004) (MSNM V3660, 3661, and 4014 [MSNM—Museo di Storia Naturale di Milano, Milan, Italy]) represent the largest sample of a Cenomanian rear-limbed snake yet known, ranging in size from a large presumed adult (Rh-E.f., 9001/9002) to a comparatively quite small presumed juvenile (MSNM V3661). As with *Pachyrhachis*, the coronoid process is very tall, the body is laterally compressed, the tail is short and paddlelike, and, in the presumed adult specimens, most of the vertebrae and ribs are pachyostotic, though this condition is not observed in the very small juvenile (MSNM V3661). Palci et al. (2013) also identified a number of important features in *Eupodophis* that are contra Rieppel and Head's (2004) anatomical interpretations and, like *Pachyrhachis* and *Eupodophis*, indicate strongly that the snake lizard

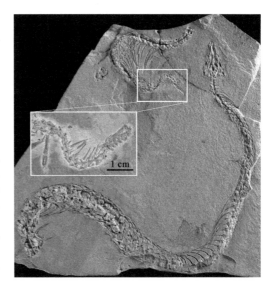

Figure 2.6. *Eupodophis descouensi*. Dorsal view of holotype (Rh-E.f. 9001, 9002, 9003) with inset box magnifying limb and girdle.

possesses a number of anatomical features that do not define modern snake lizards: like *Haasiophis*, there is no evidence from the available materials of *Eupodophis* that a laterosphenoid was present, there is a well-defined and identifiable jugal as in *Pachyrhachis*, and, similarly to *Haasiophis*, there are clear and obvious chevron bones in the tail.

## LATE CRETACEOUS—ADRIATIC-DINARIC PLATFORM

Several thousand kilometers to the north, across the Tethyan carbonate platform and the bathyal waters separating the Afro-Arabian Platform from the numerous plates that would ultimately form Italy, southern Europe, and the Balkan peninsula, was a companion carbonate platform, referred to here as the Adriatic-Dinaric Platform. *Pachyrhachis*, *Haasiophis*, and *Eupodophis* are not demonstrably present in the platy limestones outcropping or mined in modern Croatia, Slovenia, or Bosnia-Herzegovina; but again, the sediments known to produce Cenomanian snake and aigialosaurid lizards are younger, ranging from the Middle to Upper Cenomanian. From northern margins of the Cenomanian Tethys in Bosnia-Herzegovina, two snake lizard taxa have been recognized since the early twentieth century—*Pachyophis woodwardi* Nopcsa, 1923 and *Mesophis nopcsai* Bolkay, 1925. Interestingly, neither taxon is recognized to have rear limbs.

### Cenomanian—Bosnia-Herzegovina

**Pachyophis woodwardi** Nopcsa, 1923 (Figure 2.7a), **Mesophis nopcsai** Bolkay, 1925 (Figure 2.7b)

In an interesting departure from what has become the "typical" Cretaceous marine snake lizard, that is, a snake lizard with hindlimbs, there are two limbless taxa known from finely bedded, though more blocky, limestones, in quarries located near Selista (Selisca, Selisce), an eastern suburb of Bilek (Bileca), in East Bosnia-Herzegovina about 40 km inland (NE) from Dubrovnik, Croatia. Interestingly, even though there are similar aged sediments (though they are platy limestones) on the island of Hvar, Croatia, that have produced spectacular specimens of aigialosaurs and pontosaurs, to date, there are no known Cenomanian snake lizards from those rocks. Unfortunately, the skull of *Mesophis* is not preserved, and the partial skull fragments of *Pachyophis* do not allow confident assignment of the specimen to any of categories of modern snake lizard skull types.

The first snake lizard described was *Pachyophis woodwardi* Nopcsa, 1923 (Figure 2.7a) and the second, in honor of Nopcsa, was *Mesophis nopcsai* Bolkay, 1925 (Figure 2.7b). The type of *Pachyophis*

*woodwardi* (NMW A3919 [Figure 2.7a] [NMW— Natural History Museum, Vienna, Austria]) was redescribed by Lee et al. (1999a), who noted that the snake lizard appeared to lack rear limbs, had a very small head (though it is only partially preserved), and showed extreme pachyostosis of the vertebrae and ribs, to the point for the ribs (and perhaps this preservational) that the intercostal spaces are almost obliterated. The body appears to have been laterally compressed, even though the ribs throughout their length are very round. Compared to *Pachyrhachis*, *Pachyophis* has more teeth on the dentary (~23), making it more similar to *Haasiophis* (~26) than *Pachyrhachis* (~12).

The counterpart of the type and only specimen of *Mesophis nopcsai* (NHM-BiH 2039 [NHM-BiH: Natural History Museum, Bosnia-Herzegovina, Sarajevo]) (Figure 2.7b), was rediscovered by colleagues of mine from Croatia (Jakov Radovcic and Bosnian friends of his) when the collections were unpacked from their hiding place after the Balkan conflict in the early 1990s. I was able to examine the specimen in person in February 2005 in the collections of the Natural History Museum of Bosnia-Herzegovina, Sarajevo (the holotype part remains lost, and is thought to be in a private collection). There is no skull on the specimen, though the body is articulated and preserved for the most part in ventral view and preserves only a small number of postcloacal vertebrae. Overall, the skeleton appears to show a lesser degree of pachyostosis then does *Pachyophis*, but not all aspects of any one precloacal vertebra are visible, so such a comparison is difficult to make. There appear to be no limbs preserved on the specimen, and so, like *Pachyophis*, it appears to represent another individual of a legless group of aquatic Cenomanian snakes (Rage and Escuillié, 2003).

## LATE CRETACEOUS—GONDWANAN AFRICA

### Maastrichtian—Madagascar

**Menarana nosymena** Laduke et al., 2010—Type I Anilioid (not figured)

UA 9684 (UA—Université d'Antananarivo, Antananarivo, Madagascar), the "type" of *Menarana nosymena*, was described as partial postcranial skeleton of articulated and associated vertebrae and some ribs, but in truth, it is five vertebrae and a cranial fragment (basicranium), and then a larger number of referred specimens. The type material is

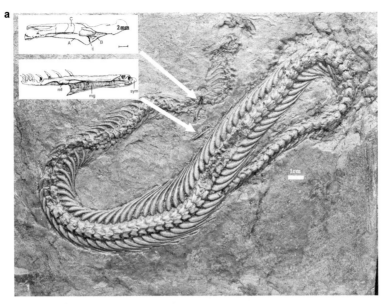

**Figure 2.7.** Limbless Cenomanian snake lizards. (a) *Pachyophis woodwardi* (Holotype, NMW A3919) (From Lee et al. 1999a). Inset boxes: Line drawings of left dentary as preserved in medial view, and jaw fragment tentatively identified as part of the right mandible, preserved in lateral view. (b) *Mesophis nopcsai* (Holotype [counterpart] NHM-BiH 2039). (Abbreviations: mg, Meckelian groove; mf, medial flange of dentary; sym, symphysis; ij, intramandibular joint; A, putative right angular; B, putative right splenial; C, putative right compound bone.)

all presumed to be from the same individual based on proximity of discovery (within a 2-meter-square area according to Laduke et al. [2010]). While this assertion is problematic, the materials considered to represent the type are most certainly those of a single "kind" of madtsoiid snake lizard and are clearly not a chimera of non-madtsoiid snake lizards. To be clear, though, I consider such type specimens chimeric and, as such, it creates significant problems for subsequent analyses,

such as phylogenetic studies, when subsequent authors treat such a taxon as M. *nosymena* as a valid terminal taxon. For the purposes of the anatomical discussion here of M. *nosymena*, I consider the braincase fragment informative and am treating it here as the only valid specimen diagnosing this taxon. There is nothing to link this fragmentary braincase element to the five other vertebral elements other than proximity in a 2-meter-square quarry grid.

Laduke et al. (2010) described the braincase fragment as composed of the basioccipital, portions of the left and right prootics and exoccipital-opisthotics (fused in many squamates and often referred to as the otooccipital [e.g., Evans, 2008]), as well as the parabasisphenoid. The elements are all fused together, thus obscuring exact sutural relationships between them, though traces are easily found on the internal surfaces and somewhat on the external faces as well. The important features preserved in the braincase that link it to other Mesozoic Gondwanan snake lizards, such as the unnamed Valanginian snake lizard from South Africa (Ross et al., 1999), *Sanajeh*, *Dinilysia*, and *Najash*, is the very large size of the otic capsule region (i.e., very large fossa for the lagena, enlarged recessus scalae tympani, etc.), suggestive of an enlarged fenestra ovalis and otic capsule. From the size of the "foramen" implied by the large and posteriorly directed opening of the recessus scala tympani, it appears that the "foramen" would be identifiable as a fenestra pseudorotunda/rotunda and that there was no crista circumfenestralis (Type 1, *sensu* Palci and Caldwell, 2014), similar to *Dinilysia* and *Najash*.

Rogers et al. (2000) concluded their remarks by noting that the pre–end Cretaceous faunas of Madagascar are not the source for the modern Malagasy faunas, and I think this is true. Where those authors did not expand further by considering the Gondwana source of the vertebrate faunas, I will—in terms of snake lizards, the similarities between arid mid-Mesozoic environments and the "madtsoiid" snake lizards that seem to populate them long after the break-up of Gondwana cannot be mere evolutionary coincidence. At least this mid-Mesozoic lineage of snake lizards, large-bodied, big-mouthed animals, lacking the modern macrostomatan gape characteristics, most surely were highly adapted to arid environments regardless of where they occurred in their Gondwanan habitats, and represent an ancient lineage of snake lizards of which we are only now beginning to garner the slightest understanding. Gondwana is most certainly a key continental player in the mid-Mesozoic origins of snakes, and just how "key" remains to be seen; in truth, the origins and radiation of snake lizards may well be tied to a completely unknown paleobiogeographical source of early lizard evolution—Pangaea (see Simões et al., 2018a).

## LATE CRETACEOUS—GONDWANAN SOUTH AMERICA

### Cenomanian—Venezuela

***Lunaophis aquaticus*** Albino et al., 2016 (not figured)

The holotype specimen (MNCN-1827A-G [MNCN—Museo de Ciencias Naturales de Caracas, Venezuela]) is composed of eleven vertebrae preserved in a small block of black shale. While known only from these few vertebrae, this snake lizard appears to be a laterally compressed, aquatically adapted animal that is quite different in terms of vertebral morphology from all of the other known forms of Cenomanian marine snakes from the Tethyan Realm. The vertebrae are not at all like those of simoliophiids such as *Simoliophis*, *Pachyrhachis*, *Haasiophis*, *Eupodophis*, *Pachyophis*, or *Mesophis*. Instead of the large, squareish pachyostotic centrum body characterizing simoliophiids (de Buffrénil and Rage, 1993), the centrum of *Lunaophis* is non-pachyostotic, long, and narrow, and the paradiapophyses are low, anterior, and directed ventrally. The neural arch, neural arch lamina, and prezygapophyses are all inflated and much larger than the centrum—the prezygapophyses and proximal neural arch lamina seem to be the region of most pronounced pachyostosis or osteosclerosis. Most of the vertebrae have either lost the neural spines or they are very low, but there is a one well-preserved vertebra with a long, thin, posteriorly directed neural spine. The anatomy of the vertebrae of *Lunaophis* strongly suggests that there is another non-simoliopiid lineage of Cenomanian-aged marine snake lizards. Only the recovery of cranial material of this taxon will help address this very intriguing question.

### Cenomanian—Argentina

***Najash rionegrina*** Apesteguía and Zaher, 2006— Type I Anilioid (Figure 2.8a–f)

The original description of *Najash rionegrina* Apesteguía and Zaher, 2006 included the holotype specimen (MPCA 390–398 [Museo Paleontológico "Carlos Ameghino," Cipolletti, Río Negro]), a short but articulated section of the axial skeleton that includes the sacral region and a loosely articulated and associated series of limb and girdle elements, and referred materials (MPCA 380–383, 385) include a second and much longer

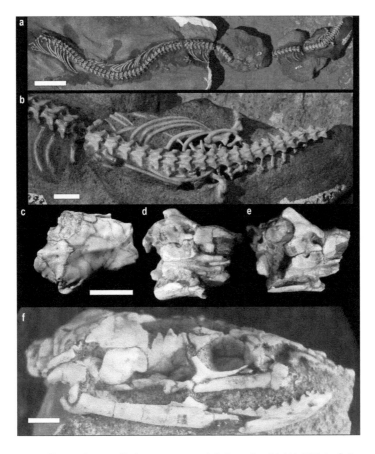

Figure 2.8. *Najash rionegrina.* (a) Dorsal view of holotype postcranial skeleton (MPCA 390-398) (scale bar = 5 cm); (b) holotype detail of pelvic girdle, leg, and anterior caudals (scale bar = 1 cm); (c) dorsal view, referred skull posterior to the left (MPCA 385); (d) ventral view, referred skull posterior to the left (MPCA 385); (e) right lateral oblique view, referred skull posterior to the left (MPCA 385); (f) MPCA 500, new and most complete known skull (scale equals 5 mm).

string of articulated vertebrae with an associated dentary (with at least eleven large tooth sockets) and an isolated and fragmentary skull with at two associated presacral vertebrae. The concept of *Najash* as created by Apesteguía and Zaher (2006) was then used by those authors and many others since (e.g., Wilson et al., 2010; Longrich et al., 2012) as a complete taxon concept for use as a terminal in phylogenetic analyses. Palci et al. (2013) cautioned that the holotype and referred materials used to create that concept were in a fact a loosely justified chimera without overlapping components and that any use in phylogenetic analysis should be restricted to the holotype postcranial skeleton.

As of the writing of this book, I have been involved in three field seasons near Cerro Policia and the La Buitrera Locality in the Candeleros Formation, northern Patagonia. The sheer volume

of new snake lizard skeletons, skulls, body sizes, and diversity is astonishing—this is truly the densest and most complete fossil preservation of any Mesozoic snake lizard yet known, a true snake lizard lagerstatten! Garberoglio et al. (in review, and 2019a,b) have slowly begun the process of working through this wealth of data and have countered Palci et al. (2013) with additional data indicating that the original concept was indeed a lucky guess—the postcranium and head associated by Apesteguía and Zaher (2006) while a chimera of individuals, likely represents the same morphospecies of snake lizard.

The original braincase referred to *Najash* by Apesteguía and Zaher (2006) (Figure 2.8c–e) is broadly similar to *Dinilysia patagonica* and to modern snake lizards such as *Cylindrophis* and *Anilius*. The otoccipital region is broad and expanded and supports two very large otic capsules with large stapedial footplates and fenestra ovale. Apesteguía and Zaher (2006) stated in their original diagnosis and description of the taxon that the crista circumfenestralis was present in *Najash*; later, Zaher et al. (2009) revised the diagnosis of Apesteguía and Zaher (2006) and stated that the crista circumfenestralis is not present, though they did argue in Zaher et al. (2009) that it is present but only weakly developed, which is not the same thing as absent. That contradiction aside, Zaher et al. (2009) continued to argue for a weak crista circumfenestralis because of the presence of an anteriorly positioned crista prootica on the margin of the fenestra ovalis; Zaher and Scanferla (2012) argue that *Dinilysia patagonica* also possesses a crista circumfenestralis based on the same argument (I will discuss this at length in Chapter 3). Without exploring in more detail the morphology of the referred skull of *Najash*, and in referring to the detailed anatomical work presented by Palci and Caldwell (2014) on the morphology of the crista circumfenestralis in snake lizards specifically, and all lizards generally, the presence of a crista circumfenestralis in *Najash*, based on the presence of only a small cristalike margin of the crista prootic along the rim of the fenestra ovalis, is nonsense. In the alphabet, the letter "O," lowercase or uppercase, is always and only the letter "O," when the line drawn to represent that letter forms a complete circle or oval or however you wish to squish it, compress it, or render it—unless you choose not to connect the beginning of your curve to its ending, in which case, you have a letter "C," not an "O." You can call it an "O" all you want, but the fact is, the rendered two-dimensional object is a "C." So it is with the crista circumfenestralis—the crista prootica and crista circumfenestralis are not morphological synonyms; quite the opposite, in fact. The crista circumfenestralis is a "fusion structure" of the crista prootica, crista tuberalis, and crista interfenestralis (see discussion in Chapter 3). For the crista circumfenestralis to be present, all three crests must fuse to form a tube of thin bone that rises up around the columellar shaft of the stapes, and in the process of doing so, because one of the contributing cristae of bone is from the basal tuber of the basioccipital, covers over the fenestra rotunda and creates a common space, or recessus scalae tympani, for the movement of perilymphatic fluid in and out of the perilymphatic space—the reentrant fluid

circuit of snake lizards. When one of more the cristae are absent, when a tube does not form, where there is no recessus scalae tympani for the formation of a reentrant fluid circuit, then the crista circumfenestralis is absent (Apesteguía and Zaher, 2006; Zaher et al., 2009; Zaher and Scanferla, 2012). Knowledge claims of something real only survive falsification by surviving falsification—the claimed presence of a crista circumfenestralis in *Najash* does not survive such testing. Not all modern snake lizards have a crista circumfenestralis, and they are not lesser snake lizards for the absence or presence of the morphology. Numerous factors, evolutionary and ontogenetic, can affect the presence or absence of such a feature, and the only means of coming to understand those patterns and processes is by being true to the data.

Getting back to the referred skull of *Najash*, what is preserved are portions of the crista interfenestralis and crista tuberalis; as would be expected in a snake lizard without a crista circumfenestralis, the juxtastapedial recess is widely open posteriorly. The basipterygoid processes are well developed and articulate with a distinct facet in the pterygoids. The prootic and posterolateral extension of the parietal form a deep and narrowly elongate fossa for the supratemporal (similar to *Dinilysia*, *Sanajeh*, *Anilius*, and *Cylindrophis*), though that latter element is missing. The prootic is also exposed dorsally between the supratemporal, exoccipital, and supraoccipital, similar to *Dinilysia*, *Sanajeh*, *Cylindrophis*, and *Anilius*. Unlike most other snake lizards, fossil and or modern, the exoccipitals do not meet dorsally, though this condition has been noted in *Haasiophis*.

The most exciting work arising from new fossil materials of *Najash* answers a long-standing debate on the morphology and evolution of the orbit in all snake lizards, not just *Najash* and related fossil forms, and that is the identification of the jugal in a new specimen of *Najash*, which is undoubtedly the best-preserved skull of any known Mesozoic snake taxon (MPCA 500; Figure 2.8f). As this work is in review (Garberoglio et al., in review) as of the writing of this treatise, it is not discussed in more detail here (though it was presented in abstract form by Apesteguía and Garberoglio, 2013), but I do note that this new fossil data yet again provides proof that the fossil record is the only data set that can conclusively address primary homolog problems in the evolution of phenotype.

THE ORIGIN OF SNAKES

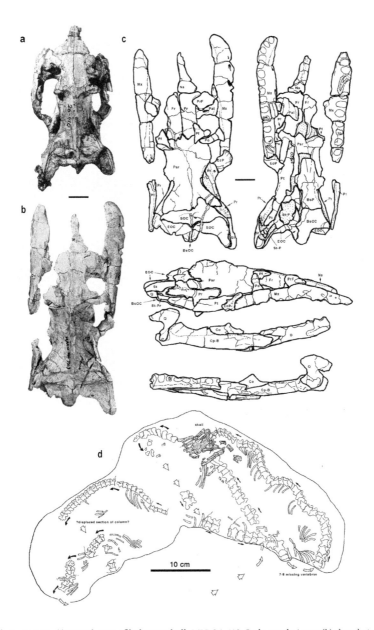

Figure 2.9. *Dinilysia patagonica*. (a) Dorsal view of holotype skull, MLP 26-410. Scale equals 1 cm; (b) dorsal view of skull of MACN–RN-1013; (c) line drawings of MACN-RN-1013 in, dorsal, ventral and right lateral views, and lateral and medial views of right mandible (From Caldwell and Albino, 2002, Fig. 1); (d) dorsal view of articulated skeleton, MACN-RN-974. (Abbreviations: BsOC, basioccipital; cen, centrum; ns, neural spine; Co, coronoid; CpB, compound bone; D, dentary; EcP, ectopterygoid; ex-col, extracolumella; Fr, frontal; fld, foramen for the lachrymal duct; fPal-Mx, Palatine-Maxillary foramen; fRO, fenestra rotunda; Mx, maxilla; Na, nasal; Pal, palatine; Par, parietal; PrF, prefrontal; Po, postorbital; Pf, postfrontal; Pr, prootic; Pt, pterygoid; Q, quadrate; SOC, supraoccipital; BsP, basisphenoid; St, supratemporal; St-F, stapedial footplate; St-P, stapedial shaft/process; VI, 6th cranial nerve [abducens]; X, 10th cranial nerve [vagus].)

## Upper Santonian/Lower Campanian—Argentina

**Dinilysia patagonica** Smith-Woodward, 1901— Type I Anilioid (Figure 2.9)

Here, I devote a great deal of text and thought to the discussion of *Dinilysia patagonica* Smith-Woodward, 1901, for the very simple reason that it was the first described Mesozoic snake lizard known from something more than a couple of isolated vertebrae (Sauvage, 1880; Marsh, 1892). The type material of *Dinilysia patagonica* MLP 26-410 (MLP—Museo de La Plata, La Plata, Argentina) described by

Smith-Woodward (1901) is a surprisingly complete (though less complete today than it was 120 years ago on first description), three-dimensionally preserved skull with a large number of associated vertebrae (still largely undescribed), and yet, until the 1970s, its description caused barely a "hiccup" in the study of snake lizard evolution and origins (Bellairs and Underwood [1951] devoted three sentences to mentioning it), if it got mentioned at all (e.g., Underwood, 1967). Why it was for so long ignored remains a great mystery to me—even a single data point is still data, particularly when it is as unique as a fossil snake lizard skull. My suspicion is that it was ignored because the specimen was from Gondwanan Argentina long before anyone thought of Argentina, and plate tectonics, and the southern continents as "Gondwana" (i.e., the ink was still wet on von Wegener's [1912] paradigm shift of "continental drift" in the early twentieth century). And because Cope (1869) and Nopcsa (1908, 1923), and fossils, clearly had been discounted as data and as ancestral origins models, a single fossil snake lizard skull was of no fundamental consequence to snake evolution when the living assemblage clearly demonstrated everything you needed to know about what was primitive and ancestral (see Chapters 3, 5, and 6). I also consider it probable that eurocentrism influenced the lack of interest in the taxon, but really cannot point to any concrete evidence to support the notion other than the fact that it was even ignored by fossil snake lizard workers (e.g., Nopcsa, 1923, 1925). The European and North American scholars of the time were embroiled in the Cope (1869)-generated debate on aquatic origins for snake lizards and their kin (see Janensch, 1906; Camp, 1923; Bellairs and Underwood, 1951), and perhaps because *Dinilysia* was clearly not an aquatic snake, the information contained in its remains and provenance were of no consequence to the scholarly debate of the day. Maybe? And finally, I have considered that the tone of the debate was changing, as I said earlier in this treatise, from a focus on fossils to one guided by the neontological data set of soft tissues (Camp, 1923; Walls, 1942; Bellairs and Underwood, 1951), modern snake lizard diversity, biogeography, and so on—for me, this one seems more likely to have been a powerful driver of the "Isolation of *Dinilysia*," particularly when coupled with the other complicating factors of eurocentrism, marine origins of snake lizards debate, and the relevance of Gondwana before there was a Gondwana.

Ultimately, though, and regardless of reason, *Dinilysia* was truly ignored, apart from the most occasional of mentions, until Estes et al. (1970), followed by Frazetta (1970), provided a serious and detailed description and revision of the type specimen of this very important late Mesozoic fossil snake lizard. Surprisingly, even Estes et al.'s (1970) efforts ended up in essentially the same limbo as did those of Smith-Woodward (1901)— apart from the occasional honorable mention (Hoffstetter, 1967; McDowell, 1974; Rage, 1977, 1984; Groombridge, 1979), *Dinilysia* was elevated to pariah status and was completely ignored yet again in major studies of snake lizard evolution. Today, which fortunately stands as a complete contrast to history, *Dinilysia patagonica*, and similar snake lizards referred to here by me as "madtsoiids" (i.e., *Najash, Menarana, Sanajeh*, and the unnamed South African snake lizard), are on the vanguard of a new understanding of early to mid-Mesozoic snake lizard evolution, not to mention the increasing presence of "madtsoiid-like" snake lizards in non-Gondwanan Laurasia (e.g., Romania [Folie and Codrea, 2005; Vasile et al., 2013]). I am taking some time here to consider the long history, anatomy, and depositional environment of the Patagonian snake lizard, *Dinilysia patagonica* Smith-Woodward, 1901, as it deserves some long-overdue ink.

## A Brief History of *Dinilysia patagonica*

Beginning in 1890 and continuing until roughly 1896, Dr. Santiago Roth, along with a number of other researchers from the Museo La Plata, La Plata, Argentina, initiated a broad survey of the geology, topography, and biology of the eastern slopes of the Argentine portion of the Andes. Roth spent some of this time working in the Argentine Patagonian provinces of Neuquén and Río Negro (Smith-Woodward, 1901). In particular, he focused much of his work on outcrops in and around the city of Neuquén, located at the confluence of the Río Neuquén and Río Limay which together form the eastward flowing Río Negro (Figure 2.1). Roth, working with Professor Francisco Moreno and a number of other geologists and paleontologists, made some of the first collections of fossil vertebrates from what is now referred to as the Bajo de la Carpa Formation, Rio Colorado Subgroup, Neuquén Group (Leanza et al., 2004; Sánchez et al., 2006), at or near Boca del Sapo on the north side of what is now the city of Neuquén. From

THE ORIGIN OF SNAKES

Neuquén, the specimens made their way to their present location at the Museo La Plata, La Plata, Argentina. Under the direction of Professor Moreno, the opportunity to study and describe the vertebrate fauna from Neuquén was offered to a number of researchers at the then British Museum of Natural History, that is, R. Lydekker and A. Smith-Woodward. Lydekker (1893) described the dinosaurs, while Smith-Woodward (1896, 1901) described the notosuchian crocodiles, turtles, and, of consequence here, the ophidian lizard specimen that would become *Dinilysia patagonica* Smith-Woodward, 1901 (MLP 26-410), first mentioned by him in the introduction to the 1896 description of the notosuchian crocodiles. For almost 70 years, Roth's specimen, the holotype of *Dinilysia patagonica* (MLP 26-410), and Smith-Woodward's (1901) description remained the only known material and descriptive characterization of this Gondwanan Cretaceous snake lizard. Based on the published literature, and perhaps from seeing the specimen in person in La Plata, though on this point I speculate, Romer (1956) named the family Dinilysiidae.

It was not until 1970, based on further preparation of the holotype specimen, that Estes et al. (1970) published a redescription of the skull of *Dinilysia*. As was noted by Estes et al. (1970) and reiterated above, this extremely well-preserved Mesozoic snake lizard had failed to generate anything more than a mention or two in the literature. Estes et al. (1970) cited key studies in the active debate on snake lizard ancestry (Bellairs and Underwood, 1951; McDowell and Bogert, 1954; Underwood, 1957), all of which had largely ignored Smith-Woodward's (1901) superb specimen of *Dinilysia*. Again, as with Smith-Woodward's (1901) description, even the Estes et al.'s (1970) redescription of *Dinilysia* failed to ignite any wide spread interest in the anatomy and relationships of this snake lizard with respect to all other snake lizards (i.e., only three studies can be found that cite Estes et al. [1970]—Rage [1977], Rieppel [1979a,b], and McDowell [1987]).

Also published in 1970 was Frazetta's (1970) biomechanical and functional morphological study of *Dinilysia* based on his observations of the holotype skull (Frazetta was a collaborator on the Estes et al. [1970] publication); he concluded that *Dinilysia* possessed a cranial apparatus whose kinematics were similar to that of extant boiids, but still less flexible, and that such snake lizards

were likely aquatic and fed on "cold-blooded vertebrates." The very first description of the postcranial remains associated with the holotype specimen was given by Hecht (1982) and even then he only described two vertebrae in detail, one from the middle and the other from the posterior precloacal region of the column. The remaining vertebrae have never been figured or characterized.

Almost 100 years had passed since the original collection made by Roth before a new specimen (MLP 79-II-27-2) was described and assigned to *Dinilysia patagonica* (Rage and Albino, 1989). These two authors described eighteen vertebrae and associated ribs collected from sediments at the holotype locality of Boca del Sapo, Neuquén city. Rage and Albino (1989) were able to modify the characterizations of *Dinilysia*'s vertebral morphology as given by Hecht (1982), but again, not much additional work was stimulated by these studies.

Some 70 years later after Smith-Woodward's (1901) description, and at about the same time as Estes et al. (1970) were publishing their study, three more fragmentary skulls were collected (MUCP v 38 [MUCP—Museo Museo de la Universidad del Comahue, Neuquen, Argentina]; MLP 79-II-27-1 & MLP 71-VII-29-1) from outcrops just to the north of the campus of the Universidad Nacional de Comahue in Neuquén. These new specimens of *Dinilysia* (MLP 71-VII-29-1, MLP 79-II-27-1, and MUCP-V-38) were described by Albino (1989: unpublished PhD thesis) in her study on the fossil snake lizards of Argentina. She compared these specimens to the holotype skull of *Dinilysia*, noted a number of differences, and concluded that they could belong to a different taxon.

However, the largest and most substantial collection of material assignable to *Dinilysia patagonica* was made by a field crew led by Professor Jose Bonaparte in 1994, all of which went to collections at the Museo Argentino de Ciencias Naturales "Bernardino Rivadavia," Buenos Aires, Argentina (MACN); these new materials came from a new locality nearly 50 kilometers away from Boca del Sapo, at Paso Cordoba (Tripailao Farm Locality) on the south side of the Rio Negro River in Rio Negro Province across from the city of General Roca, Neuquen Province. The sediments at all the known *Dinilysia* localities are from the Bajo de la Carpa Formation (Santonian–Campanian), though the quality and quantity of specimens to be found at Paso Cordoba eclipse any and all other localities. Bonaparte's fieldwork produced a large number of exceptionally well-preserved skulls and skeletons of *Dinilysia* showing osteological features missing in the holotype (MACN RN-976, 1013–1018, 1021). In 1998, I gained access to these materials through Bonaparte and via an agreement to collaborate with Dr. Adriana Albino, and proceeded forward with a broad analysis of the anatomy, phylogeny, and paleobiology of *Dinilysia patagonica* (Caldwell and Albino, 2001, 2002); subsequent studies based on these materials involved a paleobiological study (Albino and Caldwell, 2003) and an examination of the dental histology of *Dinilysia* (Budney et al., 2006), and finally an investigation of *Najash rionegrina* (Palci et al., 2013). In a decade-old review of the contribution of fossils to the understanding of snake lizard evolution (Caldwell, 2007a), the cranial anatomy of *Dinilysia* was a key element of the conclusions drawn, in terms of missing history, but also in terms of misinterpreted anatomy. Scanferla and Canale (2007) described new, though isolated, vertebral materials from the Anacleto Formation (Campanian), as the youngest known *Dinilysia* specimens (UNC-CIP 1 [UNC—Universidad National del Comahue, Neuquén, Argentina]). In 2001, a National Geographic–supported field research project to the Paso Córdoba area produced another new skull (MPCA-PV-527 [MPCA—Museo Paleontológico "Carlos Ameghino," Cipolletti, Río Negro]) that resulted in a study produced by Caldwell and Calvo (2008) where they characterized a number of new and important features of *Dinilysia*. Since the primarily descriptive work of Caldwell and Calvo (2008), the summation study of this kind would be the distillation of Agustin Scanferla's dissertation study of *Dinilysia* published as Zaher and Scanferla (2012). That work was a monographic survey of the specimens collected by Bonaparte in 1994, first surveyed by Caldwell and Albino's collaborations, that also integrated the holotype into their monograph and the new material collected by Caldwell and Calvo in 2001.

Everything published on *Dinilysia* since 2012, has been more synthetic in intent and hypothesis construction, using the morphology and anatomical details of *Dinilysia* to revise and refine our understanding of snake lizard evolution and phylogenetic relationships. Palci and Caldwell (2013) revisited the problem of the presence or absence of a crista circumfenestralis in *Dinilysia*, revising the assertions of Zaher and Scanferla (2012) that it was present, and examined the

evolution of the crest system in all other snake lizards. Scanferla and Bhullar (2014) published their study of postnatal ontogenetic variation amongst the known specimens of *Dinilysia*, followed by Scanferla (2016), who used *Dinilysia* as a key taxon in his study of the evolution of macrostomy in snake lizards. Yi and Norell (2015), argued based on CT scan data of the inner ear of *Dinilysia* and a morphometric analysis of landmarks on those reconstructions that *Dinilysia* was a burrowing snake lizard; this hypothesis has been rejected by Palci et al. (2017, 2018) on both methodological and empirical grounds and also finds no support from taphonomy or functional morphology (see below and Chapter 3). The most recent examination of new data on *Dinilysia* is an extremely detailed description of surficial brain details in *Dinilysia*, looking for common anatomies with other modern snake lizards, based on a beautifully preserved natural endocast of the braincase by Triviño et al. (2018).

One hundred and seventeen years later, one of the best-preserved fossil snake lizards in the entire world, eclipsed perhaps only now by the quantity and quality of new but undescribed *Najash* materials, is deservedly being recognized as one of the most important fossil snake lizards known to science. The pace of investigations is finally taking advantage of the beautifully preserved materials that present key bridging information between other more ancient snake lizards and subsequent later Mesozoic snake lizard evolution.

*Dinilysia* is a large-bodied snake lizard (~2 meters in length, if not more), known from multiple specimens of varying body sizes, including very small individuals that provide surprisingly detailed ontogenetic data (Scanferla and Bhullar, 2014). In broad morphological terms, *Dinilysia*, and snake lizards closely related to it, are non-macrostomatan snake lizards; that is, they do not possess the mobile cranial skeletons (e.g., palatines and pterygoids are akinetic, maxillae are fixed and immobile, the frontal-parietal suture is complex and interdigitating, etc.) characterizing almost all modern snake lizards considered to be within the clade Macrostomata (Müller, 1831) or snakes with big gapes (see also Rieppel [1988a]) (this character complex and its distribution within snake lizards will be explored in detail in Chapter 3). The literature describing the anatomy of these now numerous specimens is exhaustive in its detail and need not be reviewed in detail here (Estes et al., 1970; Caldwell and Albino, 2002; Zaher and Scanferla, 2012; Scanferla and Bhullar, 2014; Palci and Caldwell, 2014). Critical anatomical features defining *Dinilysia*, which also bear on Gondwanan madtsoiid snakes generally, include an undivided olfactory canal running through the frontals (i.e., median frontal pillars absent); the absence of a median flange on the nasal; the absence of a laterosphenoid pillar separating the trigeminal foramen into two foramina; the basisphenoid processes are prominent and strongly oriented lateroventrally; the jugal is present and is without a contact with the ectopteryoid; the crista circumfenestralis absent and there is a fenestra rotunda; there is a short supratemporal firmly sutured within a trough formed by the prootic, supraoccipital, and parietal; the coronoid has a long ventromedial process contacting the angular; the supraoccipital is excluded from the margin of the foramen magnum; there is a long tall parietal crest; the teeth are set in shallow sockets; the quadrate is a large inverted "j"-shaped element and the stapes/columella is very robust and bears an extremely large and expanded stapedial footplate; and there is an extracolumella/intercalary element that contacts the quadrate suprastapedial process to which the columella of the stapes articulates (contra Scanferla and Bhullar [2014], there is no independent stylohyal element in addition to the intercalary element).

With respect to the postcranial skeleton, there is currently no evidence for the presence of a pelvis and rear limb skeleton, but there is also only a single relatively complete postcranial skeleton that, while illustrated, remains largely undescribed (see Caldwell and Albino, 2002). The anterior precloacals of a second specimen were described by Caldwell and Albino (2002) and in great detail by Caldwell and Calvo (2008), and referred to by the latter authors as "cervicals?" These are truly unique compared to modern snake lizards, as there are unfused intercentra on the hypapophyses of the third to fourth vertebrae. The next six cervical vertebrae possess fused hypapophyses/intercentra. From the articulated specimen, MACN RN-976, the anterior 1/3 of the known precloacals bear prominent ventral hypapophyseal keels (Caldwell and Albino, 2002).

## LATE CRETACEOUS—GONDWANAN ASIA

### Cenomanian—Myanmar

**Xiaophis myanmarensis** Xing et al., 2018 (Figure 2.10)

The articulated postcranial skeleton of *Xiaophis myanmarensis* (DIP-S-0907 [DIP = Dexu Institute of Paleontology, Beijing]) is 47.5 mm in total length, making it the tiniest fossil snake known, and like an embryo or early-stage neonate (Xing et al., 2018). The vertebrae are also very small (anterior precloacals ~0.5 mm, and caudals ~0.35 mm). There are 97 preserved vertebrae, of which the anterior 87 are in articulation with each other and with their ribs; after several gaps, there are at

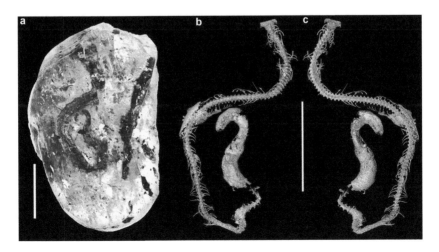

Figure 2.10. *Xiaophis myanmarensis*, overview of amber clast with synchrotron x-ray μCT image of articulated embryonic/neonate snake skeleton (DIP-S-0907). (a) Amber clast with included skeletal material; (b) dorsal view of skeleton, synchrotron x-ray μCT image; (c) ventral view of skeleton, synchrotron x-ray μCT image. Scale bar = 10 mm. (Images courtesy of copyright Lida Xing; modified from Xing et al., 2018.)

least 10 preserved caudal vertebrae. Three masses obscure some sections of the postcranium and may well be mineralized remnants of decayed tissues. The first preserved vertebra is exposed at the surface of the amber clast, indicating that many more vertebrae and the skull were lost due to preservational factors. Comparisons to more complete and contemporaneous snake lizards, such as *Haasiophis terrasanctus* (155 precloacals; largest precloacal at 70th–80th vertebra), indicate that DIP-S-0907 could be missing more than 70 vertebrae, and, of course, the skull. The preserved vertebrae show a number of interesting features, many of which are developmental and indicate that a number of characteristic snake lizard features (Xing et al., 2018) (e.g., zygosphenes and zygantra developed late in ontogeny in this ancient snake lizard as they do in modern neonate snakes, late closure of the notochordal canal, e.g., *Cylindrophis*) have been conserved in snake lizard developmental evolution for at least 100 million years. Importantly, though, the details of the vertebral morphology, regardless of ontogenetic stage, were found by Xing et al. (2018) to support recognition of *Xiaophis* as potentially a member of the Gondwanan snake lizard assemblage/clade, generally referred to as the Madtsoiidae.

The type specimen (DIP-S-0907) and a second amber clast specimen (DIP-V-15104) not associated at all with the type, preserve the first known details of ancient snake lizard integument (Xing et al., 2018). DIP-V-15104 appears to represent a shed skin of a larger individual, and the scales are diamond-shaped or ovoid-diamond-shaped, with rows that converge ventrally; there are light and dark areas that were interpreted as representing color patterning, along with circles or rings of dark patterning. DIP-S-0907 preserves a large number of scales in specific regions along the postcranial skeleton that indicate that the scales are imbricated, diamond-shaped, and thin. While taphonomy could well be a factor in affecting body shape in DIP-S-0907, it is very clear that the body was tubular toward the tail and perhaps blunt at its tip.

## Maastrichtian—India

**Sanajeh indicus** Wilson et al., 2010—Type I Anilioid
(Figure 2.11)

As described by Wilson et al. (2010). the skull and remains of the vertebral column of the type specimen of *Sanajeh indicus* were found in articulation (GSI/GC/2901-2906 [GSI/GC—Geological Survey of India, Geological Collections, Jaipur, India]); the skull was roughly 95 mm in length, and the authors estimated the snake lizard to be approximately 3.5 meters long. The braincase is elongate, and as preserved shows important similarities to both *Dinilysia* and *Najash*: the trigeminal foramen is undivided (i.e., there is no laterosphenoid), the otic capsule is enlarged, and the stapedial footplate is extremely large, covering an equal-sized fenestra ovalis; the supratemporal is not preserved (contra Wilson et al., 2010), but rather the element identified as the supratemporal is the stapedial footplate, but a groove in the parietal, prootic, and supraoccipital is present where the supratemporal would have articulated in life. There is a narrow crista interfenestralis separating the fenestra rotundum and the fenestra ovalis (as in *Dinilysia* and *Najash*), but the posterior border of the fenestra rotundum is open and bordered by a thickened crista tuberalis—consistent with Palci and Caldwell's (2014) identification of snake lizards lacking a crista circumfenestralis, *Sanajeh* lacks one as well, showing a similar condition to that observed in *Dinilysia* (Type 1). There is also a well-developed sagittal crest on the ventral aspect of the braincase that is not common to *Dinilysia* and *Najash*. As reported by Wilson et al. (2010), the maxilla is nearly complete and has a relatively short nares, whose shape morphology resembles that of anilioids. There is only a single mental foramen located anteriorly on the dentary; the teeth are broad and slightly recurved, similar to anilioids and *Dinilysia*. The preserved vertebrae are all from the precloacal portion of the column and show typical madtsoiid features including small parazygantral foramina on the posterior surface of the neural arch. The prezygapophyses lack accessory processes, and the rib articulations extend laterally beyond the margins of the prezygapophyses. Mohabey et al. (2011) subsequently described a second species of Upper Cretaceous snake from India, though this was only a vertebral form taxon, *Madtsoia pisdurensis*.

## IMPORTANT VERTEBRAL FORM TAXA

The balance of this chapter has focused on whole or relatively complete body fossils of snake lizards. But it is also a fact, and an important one at that, that the bulk of the snake lizard fossil record,

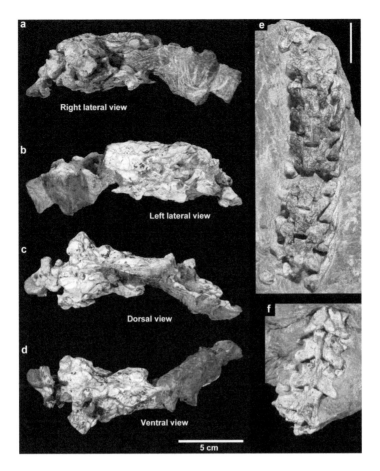

Figure 2.11. *Sanajeh indicus.* (a) Right lateral view skull and cervical vertebrae; (b) left lateral view skull and cervical vertebrae; (c) dorsal view skull and cervical vertebrae; (d) ventral view skull and cervical vertebrae; (e) ventral view, anterior precloacal vertebrae; (f) dorsal view, anterior precloacal vertebrae. (Photographs courtesy of copyright Jeff Wilson.)

particularly through the Cenozoic, is known from literally tens of thousands of isolated and disarticulated vertebrae (Rage, 1984; Holman, 2000). It is also true that most of these vertebral form taxa have been described as species from nothing more than a single vertebra, for example, the first described and named Mesozoic snake from anywhere in the world, *Simoliophis rochebrunei* Sauvage, 1880 (Figure 2.12a,b) from the south of France, to the first snake lizard described from the Mesozoic of North America, *Coniophis precedens* Marsh, 1892 (Figure 2.12c,d).

I have chosen to highlight two of these vertebral form taxa, as they are as iconic as they are uninformative in the study of snake lizards, but yet have remained important as enigmatic and controversial data points in the paleontological study of snake lizard evolution. As with all form taxa, an individual vertebra does preserve enough

information to allow recognition of the bone as a skeletal element from a "snake lizard." However, despite the number of genera and species named around isolated vertebrae, in my opinion, such specimens do not preserve enough information to merit alpha taxonomy at the generic or species level (cf. Holman, 2000) and should only be recognized as a form taxon and left in open nomenclature. Many of these form taxa have been increasingly overconceptualized, to the point that they become misleading chimeras that take on seeming species reality/validity as obligatory taxonomic units (OTUs) in phylogenetic analyses (e.g., Longrich et al., 2012). Again, and in my opinion, the only real value of form taxa is in identifying broadscale ecosystem features of faunal assemblages in time and space; that is, there were snake lizards present during the Early Maastrichtian in estuarine environments along the western shore of the Bearpaw Sea. Beyond that, there is no information contained in form taxa, in particular isolated vertebrae, that reveals anything about the origins and evolution of animals such as snake lizards. The variation in morphology, and thus function, along the axial skeleton of a snake lizard with 250+ vertebrae, from the atlas-axis complex through the precloacal series (cervicals + dorsals), to the sacrals/cloacals and postcloacals through to the caudals and terminal caudal, is so great that it is possible to identify about 20 individual form taxa from the disarticulated vertebrae of any one individual snake lizard, fossil or modern. As such, form taxa should not be used for anything more than what they represent—in the case of snake lizards, statements on isolated vertebral form linked to a "region" of the axial skeleton and statements on the broad diversity and composition of fossil vertebrate assemblages. That said, that information is informative and needs to be in the literature and understood for the data as preserved. The following outlines two of the most important vertebral form taxa from the Mesozoic that both shaped early perceptions of Mesozoic snake lizards in terms of both axial skeletal form and ecologies and habitats.

### Cenomanian—France

**Simoliotphis rochebrunei** Sauvage 1880 (Figure 2.12a,b)

When Sauvage (1880) described *Simoliophis rochebrunei*, he did not identify a type specimen for the taxon, and there was no type locality designated either. The collection of midtrunk vertebrae upon which the concept of the taxon was built has since been lost. Rage et al. (2016) recently revised *S. rochebrunei*, reviewing the history of the taxon, the sedimentary geology of the "type area" and the various productive localities, as well as illustrating and describing a large number of new specimens. Importantly, they also diagnosed the taxon against their designation of a neotype specimen (MA RND 11, neotype [MA—Museé Angouleme, Angouleme, France]), an isolated posterior midtrunk vertebra similar to those illustrated by Sauvage (1880) (Figure 2.12a,b).

In brief, the neotype vertebra is well-preserved and massive, displaying marked pachyostosis on nearly all observable cortical bone morphologies. The effects of pachyostosis even appear to have produced light and finely wrinkled texture on the external surface of the cortical bone. The vertebra is slightly taller than wide; the neural spine is tall and thickened, with a flat dorsal margin that is slightly depressed, likely for a neural spine epiphysis; the articular facets of the prezygapophyses are inclined at about 30 degrees from the horizontal and are high up on the centrum body such that the lateral tip projects above the zygosphenes (the accessory processes typical of modern snakes are absent). The neural canal is small, relatively speaking, and shows a triangular shape with a basal crest, giving an overall trefoil shape to the canal. The zygosphenal lamina is tall, depressed centrally, and the zygosphenes are prominent and face laterally. The cotyle is circular and rather shallow and small relative to the size of the centrum, as the condyle compared to the inflated centrum (effect of pachyostosis and osteosclerosis?). In ventral view, the centrum is very "square" in general outline and flat bottomed. The paradiapophyses face laterally and clearly formed a dorsal diapophysis and ventral parapophysis. The general condition of the mid to posterior precloacal vertebra is typical of the vertebrae observed in *Pachyrhachis, Eupodophis, Haasiophis, Mesophis,* and *Pachyophis*. While there is clearly a great deal of columnar variation in all of these snakes, the mid to posterior precloacals show these common features as described for this vertebra. The study of Rage et al. (2016) certainly illustrates and describes more vertebral variation than can be typified by a single neotype vertebra for *S. rochebrunei*, but this neotype specimen does nicely encapsulate the condition of the vertebrae in a diverse and widely distributed group of Cenomanian snake lizards.

### Maastrichtian—Wyoming/Montana

**Coniophis precedens** Marsh, 1892

*Coniophis precedens* Marsh, 1892 (Figure 2.12c,d) is one of the most overinterpreted snake lizard form taxa currently in existence. In contrast to recent claims by Longrich et al. (2012) of having successfully divined a complete skeleton of this species from their association of numerous disarticulated axial elements and a small number of isolated cranial elements, *C. precedens* can still only be recognized from the original isolated vertebra described by Marsh (1892) (Holotype vertebra, USNM V2134 [USNM—U.S. National Museum of Natural History, Washington, DC]). The intriguing aspect

THE ORIGIN OF SNAKES

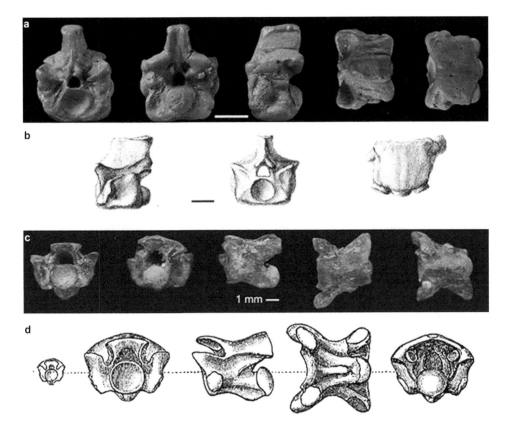

Figure 2.12. Type/neotype vertebrae of *Simoliophis rochebrunei* (MA RND 11) and *Coniophis precedens*. (a) *Simoliophis rochebrunei*, anterior, posterior, left lateral, dorsal, and ventral views of neotype vertebra (MA RND 11); (b) *Simoliophis rochebrunei* line drawings of lost holotype, left lateral, anterior, and ventral views (From Sauvage, 1880); (c) *Coniophis precedens*, holotype vertebra (NMNH V2134) in anterior, posterior, left lateral, dorsal, and ventral views; (d) *Coniophis precedens*, holotype vertebra (NMNH V2134) in anterior, left lateral, ventral, and dorsal views (From Marsh, 1892).

of the form of the *C. precedens* holotype remains is that it is so generalized that literally thousands of snake vertebrae, from a stunningly wide range of geologic ages and paleogeographic locales, have been assigned to the form genus *Coniophis*. The form-type of the vertebra is recognized in its oldest form, from the Cenomanian of Utah, *Coniophis* sp. (OMNH 33250 and 33251 [OMNH— Oklahoma Museum of Natural History, Norman, Oklahoma]) (Gardner and Ciffelli, 1999), the Campanian (Upper Cretaceous) of Alberta (Fox, 1975), through to the Middle Eocene of Wyoming (with numerous Santonian to Maastrichtian records), the Maastrichtian of Bolivia (de Muizon et al., 1983), India (Rage et al., 2004), and the Sudan (Rage and Werner, 1992), as well as the Upper Eocene of France. By comparison to any other genus of fossil snake lizard, let alone species, this is an extremely long-lived and geographically

widespread taxon of snake—likely too long and too widespread, in fact.

The preserved anatomy of the type vertebra contains diagnostic morphologies to allow recognition of the element as belonging to a snake lizard: zygosphenes and zygantra are present and well developed; the neural canal, like *Simoliphis rochebrunei* displays trefoil outline. The condyle is taller than it is wide and has a rounded ventral surface, while the centrum is cylindrical and possesses a long ventral keel that is deeper posteriorly than it is anteriorly. The synapophysis appears to be single headed, not developed into a paradiapophysis, though the synapophysis is positioned at the ventral margin of the centrum. There are no accessory processes on the prezygaphophyses; the neural arch is low; and the spine, while present, is short and present a posteriorly blunt tuberclelike tip, possibly for an epiphysis. In general terms, the type vertebra is a small, generalized vertebral element showing key snake lizard features, but little more than that.

## A FOSSIL SNAKE LIZARD WITH FOUR LEGS—OR NOT?

Martill et al. (2015) recently described *Tetrapodophis amplectus* (Figure 2.13a,b) as a fossil lizard from the Lower Cretaceous of the Crato Formation, Brazil, as the first known four-legged snake. Caldwell et al. (2016) and Lee et al. (2016) presented contrary evidence to both the identification of *Tetrapodophis*'s skeletal remains as those of a snake lizard and to the ecomorph conclusions given by Martill et al. (2015) that *Tetrapodophis* was a terrestrial, fossorial/burrowing snake lizard. As a lead author and collaborator on both of those rebuttals, and as the one of only two investigators outside of the Martill team to see the specimen personally (Dr. Robert Reisz is number two), I maintain this position, that is, that *Tetrapodophis* is not a snake lizard. As a result, I would not normally consider including it in a book on snake lizards. However, the specimen and the story as devised by Martill et al. (2015) was published in *Science* and

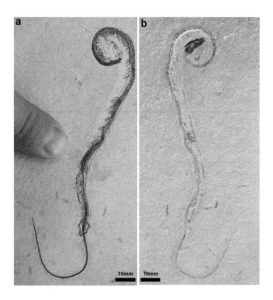

**Figure 2.13.** Part and counterpart of the holotype of the lizard *Tetrapodophis amplectus* (BMMS BK 2-2). (a) Part, containing most of the skeletal remains except the skull and right forelimb; (b) counterpart, preserving the skull in right lateral view, forelimb and isolated ribs and transverse processes. Scale bar as noted, additional scale, author's left index finger in (a).

received a great deal of attention for what it was claimed to be—"a four-legged fossil snake." Sadly, it also received a great deal of attention for how, where, and when it had been acquired, as fossil collecting under Brazilian legislation is illegal without permit, and this specimen had no such permit, and, for the fact that it was and remains in a private collection that is only publicly accessible by the graces of the private collector, which violates the ethical standards regarding data access (the specimen is housed in a private museum in Solnhofen, Bavaria, Germany).

While the Lee et al. (2016:199) response is complete and published, and concluded that whatever *Tetrapodophis* might be (snake lizard, anguimorph lizard, or space alien), it is not conclusively a fossorial animal, and in fact shows numerous adaptations and proportions consistent with being a surface or aquatic animal (see Chapter 4 as well):

> *Tetrapodophis* therefore represents an enigmatic mélange, as do many elongate, limb reduced tetrapods, of characters that can be interpreted as either aquatic or fossorial. These ecomorphological results also have potential implications for both the phylogenetic affinities of *Tetrapodophis* and for the ecology of snake origins.

Unfortunately, the full descriptive revision is not yet completed and published (Caldwell et al., in prep.), and of course this is where the substance of any real rebuttal resides—within the osteological data of the fossil itself. I will therefore present a brief overview of the osteology of the specimen, consistent with the information discussed in the abstract I co-authored and presented from the platform at the Society of Vertebrate Paleontology meeting in 2016, Salt Lake City, Utah. Following the presentation format of this chapter, I will also provide a brief overview of the depositional environment of the Crato Formation, Brazil.

***Tetrapodophis amplectus*** Martill, Tischlinger and Longrich, 2015 (Figure 2.13)

Caldwell et al. (2016) reported that the one and only specimen of *Tetrapodophis amplectus* is remarkably small (195 mm TL), but is composed of a surprisingly large number of vertebrae (~160 presacrals, and ~112 caudals). The body is very long compared to the tail, but the tail is still very long as well; proportions mean nothing compared to number of myomeres, particularly when considering snake lizard allometry in the trunk and tail. For example, a scolecophidian has a short tail—short compared to precloacal length in addition to less than 20 caudal vertebrae. *Tetrapodophis* is also still in possession of all four limbs, though the front limbs are significantly smaller than the rear limbs, and in any case, all four limbs are exceptionally small compared to a typical limbed lizard.

Martill et al. (2015) argued that the skeleton displayed twenty-four osteological features that were clearly shared with all other snake lizards; of those twenty-four features, they tested thirteen in their phylogenetic analysis of snake lizard affinities. An additional character was "behavioral," what Martill et al. (2015:417) termed a, "snake-like feeding strategy in which proportionately large prey are ingested whole." With the exception of the "snake-like feeding strategy," my first-hand observation of the specimen resulted in counterobservations and interpretations to each morphological character presented by Martill et al. (2015). As was reported in the abstract (Caldwell et al., 2016), the snout and braincase are equal in length contra the claim by Martill et al. (2015) that the snout is short, the orbit placed far forward, and the braincase long. As supporting evidence not requiring study of the specimen, see Martill et al. (2015; fig. S1A-C) and note the position of the untoothed distal ramus of the dentary above the supposed angular-splenial joint. In a snake lizard, the posterior portion of the dentary is toothed and significantly overhangs the angular splenial joint—in *Tetrapodophis*, this is not the condition; rather, the posterior ramus of the dentary, which appears to have tooth sockets almost to its terminus, thins and ends at the superior eminence of the splenial, nearly identical to the anatomy observed in the dolichosaur *Coniasaurus crassidens* (Caldwell and Cooper, 1999). In terms of orbit position, the splenial eminence sits below the coronoid process of the surangular, which is crowned by the coronoid, the apex of which is at the posterior border of the jugal, ergo, the orbit. Therefore, the orbit of *Tetrapodophis* is at the center of the skull, not located anteriorly. This level of critical assessment finds counterobservations to all twenty-four characters: the mandible is straight, not curved, there is no subdental ridge, an intramandibular joint is not preserved, nor does it appear to present where Martill et al. (2015) place it. The teeth are not snakelike but rather are taphonomically

displaced and are neither recumbent nor curved. The high seemingly precloacal vertebral counts are not exclusive to snakes (e.g., dibamids ~135; amphisbaenians ~175), though they are higher than is currently known for any other group of lizards except snake, amphisbaenian, and dibamid lizards. Claims to the contrary, there are no zygosphenes visible; even in the figures as presented, they are clearly not present. The neural spines are tall where observable, though for most of the specimen, they are not. Rib heads are not tubercular, but rather are quite simple, like typical lizards. The so called "lymphapophyses" are actually expanded sacral processes, and the claimed belly scales are not present in any form (i.e., as impressions, as permineralizations, etc.). Caldwell et al. (2016) reported on a number of new anatomical observations: a high cervical count, the presence of a complete jugal and probable postorbitofrontal, the orbit position discussed above, an elongate retroarticular process, enlarged first metapodials, reduced carpal and tarsal ossification, intercentra in the neck and tail, and reduced limb articulation surfaces. Importantly, and as mentioned above, the limbs are very small, lightly constructed, have elongate phalanges and weak girdles, and show delayed mesopodial ossification similar to a variety of aquatic reptiles (Caldwell, 1997). As was argued by Lee et al. (2016), the limbs were ineffective paddles, and the tail was long, which was interpreted as indicating that *Tetrapodophis* employed anguilliform locomotion in water or on land. I will add here that it clearly had large eyes, and, from first-hand examination of this tiny animal, could only make out small fishlike bones in its gut, and certainly not the bones of larger mammals that it had constricted and eaten whole; in fact, apart from its gut contents, there was nothing about the poscranium to suggest constriction capabilities.

## SUMMARY

So, what does all of this mean? Here is what has been reviewed: (1) seventeen fossil snake lizard taxa from the mid to late part of the Mesozoic, whose fossil remains have been found in a wide variety of depositional environments originally located in Gondwana, Laurasia, Eurasia, and across ancient marine platforms found between those land masses and along those ancient shorelines; (2) the anatomy of ancient snake lizards of a largely unique and unknown variety, whose sistergroup relationships with crown snake lizards are poorly known at best or poorly accepted in the alternative (I have my hypotheses; see Chapter 6), and who may only be represented today by a small number of crown-group snake lizards who possess skull types they inherited from their possible ancient snake lizard ancestors—Type I Anilioid, and Type II Booid-Pythonoid. In fact, what becomes obvious at this level of comparison is that once again, top–down categories have been constructed that reverse the accurate perspective, which is to start 170 million years ago and look forward, not backward beginning with the modern. In this sense, the categories should really be renamed from Type I Anilioid, to Type I *Najash-Dinilysia*–like, and Type II Booid-Pythonoid to Type II *Pachyrhachis-Haasiophis*-like. While the categories created in Chapter 1 are atemporal in the sense that they are cone-shaped clusters created using only modern snake lizards, it is clear that the essential characteristics of the modern fauna did not arise by spontaneous generation and so a temporal component becomes necessary when the goal is an evolutionary understanding of skull and skeletal anatomy and morphology.

While this summary is a lot to read, digest, and synthesize, it really is little more than the tiniest tip of the iceberg on the real data set. To be better informed, I recommend first reading the complete, original literature, and then rereading it—if you are a student of snake lizard phylogeny, do not, and I repeat do not, stop there, as this is the usual mistake: use the literature as a primary data source to be mined for metadata as character states and then code a data matrix for phylogenetic analysis. The only legitimate and scholarly next step after reading the literature is the old-fashioned empirical method—study each specimen in person. See them, draw them, understand them, go into the field and collect them, and then, slowly, properly and painfully, construct homolog concepts and characters and states, and then systematize them with original, and if need be, corrected data and homologs. There is no replacement for empirical observation.

For the more casual student of snake lizard evolution, I hope that some clarity has been achieved in pointing out how different these ancient snake lizards are from modern snake lizards in terms of their basic anatomical organization and what that suggests about their sistergroup relationships with each other and with their antecedent clades.

It is imperative in this sense to remember that the modern diversity of snake lizards is nothing more than a flickering possibility in the crystal ball of evolution, when, 70 million years before we have the first fossil evidence of vipers and higher colubroid snake lizards, *Pachyrhachis problematicus* was darting in and out of reef crevices on a giant platform reef archipelago. Or better yet, that from rocks that are 140 million years old, we have fossil evidence of the currently oldest known snake lizards from islands in the ancient Boreal Sea that covered what is now southern England, showing that snake lizard evolution was well underway 80 million years before we find a single, isolated vertebra that can be assigned to the presumed most basal/primitive snakes, the Scolecophidia. To reiterate, from the morphology of *Eophis woodwardi*, we would never predict the highly specialized morphology of a modern scolecophidian such as *Typhlops jamaicensis*, or predict the anatomy and habits of *Bungarus fasciatus* (Banded Krait) from the anatomy and paleoecology of *Dinilysia patagonica*.

By the end of this book, I want to make it clear that we cannot understand modern snake lizard phylogeny if we cannot first organize the base of the tree in the absence of the modern fauna— the data presented in this chapter are what allow us to develop that pattern of ancient sistergroup relationships; nothing else will do. The questions to be asked from the data presented here are not how ancient snake lizards fit into a phylogeny of the modern fauna (often done with a pencil and straight edge; see Vidal and Hedges [2004]), but rather are the reverse, "How do modern snake lizards nest into a phylogenetic topology, the backbone of which is constrained by ancient snake lizard phylogeny?" With that question settled, we would actually be making some progress.

The ancient Mesozoic animals profiled here are simply not very much like their crown snake lizard relatives at all—they are remarkably different. So different that, to be honest, the concept of modern snake lizard cannot be backed down a phylogenetic tree to include the ancient forms as typical of any modern kind of snake lizard. These animals, and the morphologies they actually display, redefine "snakeness" and should not be expected to fit in neatly into a phylogeny of the crown clades because that phylogeny cannot be correct unless it reconceived, and its foundational clade points redefined, by the anatomies and morphologies of its most ancient sistergroups.

"Snakeness," or better put, "snakelizardness," cannot be defined in an evolutionary sense by reference to the crown. As I have argued, and will continue to argue, in the Late Mesozoic, the snake lizard features of the crown clades were not even an evolutionary experiment yet; snake lizard phylogeny as a pattern can only be recovered at its most ancient and foundational by considering only the characteristics of Mesozoic snakes. Had we been running the experiment of predictive phylogeny, it would not be possible to anticipate the evolutionary trajectories of the modern descendent clades using only the morphology of the Mesozoic assemblages—likewise, we cannot infer the past using the data of the modern. It is simply too new and too derived to define the ancestral snake lizard. The only place we can go is to the fossil record. It is our only source for data on those ancient and unpredictable anatomies. The next chapter reconstructs the ancient snake lizard from the observed anatomies present in the cast of characters reviewed in this chapter.

# CHAPTER THREE

# The Anatomy of Ancient Snake Lizards

The words we use to describe objects, things, and kinds of things do not somehow make those kinds of things more real, nor their subsequent homology hypotheses more true, just because we have created a word to describe them, like "postorbital" and "jugal," or "apple" and "banana." I consider it true that an object is "real," but the word we have created for that object is only an idea. Thus as ideas, words express a concept that is intended to mean something and to convey that meaning from one person to the next with the intention of constancy of meaning. Following this logic, words mean everything because they express ideas and not truth or realness (the latter is what the object expresses), and therefore we need to use the meaningful words, and to commonly and precisely use them, when we create the idea and its meaning that the word is intended to represent. Therefore, when we say an object is an apple, we also mean it is not a banana; no matter how many times you might call the apple object a banana, it remains an apple, and not only as an object, but also as an idea or concept. The same is true for the objects "jugal" and "postorbital."

Because words reflect the epistemic and ontic framework of what we think we know about what is real about those objects and kinds, and how each of those objects is related to others of its kind, kinds that are different, and all kinds as a whole, we cannot afford to get it wrong, or worse than that, knowingly remain wrong. Therefore, when we detect an error, we need to correct it as soon as possible. Make no mistake, it is exceedingly difficult to create words to frame ideas that realistically and meaningfully describe this multidimensional complexity (three dimensions of shape plus the vector of time). After all, we are trying to describe the "real world," what we think we know of it, and the objects that are in it and have been in it—but, nevertheless, we must and do continue to try. In doing so, we run into a number of obvious and not so obvious difficulties. First, many words have different meanings, and when combined together as adjective and noun, will likely confer secondary meanings to objects and groupings of similar kinds of objects that may or may not have been intended (e.g., "snake" versus "ancient snake lizard" versus "ancestral snake lizard," or "jugal" versus "postorbital"). A second difficulty, creating immense confusion, is when a word is applied incorrectly to a second object or a kind, after having been carefully applied to the first object or kind, and then imbues that thing or grouping of things with incorrect meaning, which may well, in its incorrect form, be secondarily applied in the search for additional meaning (e.g., identifying a bone as "x," when indeed it is "y," and then using interpretations of "x" as metadata to construct secondary hypotheses such as character states and then phylogenies). And yet a third problem, arising form the second, and commonly encountered in our modern approach to naming objects or parts of objects, in relation to time and change over time (evolution), is that we use words from previously described objects and their parts to connect a new object and its parts in a chain of homologies to previously know objects (e.g., bone "x" in object one changed to bone "y" over time in object two). If this transformational hypothesis created during the naming process is incorrect, solving the dilemma of object part naming, of identification, becomes extremely difficult to correct.

To be more specific to the goals of this book, when speaking about the data set that is the anatomy of snake lizards, ancient or modern, we have for a long time now used a very specific anatomical lexicon of terms to name each bone of the skeleton, each muscle, each nerve, blood vessel, tissue, molecule, and so on. For the most part, these terms have been inherited by we anatomists of non-human tetrapods from those who invented that terminology by reference to the organization of human anatomy—in other words, the anatomical terms suffer a top–down effect as their original reference was to connectivity, topological relationships, and ontogeny in highly specialized human mammals, and not to birds or fishes or snake lizards, which do not look much like humans at all. Human anatomy and rat anatomy is quite comparable, but not so much for humans and snake lizards. Even less so if one is hoping to test by the Test of Congruence (*sensu* Patterson, 1982), a character suite of primary homolog concepts. In essence, what this top–down approach to simple element identification does is exacerbate the essentialist categories of archetypes—in this case, the mammal archetype versus the lizard archetype and their essential differences.

Disregarding for the time being the problem of top–down anatomical nomenclature, the empirical problem facing comparative anatomists, paleontologists, or even a molecular biologist, is to correctly identify a bony element (e.g., the jugal versus the postorbital) or a molecule (e.g., adenine versus guanine) when beginning the comparative study of first one specimen, and then another, and another, and so on. If terms applied to structures in humans do not apply, then existing terms must be modified or added to account for differing anatomies and morphologies in non-human vertebrates. From there, identifications must be done carefully and with strict adherence to connectivity, topology, and ontogeny—the same criteria at the root of Patterson's (1982) Test of Similarity. If this first crucial step of identification of an element is not done correctly, i.e., the best that it can be, then the cascade effect is that every subsequent bony, muscular, cellular, or molecular identification will be given incorrectly, because

the next one is identified by reference to the first one, and so on.

While Patterson's (1982) Test of Similarity calls for the application of the tests of connectivity, topology, and ontogeny to assist in the process of homolog concept construction and state assignments, these same criteria apply to the process of identifying the constituent parts of an organism, and of course, the whole organism itself. That the Test of Similarity follows as the only approach to developing homolog concepts and characters and states is simply a logical outcome of this very empirical process of comparative anatomy. As noted, it is easy for a simple mistake in element identification to have a catastrophic cascade effect on subsequent anatomical identifications.

For example: Step One, locate and identify the humerus; Step Two, locate and identify the preaxial element distal to the humerus as the radius, and the postaxial element distal to the humerus as the ulna; Step Three and onward, the carpals, phalanges, and unguals, all follow in a logical progression. Identify the humerus as the radius, scapula, or, worse, as a femur, and the system falls apart because logical connectivity and topologies would suggest you now have new bones in the limb or new morphologies replacing old morphologies, leading to highly modified versions of what one would expect to be present. It is easy to imagine that mistakes of this kind have been and are common in paleontology—to be fair, when looking through time at anatomies and morphologies that have been highly modified by evolution, through great tracts of geological time, it would be expected that it is actually easy to get it wrong. Still, mistakes are still wrong, inaccurate or incorrect and bear highlighting here within the context of ancient and modern snake lizard anatomy and the studies that have identified and described those seemingly novel or incorrect anatomies.

There is also another level of mistake common to phylogenetic studies, where the errors made by comparative anatomists are amplified to the level of catastrophes—this is when these anatomical misidentifications are now converted to the metadata statements of character states in a homolog concept permitting them to be recovered as synapomorphies supporting clade structures in cladograms (Patterson, 1982; Simões et al., 2017, 2018b). And here penultimate problem—hypotheses on the pattern and process of morphological evolution built around errors of anatomical identification. But, as though it could not get worse, these now severely confused and conflated errors rise to the final ignomy and become enthroned as veritable absolute truths, the overthrow of which, due to the mass of their historical burden, requires ten times the effort to falsify than was required to propose them in the first place. To illuminate the level of this problem and to try to alleviate the burden it causes through multiple orders of hypothesis construction, the remainder of this chapter focuses on some of the problematic anatomies that have plagued the science of snake lizard morphological and anatomical interpretation, particularly as concerns modern snake lizards. As will become clear, the fossil record is the only data set that has been able to provide resolution on problematic anatomies first identified via reference to only modern snake lizards. Numerous authors have written extensive monographic studies on the organization and anatomy of lepidosaur skulls, inclusive of sphenodontians (Romer, 1956), non-snake lizards (Romer, 1956; Evans, 2008; Gans and Montero, 2008), and, more to the point, snake lizards (Cundall and Irish, 2008; McDowell, 2008), and I will draw heavily from these works, pointing out where I agree and disagree with the conclusions and interpretations given. The goal of this chapter will be, in the final analysis, to create a concept of ancient snake lizards that can be deduced from the available data of the fossil record compared to the data presented by modern snake lizards, and from ancient and modern non-snake lizards. This requires a critical reanalysis of the literature and existing claims of the essential characteristics, anatomies, and morphologies of snake lizards.

## THE SKULL

### Circumorbital Region: The Jugal

The historical burden of the element identities of the circumorbital region in snake lizards, both modern and ancient, is long, complex, non-linear, and not very logical in terms of methods and perspectives used to identify the element and then to proceed to hypothesize and "discover" homologies between taxa (Figures 3.1 through 3.4). As one must

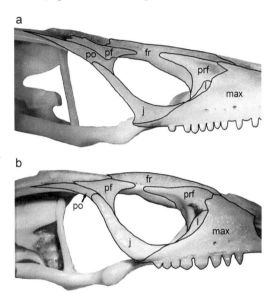

**Figure 3.1.** The orbital region of two "typical" non-snake lizards, showing variation on jugal and postfrontal contact. (a) *Diploglossus millepunctatus* (MCZ 130071), lateral view; (b) *Ophisaurus apodus* (MCZ 2094), lateral view. (Modified with permission from Palci et al. 2013. *Journal of Morphology*, 274: 973–986.) (Abbreviations: fr, frontal; j, jugal; l, lacrimal; max, maxilla; pf, postfrontal; po, postorbital; prf, prefrontal.)

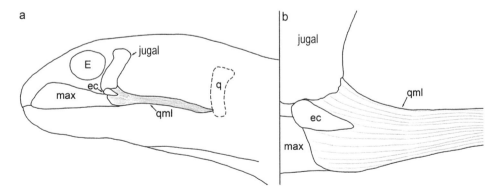

**Figure 3.2.** *Calabaria reinhardtii* (UAMZ R937). (a) Line drawings of dissection of the head and (b) enlargement of posteroventral circumorbital region. (Modified with permission from Palci et al. 2013. *Journal of Morphology*, 274: 973–986.) (Abbreviations: E, eye; ec, ectopterygoid; max, maxilla; jugal, jugal; q, quadrate; qml, quadratomaxillary ligament.)

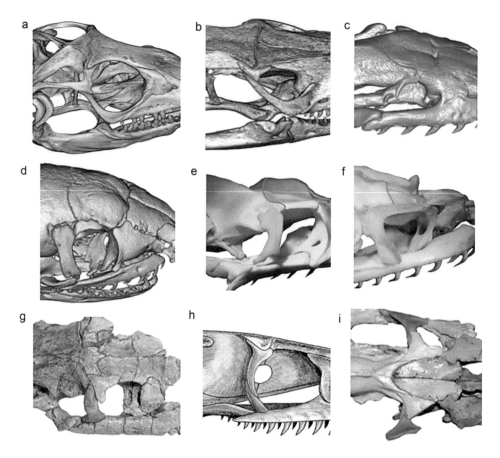

**Figure 3.3.** Right orbital region of nine lizards. (a) *Tupinambis teguixin*, dorsolateral view (FMNH 22416; CT scan); (b) *Lanthanotus borneensis*, dorsolateral view (YPM 6057; CT-scan); (c) *Cylindrophis ruffus*, dorsolateral view (USNM 297456; CT-scan); (d) *Calabaria reinhardtii*, dorsolateral view (FMNH 117833; CT-scan); (e) *Corallus caninus*, dorsolateral view (ZFMK 21667); (f) *Acrochordus javanicus*, dorsolateral view (AMNH 155254); (g) *Dinylisia patagonica*, dorsolateral view (MACN RN-1013); (h) reconstruction of *Pachyrhachis problematicus*, lateral view; (i) *Yurlunggur* sp., dorsal view of skull, right maxilla in lateral view (QMF 45391); interpreted jugal, postorbital, and postfrontal are highlighted in red, yellow, and green, respectively; the frontoparietal suture is highlighted in blue. An element that may result from fusion of jugal and postfrontal in *Acrochordus* is outlined in blue. (Modified with permission, from Palci et al. 2013. *Journal of Morphology*, 274: 973–986; CT-scan data for *Tupinambis*, *Lanthanotus*, and *Calabaria* are courtesy of M. Kearney and O. Rieppel, Deep Scaly Project.)

begin somewhere, I will begin this discussion by highlighting the arguments surrounding the postorbital margin elements of *Dinilysia patagonica*, not because it is the first snake lizard to ever enter the discussion, but because it is an example close at hand and of key importance to the problem, as it was the first fossil snake lizard to have an element in the posterior region of the orbit identified as a postfrontal; this was followed by recharacterization as probably a jugal, then a postorbital, and quite recently, as a jugal.

In the original description of *Dinilysia patagonica*, Smith-Woodward (1901) illustrated the holotype skull (MLP 26-410) as possessing a long vaguely inverted t-shaped element along the posterior

margin of the right orbit (it was missing on the left side) as the postfrontal. When Estes et al. (1970) redescribed the same specimen some 70 years later, that postfrontal element was now broken away, leaving only dorsal and ventral fragments of the element in its place, which they identified as the postorbital (more dorsal fragment) and the jugal (ventral fragment). In addition, Estes et al. (1970) also delimited a small triangular chip of the bone just anterior to their postorbital, which they called the postfrontal. Describing new material collected in 1994 and 2001, Caldwell and Albino (2002) referred to Estes et al.'s (1970) description of both a postorbital and jugal, but now had new materials of *Dinilysia* with complete elements (cf. MLP 26-410

THE ORIGIN OF SNAKES

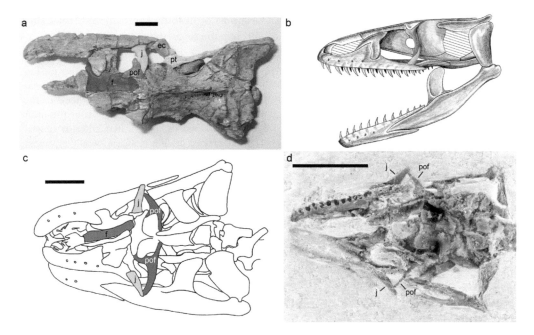

**Figure 3.4.** Topological relationship between postfrontal, jugal, and ectopterygoid in snakes. (a) Dorsal view of the skull of the fossil snake *Dinilysia patagonica* (MACN RN-1013). Note the topological relationship between postfrontal (highlighted in red), jugal (highlighted in green), and ectopterygoid (highlighted in blue) and the posteromedial process of the ectopterygoid contacting the pterygoid (highlighted in yellow). Also note the relative length of the frontal (highlighted in violet), comparable to the length of the same element in *Pachyrhachis*. (b) Reconstruction of the skull of *Pachyrachis problematicus* in lateral view; postfrontal is highlighted in red and jugal is highlighted in green. (c) Interpretive drawing of the skull of *Pachyrhachis problematicus* (HUJ-Pal EJ 3659) in dorsal view. (Modified from Caldwell, 2007a.) Note that the jugal (highlighted in green) is located anterior to the ventral end of the postfrontal (highlighted in red). The elongate frontal is highlighted in violet. (d) Photo of the skull of *Eupodophis* sp. (MSNM V 3661) showing the same topological relationships of jugal and postfrontal as in *Pachyrhachis*. Note the finished (i.e., unbroken) posterior margin of the jugal. Scale bars equal 10 mm in (a) and (c), 5 mm in (d). (Modified with permission from Palci, A. et al. 2013. *Journal of Vertebrate Paleontology*, 33: 1328–1342.) (Abbreviations: ec, ectopterygoid; f, frontal; j, jugal; pof, postfrontal; pt, pterygoid.)

with MACN RN-1013, Figure 2.9, Chapter 2), and settled on an identification of that bone as the postorbital following the convention applied to modern snake lizards, but noted there were severe problems with a postorbital identification that were easily solved by recognizing the element as a jugal. McDowell's (2008) assessment of *Dinilysia* referenced Estes et al.'s (1970) jugal and postorbital fragments, agreed with Underwood's (1967) assessment that it was a single bone, not fragments of two bones, and stated in general:

> Rieppel (1988a) assumes that such a single bone would be the postorbital…However, when compared to lizards, such a bone seems unusually low on the skull to be a postorbital and is much more likely to be a jugal
>
> MCDOWELL, 2008:588

In contrast, Zaher and Scanferla (2012) settled on an identification of the *Dinilysia* posterior orbital element as a postorbital, not a jugal. Palci and Caldwell (2013), in their large-scale review of the circumorbital series in all lizards, including snake lizards, revised the identification of the element in *Dinilysia* as a jugal, not a postorbital; they did so based on the application of principles of connectivity and topological relations to both osteological and ligamentous structures—is the debate solved? Yes, I think so, and the following addresses the "Why I think so."

The discussion around the identities of the circumorbital bones of *Dinilysia* highlights the debate that has gone on for more than 120 years regarding the identities and homologies of these dermal bones in all lizards, including those with reductions in the circumorbital series such as snake lizards, dibamid lizards, and amphisbaenian lizards (as examples, see Boulenger [1896]; Romer [1956]; Frazzetta [1966]; Underwood [1976]; Parker and Grandison [1977]; Rieppel [1977]; Bellairs and Kamal [1981]; Kluge [1993a,b]; McDowell [2008]; Palci and Caldwell [2013]; Palci et al. [2013]). Arising from these long-standing debates, the earliest convention was to identify the posterior orbital element in snakes as the postfrontal (ergo, *Dinilysia* and Smith-Woodward [1901]) and to conclude that both a postorbital and jugal were in fact absent in modern snake lizards. Seventy some-odd years later, the convention had shifted such that the element along the posterior margin of the orbit was now a postorbital, the jugal was still considered absent (consistent with dibamid and amphisbaenian lizards), and when a postfrontal did appear in some modern snake lizards (e.g., Rieppel [1977, 1988a,b], or McDowell [2008]), it was considered a neomorph, not a postfrontal. The use of the term neomorph, while a common solution to such problems, reflects a transformational, not taxic, approach to identifying elements as the *a priori* assumption is clearly that the bone cannot be there and that anything "resembling" such an element cannot be that element but must be a new bone, as evolution, ontogeny, or both had already removed that element and everyone knows it. Naturally, such an approach further confounds any attempts at subsequently discovering element homologies, as the "homology" decision was made at the element identification stage and it is now a neomorph with no homologs whatsoever.

To properly understand the modern version of the problem of historical burden on this question, it is easy to track the origin of this neomorph concept to Rieppel's (1977) conclusions on the problem when he revisited and revived Baumeister's (1908) arguments that the anterior processes of the parietal in snake lizards represent the fusion of the postfrontals to the parietals. Consistent with Baumeister (1908), if the postfrontals are lost to an ontogenetic and evolutionary fusion event, any new element would indeed be a neomorph, and the remaining element connected to and in a topological relation to the "parietal-postfrontal" fusion process would be a postorbital. Zaher and Scanferla (2012:13) noted that Rieppel's (1977) transformational approach was influential and has been broadly accepted ever since (i.e., an excellent

example of historical burden) and that he based his hypothesis on the observation that the:

> …postfrontal bones of uropeltid snakes tend to fuse with the anterolateral border of the parietal to form an anterior finger-like process of the parietal along the dorsal margin of the orbit posterodorsally…the similar anterior finger-like process present in *Anilius*…would represent a fused postfrontal.

Zaher and Scanferla (2012) continued by pointing out that Rieppel (1977) had extended his hypothesis of postfrontal-parietal fusion by assuming what they called "morphoclinal character evolution"; that is, the fingerlike process was progressively reduced and then lost and thus so was the postfrontal from macrostomatan snake lizards through to caenophidians.

No matter how obviously transformational, constrained, and burdened such arguments of identification might be, the received wisdom post-Rieppel (1977) was that the posterior orbital series of bones in snake lizards was populated first and foremost by only a postorbital bone. Thus, when Caldwell and Lee (1997) identified a pair of elements located along the posterior margins of the orbits of *Pachyrhachis problematicus* Haas, 1979, as the jugals, not the postorbitals, they innocently, foolishly, or boldly ignited a fierce debate (I assure you it was not meant to be antagonistic). The response was swift, with arguments being presented that the jugals of *Pachyrhachis* were not jugals, but postorbitals, or oddly fragmented and broken portions of the ectopterygoids (Zaher, 1998; Zaher and Rieppel, 1999a), or worse yet were unidentifiable elements or UEs despite their clarity of position, connectivity, and so on (Polcyn et al., 2005).

The collective wisdom and historical burden that snake lizards did not have a jugal was so thoroughly entrenched in common thought that any ad hoc argument of taphonomy, neomorph evolution, or the "we can't solve this problem of identification, but the jugal is absent," was deemed acceptable. I admit, I was expecting a reaction post-1997, but I assumed that common sense would reign when empirical evidence was presented to positively identify a jugal in an ancient snake lizard—after all, from Boulenger (1896) forward, everyone had agreed that snake lizards were indeed lizards, that lizards both modern and ancient had jugals, and that an ancient

THE ORIGIN OF SNAKES

snake lizard, by deduction, must have once had a jugal. The prediction would therefore hold that the fossil record would one day produce a fossil snake lizard with a jugal. Simple. Caldwell and Lee (1997) clearly thought that *Pachyrhachis* was a fine instance of that condition. So what went wrong?

The problem really, as I now see it, is that to identify the jugal as present in an ancient snake lizard and allow it to be called a snake lizard rather than simply another kind of lizard, would in fact destabilize the essentialist and idealized type concept/archetype that had developed around "modern snake lizard." This essentialist-driven boundary expectation of snake lizardness required this group of tetrapods to have become snake lizards long after they lost the jugal; that is, when they lost the jugal, they were still just "lizards," not snake lizards, and that presumably other elements were lost or acquired as well (e.g., Longrich et al., 2012). Clearly no real "snake" or snake lizard could be a snake lizard and still possess a jugal—such claims are heresy to the dogma of essentialism.

This essentialist archetype concept of the snake lizard circumorbital series had so dominated precladistic anatomical thought that by the time the first cladistic analysis of squamates was rolled out by Estes et al. (1988), the jugal could only be considered as an essential and defining absence character in all modern snakes. Estes et al. (1988:129) presented two characters, "16. Postorbital: (0) present; (1) absent" and "32. Jugal-postorbital bar: (0) jugal large, postorbital bar complete; (1) jugal reduced or absent, postorbital bar incomplete." For all modern snake lizards, Estes et al. (1988) coded Character 16 as "0" for their Serpentes, and Character 32 as polymorphic "0,1" for Serpentes based on Estes et al.'s (1970:149) identification of a "jugal" fragment in *Dinilysia*. Despite the fact that Estes et al. (1988) did not include any fossil taxa in their analysis, they did introduce a polymorphic coding for Serpentes into their analysis by including the state "1" based on *Dinilysia*. Despite this head nod to the fossil record, subsequent authors ignored Estes et al.'s (1970) jugal identification in *Dinilysia* and proceeded forward with Estes et al. (1988) at the core of their character matrices, that is, with the idea that a snake could not be a snake if it possessed a jugal, and scored it as such in everything they coded (e.g., Caldwell, 1999a; Longrich et al., 2012; Zaher and Scanferla, 2012;

Martill et al., 2015; Reeder et al., 2015). Subsequent workers even went so far as to recode *Pachyrhachis*, *Dinilysia*, and *Eupodophis* to follow that essentialist scheme (e.g., Gauthier et al., 2012; Longrich et al., 2012; Zaher and Scanferla, 2012; Martill et al., 2015; Reeder et al., 2015). As an example of the construction of the character and its states, I highlight Gauthier et al. (2012:128):

> 142. Jugal: (0) present (*Oplurus cyclurus*, lateral close-up of anterior skull); (1) absent (*Feylinia polylepis*, lateral closeup of anterior skull).

While it is unclear how the above quote links to a character delineated by Estes et al. (1988), who had no distinct, "jugal absent" character, it is clear that based on Rieppel's (1977) transformational model, the essentialist archetype concept for all snake lizards as lacking a jugal became the standard character state entry into any and all taxon-character matrices for lizard phylogenetic analysis (see Caldwell [1999a] as an example of the non-critical use of a data set that deserves serious criticism).

Now, in all truth, and to be fair to the data set provided by modern snake lizards, the identification of a jugal versus a postorbital, postfrontal, supraorbital, and so on, is neither clear nor straightforward. However, to also be honest about the scholarly discourse on the question, it has, as I said above, been muddled, confused, and hardly empirical. Summing up everything from Boulenger (1896) to Rieppel (1977), Palci and Caldwell (2013:973) stated:

> …the concept that the jugal is absent in all snakes as assessed against living forms, and therefore any jugal-like element in a fossil snake cannot be a jugal because in all snakes the jugal must be absent (i.e., absence of a jugal is seen as a synapomorphy of snakes).

While this logic string is clear, the root problem is that it is top–down in its construction regarding both evolution and time, which proceeds from now into the future, or past to the present, not present to the past. And, the logic string is also circular, not to mention erroneous regarding its confusion and conflation of primary and secondary homologies (synapomorphy) (this top–down problem is a central thesis of this book, which is why the fossil record is so critical to solving these problems).

This is also why, when the fossil record provides insight (i.e., *Pachyrhachis*), it is so confusing to see the past being fitted into the present under the duress of ad hoc arguments—taphonomy, neomorphs, mistaken comparative anatomy on the part of the heretical proposer, and so on. When coupled today, as such identifications are, with the role of serving as character states in the search for synapomorphies and homologs, we must be even more vigilant of the dogma of element identifications. As Palci and Caldwell (2013) noted in citing de Pinna (1991), and as has recently been reiterated by Simões et al. (2017), primary homology statements (i.e., conjectures of homology) should only be based on anatomical similarity and the test of conjunction, and must be devoid of preconceived ideas of phylogenetic relationships and evolutionary processes. While such statements are easy to write, in practice, it is extremely difficult to get it right because so much is being imagined yet described as empirical (see Rieppel and Kearney, 2002).

In non-snake lizards, there are nine elements that frame the orbit: the frontal, prefrontal, lacrimal (when present), maxilla, jugal, ectopterygoid, postorbital, postfrontal, and, in a few species, the supraorbital or palpebrals (Figures 3.1a,b and 3.2a,b). Palci and Caldwell (2013) did not include the ectopterygoid because it is not usually visible in lateral view, but I am including it here as it is present and has been a point of controversy in identification and homolog arguments. The key elements in this debate on the big picture "lizard" homologs of the posterior orbital elements in snake lizards are the postfrontal, the postorbital, and the jugal; the critical issues of connectivity and topological relationship (see Patterson, 1982) require consideration of the ectopterygoid, maxilla, frontal, parietal, and squamosal (Figures 3.1 and 3.2).

As the postorbital is the element that had long won the debate in modern snake lizard anatomy and has burdened the discussion ever since (Boulenger, 1896; Baumeister, 1908; Rieppel, 1977), then it is the postorbital that must be knocked down from its throne, if in fact an argument can be made for a jugal. The following is not a detailed review of the history outlined above, but rather a presentation of a common-sense logical string of connectivities and topologies that I think make it clear that the best fit of the data is that the remaining posterior orbital element in fossil and modern snake lizards is a jugal.

I begin by examining the connectivities and topological relationships of the postorbital bone in non-lizard reptiles, for example, sphenodontians, and use the living tuatara as the start point. In the tuatara, the postorbital is a triradiate bone sending one process dorsally to contact the postfrontal (which excludes it from contact with the parietal), a ventral process that contacts the jugal, and a posterodorsal process that overlaps the squamosal on the latter element's lateral surface. Of consequence to this discussion is that in *Sphenodon*, the jugal also is triradiate, sending a process dorsally to the postorbital, anteriorly to the maxilla (and with a medial contact with the ectopterygoid), and posteriorly to meet the quadratojugal and descending process of the squamosal, thus framing a lower bar to the lower temporal fenestra.

As lizards of all kinds, including snakes, reduce the bony processes of elements such as the jugal and squamosal in order to "open up" the lower portion of the lower temporal fenestra, the jugal loses its connection to the quadrate complex; coupled with the evolution of a streptostylic/mobile quadrate in squamates (see Simões et al., 2017), the quadrotojugal is lost, as is the long descending process of the squamosal. Increasing the size of the temporal fenestra is considered to permit enlarged muscle masses in those spaces, thus accommodating muscle bulging during contraction and increased bite force and performance. In lizards, including snake lizards, there is still a ligament band that spans the gap created by loss of the lower temporal bar, referred to in snake lizards as the quadratomaxillary ligament (Figure 3.3) and in non-snake lizards as the quadratojugalmaxillary ligament (though this likely should change for the former considering the jugal is in fact quite often present in snake lizards [Palci and Caldwell, 2013]).

Typical non-snake lizards possess an upper temporal fenestra separated from the lower temporal region by a bony bar of bone formed by the postorbital and its posterior ramus, and a ramus of bone from the squamosal, which underlaps the postorbital. The squamosal articulates with the supratemporal and the quadrate, often with a contact point with the parietal; the opisthotic process/paroccipital process is usually just below the supratemporal but never in contact with the squamosal. The postorbital bone bears two rami, the one mentioned with the squamosal, and a more anterior one that contacts the jugal, postfrontal, and in some cases the parietal. While the descending process of the postorbital can be large in non-snake lizards, the element is

THE ORIGIN OF SNAKES

most certainly a principal component of the lateral margin of the supratemporal fenestra and by no means dominates the posterior margin of the orbit even though it commonly has a large ramus in that region. Similar to the lower temporal region, in lizards such as snake lizards, the upper part of the cheek becomes even more emarginated, involving significant loss and reduction of a variety of elements. First, the squamosal disappears in snake lizards, with the suspensorium shifting to a quadrate articulating with a supratemporal only. Therefore, the anterior process of the squamosal, which would normally articulate with the posterior postorbital ramus, is now missing.

Going back to Sphenodon, an important baseline connection and topology in this logic string concerns the position of the postfrontal—a triangular wedge of bone that clasps the frontoparietal suture. In almost all non-snake lizards, this connection and topology is clear and apparent to the observer, though there are odd examples where identifications would suggest otherwise (see Evans, 2008; regarding agamid lizards, the jugal bones often contact the squamosal, as well as satisfying all connections and topologies of typical jugals, while the element I would call a postfrontal is considered the postorbital; I would argue the postorbital has been "lost" in a dermatocranial fusion in many cases to form a postorbitojugal, or sometimes the postfrontal to form a postoribtofrontal, but that the jugal and postfrontal remain based on their topologies and connectivities). Returning to the postfrontal, when a triangular wedge of bone of varying size with no elongate posterior or ventral processes/rami clasps the fronto-parietal suture, I would argue that the logical first identification is "postfrontal" (Figures 3.1 and 3.2). When an element articulates with the postfrontal and descends toward the maxilla and ectopterygoid, but does not send a posterodorsal process to contact the squamosal, that element is not a postorbital, but a jugal, and that element is likely to also be part of a three-point insertion for the quadratomaxillary ligament. I would further suggest in this logic string, where a squamosal is absent in modern snake lizards, and for that matter in all known fossil snake lizards, that it is likely that a postorbital would also be absent as they are both elements of the upper temporal bar, which is lost in snake lizards. This leaves two elements in the posterior margin of the orbit—a postfrontal and a jugal (Figure 3.3).

Observing modern and fossil snake lizard circumorbital anatomy in a purely taxic manner,

focusing on essential differences compared to non-snake lizards and without conceptualizing transformations and mutability in the least, highlights a number of peculiarities: (1) no upper temporal bar of bony tissue; (2) the postorbital region of the skull is extremely elongate; (3) though the preorbital region is comparatively short, the maxillary bone extends a long toothed ramus beneath the orbit; (4) the toothed palatal bones are also elongate posterior to the orbit, allowing the ectopterygoid bone to overlap the posterior ramus of the maxilla posterior to the orbit (this means that the usual non-snake lizard contact of the ectopterygoid to the medial surface of the jugal maxillary ramus is now absent). These peculiarities of the snake lizard skull create essential differences in the skulls of modern and fossil snake lizards that obfuscate the problem of element identifications, as several bony element groups are both highly modified and obviously absent. However, the fossil record of ancient snake lizards has solved a couple of these problems in the best way possible—new empirical observations—and we need not "juggle the jugal" anymore (Caldwell, 2007a).

The data from Dinilysia patagonica are unequivocal regarding the logical flow of connectivity and topology—postfrontal and jugal (Figures 3.3g and 3.4a); Estes et al. (1970) were correct, except that there is no postorbital, as were Caldwell and Albino (2002) in following Estes et al. (1970) and as was McDowell (2008). The same statement is certainly verified by reference to observable anatomy for both Pachyrhachis (Figures 3.3h and 3.4b,c) and Eupodophis (Figure 3.4d) and for the much younger Cenozoic madtsoiid snake lizard, Yurlunggur (Figure 3.3i) (Scanlon, 2006—who gave the same rationale I presented above). The fossils and their clear data do not require ad hoc arguments of neomorphs or special taphonomy (Zaher and Rieppel, 1999a,b; Rieppel et al., 2003a; Rieppel and Head, 2004; Zaher and Scanferla, 2012) to explain them in order to pigeonhole them into the ironclad morphological disparity of the neontological data set.

As an example, Rieppel and Head (2004:12) wrote that for the three new specimens they were describing Eupodophis descouensis Rage and Escuillé, 2000 that:

> The ectopterygoid of specimen MSNM V 3661 closely resembles that of Pachyrhachis…However, as will be discussed in more detail below, the

position of the ectopterygoid almost entirely in front of the postorbital results from the unusually elongate frontals that carry the postorbital backwards.

On the other hand, following the logical series I outlined above, Palci et al. (2013) recognized both jugals and ectopterygoids, and, as was noted by Caldwell (2007a), Rieppel and Head's (2004) ad hoc arguments also required the empirical observation of the length of the ectopterygoid to be 1.4 times longer than the element actually is.

The same was true of Caldwell and Lee's (1997) and Lee and Caldwell's (1998) identification of right and left jugals in *Pachyrhachis problematicus* (Figures 3.3h and 3.4b,c), along with right and left postfrontals (as revised from postorbitofrontals; Palci and Caldwell, 2013), and large, elongate, spatulate ectopterygoids (best observed on HUJ-PAL 3775). Zaher and Rieppel (1999a) immediately challenged the identification of the jugals and postfrontal and argued that the jugals were symmetrically broken fragments of ectopterygoids (thus ignoring the actual ectopterygoids in order to make their ad hoc taphonomy argument—which was by necessity followed by Rieppel and Head [2004]; for *Eupodophis*). Just to make sure that ad hoc arguments ruled when refuting reasoned arguments based on connectivity and topology, Polcyn et al. (2005) followed Rieppel and Head (2004) and Zaher and Rieppel (1999a) by stating, "Caldwell and Lee's identifications are implausible," for the jugal and postfrontal, though they agreed with Caldwell and Lee (1997) regarding the identity, size, and shape of the ectopterygoid, (contra Zaher and Rieppel, 1999a; Rieppel and Zaher, 2000a; Tchernov et al., 2000; Rieppel et al., 2003a; Rieppel and Head, 2004). Not to be outdone by Rieppel and Head (2004) and Zaher and Rieppel (1999a), Polcyn et al. (2005) went a step further and refused to provide identities for five elements, what they termed the unidentified elements (UEs), which correspond to Caldwell and Lee's (1997) right and left jugal and right and left postfrontal (two fragments of the right postfrontal to make five UEs). While being conservative in identifying ancient snake lizard anatomies might be well advised, particularly when you wish to avoid conflating the problem of primary homology statements, Polcyn et al.'s (2005) refusal to identify these right and left bilaterally symmetrical elements was clearly nothing more than the dogmatic

maintenance of the historically burdened assertion that in snake lizards, fossil or modern, there is no jugal. How do I know this? In their abstract, Polcyn et al. (2005), in the absence of identifying their five UEs, state: "There is no jugal." (Polcyn et al., 2005: Abstract). Simply put, you cannot know something is not there if your comparative anatomy is unable to identify elements that are clearly present. Such logic has continued to be applied even when an element is clearly present in modern snake lizards. Rieppel et al. (2009) described in detail the cranial anatomy of *Liotyphlops* from CT scans, noting the presence of a tiny and very thin "pe" element, a "postorbital element," suspended in tissue (intramembranous bone) and not in articulation with any element. The "pe" frames the posterior wall of the orbit from the posterior margin of the prefrontal at the midpoint of the frontal, and angles toward the maxilla. As noted repeatedly above, among lizards and their kin, these are the topological relations and connectivities of the jugal, not a mysterious "pe" or "UE" element of uncertain identity. As Dr. David Reid, a sports medicine physician I took a course with back in the mid-1980s, most impactfully said to me one day (in reference to orthopedic surgical problems), "Michael, when in North America, if you hear hoof beats, think horses, not zebras."

Herein lies the problem with special arguments—they are inconsistent with evidence, if not made in the absence of it, and require ad hoc argumentation. As I said much earlier in this chapter, the fossil record is the only data set that reveals the patterns observed in the modern; the modern does not reveal its complete and accurate history, it only reveals its current state of being. The fossil record should be used to confidently revise, redefine, and rediagnose what we mean when we create mutability criteria around modern snake lizard anatomy and morphology, and not what has been done, which is to think in terms of Darwin's evolution and then to draw immutable boundaries around modern archetypes for groups of things merely because we think we have it figured out.

As Caldwell (2007a) wrote:

> Debate on the nature, identity, and delimitation of a character, as pedantic as that debate might become, is the only mechanism of falsification available in the testing of cladistic statements… the debate on the jugal, ectopterygoid and postorbital, represents an essential character delimitation debate on the sistergroup relationships of snakes.

THE ORIGIN OF SNAKES

And, as a final note on this question of the jugal versus the postorbital, or the postfrontal for that matter, I refer back to the newest ancient fossil snake lizard data from *Najash rionegrina* (Chapter 2), where a new specimen, a complete, three-dimensionally preserved skull, has set the record straight on the identity of the bones of the orbit (Apesteguía and Garberoglio, 2013), though at the time of the writing of this treatise, the work remains in press—the jugal is most certainly present, thus settling the long-standing debate (F. Garberoglio, personal communication).

## The "Crista Circumfenestralis"

The debate around the jugal versus postorbital has been made complex because of a long history of presumptions, ad hoc arguments, and circular reasoning, all folded into an essentialism derived by reference to a modern snake lizard archetype. However, the debate on the "crista circumfenestralis" and other structures and homologs in the otoociput of the braincase, while complex for the same reasons, is also difficult because the anatomy is exceedingly complex (Figures 3.5 through 3.7). Admittedly, the same top–down essentialism

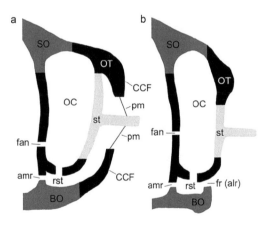

**Figure 3.5.** Schematic diagram illustrating the difference in the otic region of a snake lizard and that of a typical lizard (medial is to the left). (a) Schematic cross-section of the otic region of a snake; (b) schematic cross-section of the otic region of a lizard. (Modified with permission from Palci, A., and Caldwell, M. W. 2013. *Journal of Morphology*, 274: 973–986.) (Abbreviations: alr, apertura lateralis recessus scalae tympani; amr, aperture medialis recessus scalae tympani; BO, basioccipital; CCF, crista circumfenestralis; fan, foramen for auditory nerve; fr, fenestra rotunda; OC, otic capsule; OT, otoccipital; pm, pericapsular membrane; rst, recessus scalae tympani; SO, supraoccipital; st, stapes.)

noted for the jugal-postorbital debate is present in the debate on the "crista circumfenestralis" (Hasse, 1873; Gaupp, 1900; de Beer, 1937; Bellairs and Kamal, 1981; Rieppel, 1985), but in reality, the real difficulty is the complexity of the anatomy. To navigate this complexity and get to a discussion of the core problem, which is the use of the term "crista circumfenestralis" to reference a morphology of consequence to the evolution of snakes, it is necessary to first communicate, as simply as possible, the anatomy of this region of the braincase. I will then follow that introduction of anatomy and morphology with a discussion of the role these structures have played in the debate on snake lizard ingroup and outgroup phylogeny.

### *"Crista Circumfenestralis I"*

From my perspective, and after long consideration of this supposed morphological feature (Caldwell, 2007a; Caldwell and Calvo, 2008; Palci and Caldwell, 2014), I will submit that the "crista circumfenestralis" does not exist in snake lizards. I will also follow that point by stating that the "crista circumfenestralis" is essentially a synonym of the "bony juxtastapedial recess" and that the latter term better conveys what is being observed and characterized.

Baird (1960:929) first coined the term "circumfenestral crest" to describe a "tubelike" structure based on his observations of a small subset of snake lizards—derived colubroids. He noted that the colubroid "circumfenestral crest" forms from the crista prootica, located along the prootic margin of the fenestra ovalis, which is in contact with a rim or crest of bone formed by the fusion of separate crests from the otoccipital bone (the fused chondrocranial bones, the exoccipital-opisthotic Cundall and Irish [2008]), the crista interfenestralis and crista tuberalis. Baird (1960:951) also made the following observations from his examinations of the putatively basal snake lizards, the scolecophidians *Tyhplops* and *Leptotyphlops*:

These relationships suggest that the circum-fenestral crest has been elaborated in this form and has completely enclosed the juxtastapedial fossa and sinus, except for the small opening traversed by the shaft of the columella.

And further on page 951:

The external opening in the otic capsule of *Leptotyphlops* would, therefore, not be the

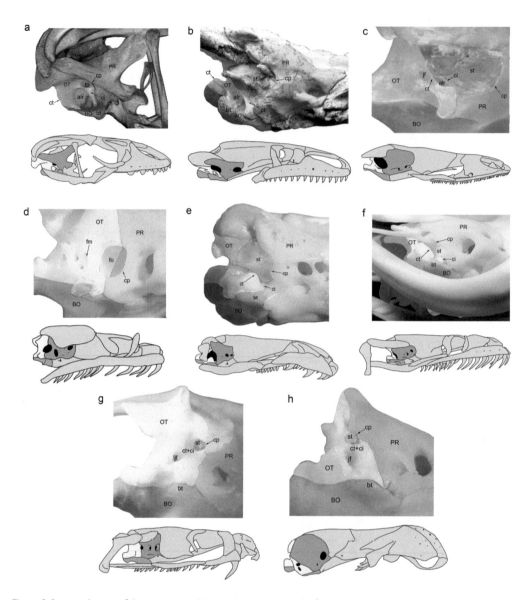

**Figure 3.6.** Lateral views of the otic regions (Type I—Type IV variants) of a varanid lizard and seven snake lizards (skull diagrams have quadrates, stapes, and in some cases pterygoids omitted to better illustrate the otic region). (a) *Varanus exanthematicus* (TMP 1990.7.33); (b) *Dinilysia patagonica* (MACN 1014); (c) *Xenopeltis unicolor* (USNM 122782); (d) *Acrochordus javanicus* (AMNH R-89839); (e) *Anilius scytale* (MCZ 17645); (f) *Python molurus* (ZFMK 5161); (g) *Bungarus fasciatus* (AMNH 56198); (h) *Rhinotyphlops schlegelii* (MCZ 38551). (Modified with permission from Palci, A., and Caldwell, M. W. 2014. *Journal of Morphology*, 275: 1187–1200.) (Abbreviations: alr, apertura lateralis recessus scalae tympani; BO, basioccipital [highlighted in red]; bt, basioccipital tuber; ci, crista interfenestralis; cp, crista prootica; ct, crista tuberalis; fm, fissura metotica; fo, fenestra ovalis; jf, jugular foramen; OT, otoccipital [highlighted in green]; PR, prootic [highlighted in blue]; st, stapes.)

fenestra ovalis, but the constricted mouth of the juxtastapedial fossa; Haas' (1930) description of *Typhlops* allows speculation that the same relationships may exist in that form.

As is clear from Baird's (1960) original characterization, the "circumfenestral crest" forms

a recess around the footplate and shaft of the stapes or columella, reflecting modifications to the otic capsule and lateral aperture of the recessus scala tympani; the result of these bony modifications is to reroute the reentrant fluid circuit such that perilymph is now moved out of the bony labyrinth of the otic capsule and onto the exterior surface of

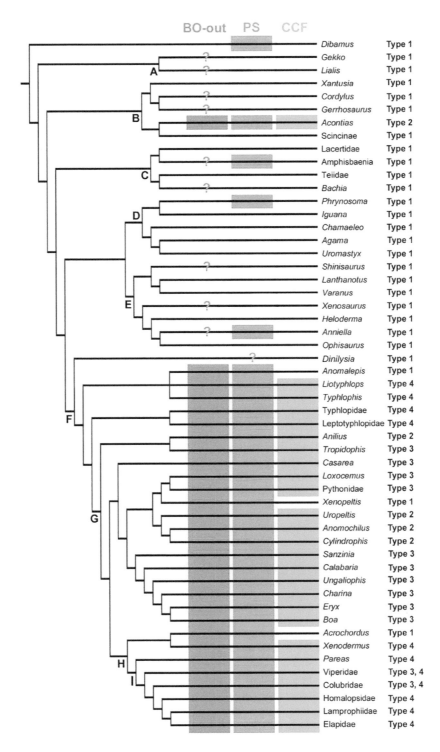

Figure 3.7. Distribution of the "crista circumfenestralis" (CCF), periotic sac (PS), and exclusion of the basioccipital from the juxtastapedial space (BO-out), within a phylogeny of Squamata (phylogenetic relationships from Pyron et al. 2013; position of *Dinilysia* according to Gauthier et al., 2012). Note correlation between CCF and BO-out. Taxa where the basioccipital is fused to the otoccipital are marked with a "?" under BO-out. (A) Gekkota; (B) Scincoidea; (C) Lacertoidea; (D) Iguania; (E) Anguimorpha; (F) Ophidia; (G) Alethinophidia; (H) Caenophidia; (I) Colubroidea. (Modified with permission from Palci, A., and Caldwell, M. W. 2014. *Journal of Morphology*, 275: 1187–1200.)

the columella and what previously was the lateral wall of the bony otic capsule. The space, or the juxtastapedial recess/pericapsular recess, now exists between the previous lateral wall of the otic capsule and interior surface of what Baird (1960) called the circumfenestral crest. Baird's (1960) circumfenestral crest does not exist as a single crest of bone, nor does it function as a crest, but rather as a housing for the pericapsular sinus and its perilymphatic fluid.

By the usual definition, as bony structures, cristae are lines, crests, or ridges to which muscles are attached via collagenous ligament fibers at either their insertion or origin; in long bones, such cristae are limited to a single endochondral element, but on cranial bones, they may well extend across or along more than a single element; for example, the sagittal crest runs along the midline of both the parietal and supraoccipital and serves as the point of origin for a wide variety of muscles. The "crista circumfenestralis," in contrast, is not a crista for muscle insertions nor origins, but, as noted, forms a partial dome over the footplate of the columella with a hole/foramen through which the columellar shaft emerges to enter the middle ear cavity (a membrane seals the opening around the shaft of the columella). In the highly specialized form where the tube is complete, such as in colubroids and scolecophidians, the circumstapedial opening is very small. The matter of muscles attachments will become important shortly concerning the origins of the crests that converge to form the juxtastapedial dome.

But, and more to the point, the "circumfenestral crest" or "crista circumfenestralis" of scolecophidians and derived colubroids is not found to be a common anatomical feature in many other modern snake lizards (McDowell, 1979; Rieppel and Zaher, 2001; Palci and Caldwell, 2014) that in any way deserves *a priori* to be considered a primary homology assignment of "present" (e.g., Longrich et al., 2012; Zaher and Scanferla, 2012). For fossil snakes, contrary to claims of its presence, each of those claims have systematically been falsified by subsequent investigations (for an overview, see Palci and Caldwell, 2014). I do not wish to "throw the baby out with the bathwater," so if there is to be any value in recovering phylogenetically informative data in the otoccipital region of lizards, it is critical to know what we are talking about and then atomize it effectively or discard it altogether. The following is a brief and necessary overview of the otoccipital region of fossil and modern snake lizards.

## Bones of the Otoccipital Region

The bones of the otoccipital region are few in number, but exceedingly complex in terms of their morphology. It is critical to remember that all the bony structures in the otoccipital region form first in cartilage; that is, they are all part of the developmental component of the skull known as the chondrocranium. However, just to make sure the origin of the cartilaginous precursors of the bony chondrocranium have an even more complex developmental history, there are cartilaginous elements of the braincase that form long before there are recognizable precursors of the cartilaginous chondrocranial elements. I will ignore this level of braincase complexity here, but recommend, for a brief but dense review, the introductory paragraphs in Rieppel and Zaher (2001) or, for a denser and more lengthy treatise on the subject, de Beer (1937), Kamal (1965), and Rieppel (1985). The end result of chondrocranial chondrification is that cartilage precursors of the bony otoccipital region form what will ossify into the anterolaterally positioned right and left prootics; the posterior and posterolaterally positioned right and left otoccipitals (in snake lizards, these paired elements are considered the fusion of the opisthotic and more posterior exoccipitals, both of which found as separate elements in most other lizards); the single basally positioned basiooccipital; and the more anteriorly and basally positioned single element, the basisphenoid. The roof of the braincase, above the occiput, is formed by the right and left otoccipitals that meet above the foramen magnum, a single supraoccipital anterior to those two elements, and the non-chondrocranial element, the parietal (derived from dermatocranial ossification without a cartilage precursor).

For snake lizards, this means that one side of the braincase, including the floor and roof, is composed of exactly four bones: prootic, otoccipital, basioccipital, and supraoccipital. This should not be too complex, and as the number of bones goes, it is not. However, this region of the braincase houses the apparatus of the inner ear complex, which forms a large, bulging, and dense labyrinth of tubes and cavities (also called the bony labyrinth or otic capsule) crossing through, within, out of, and into again the bony tissues of the prootic, otoccipital, and basioccipital. At the same time these four bones (five in other lizards where the exoccipital and opisthotic are not fused)

THE ORIGIN OF SNAKES

are supporting the labyrinth, they are also being pierced by a vast and complex array of foramina for cranial nerves, blood vessels, and ducts. The division of the fissura metotica in lizards, retained in snake lizards, also makes the region more complex, as the anterior division houses the glossopharyngeal nerve (Cranial Nerve IX), while the posterior division houses the vagus nerve (Cranial Nerve X); the anterior division becomes the recessus scala tympani, while the posterior one becomes the vagus/jugular foramen.

## Cranial Nerves

The bony and cartilaginous braincase surrounds the brain (central nervous system) and its paired and bilaterally symmetrical cranial nerve trunks (twelve in higher vertebrates, including all lizards, though only eleven can be identified in snake lizards), as well as a wide variety of sensory structures linked to the senses of smell, taste, sight, touch, and hearing. Two of these cranial nerve systems arise from the forebrain, for example, Cranial Nerve I, the paired olfactory nerves, and Cranial Nerve II, the paired optic nerves. The other ten (or nine) paired right and left cranial nerve tracts in lizards arise from the brain stem, or hindbrain, and thus are clustered in the more posterior and ventral portions of the braincase, in particular in and around the otoocciput, where the exits or foramina are to be found just anterior to the otic capsule of the otoocciput (Cranial Nerve VII, facial nerve, piercing the prootic bone), inside the otic capsule but not exposed on the external surface of the braincase (Cranial Nerve VIII, vestibular cochlear nerve, which, critical to this discussion, pierces the bony wall of the otic capsule, usually through the opisthotic bone), and posterior to the otic capsule (Cranial Nerves IX–XII, respectively, the glossopharyngeal nerve, the vagus nerve, the "spinal" accessory nerve, and the hypoglossal nerve, also within the opisthotic bone). These seven nerve trunks/tracts are clustered anterior and posterior to the otic capsule and exit through a wide variety of variably positioned foramina, some of which are shared between various cranial nerves and vasculature and others that are not shared, but singular. There are also a number of blood vessels, both arteries (going into the brain) and veins (exiting the braincase), that have variable but independent foramina perforating the otoocciput in this region of the brain case.

## Perilymph and Endolymph

The brain and spinal cord are surrounded by cerebrospinal fluid (CSF), which is produced by choroid plexus cells located in the ventricular spaces within the brain. The CSF is replenished several times a day, circulates throughout the cerebrospinal space, is critical for immunological and mechanical protection of the brain, and enters the otic capsule via the medially located perilymphatic duct. Importantly, this is the same fluid that is moved in and out of the bony labyrinth/otic capsule (reentrant fluid circuit) through the lateral aperture of the tympanic recess and onto the outer surface of the otic capsule; this fluid circuit is recognized in numerous modern snake lizards, regardless of whether they possess Baird's (1960) bony "crista circumfenestralis" (e.g., *Acrochordus javanicus* [McDowell, 1979; Rieppel and Zaher, 2001]).

Movement of the perilymph is key to understanding the problem of the "crista circumfenestralis" as a misused primary homolog with simplistic states of absent and present. The character of interest should be (see Rieppel, 1988a,b) the fluid circuit of perilymph, whether snake lizards all have a common fluid circuit, and whether it is shared with any other clade of lizards. Rieppel (1988a,b) concluded snake lizards were unique in the organization of their perilymph fluid circuit. Among modern snake lizards, it is possible to verify the flow circuit and confirm, in the absence of any bony structures whatsoever, the presence of a reentrant fluid circuit; in fossil snake lizards this is not possible to ascertain with confidence in the absence of the necessary bony structures. As was demonstrated by McDowell (1979) for *Acrochordus*, that fluid circuit exists in the absence of the "crista circumfenestralis," with the perilymphatic sac extending out on the external surface of the otic capsule through what McDowell (1979) and Rieppel and Zaher (2001) considered the jugular foramen (the latter authors argue that the lateral aperture of the recessus scala tympani [fenestra rotunda] has been lost). In non-snake lizards (though there are exceptions [see Palci and Caldwell, 2014]), the fenestra rotunda is separated from the fenestra ovalis by the crista interfenestralis and in non-snake lizards is covered by a thin membrane that bulges laterally in compensation to pressure waves created in the perilymphatic space by movement of the stapedial footplate.

As noted, for modern snake lizards, even in the absence of "crista circumfenestralis," the perilympahtic sac extends out of the tympanic recess through the lateral aperture of the recessus scala tympani/fenestra rotunda and onto the lateral surface of the bony otic capsule, surrounding the footplate of the columella. In a large number of species of modern snakes, mostly colubroids and scolecophidians, the crista interfenestralis fuses to the crista tuberalis and forms a bony dome over the extruded perilymphatic sac, the juxtastapedial recess. Because of the presence of the domelike "crista circumfenestralis," the fenestra rotunda is now captured "inside" of this complex, and is often referred to as the fenestra pseudorotunda. The membrane around the shaft of the columella is the pericapsular membrane. Detecting the existence of a soft tissue reentrant fluid circuit, regardless of the exit foramen of the perilymphatic sac, is impossible in a fossil snake lizard. The problem is, this is really the character we wish to understand, and not the variety of bony tissues that may or may not form a protective juxtastapedial recess around, over, and beside it. Whether the perilymphatic sac actually extrudes out of the fenestra rotunda may well be the relevant snake lizard synapomorphy—it is certainly not the crista circumfenestralis, as it is not common to all snake lizards, nor is it the juxtastapedial recess, which appears to be a synonym for the former.

### Inner Ear Complex-Membranous Labyrinth

Contained within the otic capsule/bony labyrinth created by the prootic, otoccipital, and basioccipital lies the inner ear complex, or the membranous labyrinth of the inner ear. This entire complex of interconnected soft tissue structures is significantly smaller than the bony labyrinth and is suspended in that space by the perilymphatic fluid (see above). The membranous labyrinth is also fluid filled—endolymph—and that fluid moves through and past the various sensory structures (three receptor crests in the semicircular canals; three maculae in the saccule, utricle, and lagena; and two different papillae, one auditory in the cochlea), all of which support the sensory functions of hearing (auditory system) and equilibrium (vestibular system). The auditory system in non-mammalian amniotes is composed of the lagena (a more simplified version of the coiled cochlea and cochlear duct of mammals), while the vestibular system is

composed of the saccule, utricle, three ampullae, and the three semicircular canals, each of which extends out and back to its respective ampulla (horizontal, superior, and posterior canals). This entire complex is innervated by Cranial Nerve VIII, the vestibulocochlear nerve, which bears two branches: the vestibular branch I, with nerve fibers running from the saccule, utricle, and the three ampullae of the semicircular canals; and the cochlear branch II, which is connected to the cochlea and organ of Corti in the lagena. The hair cells within the organ of Corti respond to pressures waves in the endolymph created via transduction through the basilar membrane of the lagena from the perilymph and physical movements of the columellar footplate. Within the saccule and utricle are found one or perhaps many endolymphatic infillings known as otoliths, which have a defined and characteristic structure in various vertebrates including snake lizards (e.g., de Burlet, 1934), and in many fossil and modern snake lizards can be extremely large (e.g., *Dinilysia*; see Zaher and Scanferla [2012]).

### Lateral Wall of the Otic Capsule

Broadly speaking, there are two general conditions of form of the lateral otic capsular wall: (1) the presence of a bony dome ("crista circumfenestralis," or my preferred term, the bony juxtastapedial recess) pierced by the stapedial shaft, where neither the fenestra ovalis nor fenestra rotunda are visible in lateral view (nor the footplate of the columella), and the only other visible foramina are those associated with Cranial Nerves IX, X, XII, and the jugular vein (Type 4 of Palci and Caldwell [2014]); and (2) the absence of a complete bony juxtastapedial recess forming a tube around the stapedial shaft (crista circumfenestralis), the obvious presence of a separate (unfused) crista prootica, crista interfenestralis, and crista tuberalis, where the footplate of the columella is visible in whole or in part, as is the fenestra rotunda (lateral aperture of the recessus scala tympani), and, of course, the foramina associated with Cranial Nerves IX, X, XII, and the jugular vein are posterior to the crista tuberalis (Types 1–3; Palci and Caldwell [2014]). (There is significant variation among the several thousand snake lizards in Types 1–3 in terms of the degree to which these numerous features are developed, or not, and thus do not lead to the presence of juxtastapedial recess.)

THE ORIGIN OF SNAKES

## The Middle Ear

The medial portion of the middle ear of snake lizards contributes a single element to the form and function of the otic capsule and inner ear, and this is the columella and its footplate, the latter of which rests along the rim, or perhaps slightly below it, of the fenestra ovalis. That footplate is in contact with the perilymph within the pericapsular space above the lagenar portion of the membranous labyrinth of the inner ear. Soundwave-induced movements of the columella and its footplate pass through the perilymph and via transduction create pressure waves in the endolymph fluid within the lagena/cochlea. The footplate in snake lizards is variable in size, from very tiny in scolecophidians (Figure 3.6h), to massively large in *Dinilysia patagonica* (Figure 3.6b), to moderately sized in an elapid such as *Bungarus* (Figure 3.6g).

The more lateral component of the middle ear complex in snake lizards varies somewhat between many modern snake lizards and in particular Mesozoic-aged fossil snake lizards. In modern forms such as macrostomatans (e.g., pythons, boas, colubroids), the shaft of the columella is extremely elongate and thin, crossing the middle ear cavity to contact the shaft of the quadrate near the middle of that element; the contact of the distal tip of the columella with the quadrate is often with a small process on the latter (stylohyal process; see below).

A second form of middle ear organization is observed in non-macrostomatan snake lizards such as *Anilius*, *Cylindrophis*, and *Xenopeltis*, where the columellar shaft is short, robust, and posterodorsally inclined toward the suprastapedial process of the quadrate. The columella does not directly make contact with the suprastapedial process, but rather articulates with a small, subrounded ossicle of bone (the stylohyal, or intercalary bone, which could be derived from the extracolumella) that is fused/articulates with the suprastapedial process.

A third morphology is recognized for the lateral component of the middle ear complex, observed in some scolecophidian snake lizards, where the columella can have a ligamentous connection with the quadrate (Rieppel, 1980b).

## The Otic Capsule of Fossil Snakes

The fossil record does not preserve any evidence, ever, of fluids such as the perilymph, or of soft tissues such as nerves or the actual membraneous labyrinth (Figure 3.5). However, because the cartilage precursors of the otoccipital bones form around those soft tissue structures, and in turn ossify, and because fossilization can preserve even the cell spaces occupied by osteocytes, the fossilized bones of the otoccipital region can superbly preserve the internal and external anatomy and morphology of the otic capsule. Though the bony labyrinth is much larger than the membranous labyrinth it houses, this bony labyrinth still reflects the basic organization of membranous one (e.g., micro CT scan images of the bony labyrinth [Yi and Norell, 2015; Palci et al., 2017, 2018]). The same is true for foramina—though they are only parts of the conduit system of ducts, nerves, and blood vessels, for the parts they preserve, they are invaluable data points. As noted, the same is true for the exterior of the otoocciput and the various crests of the prootic, otoccipital, and basioccipital, which have been superbly well preserved in a number of snake lizards. The missing components are of course the presence/absence, course, and morphology of a possible periotic sac and pericapsular space when there is no bony juxtastapedial recess, as compared to the condition described for the modern snake lizard, *Acrochordus javanicus* (McDowell, 1979; Rieppel and Zaher, 2001).

To date, there are three fossil snake lizards from the Mesozoic that preserve the otoccipital region in enough detail to be able to describe whether there is a juxtastapedial recess formed around the shaft of the columella as characterized by Baird (1960) when he defined his "circumfenestral crest": *Dinilysia patagonica*, *Najash rionegrina*, and, to a lesser degree, the unnamed South African snake lizard. It is tempting to add *Sanajeh indicus* to this list, but I have not personally observed the specimen, and the original description is lacking in details regarding the fenestra ovalis and stapedial footplate; for the time being, it is excluded from this discussion and description of anatomy.

## Constrictor Internus Dorsalis Group

I have never read a discourse considering muscle origins/insertions and the evolution of myology in relationship to the otic capsule and the "crista circumfenestralis." So I am going to explore what I consider an important group of muscles, the constrictor internus dorsalis (CID) muscles, one of which in particular (the musculus protractor pterygoideus) is of relevance to the morphology and evolution of the "crista circumfenestralis" in snake lizards as compared to typical lizards, but is also a key muscle to modern snake lizard functional morphologies associated with prey capture and ingestion (see Kley [2001]).

The CID is composed of three or four muscles originating on elements of the braincase (bone, cartilage, and fascia), that are dorsal to the palato-pterygoid complex of bones and are commonly innervated by a branch of the Trigeminal Nerve ($V_4$): the musculus levator pterygoideus, m. protractor pterygoideus, m. levator bulbi dorsalis and m. levator bulbi ventralis. In very broad terms, in non-snake lizards, when contracted, the first two CID muscles "constrict"/"stabilize" the braincase against the palatopterygoid bones, while the last two open the lower eyelids (contraction of the m. levator bulbi dorsalis opens or depresses the lower eyelid, while the m. levator bulbi ventralis is a "tensor" of the palatal membrane [Haas, 1973]). The remainder of the CID group does not elevate or protract the pterygoid complex in a sphenodontian such as the Tuatara, nor in a typical lizard such as an iguana, as these elements are not kinetic in the first place, but rather serves to resist torque forces being exerted against the braincase via the palate and upper jaws during head twisting and biting by stabilizing the braincase against the pterygoids and thus the palate and maxillae (Ostrom, 1962; Haas, 1973; Holliday and Witmer, 2007; Jones et al., 2009). The m. levator bulbi dorsalis originates from a tendon on the epipterygoid, while the m. levator bulbi ventralis originates on the anteroventral margin of the membranous braincase (Haas, 1973); according to Haas (1973), the latter inserts on the palatal membrane just anterior to the insertion of the m. levator pterygoideus. The musculus levator pterygoideus originates on the "ventrolateral surface of the cartilaginous orbitosphenoid medial to the dorsal extremity of the epipterygoid" (Ostrom, 1962, p. 733) and inserts on the medial surface of the epipterygoid base and the dorsal and medial surfaces of the pterygoid.

In non-snake lizards, the muscle of consequence to the evolution and morphology of the "crista circumfenestralis" of snake lizards is the musculus protractor pterygoideus, which in a non-snake lizard originates from the lateral surface of the braincase (dorsal surface of the basisphenoid-basipterygoid process, the anteroventral part of the prootic along the prootic crest or crista prootica (note: this is not the same crista prootica as defined on the prootic

margin of the fenestra ovalis), and the ventral part of the orbitosphenoid) and inserts along the dorsal crest of the quadrate ramus of the pterygoid (Anderson, 1936; Ostrom, 1962; Haas, 1973).

In snake lizards, the musculus levator bulbi dorsalis is lost with the loss of the lower eyelid, and, of course, eyelids in general in snake lizards. The m. levator bulbi ventralis considered "restructured" (Haas, 1973) as m. retractor pterygoideus), now originates not on the epitpterygoid, which is absent in snakes lizards, but rather on the parietal, and inserts on the pterygoid. Likewise, the origin of the m. levator pterygoideus on the ventrolateral surface of the orbitosphenoid in typical lizards, with an insertion on the dorsal crest of the pterygoid and foot of the epipterygoid, shifts in snake lizards to an origin on the ventral surface of the descensus parietalis (descending flange of the parietal) and an insertion on the midportion of the pterygoid anterior to the ectopterygoid process of the pterygoid and immediately anterior to the large insertion of the m. protractor pterygoideus. It is the attachments of origin of the m. protractor pterygoideus that are of consequence in snake lizards to this discussion of the "crista circumfenestralis." In snake lizards the prootic crest is lost as an elongate crista of muscle origins, running anteriorly from the margin of the fenestra ovalis to the superior rim of the trigeminal notch, along with the orbitosphenoid bone and membranous braincase. The m. protractor pterygoideus of snake lizards now originates on the basisphenoid and inserts along a long crest on the dorsal surface of the quadrate ramus of the pterygoid—in other words, the origin of the muscle changes but the insertion does not. Critically, for the crista prootica and its relationship to the fenestra ovalis and involvement in the "crista circumfenestralis," the constraints on the otic capsule and crista prootica that would have been imposed by this muscle origin site are now gone. Combining all of these chondrocranial and dermatocranial modifications together, along with musculoskeletal evolution, results in an otic capsule and crista prootica that is not under stress due to muscle attachments and can specialize around specific sensory system evolution linked to hearing and balance/equilibrium.

Most importantly for snake lizard maxillary and pterygopalatal mobility, these extreme modification of the CID muscle origins and insertions, and highly modified functions, permit extreme kinesis of the maxilla, pterygoid, ectopterygoids, and palatines in Type IV Viperid-Elapid-Scolecophidian skulls (Bolt

THE ORIGIN OF SNAKES

and Ewer, 1964; Cundall, 1983; Greene, 1983; Kley, 2001; Cundall and Irish, 2008), where the m. *protractor pterygoideus* is the key muscle involved in the elegant mechanical system that results in protraction of the mobile maxilla and the extension and erection of the fangs in Type IV modern snake lizard skulls. The m. *levator pterygoideus* is also essential to this process, as it too levates and stabilizes the pterygoid at the ectopterygoid joint by retaining some of its ancient CID function. Whether this mechanical elegance is accomplished by right and left synchrony as it is in viperids-elapids (Boltt and Ewer, 1964), or by left and right asynchrony as in typhlopids and presumably anomalepidids (Kley, 2001), or by "pterygoid walks" in Type I and Type II modern snake lizard skulls is of no consequence, as the framework musculature, osteology, and innervation are the same. The key anatomical feature here is that the crista prootica has lost its function as a point of origin for the protractor pterygoideus, thus freeing the latter to evolve highly specialized functions in feeding. It is critically important to reiterate here that the crista prootica of the snake lizard crista circumfenestralis is not the same bony crest associated with the m. *protractor pterygoideus*.

## "Crista Circumfenestralis II"

Summing up this examination of the "crista circumfenestralis," I can only conclude that it does not exist in snake lizards, but rather is a feature localized in the groups that Baird (1960) first postulated it to be present in—at least some caenophidians. Even there, the use of the term "crista" should be abandoned and the older reference to a bony juxtastapedial recess used to refer to such a space only when that dome of bone, produced by cartilaginous contributions arising from the chondrocranial prootic and exoccipital-opisthotic (otoccipital), is present and surrounds the columellar footplate and the shaft. If the various cristae do not form a complete juxtastapedial recess, then they do not, and there is no need to create a character for phylogenetic analysis to conclude, via the Test of Congruence (see Patterson, 1982), that yet another synapomorphy of all snake lizards has been recovered. A complete juxtastapedial recess is only ever observed in Type III and Type IV modern snake lizard skulls.

MATTERS OF CHARACTER

As with all data, the problem is how to transform the data from a suite of empirical observations into some form of coherent, testable knowledge claims (i.e., hypothesis bounded) that are true to the data but at the same time inform us in a meaningful way (i.e., the ontologic informs the epistemic). In systematics generally, we try to make that transformation of data into knowledge and testable hypotheses via the process of constructing primary homologies (de Pinna, 1991) and ultimately testing them via the Test of Congruence (*sensu* Patterson, 1982).

The problem, of course, and this will remain a consistent theme through this treatise, is to acknowledge the constraints imposed by the top–down essentialist archetype concept derived from the singular reference to the condition of modern snake lizards. Similar to the problem outlined for the circumorbital series, the same historical burden and top–down approach to essential anatomies required to be a "snake" as opposed to a "lizard" has infected the anatomical identification of the so-called crista circumfenestralis as a synapomorphy of all snake lizards, modern and fossil. Examples from the literature are easy to find, and because *Dinilysia patagonica* has been such an important Mesozoic snake lizard taxon in recent years, I have elected to use it, and the literature dedicated to it, to highlight the depth and pervasiveness of the problem. From the abstract of Zaher and Scanferla (2012:194):

> We also reinterpreted the structures forming the otic region of *Dinilysia*, confirming the presence of a crista circumfenestralis, which **represents an important derived ophidian synapomorphy**. (boldface mine for emphasis)

Their reinterpretation of Estes et al. (1970), Caldwell and Albino (2002), Caldwell and Calvo (2008), and Caldwell (2007a) was justified by the following anatomical identifications and conclusions. It is also critical to understand for Zaher and Scanferla (2012), and numerous authors before them, the presence of a crista circumfenestralis is an "important derived ophidian synapomorphy" that must be present in *Dinilysia* because *Dinilysia* is a "snake" (the ancient metaphor of the circle created by the snake eating its own tail is dizzyingly accurate).

So, to defend their reinterpretation, Zaher and Scanferla (2012:213) noted: (1) that Estes et al. (1970) and Caldwell and Albino (2002) were mistaken in claiming the crista circumfenestralis was absent because the specimens they examined were broken; (2) that MACN 1013 possesses the cristae prootica, interfenestralis, and tuberalis, which means the crista circumfenestralis is present; (3) the stapedial footplate is recessed posterior to a weak crista prootica; (4) the crista tuberalis form a posterior projection or shelf; (5) the crista tuberalis fails to enclose the juxtastapedial recess, which remains wide open posteriorly (and that this was a shared feature of scolecophidians, anilioids, *Xenopeltis*, and *Loxocemus*), as taken from Rieppel et al. (2003b); and (6) the crista tuberalis is formed only by the otooccipital.

To rebut, I here follow the same arguments provided by Palci and Caldwell (2014): (1) while it is not true that Estes et al. (1970) had access to any other specimens, I can confirm that I personally prepared the specimens MACN 1013 and 1014 through 1998 and 1999, and observed the conditions of the three cristae and that even if slightly broken, I can confirm that they do not contact each other to form a composite crest coherent with the condition originally circumscribed by Baird (1960), such that, for example, the crista prootica is never in contact with the crista tuberalis, as per Caldwell and Albino (2002) (Figure 3.6); (2) all lizards I am aware of possess separate cristae prootica, interfenestralis, and tuberalis, and this is not evidence of the presence of a crista circumfenestralis (Figure 3.6); (3) while it is true that the stapedial footplate is recessed posterior to a weak crista prootica in *Dinilysia*, it is also true in *Varanus* and other lizards, and, again, there is no crista circumfenestralis; (4) it is also true that in *Dinilysia*, the crista tuberalis forms a posterior projection or shelf, but such a shelf does not constitute a crista circumfenestralis, nor does it build a juxtastapedial recess; (5) again, it is true that in *Dinilysia*, the crista tuberalis fails not just to enclose, it also fails to *create* the juxtastapedial recess; it is also true that this condition is observed in anilioids, *Xenopeltis*, and *Loxocemus*, but it is not true that this is observed in scolecophidians where a juxtastapedial recess is clearly present as illustrated by Palci and Caldwell (2014), which stands in sharp contrast to both Zaher and Scanferla's (2012) coding of the same for scolecophidians in their Character 80 (see below); Palci and Caldwell (2014) also contradict the literally verbatim claim made in Rieppel et al. (2003b:820), "Indeed, the crista tuberalis does not close the juxtastapedial recess posteriorly in scolecophidians, *Dinilysia*, anilioids, *Xenopeltis* and

*Loxocemus*," which is the same statement given by Zaher and Scanferla (2012); and (6) it is true that in *Dinilysia* the crista tuberalis is formed only by the otooccipital, but it is also true that the same is present in *Varanus*, as an example.

As Palci and Caldwell (2014) showed, *Dinilysia* does not possess a crista circumfenestralis, and as I hope I have made clear in this section of Chapter 3, the crista circumfenestralis does not exist in most snake lizards, and in those that possess it, it is really a synonym for the juxtastapedial recess. The lingering if not overarching problem, of course, is that variations on the theme of "circumfenestral crest" are consistently recycled in the literature as primary homologies in multiple and exceedingly dependent character state complexes for phylogenetic analysis. For example, characters 80–83 from Zaher and Scanferla (2012) are virtually identical to characters 34–37 from Rieppel et al. (2003b). Characters 80 and 81 from Zaher and Scanferla (2012) are given as examples here:

80. Juxtastapedial recess defined by crista circumfenestralis absent (0), present but open posteriorly (1), or present and closed posteriorly (2).

81. Crista circumfenestralis exposes most of stapedial footplate (0), converges upon stapedial footplate (1).

As can be readily grasped from reading such character descriptions and primary homolog constructs, the presence of crista circumfenestralis is a given—the existence of this structure and its homologies are never tested. In doing this, dependent characters have been created to test conditions of crests and recesses where none form a crest system nor a recess for the perilymphatic sac. Ultimately, what this means is that this supposed snake lizard synapomorphy is an untested homology supported by four characters with nine states, yet I would argue it is not real. For me this chaotic approach to the configuration of the juxtastapedial recess should be abandoned and these characters discarded. I would also recommend that the appropriate empirical approach is to determine the presence or absence of the perilymphatic sac's extrusion out of the fenestra rotunda and onto the footplate of the columella. This is likely to be a key synapomorphy of snake lizards, followed by a derived state in colubroids and scolecophidians (linking Type III—Colubroid and

THE ORIGIN OF SNAKES

Type IV Viperid-Elapid-Scolecophidian) where a true juxtastapedial recess is formed. Such a condition is not known for any of the Mesozoic snake lizard fossil taxa (see Chapter 2 for details on the unnamed South African cranium, *Najash, Dinilysia, Sanajeh*, probably *Menarana*); I predict it is unlikely to be observed in any Type I—*Najash-Dinilysia*-like skull, consistent with the modern Type I—Anilioid skull, nor in a Type II *Pachyrhachis-Haasiophis*-like skull consistent with the general absence of a closed juxtastapedial recess in Type II—Booid-Pythonoid skulls. The linkage to the categories created by Palci and Caldwell (2014) for the presence/absence and completeness of the "crista circumfenestralis" in snake lizards is nearly one to one (see Figure 3.7).

## Intramandibular Joint

The intramandibular joint has been characterized for some time as a single homolog in the literature (for lizards such as the varanoid lizards, that is, *Lanthanotus*; see McDowell and Bogert [1954] and Gauthier [1982]; for mosasaurian lizards see Cope [1869]; Williston [1898]; Camp [1923, 1942], Gregory [1951]; Russell [1967]; Lee [1997a]; Caldwell [1999a,b]; and for snake lizards see Rieppel [1988a,b]; Caldwell and Lee [1997]; Lee and Caldwell [1998]; Rieppel and Zaher [2000a]; and Lee and Scanlon [2001]) (Figures 3.8 through 3.12). However, as with all dictums of received wisdom, this historical burden of common usage

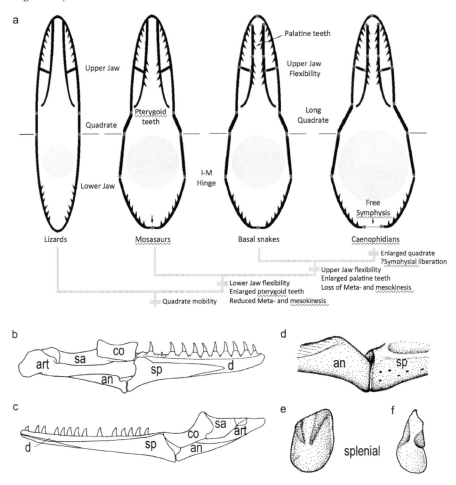

**Figure 3.8.** (a) Lee et al. (1999a,b) diagram of the evolution of snake feeding systems in the context of a sistergroup relationship with mosasaurs (? indicates uncertainty of the timing of appearance of this feature within Caenophidians). (b) Medial view of left mandible of the plioplatecarpine mosasaur, *Platecarpus* sp. (Modified from Gregory, 1951.) (c) Medial view of right mandible of the mosasaurine mosasaur *Clidastes* sp. (Modified from Gregory, 1951.) (d) Close-up of angular-splenial joint of *Platecarpus* sp. (e) Joint surface view of right splenial, from *Platecarpus* sp. (f) Joint surface view of right splenial, from the snake lizard *Cylindrophis ruffus*. ([a, d-f] Used with permission, original illustration, copyright M. S. Y. Lee et al., 1999a,b.)

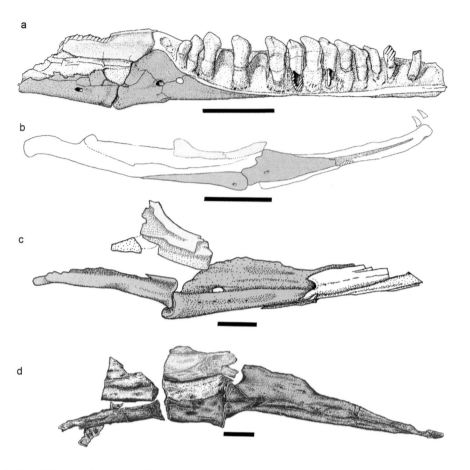

**Figure 3.9.** Medial views of mandibles of basal pythonomorphs. (a) Left mandible *Coniasaurus crassidens* (NMH R 3421). (Modified from Caldwell and Cooper, 1999.) (b) Left mandible *Pontosaurus lesinensis*. (Modified from Pierce and Caldwell, 2004.) (c) Fragments of right mandible, image reversed, holotype *Dolichosaurus longicollis*. (Modified from Caldwell, 2000b.) (d) Fragments of left mandible, holotype of *Komensaurus carrolli*. (Modified from Caldwell and Palci, 2007.) (Red = splenial; purple = angular; green = coronoid.) Scale bars = 1 cm.

is by no means a verification of the quality and reliability of the homolog concept at the root of the character. It is also not clear if there even is such a "joint" in the mid-mandibular region, or if it is in fact a collection of bony contact surfaces that collectively result in flexibility, if not real kinesis, but individually do not define a joint of any kind (Figures 3.9 through 3.12).

The key question I will address in this examination of the intramandibular joint in snake lizards and their kin is whether such a joint even exists (Figure 3.8a–f)—that is, is this intraramal portion of the mandible and its six independent elements, a single joint or multiple joints formed by widely spaced and non-overlapping sutural surfaces, and thus not a single joint at all? The second question, assuming the answer to the

first one is at least a partial yes (a joint is present between at least two bones even if not all), is whether it is one character or multiple characters. The third question, or problem, is to define the/ these testable primary homologs and homolog concepts among the bones of the intramandibular region (see Simões et al., 2017).

Determining the "jointness" of this joint and the primary homologs to be conceptualized in the mandible is important for a number of reasons, not the least of which is the linkage of this mandibular system to the much-studied mechanics of modern snake lizard feeding systems. Researchers specifically, and the global public generally, are fascinated (the latter group morbidly so) by the capacity of most modern snake lizards to ingest prey the same size, or significantly larger, than

THE ORIGIN OF SNAKES

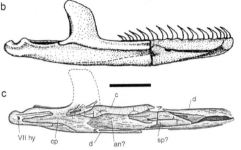

Figure 3.10. *Pachyrhachis problematicus* holotype (HUJ-Pal EJ 3659). (a) Ventral view; (b) reconstruction of right mandible, lateral view (Modified from Lee and Caldwell, 1998); (c) stippled line drawing, left mandible medial view (Modified from Rieppel and Zaher, 2000a).

themselves. To highlight the complete literature on the subject is impossible, though a few stand out both in terms of diversity and linkages to ancient snake lizards and related modern forms, including the research output of Underwood (1957), Gans (1961), Greene (1983, 1984), Gartner and Greene (2008), Kardong (1986), Kardong and Berkhoudt (1998), Kardong et al. (1986), Cundall (1983, 1995), Cundall and Gans (1979), Cundall and Greene (2000), Deufel and Cundall (2010), and the work of Kley (2001) and colleagues, for example, Kley and Brainerd (1996, 1999), on scolecophidians.

When we broaden the question and data set beyond the modern fauna to include ancient snake lizard anatomies, and to then consider the origin of the "intramandibular joint," the question becomes of relevance to paleontologists and herpetologists alike, and to the goals of this book. I begin with a survey of the elements at the "joint" plane in lizards, including snake lizards, and then examine the kinds of joints currently recognized in vertebrate animals and what kinds of joints are present midmandible in lizards

possessing an "intramandibular joint." I continue by examining form and function, top down, from Type IV and III snake lizard skull types, to those of ancient snake lizards, and conclude by examining the specifics of the debate on the intramandibular joint and its constituent elements and the primary homolog concepts that have been developed around them.

### Elements at the Intramandibular Joint

Atomizing the intramandibular joint, or "intraramal joint" (*sensu* Rieppel, 1988a,b), of lizards to its constituent dermatocranial and splanchnocranial parts identifies the following elements: (1) non-snake lizards present an unfused postdentary arcade of bones (Figures 3.8b–d and 3.9a–d). The prearticular and articular may fuse into a single medial element reminiscent of a "compound bone" in snake lizards (see below), such that at the "intraramal joint" if present, or in the "intraramal region" if a "joint" is not obvious, the articular/prearticular crosses the plane of the joint surface. They also possess surangular, coronoid, angular, splenial, and dentary bones, as well as a Meckel's cartilage (part of the splanchnocranial skeleton). In non-snake lizards, the Meckel's cartilage is internal to the surangular-articular-prearticular, crosses the intraramal joint as it does in snake lizards, and is covered posteriorly by the splenial, but in some non-snake lizards may be exposed anteriorly along the medial face of the dentary; in other groups of non-snake lizards, the splenial fully covers the exposed portion of the canal anterior to the intraramal region and the remainder of Meckel's cartilage continues anteriorly within the dentary and is not exposed anterior to the splenial. (2) Snake lizards present a dentary, splenial, angular, coronoid, and compound bone (which is a fusion of the surangular, articular, and prearticular, and also crosses the plane of the joint) (Figures 3.10 through 3.12), plus the Meckel's cartilage that is also continuous across the joint and is enclosed within the compound bone and dentary. Meckel's cartilage is also exposed ventromedially in snake lizards, as the splenial does not fully enclose the anterior portion of the Meckelian canal along the dentary (Figures 3.10 through 3.12).

Regardless of whether we are examining snake lizards or non-snake lizards, all of these animals

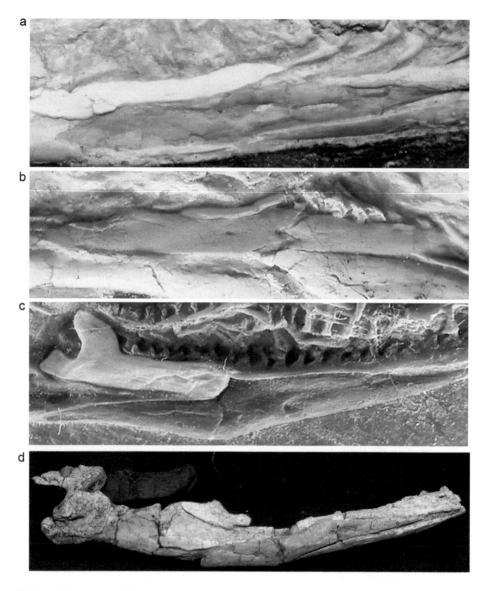

**Figure 3.11.** Medial view, mandibles/intramandibular joints of ancient snake lizards. (a) *Pachyrhachis problematicus* (HUJ-Pal EJ 3659), left mandible, close-up; (b) *Pachyrhachis problematicus* (HUJ-Pal EJ 3659), right mandible, close-up, image reversed for comparison; (c) *Haasiophis terrasanctus* (HUj-Pal EJ 695), close-up; (d) *Dinilysia patagonica* (MACN-RN-1013), entire mandible. (Red = splenial; purple = angular; green = coronoid.)

retain a continuous, flexible rod of cartilage, Meckel's cartilage, that crosses the plane of the joint, and a much less flexible compound bone, or prearticular/articular + prearticular, that also crosses the plane of the "joint" surface (Figures 3.8 through 3.12). It is true that the narrow diameter of the cartilaginous rod of Meckel is flexible within the mandible as a whole, but it is also true that functionally, it will prevent the intraramal/intramandibular joint from flexing too acutely, but also allows flexion at the bony contact surfaces that would otherwise risk dislocations or fractures. Before examining the specifics of modern and ancient snake lizard intraramal element morphologies and comparing them to those of typical modern and ancient lizards, it is relevant to explore briefly the anatomical nature of joints in general and the kind of "joint," if it is a joint, represented by the intramandibular/intraramal region of those lizards (snake and otherwise) that appear to possess one.

THE ORIGIN OF SNAKES

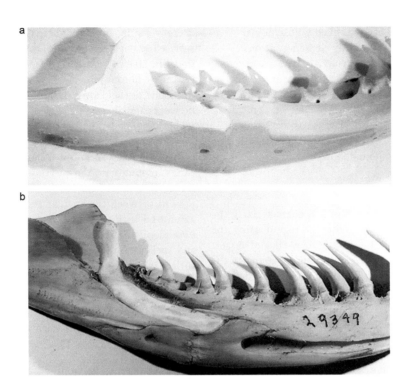

Figure 3.12. Medial view, mandibles/intramandibular joints of snake lizards. (a) *Cylindrophis ruffus* left mandible, close-up (AMNH R-85647); (b) *Eunectes murinus* (AMNH 29349), left mandible. (Red = splenial; purple = angular; green = coronoid.)

## What Is a "Joint?"

In the broadest of terms, a joint between two or more bony elements generally functions to permit or prevent movement at that joint surface. As joints function around either facilitating or restricting movement of one or more bones relative to each other, there are a rather large number of observable joint types that exist to meet specific functional requirements. There are three types of bony joints or articulations in the skeleton of vertebrates:

1. Fibrous joints, which are generally subdivided into three categories: (i) sutural joints, which are uniquely found only between skull bones, vary depending on the shape of the bony margins at the sutures (planar serrate, denticulate, squamous, limbous, and a schindylesis-shaped sutural surface); (ii) syndesmoses, where large fibrous ligaments connect two bones together; (iii) gomphoses, where ligaments (e.g., the periodontal ligament anchoring a tooth in its alveolus) fix a bony structure into a socket or space in another bony structure.

2. Cartilaginous joints, where two bones are held together by an intervening cartilage to form a point of articulation that is not necessarily mobile and can be subdivided into two joint types: (i) synchondroses (primary cartilaginous joints), where a plate of hyaline cartilage links two bony elements in a temporary but strong and immovable joint (e.g., cartilage plate between epiphyses and diaphysis of a growing long bone); (ii) symphyses (fibro-cartilaginous joints or secondary cartilage joints), where bony surfaces are covered by a hyaline cartilage sheath and the bones are connected throughout life by fibro-cartilage (e.g., pubic symphysis, intervertebral joints).

3. Synovial joints, or diarthroses, create a connection between bony elements where a fibrous joint capsule (outer layer articular capsule, inner layer synovial membrane) crosses the space between the bones and wraps continuously around the joint space, leaving a cavity, the synovium, between the two bony elements. This

synovial cavity is filled with synovial fluid, and the bony element portions exposed in the synovial cavity are covered with hyaline cartilage. There are seven broad categories of synovial joints: (i) plane joints; (ii) hinge joints; (iii) pivot joints; (iv) condyloid joints; (v) saddle joints; (vi) ball-and-socket joints; and (vii) compound joints.

## Typical Lizards

Rieppel and Zaher (2000a) characterized the typical lizard midmandibular region using the same taxa examined by McDowell and Bogert (1954)—*Varanus* sp. and *Lanthanotus borneensis*. For the purpose of brevity, I will summarize the detailed anatomy presented by these authors. In *Varanus*, and most other typical lizards, the elements of the mandible all heavily overlap each other, particularly on the medial face of the mandible. In particular, the splenial is a thin sheet of bone that overlaps a portion of the dentary dorsally and ventrally, and significantly overlaps an elongate and thin anterior process of the coronoid that itself is internal to dentary and superior to the Meckelian cartilage in the canal; the splenial also overlaps the elongate prearticular process of the articular bone as it too enters the dentary ventral to the Meckelian cartilage, and overlaps an anterior process of the angular, ventral to the Meckelian cartilage, that forms a floor to the Meckels canal ventrally. As noted by Rieppel and Zaher (2000a.9), there is an important difference between modern snake lizards and typical lizards concerning the exposure of Meckel's cartilage on the medial face of the surangular in the adductor fossa in modern non-snake lizards:

> In squamates other than snakes, Meckel's cartilage is exposed at the bottom of the adductor fossa for approximately half of its length, and fibers of the posterior adductor insert into it.

This feature in modern non-snake lizards, as compared to modern and fossil snake lizards, will be relevant to the anatomical features of the mandibles of fossil pythonomorphs (see below).

In lateral view, the dentary describes a low-angle "w-shape" where it overlaps the coronid, surangular, and angular. The complex anterior processes of these elements are not at all visible in

lateral view. In summary, there is no kinetic joint surface present between any of the elements of the midmandible in *Varanus*; all these joint surfaces in *Varanus* are fibrous joints where bone are relatively immobile; are held together by dense collagen fiber networks and Sharpey's fibers; and demonstrate a wide variety of denticulate, squamous, and planar bony contacts.

The cryptic varanoid *Lanthanotus borneensis* varies in comparison to *Varanus* and other typical lizards in terms of the kind of bony articulation present between the angular and splenial bones as viewed in medial and ventral aspect—these two elements appear to form a shallow ball and socket joint surface, or perhaps saddle joint surface where the splenial is recessed and receives a shallow anterior projection of the angular. Most notably, by comparison to almost all other typical lizards, *Lanthanotus* presents an extremely reduced splenial with few if any overlapping articulations as observed in *Varanus*. Otherwise, though the coronoid is less complex by comparison to *Varanus*, it still strongly overlaps the dentary laterally and medially, and the surangular and prearticular process of the articular extend into the Meckelian canal around and below the Meckel's cartilage (as was illustrated by Rieppel and Zaher [2000a]), the angular also extends an anterior process lateral to the significantly reduced splenial in order to form a floor for the Meckelian canal. By comparison to *Varanus*, *Lanthanotus* presents a joint surface between the angular and splenial that is potentially not limited to a fibrous joint system, but rather could be a symphysis-style cartilage joint (see above); despite the weak ball-and-socket-like morphology displayed at this joint, it has never been reported in either *Lanthanotus* or modern snake lizards that the angular-splenial articulation was synovial.

If one element in the mandibular set of typical lizards strongly resembles any element at all in modern and fossil snake lizards, it is most certainly the splenial, both in terms of size and the lack of overlapping articulations. The angular is also very similar to that of many modern snake lizards, including the socket or saddlelike face it presents to the splenial, and the lateral process it projects below the Meckel's cartilage (cf. *Lanthanotus* and *Cylindrophis* as illustrated by Rieppel and Zaher [2000a; figs. 3 and 6]). It is fair to conclude that there is no intramandibular joint (i.e., one or more bony joints permitting movement or flexion between elements at their joint surface) in modern

THE ORIGIN OF SNAKES

non-snake lizards, with *Lanthanotus* as the exception where a jointed articulation is present between the angular and splenial.

## Pythonomorph Lizards

Though Rieppel and Zaher (2000a) built on the detail provided by McDowell and Bogert (1954) regarding modern snake lizard mandibular element morphology, they did not illustrate comparable details of each mandibular element for even a single mosasauroid, presenting instead a single black-and-white photograph, in lateral view, of a poorly preserved mosasaur left mandible referred to the genus *Platecarpus* (Figures 3.8 and 3.9). In consideration of the rich details provided for eleven species of modern snake lizards (basal alethinophidians and basal macrostomatans), as well as *Varanus* and *Lanthanotus*, it is odd that in their characterization of a key taxon in the broad argument they were critiquing—that snake lizards are the sistertaxon of mosasauroid lizards in a global analysis of lizard phylogeny as hypothesized by Caldwell and Lee (1997), Lee and Caldwell (1998), and Lee et al. (1999a)— that they said little to nothing except to refer the reader to other works (e.g., Camp, 1942; Russell, 1967; Bell, 1997). In addition to noting the absence of a detailed consideration of the mosasaur mandible by Rieppel and Zaher (2000a), that study must also be criticized for the complete absence of any discussion of the mandibular anatomy of even more basal pythonomorphs. There are literally thousands of mosasaur specimens in collections around the world, along with dozens of basal mosasauroids and dolichosaurs for which materials and existing illustrations would have provided comparable empirical evidence to the anatomies they illustrated for modern snake lizards. My criticism here is pointed because Rieppel and Zaher's (2000a) study was a harsh critique of the pythonomorph-snake lizard sistergroup hypotheses of Caldwell and Lee (1997), Lee and Caldwell (1998), Caldwell (1999a,b), and Lee (1997a, 1998)—and as Popper (1959) argued, a replacement hypothesis must be superior to the one it claims to refute and thus to replace, and such refutations depend on superior data analysis, not just passionate criticism. A review of a dozen modern snake lizard mandibles does not replace the fact that the only data of consequence in such a discussion and refutation are those of the ancient fossil snakes and their contemporaneous proposed sistertaxa (pythonomorphs), as these animals and their osteology were the data linking these ancient lizards to ancient snake lizards and thus to modern snake lizards.

MOSASAURS

The pythonomorph condition of form in the midmandibular region was fist characterized by Cope (1869), who also coined the taxon name Pythonomorpha in that study, but notably did not create the term "intramandibular joint," but rather in his list of characters shared by his Pythonomorpha, described his character #7 thusly, "The subarticular and splenial elements of the mandible are connected by articular faces" (Cope, 1869:253) (Figure 3.8b–e). He went on to say, "The articulation of the splenial…is highly characteristic of the boaeform serpents of the genera *Loxocemus* and *Eryx*, though it does not occur in *Boa* proper, nor in many other serpents. This has allowed of considerable motion…" (Cope, 1869:255). For Cope (1869), there was no intramandibular joint, though there was a joint of sorts formed by the articular faces of the angular and splenial.

Examining a broader group of pythonomorphs than was available to Cope at the time demonstrates a remarkable uniformity between basal pythonomorphs such as dolichosaurs and higher mosasauroids such as aigialosaurs and mosasaurs (Caldwell, 2012) (see also Figures 3.8 and 3.9), and most certainly displays a wide variety of joint types between specific elements (see below), as well as an overall plane of flexion between the postdentary bones and the dentary-splenial. However, the degree of separation between elements superior to the angular and splenial, that is, the development of a plane of flexion above the angular-splenial joint, is highly developed in derived mosasaurs, but not in more basal pythonomorphs (*sensu* Lee, 1997a). A key feature of all pythonomorphs, linking them to modern and fossil snake lizards, that does not include the "intramandibular joint" around which they differ significantly from modern non-snake lizards, concerns the exposure of Meckel's cartilage in the mandibular fossa. Rieppel and Zaher (2000a) cited Russell (1967:52) but did not cite him with an actual quote:

> The Meckelian fossa is located under the posterior half of the coronoid suture and is walled dorsally and laterally by the surangular and medially by the prearticular.

In other words, though mosasaurs specifically, and pythonomorphs generally, do not possess a compound bone, they do possess a unique morphology of both the surangular and articular-prearticular bones compared to non-snake lizards that covers over the Meckelian cartilage, even though the elements do not fuse to form a compound bone—like the compound bone, the surangular and articular-prearticular of pythonomorphs form a tube around Meckel's cartilage anterior to the mandibular fossa. Thus, pythonomorphs are more similar to snake lizards regarding this feature than they are to other non-snake lizards. Similar to the compound bone and its extreme anterior penetration of the dentary bone (note: this does not reflect the autapomorphic nature of the lateral dentary excavation, but is merely a descriptor of the anterior extent of the compound bone of snake lizards), the articular-prearticular bone of mosasaurs is enlarged, elongate, and dorsoventrally deep compared to the prearticulars and prearticular processes of other modern non-snake lizards. The massive size of the mosasaur prearticular-articular is the principal suspension point for the dentary, as it inserts into the Meckelian canal of the dentary and is also a bony buttress to limiting dislocative forces within the intraramal zone of the mandible.

The massiveness of the articular-prearticular process is easily observed in *Platecarpus* and *Clidastes* (Figure 3.8b,c) and is essential to mandibular stability, as the coronoid either does not contact or only weakly contacts the dentary dorsally and sits perched on the surangular, forming a fibrous schindylosis joint atop the coronoid process. The dentary does not send a dorsal and posterior process across the intramandibular joint plane, but rather ends quite squarely at that plane as does the surangular. As noted, the articular-prearticular, along with the Meckelian cartilage it covers, crosses the joint plane, with the former extending a robust bony process quite deeply into the Meckelian canal of the dentary. The angular and splenial (see Lee, Bell and Caldwell, 1999a) form a symphysis-style joint and articulate in a weak ball-and-socket-like joint, similar to a rib head and its "socket" on a synapophysis. Among pythonomorphs where the angular-splenial joint surface can be observed, there is a great degree of subtle variation regarding the facet shapes on the splenial and angular (i.e., a simple concave

surface for receipt of an equally simple angular convexity, or concave and convex surfaces with matching grooves on the splenial and ridges on the angular of varying number and size). In broad terms, though, the splenial is recessed and receives the convex facet of the angular in a similar manner to the procoelic vertebral condition of cotyle and condyle. A joint surface is obvious between the angular and splenial, and when coupled with the lack of overlapping fibrous joint surfaces between the other elements, contributes to an intramandibular flexion constrained in its range of movement by the cartilaginous symphysis joint between the angular and splenial (likely with a hyaline cover on each bone, though it is not synovial) and by the fact the prearticular process of the articular extends into the Meckelian canal not as a thin sliver of bone, but as a large and substantive bony process alongside the Meckel's cartilage.

### DOLICHOSAURS, PONTOSAURS, AND AIGIALOSAURS

While "mosasaurs" are clearly derived pythonomorphs, there are numerous other more ancient and basal pythonomorphs that compare much better to ancient snake lizards, as they are less derived and specialized than the mosasaurs (Figure 3.9a–d). For the most part, these basal pythonomorph lizards are Cenomanian in age, though there is one taxon from the Lower Cretaceous of Japan (*Kaganaias hakusanensis* Evans et al., 2006), but the specimen does not possess mandibular material and so cannot be compared here. The exact relationships of *Tetrapodophis amplectus* remain controversial, though Lee et al. (2016) and Paparella et al. (2018) have presented data and hypotheses suggesting it is a basal pythonomorph close to dolichosaurs. It is potentially much older (Albian) than most dolichosaurs, pontosaurs and aigialosaurs, but it does have a partially preserved mandible in medial view, in particular, the splenial and posterior dentary (see Martill et al., 2015; Fig. S1A–C).

Excellent examples with well-preserved mandibular elements include *Coniasaurus crassidens* (Figure 3.9a) *Pontosaurus lesinensis* (Figure 3.9b), *Dolichosaurus longicollis* (Figure 3.9c), and *Komensaurus carrolli* (Figure 3.9d). While the preservation of the type and referred specimens of *Pachyrhachis problematicus* (Figure 3.10) make it difficult to compare the intraramal anatomy to those of

THE ORIGIN OF SNAKES

this selection of dolichosaurs and aigialosaurs (Figure 3.9), the anatomy of these marine lizards is virtually indistinguishable from that of the ancient snake lizard *Haasiophis terrasanctus* (Figure 3.11c).

All known basal pythonomorphs present a concave splenial that receives an anteriorly convex process of the angular (Figure 3.9a–d) on the medial edge, while the more lateral portions, where visible, indicate a concavity on the angular; there is a lateral anterior angular process as in *Lanthanotus*. The angulars, where visible, are all uniformly pierced by the posterior mylohyoid foramen, and the splenials present the anterior mylohyoid foramen (both foramina are for the mylohyoid nerve, an early branch of the Inferior Alveolar Nerve [$V_3$], Mandibular Branch of the Trigeminal Nerve). The intraramal region presents a clear angular-splenial joint surface, but there is no plane of flexion between the surangular-dentary and articular-prearticular as is observed in higher mosasaurs. Instead, the splenial presents a medially very tall posterior eminence above the anterior mylohyoid foramen that rises to meet the anteromedial process of the coronoid and covers the contact of the articular-prearticular and dentary. The articular-prearticular process still inserts deeply into the dentary, but it is not visible below the coronoid and superior to the splenial as it is in mosasaurs. The result of the absence of the plane of flexion in more basal pythonomorphs is that the dentary is quite firmly clasped in a fibrous joint with the coronoid and splenial medially and the surangular and coronoid laterally, but does not send a toothed posterior process to overlie the coronoid as seen in snake lizards (the anatomy observed in *Tetrapodophis amplectus* is identical to that of basal pythonomorphs, not snake lizards). The anatomy of the joint formed by the angular and splenial in basal pythonomorphs is indistinguishable from that of *Haasiophis* (Figures 3.9a–d and 3.11c), except that in *Haasiophis* the coronoid and angular meet posterior to the splenial, a condition common to ancient snake lizards and to Type I and Type II modern snake lizards (Figure 3.12).

### Ancient Snake Lizards: Type II
### *Pachyrhachis-Haasiophis*

Caldwell and Lee (1997) and Lee and Caldwell (1998) in their redescription of Haas's (1979, 1980a) specimens as a single taxon, *Pachyrhachis*

*problematicus*, naturally characterized the mandibular anatomy of both the type and referred specimens (Figure 3.10; Chapter 2), and found similarities in the anatomical features of the mandible, with modern snake lizards and with pythonomorph lizards. Careful study of the holotype skull of *Pachyrhachis* (Figure 3.10a) and the referred materials (Lee and Caldwell, 1998: figs. 5 and 6) led to the reconstruction of the lower mandible (Figure 3.10b), as given in Caldwell and Lee (1997). Today, the anatomy of the angular-splenial joint remains difficult to reconstruct, as the right mandible is crushed and split, and the angular and splenial have been turned outward and shattered (top of frame in Figures 3.10a and 3.11a,b), while the left mandible has been turned face down and the coronoid is pushed down onto and over both the angular and splenial (Figures 3.10a and 3.11a,b).

From restudy of the two specimens in 2011 and reference to my original notes and drawings from 1996, it is clear that, with respect to *Pachyrhachis*, some middle ground can be found between the interpretations of Caldwell and Lee (1997) (Figure 3.10b) and Rieppel and Zaher (2000a) (Figure 3.10c). Caldwell and Lee (1997) and Lee and Caldwell (1998) accurately illustrated the anatomy as preserved in the type and referred specimens and based their reconstruction on a composite of the preserved anatomies. Rieppel and Zaher's (2000a) illustrations of the preserved anatomy do not compare well to either Caldwell and Lee's (1997) renderings, nor to the actual specimen. On the other hand, I also cannot agree with the reconstruction produced by Caldwell and Lee (1997), particularly as concerns the articulatory surfaces of the angular and splenial and the positions of the anterior and posterior mylohyoid foramina. The right and left mandibles of the holotype are badly crushed and do not compare well to each other as a result (Figure 3.11a,b), but it appears as though the coronoids are in place, and both jaws mark the joint surface between the angular and splenial. Crushing and displacement make it difficult to compare the right and left sides, but it appears that the anterior mylohyoid foramen is just anterior to the articulation point on the left side, and at or in the articular surface on the right side. The two splenials (Figure 3.11a,b) are roughly the same length, as are the preserved remnants of the angulars. The right and left sides are difficult to compare, do not clearly support the reconstruction of Caldwell and Lee (1997), but do seem to suggest that there was a small posterior process of the splenial underlying the angular.

*Haasiophis terrasanctus*, on the other hand (Figure 3.11c), is perfectly preserved and shows that at least in this ancient Cenomanian-aged snake lizard (and in medial view) the coronoid was elongate, overlapped by the dentary, and contacted the splenial anteriorly and the angular dorsally, and that the compound bone was obscured medially by the coronoid-angular contact. Most importantly, the splenial is clearly concave medially, into which concavity fits an anterior convexity of the angular (though it is not clear how large that convexity is). The morphology of that joint surface moving laterally is not known. The anterior mylohyoid foramen opens some distance anterior to the angular-splenial joint surface, though the posterior mylohyoid foramen in the angular is slightly closer to the articulation. Again, *Haasiophis* is hard to compare to *Pachyrhachis*, but is very similar to the mandibular anatomy of modern snake lizards such as *Cylindrophis* and *Eunectes* (Figure 3.12a,b) (Type I and Type II skulls, respectively) and to *Dinilysia patagonica* (Type I *Najash-Dinilysia* skulls)—that is, a large coronoid underlying the tooth posterior process of the dentary, contacting the angular and splenial and overhanging the joint line of the angular and splenial (there are some derived features of Type I and II skulls that differ from *Haasiophis*, and these will be discussed below).

### Ancient Snake Lizards: Type I Dinilysia-Najash

The medial face of the mandible has not yet been described for the new skull materials of *Najash*, but there are numerous specimens and descriptions in the literature for *Dinilysia patagonica* (Estes et al., 1970; Caldwell and Albino, 2002; Caldwell and Calvo, 2008; Zaher and Scanferla, 2012) (Figure 3.11d). The angular and splenial form a clear and simple joint surface where the splenial is slightly concave and the angular is slightly convex. The posterior mylohyoid foramen of the angular is relatively close to the joint surface, while the anterior mylohyoid foramen is some distance forward on the splenial. The coronoid is large and unusually shaped in its dorsal and lateral faces, but at its contact with the angular above the angular/splenial joint, it contacts the splenial anterior to that joint surface. The coronoid is incomplete medially in the figured specimen (Figure 3.11d), but the facet continues forward and, as figured by Caldwell and Calvo

(2008), is overlapped by the posterior toothed process of the dentary. The compound bone is pinched out posterior to the angular and so does not contact the splenial on the medial face (it is overlapped by the splenial internally).

### Type I and Type II Modern Snake Lizards

As exemplars of Type I and Type II modern snake lizard skulls and their mandibles, *Cylindrophis ruffus* and *Eunectes murinus* meet predictions arising from the anatomy of *Haasiophis* and *Dinilysia*: they both possess a toothed posterior process of the dentary overlying the compound bone and coronoid and extending to the base of the coronoid process; the coronoid possesses a long anterior process that contacts the angular and splenial, thus excluding the compound bone in medial view from the joint plane formed by the angular and splenial; the angular and splenial, in medial view, present a relatively straight and vertical suture; the articular face-suture of the angular and splenial shows some degree of interdigitation and interfingering, particularly in *Eunectes* and presumably in older versus younger individuals of both taxa (Figure 3.12a,b).

The anatomy of extant basal alethinophidians and basal macrostomatans was a key data set for Rieppel and Zaher (2000a) in their critique of the joint plane of the angular and splenial as presented in Lee, Bell, and Caldwell (1999a). Rieppel and Zaher (2000a: figs. 6, 8, 10) presented as evidence of bone contact and complexity between the angular and splenial, histological sections from the intraramal region (i.e., *Anilius*, *Cylindrophis*, *Plecturus*, etc.) along with drawings of all elements in medial or lateral view. Histological sections unfortunately convey little to no information on the overall configuration of a joint surface, as each histological slice is roughly 5 microns or 0.005 mm in thickness. Subtleties of joint topography, that is, undulations on the joint face, can appear as vacuities or an absence of bone, as it would require 200 sections at 0.005 mm each to visualize in three dimensions a joint surface topology comprising only 1 mm of three dimensional space. The argument presented by Rieppel and Zaher (2000a) is that joint topology is not as simple as the medial or ventral aspect views of gross morphology would suggest that there are internal tabs and processes in snake lizards that are not present in mosasaurs, and that this means that the modern snake lizard angular-splenial joint is more complex than that

THE ORIGIN OF SNAKES

of mosasaurs. To this point, I provide no rebuttal, but rather concur with weak qualification—the medial aspect of the angular-splenial joint surface is slightly less complex in mosasaurs as compared to Type I and Type II modern snake lizards as it lacks observable internal processes, but is still a unique and complex joint surface.

There is no clear anatomical argument to be made that the jointed surface between the angular and splenial creates a mobile ball-and-socket joint in Type I and Type II snake lizards, even though this language is used quite consistently in the literature. For example, Character 377 from Gauthier et al. (2012:213), describes the joint in *Cylindrophis* (Type I skull) and *Calabaria* (Type II skull) as a ball and socket when it is not (I am in favor, however, of Gauthier et al.'s [2012] avoidance of coding an intramandibular joint present/absent, though it is discussed as such in their text):

> 377. Splenial-angular articulation: (0) splenial overlaps angular…; (1) with ball on splenial (below level of posterior mylohyoid foramen) fitting into socket on angular (*Cylindrophis ruffus*…); (2) with ball on angular fitting into socket on splenial (*Plotosaurus bennisoni*…); (3) flat, abutting joint (*Calabaria reinhardtii*…).

Though the trend in all modern snake lizards is toward an increasing reduction of the splenial and angular (see below for Type III and IV snake lizards), the intraramal region would be highly unstable if there were not multiple points of resistance to unconstrained outward flexion of the mandible during prey ingestion. Cundall (1995: fig. 5) presented a highly detailed analysis of the functional morphology of the mandible during feeding and prey ingestion in *Cylindrophis* *ruffus*. He found that in the intraramal region around what he referred to as the intramandibular joint, the mandible flexed at the joint to a maximum of a 45-degree angle (hyperextension 5%, flexion 40%)—that flexion was supported by the angular-splenial joint medially, and the lateral joint formed by the dentary-compound bone. The complex surfaces and sutures created by overlapping coronoid, compound bone, and dentary elements, and the tiny processes and interfingered suture lines of the splenial and angular, seek to resist rupture along the plane of flexion in these two joints. The complexities identified by Rieppel and Zaher (2000a) do not delegitimize the basic

primary homolog present in pythonomorphs and all snake lizards; rather, these complexities are the synapomorphies of subsequent adaptation and evolution.

The same is clearly very true in basal and derived pythonomorphs where the prearticular-articular evolves massive size in the latter and in both crosses the intramandibular plane of flexion, with complex articulations with the coronoid and dentary, and in support of Meckel's cartilage, in order to stabilize the joint. The same solution is clearly present in all snake lizards, with specializations and added complexities at the intraramal region in Type I and Type II modern snakes.

All that notwithstanding, the morphologies of the angular and splenial, and their particular joint surfaces, regardless of supernumerary processes and convexities versus concavities, cannot be viewed as anything but primary homologs between pythonomorphs, ancient snake lizards, and all modern snake lizards: the same bones, in the same position in all three groups, with similar morphologies, pierced by the same nerve foramina, interacting with the similar structures (an exception is the compound bone verus prearticular-articular) pass all possible Tests of Similarity. The observations of topology and connectivity must be respected, because if not, the fate of all primary homology statements becomes suspect, as the power of these tests would be rendered meaningless for nearly all primary homology statements for all characters in all data sets; for example, the single frontal of most lizards would not be the homolog of the paired frontals of varanid lizards because varanid lizards possess a suture (the list is endless where there are slight variations in the microdetails of anatomy).

### Type III and Type IV Modern Snake Lizards

The mandible of modern snake lizards grouped into Type III and IV skulls is composed of a surprisingly invariant number of elements (see Cundall and Irish, 2008): compound bone, coronoid (only in scolecophidians), splenial, angular, and dentary. The most conspicuous difference between Type III and IV snake lizards and more typical lizards is that the postdentary elements of the articular, prearticular, and surangular all fuse together to form a compound bone, the coronoid is usually lost completely, and the angular and splenial both become extremely small and often fuse together (the intraramal joint present in Type I and II skulls is thus lost). Size variation and allometric scaling of all of these elements varies widely as well, though the general trend is for the compound bone to become extremely elongated, the dentary becomes very small, and the remaining elements are lost or highly reduced in size as well. Another very obvious difference with non-snake lizards is that on the lateral surface of the dentary bone, the compound bone appears to deeply invade the dentary and is exposed laterally in a deep, shallow v-shaped fossa on the dentary (common to Type I and II as well). I say appears, because frame of reference is everything in describing a morphology such as this joint surface—an alternative descriptor is not that the compound bone has invaded the dentary, but rather that the dentary has developed a dorsal and a ventral process that extends posteriorly to clasp the medial face of the compound bone, the dorsal process of which in snake lizards is almost always toothed (exceptions of course are noted in scolecophidians where the lateral excavation of the dentary is absent, though the dorsal process is present, and the dentary is not toothed over the compound bone). Regardless of the frame of reference and the language used to describe this condition and its variants in modern snake lizards, it is not observed in typical lizards.

Among modern snake lizards, viperids, elapids, and typhlopids show the most reduced condition of the mandible (e.g., List, 1966; Kley, 2001; Cundall and Irish, 2008:521), where even the angular and splenial may be fused into a single element, the dentary, though toothed (toothless in typhlopids), is very small, and the compound bone has become extremely elongated. Not too surprisingly, two-thirds of scolecophidians show similar mandibular morphologies (though not in leptotyphlopids, which are unique among snake lizards regarding mandibular morphology and intramandibular kinesis [Kley and Brainerd, 1996; 1999; Kley, 2001]—dealt with later in this chapter) (Figure 1.5a–c). Anomalepidids and typhlopids present a compound bone that is extremely elongated, a coronoid that is enormous, and splenials and angulars that are extremely reduced, as is the dentary, which bears no teeth at all. In contrast, leptotyphlopids have a comparatively robust dentary, though it is short and bears only four to five teeth, the splenial is a chip of bone, but the angular is larger; the coronoid is present and complex in its rugosities and muscle

attachments; and the compound bone is extremely short and simplified. For leptotyphlopids, the functional role of the elongate compound bone in anomalepidids and typhlopids is assumed by an extremely elongate and virtually horizontally oriented quadrate bone. Anomalepidis and typhlophids, similar to vipers and elapids, extend their maxillary fangs in a bite phase similarly to each other (see above discussion of the evolution and function of the CID musculature, Boltt and Ewer [1964], and Haas [1973]), and so evolution has emphasized the morphology and function of the mandible to facilitate maxillary function (mandibles need to assist with ingestion and maximizing gape size). In leptotyphlophids, which feed and attack/capture prey using mandibular raking (Kley and Brainerd, 1999; Kley, 2001), the maxilla is toothless and relatively immobile, and the functional and morphological emphasis is on kineticism of the mandibular complex, including the quadrate. These highly specialized Type IV skull types are most clearly extreme specializations of the modern snake lizard skull and would not be predicted in any hypothetical framework to display characteristics of the ancient snake lizard skull. The same holds true for Type III—Colubroid skull types, where the same trends observed in Type IV skulls are observed, except that the relative length of the compound bone versus the dentary is closer to 1:1. The compound bone is large, but so is the dentary (and it can bear a large number of teeth and is a key bony structure in prey capture and prey ingestion strategies), and the angular and splenial are relatively large and form a distinct, medially positioned joint surface with each other that is not so obvious in Type IV skull type mandibles due to the extreme reduction in size of the angulars and splenials or, in fact, their fusion.

With respect to the intramandibular joint, the mandibles of Type III and IV modern snake lizard skulls (except typhlopids and presumably anomalepidids, which are secondarily, not primitively, rigid) demonstrate a high degree of flexion within the contact "zone" between the compound bone and the dentary-splenial-angular, but this does not reflect specific mobility at a synovial joint system in particular. Because of the significantly large excavation on the lateral surface of the dentary, outward pressure against the medial face of the mandible (e.g., caused by the ingestion of large prey in the mouth), the compound bone-dentary contact no doubt can flex elastically; as the

THE ORIGIN OF SNAKES

splenial-angular articulation is similarly simplified (i.e., neither sutural nor complex) and sits at the midpoint of the compound bone-dentary contact, it too likely can flex laterally to facilitate general flexion of the entire mandible under ingestion pressures. Whether Type III and IV snake lizards have a well-developed intramandibular joint is clearly not consequential to the evolutionary origins of this intramandibular kinesis and flexion, but rather is yet another example of the result of extreme specializations of the first intramandibular joint system that evolved hundreds of millions of years ago.

## Summary

Without mincing words, the "intramandibular joint" does not exist in modern snake lizards or ancient snake lizards, nor in modern or ancient non-snake lizards—Question One is answered. Therefore, primary homolog statements resulting in character codings seeking to discover homologies for such a "joint" should be abandoned, that is, joint present versus absent. What is commonly present, and similar, in pythonomorphs and all snake lizards currently known is a symphysis-type joint surface of variable morphology between the angular and splenial bones; in all snake lizards known, with the exception being typhlopid scolecophidians, there is a dentary-compound bone "joint" (see Cundall, 1995). All other intraramal elements and their articulations appear, functionally, to resist dislocation under maximum stress for the mandible to flex.

Most recently, as a means of critiquing hypotheses of sistergroup relationship between mosasauroids and snake lizards, Rieppel and Zaher (2000a) focused on the observation that in modern snake lizards, the angular and splenial surfaces were reversed in the convexity-concavity in some modern snake lizards as compared to ancient mosasaur lizards, and that the mosasaurian lizard form was more simple than that of modern snake lizards. While neither of these statements is completely incorrect, it does not negate the fact that a similar joint between two homologous bones exists in the same position between potential sistertaxa. The question is not how a mosasaur lizard might compare in such a detail to the splenial-angular joints of modern snake lizards that are 85 million years younger, but rather what anatomies are observed in such details between ancient snake lizards such as *Haasiophis terrasanctus* or *Najash rionegrina* and a contemporaneous dolichosaurian lizard such as *Dolichosaurus longicollis* (both are approximately 100 million years old). In such comparisons, we find the missing elements of the Rieppel and Zaher (2000a) data set—the empirical observation is that there are no observable differences (in both the snake lizard and dolichosaurian lizard, the splenial is concave and the angular convex). This is far more than merely germane to the question, it is in fact strongly suggestive of a primary homolog concept for the joint surface as displayed by equivalent primary homologs of the splenial and angular—the arbiter must be, in the end, the Test of Congruence. However, that test must be undertaken with an appropriate data set and not one biased prior to the assessment of primary homologs.

In addition, the intramandibular plane of flexion present in higher pythonomorphs is absent in *Dolichosaurus* and other basal pythonomorphs, as it is in *Haasiophis* where the toothed posterior process of the dentary firmly clasps both a large coronoid and compound bone. That subsequent snake lizard evolutionary adaptation and radiation diverges from that of higher pythonomorphs is no surprise. Long before there were mosasaurs, pythonomorphs were diverging away from the condition shared with an as yet unknown ancient snake lizard. What I find in the summary analysis of the character debate on the splenial-angular joint is a false dichotomy created through the comparison of non-comparable forms.

With respect to the remaining intraramal elements in snake lizards and pythonomorph lizards, it is clear that there is no single joint formed, though the compound bone—dentary "joint" (e.g., Cundall, 1995) is clearly able to facilitate extreme flexion of the intraramal zone laterally during prey ingestion, as so is the articular-prearticular of pythonomorph lizards (Lee et al. [1999a,b] noted ≤30 degrees of excursion). Again, the trend in both clades is to use the same postdentary mandibular unit for exactly the same purpose—intraramal mandibular stability—the principal difference is that higher pythonomorph evolution and ontogeny did not lead to the fusion of the postdentary elements into a compound bone, but it came close as it diverged away from the condition of all other typical lizards and evolved further around a fused prearticular-articular.

As a characteristic of ancient snake lizards, it is abundantly clear that Late Jurassic and Early Cretaceous snake lizards possessed medial excavations of the dentary (Chapter 2—*Portugalophis* and *Diablophis*), likely for a compound bonelike process (see the probable compound bone of *Diablophis* [Caldwell et al., 2015]), and also possessed posterior dentary processes that were toothed (the maxilla was also suborbital and toothed, thus corresponding to the posterior toothed process of the dentary— *Parviraptor* and *Portugalophis*). In the absence of evidence, but by logical extension from the available evidence (anatomical evidence from *Parviraptor*, *Diablophis*, and *Portugalophis*, and evidence from the sistergroup to snake lizards, the pythonomorphs), it seems likely that the splenial and angular would have formed a symphyseal joint along the medial face of the mandible. Substantial flexion may have been possible, though it is not clear if these oldest ancient snake lizards could flex as substantially as can a Type I Aniliod skull (Cundall, 1995) or by extension, as much as *Dinilysia*, *Haasiophis*, *Najash*, or even *Coniasaurus*. Contrary to the arguments of Rieppel and Zaher (2000a), there is anatomical similarity between the angulars and splenials of snake lizards and pythonomorph lizards—to such an extent, in fact, that it easily passes the Test of Similarity. I also do not agree that the atomization of convexity on one element and concavity on the other, no matter which way it goes (see Gauthier et al., 2012: Char 377), is consequential to the diagnosis of primary homolog concepts. As discussed, it is not surprising that modern snake lizards do not compare well to mosasaurs in every detail, but even here, the similarities are recognizable, unless of course the historical burden of the burrowing origin of snakes, and the amphisbaenian-scolecophidian motif, dominates the empirical observation of anatomies in the first place.

I also must point out that the transformational narrative with mosasaur lizards as intermediate to snake lizards, as presented by Lee et al. (1999a), does not match the increasing complexity presented by the fossil record—mosasaur lizards are not intermediate to snake lizards, but rather are highly derived within basal pythonomorphs. As described above, basal pythonomorphs are much more similar morphologically to such snake lizards as *Haasiophis* and *Dinilysia*, with some similarities, particularly in terms of intraramal flexion, to Type I Anilioid skulls. This means, of course, that the intraclade evolution of mandibular flexibility is

parallel, specialized, and reaches its acme in Type III and Type IV snake lizards. While scolecophidians as Type IV skulls are clearly not basal snakes in terms of the "intramandibular joint," the neat progression implied by Lee et al. (1999a) from mosasaurs to *Pachyrhachis* is now exceedingly more complex with the inclusion of Gondwanan snakes going back in time to the Valanginian, with the introduction of the Late Jurassic–Lower Cretaceous "*Parviraptor*-Group," and including the new data on basal pythonomorphs (Figure 3.9). These ancient snake lizards are significantly older (nearly 80 million years older) than the earliest mosasaurs and 75 million years older than *Pachyrhachis*, *Haasiophis*, *Eupodophis*, *Najash*, and contemporaneous pythonomorphs. Unfortunately, the Lower Cretaceous Japanese dolichosaur *Kaganais* does not preserve any information on the intramandibular region, and *Tetrapodophis amplectus* remains in limbo and as of yet has not benefited from a complete recharacterization of the error-riddled original comparative anatomy (Caldwell et al., in prep.). Still, with each new form discovered in the fossil record, these new data simply rattle the old questions harder, as the story gets more complex and demonstrates the presence of specializations within sistertaxa as old questions get answered and a dozen new questions arise.

### Intermandibular Joint: Dentary

While there has been considerable debate on the anatomy of the "intramandibular joint" and its evolution and role in snake and mosasaurian lizard gape evolution, there has been little or no debate on the intermandibular joint, or, more accurately described, the interdentary joint, as no other elements are present at the joint except the right and left dentaries (Figure 3.13a–h). In straightforward anatomical terms, the tips of the dentaries in most lizards are solidly ankylosed to each other, often at a joint surface between the two dentaries that is quite complex. In snake lizards, the general trend over time and through complexity (primitive to derived) is for the joint to evolve from a tight fibrous connection (e.g., *Cylindrophis* [Cundall, 1995]) with little capacity to spread at the tips, to the extremes of mandibular tip separation and distension seen in higher snakes (Cundall and Irish, 2008). In all modern snake lizards, the ankylosis or syndemosis of the dentary tips of typical lizards is lost, the dentary tips can separate from each other even

THE ORIGIN OF SNAKES

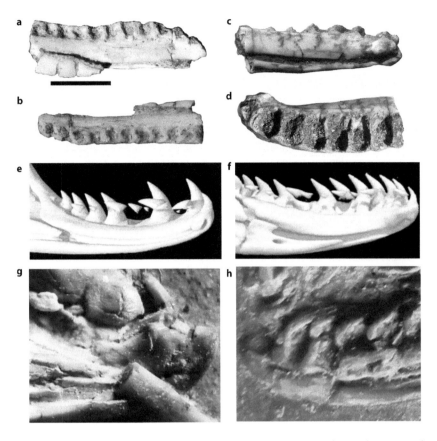

Figure 3.13. Intermandibular joint morphology in Type I Ancient Snake Lizards assigned to *Najash rionegrina* and Type I Anilioid Snake Lizards. (a–b) Left dentary cf. *Najash rionegrina* MPCA 390. (a) Medial view; (b) occlusal view. (c–d) Left dentary cf. *Najash rionegrina* MPCA 380. (c) Medial view; (d) occlusal view; (e) *Anilius scytale*, left dentary medial view (Digimorph.com); (f) *Cylindrophis ruffus*, left dentary, medial view (Digimorph.com); (g) left dentary tip, paratype, *Pachyrhachis problematicus* HUJ-PAL 3775; (h) right dentary tip, holotype *Haasiophis terrasanctus* HUJ-PAL 659.

if the distance is comparatively small, and this variation in gape size appears to be constrained around soft tissue flexion and function (Cundall, 1995). This means, of course, that the origins of the interdentary joint, like so many other anatomical features of snake lizards, cannot be understood by reference to modern forms. Unfortunately, in fossil snake lizards, the answers are not readily interpreted from the known specimens, because of the opposite problem—the soft tissues are absent. Despite claims to the contrary (e.g., Zaher et al., 2009), the bony data presented by modern snake lizards, and the bony data available from fossil snake lizards, would suggest that the known fossil snake lizards are no different than the modern forms, even if all of them are *Cylindrophis*-like or *Eunectes*-like in their anatomy (as I will show below, no known Mesozoic fossil snake lizard presents an anatomy similar to that of colubroid snake lizard). Presenting arguments for being more typically lizardlike, as opposed to snake

lizard–like, means that questions must be asked of the data being presented.

## Morphology of the Dentary Tips

As has been recognized for some time in mosasaurian and pythonomorph lizards, the right and left dentaries do not form a symphysial joint at their anterior tips (Russell, 1967; Lee et al., 1999a,b). The dentary tips are basically smooth and rounded and show no evidence of even a slight roughening, let alone a facet or sutures for a joint surface; thus, there is no bony ankylosis or syndesmosis. The dentary tips were no doubt connected to each other by a fibrous ligament pad and associated tissues, but the proportions and elasticity of these tissues are unknown (for a comparison between various modern snake lizards, see Young [1998]). The absence of a bony ankylosis/syndesmosis is a shared condition between pythonomorphs and all modern snake lizards (but not necessarily a synapomorphy). However, despite the absence of a bony joint between the dentary tips, generalizing to all pythonomorphs as being similar to all snake lizards is incorrect. The comparable condition is between basal pythonomorphs, such as *Pontosaurus kornhuberi* (Pierce and Caldwell, 2004) (Figure 3.9b), where the dentary tips present a morphology that is nearly identical to that of *Najash* (Figure 3.13a–d) or *Anilius* and *Cylindrophis* (Figure 3.13e,f) (i.e., the Meckelian canal extends to the tip of the dentary and separates the facet into dorsal and ventral faces). In contrast, the more derived pythonomorphs, such as the mosasaurine mosasaurid *Plotosaurus bennisonni* (see LeBlanc et al., 2013: fig. 2c,d), present a broad contact between the dentaries anterior to the tips, but the tips are straight and without facets or roughening, and diverge from the parts of the dentaries that are broadly in contact with each other.

The details of dentary tip morphologies in snake lizards, modern and ancient, fall into two broad categories linked to the four identified skull categories (Types I–IV). An important point to remember, however, that overprints the interdentary "joint" is that with increasing age through ontogeny (Zaher et al., 2009; Palci et al., 2013), almost all modern snake dentaries develop a noticeable curvature of the anterior portion of the dentary; this curvature effectively "points" the tips toward each other, heightening the effect of an anterior tip-medial process. Speculation on this point suggests that curvature through ontogeny is likely as much epigenetic as it is genetic and results, like points of muscular insertion and origin, from stress forces resulting from feeding mechanics. Type I Aniliod and Type II Booid-Pythonoid skulls all present a dentary tip morphology where there is a Meckelian canal bisected facet of varying size and proportions. Type III–Colubroid (excluding viperids and elapids) and Type IV–Viperid-Elapid-Scolecophidian skulls possess no facet structures at all, though the dentary tips are usually curved and often present a small, medially directed anterior tip.

Among ancient fossil snake lizards, only a handful of forms possess complete dentaries, permitting assessment of this morphology, and thus allow the development of hypotheses on the evolution of the interdentary joint. The Late Jurassic–Early Cretaceous snake lizards *Eophis underwoodi* (see Figure 2.1a), *Diablophis gilmorei* (see Figure 2.1d), and *Portugalophis lignites* (see Figure 2.1h,i) all present straight dentaries with small anterior tip processes, but no obvious sutural facet for

a syndesmosis. Unfortunately, though, only the oldest of the three, *Eophis underwoodi*, presents a perfectly preserved tip with no curvature and no evidence of facet or symphysial surface. For Type I *Najash–Dinilysia* skulls, only *Najash* presents data on this morphology, as the dentary tips for other fossil snake lizards assigned to this category are not preserved (Gómez [2011] described an anterior dentary tip for an Upper Campanian–Lower Maastrichtian snake lizard from Patagonia that, like *Najash*, appears to preserve the anterior tip, but there was nothing else recovered of this animal and it remains in open nomenclature). From observations of specimens assigned to *Najash*, two morphologies are present (Figure 3.13a–d)— one straight and without an anterior facet, and the second curved and in possession of a facet bisected by the Meckelian canal. Type II *Pachyrhachis–Haasiophis* skulls present dentaries that are straight throughout their length (Figure 3.13g,h) and, where observable, similar to *Najash* and *Cylindrophis* with respect to the Meckelian canal bisection of the anterior facet, for example, *Pachyrhachis* (Caldwell and Lee, 1997; Lee and Caldwell, 1998) (Figures 2.4a, 3.4b,c, and 3.13g), *Eupodophis* (Rage and Escuillié, 2000; Rieppel and Head, 2004) (Figures 2.6 and 3.4d), and *Haasiophis* (Rieppel et al., 2003a; Palci et al., 2013) (Figures 2.5c,d and 3.13h).

### Details of Najash rionegrina

As *Najash* has been argued to be a fossil snake lizard presenting evidence of a more typical lizard intermandibular joint (Apesteguía and Zaher, 2006; Zaher et al., 2009) (Figure 3.13a–d) beyond the recognized ligament in modern snakes (e.g., *Cylindrophis* as per Cundall, 1995), it bears further discussion here. This is in part because *Najash* is a contemporary of marine snake lizards not considered by Zaher et al. (2009) to present a *Najash*-like symphyseal "joint" (e.g., *Pachyrhachis*, *Eupodophis*, *Haasiophis*, and *Pachyophis*). Zaher et al. (2009: figs. 2a–h; 6a–d) described in detail and illustrated the holotype dentary and an isolated referred dentary of *Najash rionegrina*, presenting the argument that the preserved anatomy indicated the presence of a "symphysis" and stated the following (Zaher et al., 2009:811):

> It can be concluded that *Najash* had a tight contact, comparable with the one present in extant lizards with limited mobility between their dentaries as a result of strong ligaments

being present instead of a bony symphysis (e.g. *Varanus griseus*… and *Xantusia vigilis*).

Zaher et al. (2009) argued that the symphysis was ligamentous but very tight, as opposed to a bony syndesmosis or the presumably loose ligamentous connection in higher snake lizards and was similar to that of extant lizards. However, even a complex syndesmosis (see above details on joints presented in the discussion of the intramandibular joint) involves collagen fibers crossing the sutural space between elements regardless of the complexity or simplicity of the joint surface. While Palci et al. (2013) countered the arguments presented on the holotype construction including MPCA 390 (Figure 3.13a,b), they did not discuss or counter the suggestion by Zaher et al. (2009) that the symphysial facet of *Najash* represented a similar intermandibular joint to those of extant lizards, in particular with no discussion of pythonomorph lizard morphologies.

While I disagree with Zaher et al. (2009) and consider the evidence presented by such taxa as *Pachyrhachis problematicus* (Figure 3.13g) to present the same morphology as *Najash rionegrina* (Figure 3.13c) and *Anilius* and *Cylindrophis* (Figure 3.13e,f), I also have a general concern about extrapolating from one fossil snake lizard, in this case a single isolated anterior dentary fragment of a referred specimen, to similarities or dissimilarities with "extant lizards" in an attempt to distance *Najash* from all alethinophidians. It is common to generalize in order to find support for arguments, but in fact such generalizations are dangerously inductive and subject to immediate falsification with one counterexample—as there are nearly 10,000 species of extant lizards (typical lizards and snake lizards), it is in fact unlikely that there is not a counterexample. There are thousands of counterexamples among lizards on this morphology, revealing significant morphological disparity within lizards regarding the symphysial region (cf. Figure 3.13 with Figure 3.18 in the tooth attachment section of this chapter, or see Digimorph.com).

An alternative and more specific interpretation to Zaher et al. (2009) is that *Najash* and almost all snake lizards present an intermandibular "joint" morphology that is broadly shared in its specific and general characteristics by all anguimorph lizards, including basal pythonomorph lizards such as *Pontosaurus kornhuberi* (Figure 3.9c). Again, anatomical details are critical to such comparative interpretations, and from among the available

materials, there are two morphs present in *Najash* (Figure 3.13a,b versus 3.13c,d). Zaher et al. (2009) argued these morphs represented ontogenetic stages, but in the absence of ontogenetic information corroborating such a conclusion, the two morphs must be accepted for what they are—morphological variants—until proven otherwise. MPCA 390 presents a straight, gracile dentary with a rounded tip showing no evidence of a medial facet at the anterior tip. MPCA 380 presents a medially curved dentary with a distinct dorsal tubercle and a smaller ventral eminence, bisected by the anteriormost portion of the Meckelian canal. The condition in MPCA 390 resembles higher snake lizards such as colubroids (in very broad terms), but certainly not basal alethinophidians such as *Eunectes* (Figure 3.18b) or other boiines or pythonines where there is a pronounced medial tubercle at the tip of the dentary, the bulk of which is all dorsal to the anteriormost portion of the Meckelian canal. The most comparable modern snake lizard morphologies are those of the Type I Anilioids, *Anilius scytale* (Figure 3.13e) and *Cylindrophis ruffus* (Figure 3.13f). These two modern snake lizards present nearly identical morphologies to that of MPCA 380, where the process or eminence dorsal to the Meckelian canal is enlarged and projects medially, while the ventral eminence is smooth and reduced. The morphology of this region is similar to that of anguimorphs, such as *Pontosaurus kornhuberi* (Figure 3.9c) and the modern varanoid *Varanus exanthematicus* (Figure 3.18c). Interestingly, the Meckelian canals of *Pachyrhachis* (Figure 3.13g) and *Anilius* (Figure 3.13e) show the same dorsal curvature or "twist" as is observed in *Varanus*, thus positioning the anteriormost portion of the Meckelian canal vertically and the dorsal and ventral facets for the symphysis posteriorly and anteriorly, respectively. *Najash* (MPCA 380) (Figure 3.13c) presents a morphology more similar to that of *Cylindrophis* (Figure 3.13f).

### Implications for the Ancestral Snake Lizard

Zaher et al. (2009) presented their arguments on the nature of the symphysis to position *Najash* as the most primitive or basalmost snake, by comparing it to extant lizards but not to pythonomorph lizards. Among modern snake lizards, cf. *Najash* MPCA 380 compares well to *Cylindrophis* and *Anilius* (Figure 3.13), but is not similar to the condition observed in other fossil snake lizards such as *Pachyrhachis*, *Eupodophis*, and *Haasiophis*, which seem more similar to the condition observed in MPCA 390, cf. *Najash*, that is, the absence of any kind of eminence and/or curvature. Unfortunately, the dentary tip remains elusive for *Dinilysia*, *Sanajeh*, and *Menarana*, and so it cannot be confirmed in other Mesozoic Gondwanan snake lizards potentially sharing clade membership with *Najash*. Interestingly, the dentary tips of *Pachyrhachis*, *Eupodophis*, and *Haasiophis* are not similar to the pronounced processes observed in modern basal macrostomatan snake lizards (Figure 3.18b), but more similar to the effaced rounded tips of both pythonomorph lizards and higher snake lizards within the colubroid radiation.

What do all of the data suggest for the anatomy of the ancestral snake lizard? If *Eophis underwoodi* provides any insight on such a question, I would argue that the jaws were straight, not curved, and that in early ancient snake lizards, the symphysis was already lost. *Cylindrophis* (Cundall, 1995) presents anatomical features that make it clear that a joint can be lost, that the dentary tips can separate, and that soft tissue anatomies can still significantly restrict mobility at the joint. However, linking *Eophis*, *Portugalophis*, and the other known Upper Jurassic and Lower Cretaceous snake lizards to *Najash*, then to *Anilius* and *Cylindrophis*, and from there to modern higher snake lizards, cannot and should not be done in a straight line. There is no linear scala naturae to this transformation of morphology. In fact, the evidence demonstrates the contrary—it is a nexus of transformations with clade characteristics and convergent, or even parallel, evolutionary innovations and most strongly pedomorphic evolution characterizing the Type III and IV skull categories. A simple morphology such as the interdentary joint reducing itself from a bony syndesmosis to a non-joint linking dentary tips only by soft tissues is nearly as simple a system as can be profiled in this chapter. It is thus possible to observe the simple origins of straight dentary with a reduced, flattened contact at the tips (*Eophis*), to a "clade" of Gondwanan snake lizards inclusive of *Najash* and *Dinilysia*, through to *Anilius* and *Cylindrophis* as the few modern forms, and the Type II to Type IV categories, ancient and modern, that appear initially as more pythonomorphlike, then specialize independently of the Type I radiation, moving toward increasing reduction of the joint through pedomorphosis. I recognize this scenario is very transformational in its transitions, but it reflects that pattern of morphology as interpreted phylogenetically.

## Mental Foramina of the Dentary

Building the ancestral snake lizard from the anatomical features of ancient fossil snake lizards requires constant attention to the details, not to mention continual adherence to the historical burden of concepts developed around the essential characters of snake lizards (Figure 3.14). From Rieppel and Zaher (2000a:29), in their critique of Caldwell and Lee's (1997) assertion that there are at least two mental foramina on the dentary of *Pachyrhachis*:

> By contrast, snakes have a single mental foramen located toward the anterior tip of the dentary.

And:

> In conclusion, *Pachyrhachis* is characterized by the presence of a single mental foramen, as is characteristic of all snakes.

While a reduction of the number of mental foramina of the dentary found in typical lizards ($\geq 2$ to ~6–7) to the number found in modern snake lizards (0 in anomalepidid and some typhlopid scolecophidians and 1 in alethinophidians and leptotyphlophid scolecophidians, to be exact) is unlikely to represent a radical evolutionary innovation in snakes, the concern I raise here is that once again it is clear that there is a long-held and common certainty that one and only one mental foramen in each dentary represents the condition of "all snakes" (which according to the facts would mean if the received wisdom is correct as expressed by Rieppel and Zaher [2000a], then anomalepidid and typhlopid scolecophidians are not snake lizards). As a result of this firm belief that "all snakes," fossil and extant, must be static around all characters of snake lizards as per historical burden, Rieppel and Zaher (2000a) went to extreme lengths to counterargue Caldwell and Lee (1997) and Lee and Caldwell (1998), suggesting the latter authors had misinterpreted bubbles and/or damage as the extra foramen. While the argument is of no concern, the data actually are, and so it is instructive to examine Rieppel and Zaher (2000a: fig. 16) in detail—what they illustrated for *Pachyrhachis* was their version of the medial faces of the mandibles (Figure 3.10c), as their study was intended to be a rebuttal of characters of the intramandibular joint. They did not illustrate a lateral view of either dentary from the holotype (HUJ-Pal EJ 3659) and so provided no

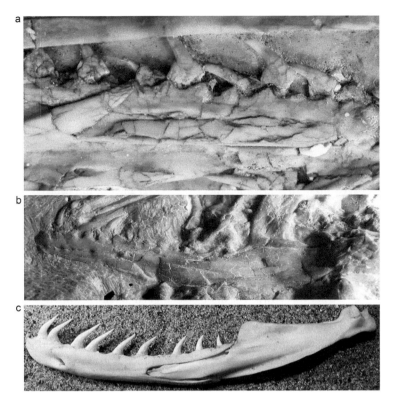

Figure 3.14. Lateral view, right and left mandibles, referred specimen of *Pachyrhachis problematicus* (HUJ-Pal EJ 3775). (a) Right mandible; (b) left mandible; (c) right mandible, *Python* sp. (UAMZ 57047). Scale bar = 5 mm.

visual support for their contrary interpretations, and they also ignored the data present in the referred specimen as they did in their critique of the intramandibular joint.

In response, Palci et al. (2013) reexamined both the type and referred specimen, and, while admittedly the dorsal surface of the holotype specimen is difficult to observe through the acrylic in which it is embedded, it remains possible to observe more than one mental foramen on the left dentary; unfortunately, the right dentary is hidden under the right maxilla. In fact, and as was ignored by Rieppel and Zaher (2000a), in support of Palci et al. (2013) and Caldwell and Lee (1997), Haas (1979:62) noted in his original description of the holotype (HUJ-Pal EJ 3659): "there are shallow alveolar depressions for about 10 teeth, caudally growing deeper in succession, and about 6 trigeminal foramina." While he later retracted that claim (Haas, 1980b:95) in his description of the referred specimen as a new species, "*Ophiomorphus colberti*," importantly, the referred specimen HUJ-Pal EJ 3775 confirms the number of foramina noted by Haas (1979) for the holotype specimen,

with the referred specimen presenting four foramina on the right dentary and five on the left (Figure 3.14a,b). As further support for Caldwell and Lee's (1997) original interpretation of at least two, Polcyn et al. (2005), based on CT scans of the holotype specimen, noted the presence of at least three foramina on the left dentary. Similar to the problem they faced with "UEs" as noted above in the discussion of the circumorbital bones, Polcyn et al. (2005) interpreted their scans as presenting only one true mental foramen, dismissing the other two or more as breakage or as taphonomic artifacts. While Polcyn et al. (2005) may or may not have agreed with Rieppel and Zaher (2000a), it seems clear that they too held to the notion that "all snakes" possess only a single mental foramen—more than one and you cannot be a "snake"; therefore, all ad hoc argumentation to will away the extra foramina was acceptable.

In contrast, and clearly supporting the "single foramen"—"all snakes" perspective, the second rear limbed snake from Ein Jabrud, *Haasiophis terrasanctus* (see Chapter 2), appears to present only a single mental foramen on both the right and left dentaries of the type and only specimen (HUJ-Pal EJ 695), as originally described by Tchernov et al. (2000) and later more completely characterized by Rieppel et al. (2003a). Close examination by me, Alessandro Palci, and Randall Nydam in 2011, which resulted in Palci et al. (2013), confirmed that no other foramina are visible on the right and left dentaries of the type, other than the far anterior foramen as described by Tchernov et al. (2000). It is interesting that in *Haasiophis* the condition observed in modern snake lizards is already demonstrated by a Cenomanian-aged snake lizard, while other contemporaneous snake lizards show a much more plesiomorphic condition of form ($\geq$2).

Unfortunately, despite the wealth of morphological data on *Dinilysia* (Chapter 2), the known material does not provide further clarification on the plesiomorphic condition for Mesozoic Gondwanan snake lizards, as complete dentaries are not yet known. The most complete dentary is a right element from MACN RN 1013, but this specimen still only preserves the posterior 1/3 of the dentary. However, just anterior to the excavation for the compound bone, there is a single large mental foramen. Considering the posterior position of this foramen, it is likely that there was at least one more anterior foramen, if not more. Regardless, at this point, only a single large foramen is known for *Dinilysia*, and in Zaher and Scanferla (2012) it was correctly coded as "?" rather than as a zero or one state.

Some insight, however, is provided by the newest materials of *Najash rionegrina* (MPCA 500) (F. Garberoglio personal communication), which very clearly possess two mental foramina on the right dentary (Figures 2.8f and 3.13a–f) consistent with published accounts of isolated *Najash* dentaries (Zaher et al., 2009; Palci et al., 2013). However, not too surprisingly, MPCA 500 also possesses a single foramen on the left dentary, and this is unequivocal (Garberoglio et al., in prep.). Similarly to *Dinilysia*, the posteriormost foramen is located just anterior to the notch for the compound bone, while the more anterior one is approximately two to three tooth positions posterior to the symphysial tip of the dentary. By analogy only, I would predict that *Dinilysia* likely possessed at least two mental foramina.

In terms of neuroanatomy and its evolution, the mental foramina of the dentary are exit points for small sensory branches of the anterior infralabial nerve (mental nerve in mammals) and a branch of the inferior alveolar nerve, which itself is a nerve trunk arising from the Mandibular Branch ($V_3$) of Cranial Nerve V, the Trigeminal Nerve. As $V_3$ descends from the posterior prootic foramen (posterior subdivision of the trigeminal foramen created by the laterosphenoid ossification in most modern snake lizards) toward the mandibular fossa of the compound bone, it enters the inferior alveolar canal at the anterior end of the mandibular fossa. As the nerve continues internal to the compound bone, it bifurcates to form two principal nerve branches: (1) posterior infralabial nerve; (2) anterior infralabial nerve. The posterior infralabial exits the compound bone from the superior alveolar foramen and extends laterally onto the dentary-compound bone, ventral to the tooth row of posterior process of the dentary (Rieppel and Zaher [2000a] refer to the superior alveolar foramen of snake lizards as the anterior surangular foramen, which is the name applied in typical lizards where there is no compound bone). The posterior infralabial innervates the posterior portions of the infralabial gland and skin in modern snake lizards where it has been identified (Auen and Langebartel, 1977), while the second main branch continues anteriorly and into the dentary to emerge from the single mental foramen in modern snake lizards to innervate the anterior portion of the infralabial gland and the surrounding skin. In both typical lizards and modern snake lizards, the infralabial glands (and supralabial glands along the maxilla and innervated by mental foramina filled by branches of the V2, the maxillary branch of the trigeminal nerve), are, in simple terms, salivary glands producing secretions that are serous, mucous, or mixed mucous-serous in nature (e.g., Oliveira et al., 2008; Martins et al., 2018). Venom glands are distinct from the infra- and supralabial glands and usually underlie or are posterior/superior to the labial glands.

The relationship of the anterior and posterior inferior alveolar nerve branches to the skin, but most critically the infralabial gland, is not unique to snake lizards among all lizards. Rather, if I were to hazard a guess, the excursion of the lateral dentary wall by the anterolateral process of the compound bone has likely mechanically excluded the exit points for the anterior infralabial nerve (see *Dinilysia patagonica* as an example) until the excursion of the dentary has been "cleared," so to speak, by the nerve tissue. Also, as has been examined in some detail, the steadily decreasing size of the dentary through Type III and Type

IV skull types means that the mental foramen appears to keep moving forward on a dentary bone that is steadily decreasing in size, which means that the anterior component of the infralabial gland will likely diminish in size with it; in typhlopids and anomalepidids, where the dentary is diminished to nothing, as are the infralabial glands, the anterior infralabial nerve is gone, as are the mental foramina (*Typhlops jamaicensis* is an exception with two mental foramina). In contrast, in leptotyphlopids where infralabial glands have been identified (Martins et al., 2018), the dentary retains a single large mental foramen for the excursion of the anterolabial nerve fibers to innervate the gland (it is also consequential that leptotphylopids feed by mandibular raking [Kley, 2001] and as such would benefit from inferiorlabial gland secretions both serous and mucous).

It is clear that because snake lizards are in fact lizards, and typical lizards have numerous mental foramina on the dentary bone, the reduction of mental foramina in modern snake lizards to either 1 or 0 is a derived state from the plesiomorphic or primitive state of $\geq 2$. Whether this vector of transformation is linear, that is, from $\geq 2-1-0$, is not yet clear, though it appears to be so even if convergent between lineages (in any case, the 0 state is certainly derived, not primitive or plesiomorphic, and is observed only in two clades of scolecophidians).

The historical burden of archetypes with essential characteristics—in this case, that to be a snake lizard you must have only one mental foramen—must be shed if we are to understand the evolution of snake lizards. Fortunately, the fossil record provides insight on this point such that *Pachyrhachis* and *Najash* can be understood to possess four to five and two, respectively, and are still in all respects "snakes," ancient or otherwise. Importantly, though, a late Mesozoic snake lizard such as *Haasiophis* already had evolved toward the state of one mental foramen present among almost all modern snake lizards. Clearly, the condition of modern snake lizards was evolving in parallel among the numerous lineages of snake lizards alive in the later Mesozoic in both the sea and land (*Najash* is already reduced to two and varying from side to side to one).

## Extracolumellar Cartilages: Middle Ear

The conceptualization of primary homologs from elements of the middle ear of modern snake lizards, followed by the Test of Congruence, has never been proposed in studies of snake lizard phylogeny (Figures 3.15 through 3.17). In part, this is because there is only a single bone, the columella, associated with one or more ossified or cartilaginous elements that have variably gone by the names extracolumella, stylohyal, intercalary. Embryologists have debated the identity of these several elements around historical embryological criteria of homology (e.g., Parker, 1878a; de Beer, 1937), but without success. Rieppel (1980b) attempted to clarify the problem by adding the pathways of nerves and blood vessels to the embryological data but found these data did not solve the problem either.

As there is only a single element varying between snake lizards and non-snake lizards, systematists have not considered the extracolumella of non-snake lizards in anything but the most simplistic terms. As an example, from Gauthier et al. (2012:148), Character 195 proposed a single binary character applied to all lizards: "195. Extracolumella: (0) present; (1) absent. Rieppel (1980a)." Gauthier et al. (2012) coded all snake lizards as "1," or extracolumella absent. Despite the embryological uncertainties, the question of course is whether this is accurate for all modern snake lizards and whether the extracolumella is actually present in ancient snake lizards, when it was lost if it was lost; and, of course if it is lost, begs the question as to the identity, de novo element or otherwise, of the "second element." It also forces a revisit to the origins of the debate on snake lizard middle ear anatomy and homologies as driven in the first instantiation by nineteenth-century embryologists, and to reference the nineteenth-century data described for mosasaurian pythonomorph lizards that bear on this question. I will begin with a brief review of the embryology of these elements, followed by a review of the extracolumella of non-snake lizards, including mosasaurian pythonomorphs. I will then outline the anatomy and morphology of the additional elements of snake lizards, then return full circle to the seeming embryological conundrum.

## Embryology: Cartilage and Arches

The developing vertebrate embryo possesses six pharyngeal arches, formed by cells from all three germ layers (ectoderm, mesoderm, and endoderm) along with the critical and morphogenesis directing influence of neural

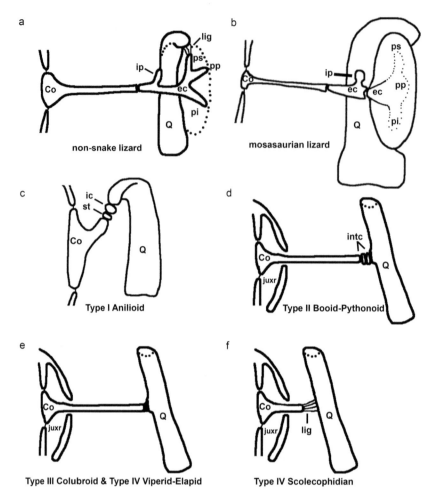

Figure 3.15. Schematic representations of middle ear anatomy. (a) Non-snake lizard; (b) mosasaurian pythonomorph; (c) Type I Anilioid modern snake lizard; (d) Type II Booid-Pythonoid modern snake lizard; (e) Type III Colubroid and Type IV Viperid-Elapid; (f) Type IV Scolecophidian. (Abbreviations: Co, columella; intc, intermediate cartilages; ec, extracolumella; ip, internal process of extracolumella; juxr, juxtastapedial recess; lig, ligament; pi, pars inferior of extracolumella; pp, posterior process of extracolumella; ps, pars superior of extracolumella; Q, quadrate.)

crest cells. Of consequence to the middle ear anatomy of snake lizards, and all lizards for that matter, are the cartilages/bones derived from the first (mandibular arch) and second (hyoid arch) pharyngeal arches.

Development of the first pharyngeal arch is complex, as it divides early to form the maxillary and mandibular processes within which develop the palatoquadrate (=pterygopalatoquadrate) and Meckel's cartilages, respectively. Mesoderm and neural crest cell synergy form the premaxilla, maxilla, most of the mandibular bones, and, of consequence for snake lizards, the quadrate bone from the posterior portion of the palatoquadrate cartilage.

The hyoid arch, or second pharyngeal arch, also gives rise to bone and cartilage critical to the anatomy and function of the middle ear. The ancient vertebrate hyoid arch cartilages include the hyomandibular, ceratohyal, and single basihyal. In mammals, as a top–down perspective on hyoid arch formation, the upper arch is known as Reichert's cartilage, which has two distinct proximal and distal portions connected by a mesenchymous section; the upper portion forms the mammalian stapes, while the lower portion forms parts of the hyoid complex; distal to the stapedial portion, mesenchyme eventually forms the stylohyal condensation that will become the stylohyoid process. For snake

THE ORIGIN OF SNAKES

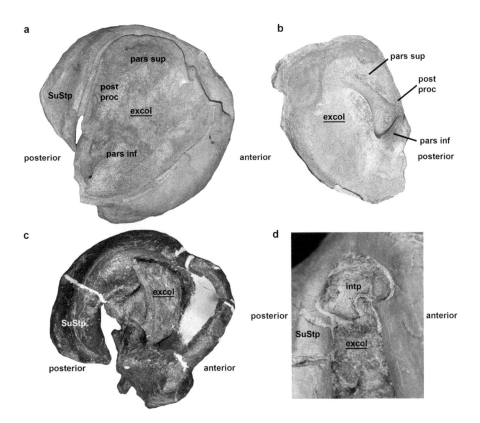

Figure 3.16. Quadrates and middle ear osteology of plioplatecarpine mosasaurids. (a) Lateral view, holotype right quadrate, *Plioplatecarpus houzeaui* (IRScNB R36); (b) "*Plioplatecarpus*" sp. Isolated right extracolumella, in medial view (IRScNB R37); (c) lateral view (image reversed for comparison) of left quadrate of *Latoplatecarpus nichollsae* (TMP 83.24.01); (d) close-up of internal process, medial surface of left quadrate, cf. *Platecarpus* sp. (Note: *excol* is underlined to emphasize that is the complete disclike structure, not just the portion proximate to the abbreviation.) (Abbreviations: *excol*, extracolumella; *intp*, internal process; *pars inf*, pars inferior of extracolumella; *pars sup*, pars superior of extracolumella; *post proc*, posterior process of columella; *SuStp*, suprastapedial process.)

lizards, Parker (1878b) concluded from his study of the colubrid *Natrix natrix*, that the homologies of the snake lizard columella auris was with the hyomandibular cartilage, and that the more inferior elements of the ceratohyal and basihyal formed the hyoid bone complex supporting the tongue (similar to the subdivision of Reichert's cartilage in mammals). Parker (1878b) also observed in *Natrix* the formation of an additional cartilage condensation located between the condensations of the quadrate and the columella, in other words, distal to the columella (similar to the stylohyoid of mammals)—he termed this the stylohyal cartilage and observed that in adult *Natrix* it fuses to the shaft of the quadrate and eventually ossifies. Kamal and Hammouda (1965a,b:187) found the same thing for the boiine *Eryx*—cells from the distal tip of the columella proliferate late in ontogeny to form a cartilaginous nodule that then may fuse to the shaft of the quadrate and ossify with increasing age of the individual (see below, and see also Kamal and Hammouda [1965b] on *Psammophis*). And, as was noted by de Beer (1937) for *Lacerta*, the formation of the columella from the hyomandibular condensation of the upper portion of the hyoid arch results in the extracolumella of Gadow (1888), what de Beer (1937:225) refers to as the "cruciform plate," and connections of those numerous processes, one of which was to the ceratohyal at an earlier stage in development. This of course is the expected position of a stylohyal cartilage—it should be at the mesenchymous interface between the hyomandibula and ceratohyal early in hyoid arch development and would strongly support the snake lizard stylohyal as the degenerate homolog of the extracolumella of non-snake lizards.

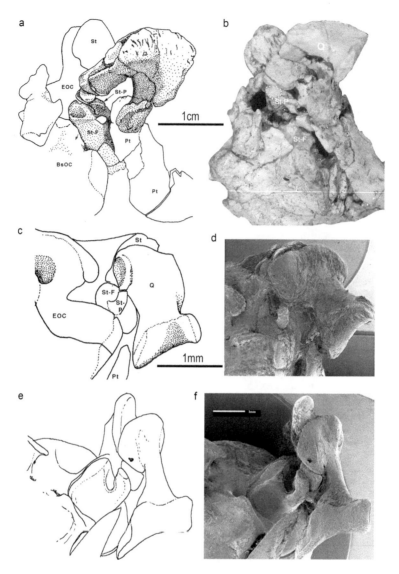

**Figure 3.17.** Right quadrate and middle ear osteology of *Dinilysia patagonica, Cylindrophis ruffus,* and *Xenopeltis unicolor.* (a) *Dinilysia patagonica,* MACN 1014 pen-and-ink rendering; (b) photograph of the same; (c) *Cylindrophis ruffus,* USNM 197456 pen-and-ink rendering; (d) scanning electron microscope (SEM) photograph of the same; (e) *Xenopeltis unicolor* (USNM 287277) pen-and-ink rendering; (f) scanning electron microscope photo of the same.

As was reviewed by Rieppel (1980b), numerous authors, using a wide variety of criteria, have argued the homologies of a stylohyal versus an extracolumella versus an intercalary cartilage in snake lizards, always with comparisons to non-snake lizards. Two opposing views based on very different information can be distilled from more than 100 years of study of snake lizard mandibular arch and hyoid arch development: (1) that the stylohyal of snake lizards is an extracolumellar degeneration of the typical lizard state, resulting in little more than the internal process (McDowell,

1967), and (2) that the stylohyal is an intercalary cartilage/bone (an extra element inserted between two others) and the cartilaginous portion of the columellar shaft is the remnant internal process (Kamal and Hammouda, 1965a,b; Rieppel, 1980b). While I will discuss the limitations of embryological homologies below, it is important to note here that the Parker (1878a,b) to McDowell (1967) opinion on "stylohyal = internal process" is a topological relation Test of Similarity (*sensu* Patterson, 1982), while the stylohyal = intercalary cartilage ignores topology and explicitly relies

THE ORIGIN OF SNAKES

on an assumption of ontogenetic constancy of an independent homolog, the course of the tympanic nerve, "Assuming a constant course of the chorda tympani, the ophidian stylohyal can be homologized with the intercalary cartilage of lizards while the cartilaginous distal portion of the ophidian stapes represents the internal process" (Rieppel, 1980b:45). I find such assumptions perplexing, especially considering that in many lizards, including snake lizards, the stapes is no longer perforate and the course of the stapedial artery has now changed—meaning that arterial and nerve courses can change independently of cartilages and bones, which form around the former pair of tissue types (nervous and vasculature) and can vary independently. Assuming constancy of the course of a nerve in order to recalibrate topological relations of splanchnocranial elements is an unwarranted use of ontogeny to ignore topology. For example, Cranial Nerve XI, the spinal accessory nerve, is absent in snake lizards, but, its absence does not mean that the bone it normally passes through, the exoccipital-opisthotic, is no longer an exoccipital-opisthotic. That element's identity is not dependent on the presence, absence, or course of the spinal accessory nerve. I would argue the same holds true for the columella and the stapedial artery, for the extracolumella (not matter how reduced), and for the path of the tympanic nerve (chorda tympani).

## The Columella and Extracolumella of Non-Snake Lizards

The middle ear of non-snake lizards is composed of a single ossified element, the columella, and a cartilaginous element of varying complexity, the extracolumella (Figure 3.15a). The columella is a simple element—beginning proximally, it presents an expanded footplate that is suspended in the fenestra ovalis of the otic capsule by the oval membrane (see above under "The Crista Circumfenestralis"); moving distally, the columella constricts to a bony rod of variable length and thickness that at its lateral terminus expands slightly to form a flattened pad or sometimes a cuplike distal tip. The simple rod of the extracolumella may well immediately articulate with the extracolumellar, or it might have a distally directed shaft/rod portion that is not ossified but rather is cartilaginous. Following that unossified shaft portion, the columella is

then followed in the chain by the cartilaginous extracolumella. The columellar-extracolumellar contact is highly variable between and within differing clades of non-snake lizards: (1) an actual articulation, (2) gradation from bone to cartilage of the columella to cartilage of the extracolumella, (3) a stiffened rod where cartilage and bone are wrapped by a collagen fiber bundle.

The cartilaginous extracolumella is somewhat more complex in its organization as compared to the columella. In most non-snake lizards, it is composed of a varying number of processes that may present as much as a four- to five-part morphology. Wever (1978) designated three different forms of organization for the extracolumella: (1) the iguanid, (2) the gekkonid and scincid type, and (3) a number of degenerate forms where processes are lost. Following Versluys (1898), Wever (1978) described four parts to the "insertion piece" of the extracolumella, that is, the portion of the extracolumella that radiates onto and across the tympanic membrane in a variety of directions, similar to the roots of a tree—the pars superior, pars inferior, anterior process, and posterior process. From observations given by Wever (1978), the anterior and posterior processes are variably present in differing groups of non-snake lizards, while the pars superior and pars inferior are generally present and are the principal "tuning fork" that transduces vibrations of the tympanic membrane to the columella. In the iguanid type, Wever (1978) noted that there is an extra process, arising from the shaft of the extracolumella before it passes the quadrate, that he termed the internal process; this process deviates anteriorly from the extracolumellar shaft to contact the shaft of the quadrate (he considered it to serve as stabilizer for the columella against lateral displacement during vibration). The remaining processes of the extracolumella arise from the cartilage postinternal process. The iguanid type is common to most non-snake lizards and the gekkonid type is specific to gekkonids, while the scincid type is found in scincids, anguids, and several species of xantusiids. What Wever (1978) referred to as degenerative forms where the extracolumella is highly reduced or modified in terms of processes are found in anniellids, chamelonids, xenosaurids, dibamids, feyliniids, and *Lanthanotus* (see McDowell [1967] for details of this important platynotan lizard).

The embryology of the "intercalary element" is made complex by numerous contradictions between the descriptions of various authors. For example, Wever (1978:414; fig. 11-18) describes an intercalary element for the Gila Monster, *Heloderma horridum*; from the posterior process of the extracolumella, the extracolumellar ligament extends toward the paroccipital process to which is attached an "intercalary element." De Beer (1937:220–221) describes a similar element, which he terms the "processus dorsalis," "Dorsal to the columella is a nodule of cartilage, in mesenchymatous connection with it, and forming the processus dorsalis. This cartilage becomes attached to the crista parotica of the auditory capsule, and is known as the intercalary or processus paroticus." Combining de Beer (1937) and Wever (1978), it seems clear that the "processus dorsalis" of the former and the "intercalary" of the latter are the same cartilaginous disc associated with the paroccipital process of the exoccipital-opisthotic and linked to the extracolumella by the extracolumellar ligament. The ligamentous attachment extending from the posterior process of the extracolumella to the intercalary element of the paroccipital has no clearly defined embryological correlate from either the mandibular or hyoid arches. The paroccipital process of the exoccipital-opisthotic develops from cartilage condensations that are not associated with the palatoquadrate cartilage or the hyomandibular-stylohyal condensations and the pharyngeal arches.

Complicating matters of embryology further, Baird (1970:200; figs. 2 and 4) describes and figures the exact opposite condition for lizards—his "dorsal process" arises from the extracolumella at or near the junction with the columella, proximal/medial to the internal process and extends dorsally toward the paroccipital process. This "dorsal process" is connected to the paroccipital process by a ligament. Baird (1970) provides no description of an intervening intercalary element between the dorsal process ligament and the paroccipital process. Baird's (1970:229; fig. 25) dorsal process in *Sphenodon* appears to articulate/contact the quadrate, not the paroccipital process, and may well be the internal process and not a "fifth" extracolumellar process. Rieppel (1978; 1983) adds more complexity to the problem of intercalary element origins by suggesting all intercalary elements in and around the quadrate are splanchnocranial condensations and that both the mandibular and hyoid arches produce such structures; it seems clear that the mandibular arch may well result in numerous

condensations linked to the palatoquadrate cartilage, but it is unclear as to the source if it is "hyoidal" in its origin.

The extracolumellar ligament that inserts on the pars superior or posterior process is important in this discussion as regards the points of origin and insertion of the ligament: (1) they originate from an intercalary cartilage/bone (extra chondrification/ossification) located on or near the cephalic condyle of the quadrate; (2) originate from the condyle in the absence of an intercalary bone/cartilage; or (3) originate from the paroccipital process of the exoccipital-opisthotic (otoccipital bone).

The non-snake lizard anatomy of the middle ear structures defines the essential ground plan of the snake lizard system, yet its complexity and descriptions are most certainly problematic, as they are neither clear nor in agreement. From a comparison of the range of presented anatomies among non-snake lizards, and by reference to any recent phylogenetic scheme for all lizards (e.g., Reeder et al., 2015), it is clear that the plesiomorphic condition is "extracolumella present and displaying numerous distal processes." The derived conditions appear convergently within almost all major clades and represent the absence of various processes to complete loss of the extracolumella; the columella proper is never lost, though the distal cartilaginous extremities appear to be.

## Mosasaurs and the Pythonomorph Middle Ear

The mosasaurian middle ear is unique among lizards on several points of anatomy, but unfortunately the same features cannot be confirmed in basal pythonomorphs due to a lack of preservation (Figures 3.15b and 3.16a–d). For higher mosasaurs though there are numerous well-preserved fossils with both the columella and the complete extracolumella preserved (internal process, pars inferior, superior, and posterior processes). The internal process appears to have ossified, while the remaining extracolumellar structures appear to be either ossified or calcified (detailed histological investigations have not been undertaken to date). The internal process has a taxon-specific ovoid to bean-shaped tip that articulates in a stapedial pit of similar but negative shape near the top of the quadrate shaft at the stapedial notch, on the internal or medial face of the quadrate (Dollo, 1903; Camp, 1942).

THE ORIGIN OF SNAKES

The internal process extends an anterior process through the stapedial notch of the quadrate to cup a convex structure from which expand at least three arms (pars inferior, superior, and posterior process) that radiate laterally and fuse into a solid mass of bony tissue to form a large disc. The disc has been variably referred to as an ossified/calcified tympanum (Dollo, 1903; Callison, 1967; Russell, 1967) or an extracolumella, at least for parts of the structure (Callison, 1967). What is abundantly clear from comparisons to other lizards and from observation of the anatomy of the extracolumella is that the disclike structure of the mosasaur middle ear apparatus is not a tympanum.

As McDowell (1967:155; figs. 1-3) noted for the earless monitor *Lanthanotus* (as should be clear from its common name, it has no external ear and thus no tympanic membrane), the extracolumella extrudes through the stapedial notch below the suprastapedial process and presents a broad, flattened ovate, disclike structure that fills the lateral face of the quadrate around the circumference of the tympanic ala. It is not the tympanum, which is lost, and it is not an ossified/calcified tympanum, as it is completely cartilaginous. Were it to become a bony structure, it would do so via ossification, as it is an endochondral tissue derived from the stylohyal of the second pharyngeal arch or hyoid arch. While it is possible for cartilages to calcify, it is likely that bone development would proceed by ossification (osteocytes replacing chondrocytes) and not the mere calcification of collagen fibers.

Therefore, as with *Lanthanotus*, mosasaurians likely did not possess a tympanum, and the skin would have covered over the space where the tympanum had been, as in snake lizards. The bony disc of the mosasaurian extracolumella includes at least three processes fused into a single disclike mass (pars superior, pars inferior, and posterior process) and a fourth, the internal process, that forms a solid attachment for the columella-extracolumella against the shaft of the quadrate. Importantly, and perhaps a key point in understanding snake lizard anatomy and homologs in this region, is that the pars superior and posterior process portions of the extracolumellar disc are directly in contact with the descending suprastapedial process while the internal process is not, nor is the pars inferior.

Therefore, when subjecting snake lizard primary homolog concepts to Tests of Similarity

relying on topological relations, these anatomies would suggest that any cartilaginous or bony structure aligned with the cephalic condyle or suprastapedial process would be, by dint of topology, an extracolumellar remnant of the pars superior and/or posterior process (e.g., *Dinilysia patagonica* [Figure 3.17a,b]). If the contact is made with the shaft of the quadrate, then the Test of Similarity relies on the consideration of a primary homolog concept constructed around the internal process. Identifying either element as an intercalary bone/cartilage makes no sense in light of the observed topologies, embryology, and so on. However, a review of modern and fossil snake lizard anatomies is required to further support this prediction and the hypotheses arising from it.

### The Columella and Additional Elements of Modern Snake Lizards

The columella of modern snake lizards presents two broad morphological categories (Figures 3.15c–f through 3.17a–f). In Type I Anilioid skulls, the columella presents an extremely large footplate, matching the very large size of the fenestra ovalis (Figure 3.6c,e); the shaft of the columella is relatively short, stout, and angled posterodorsally. Between the distal tip of the columella and the medial face of the suprastapedial process of the quadrate are one or more intervening cartilages (e.g., the uropeltid *Rhinophis* [Wever, 1978; fig. 20-11]) and a single cartilaginous element in *Anilius* and *Cylindrophis* (Figure 3.17c,d) or sometimes a single bony or cartilaginous element in *Xenopeltis* (Figure 3.17e,f).

In Type II–IV modern snake lizard skulls, the footplate is comparatively very small and enclosed within the cavity of the juxtastapedial recess (see above on the crista circumfenestralis) (Figures 3.6d,f–h and 3.15d–f). The shaft is extremely elongate and generally very thin (scolecophidians are the exception, where the shaft is usually very short, though the footplate is also small, as is the fenestra ovalis). The angle of the columella in Type II–IV skulls varies a great deal and is dependent on the position of articulation on the quadrate. For Type II–IV skulls, where there is no suprastapedial process to speak of on the quadrate, and certainly no Type I Anilioid-like element, there are one or more elements between the columella and the quadrate. The columella contacts the shaft of the quadrate and articulates with an articulatory process on the quadrate (Figure 3.15d,e). In Type II skulls, the columella and the articulating process possess cartilaginous caps, and between those caps there are intervening or intermediary cartilages (see Wever, 1978: figs. 20-7 to 20-9). In Type III and IV skulls, there are no intervening/intermediary cartilages and the columellar cartilage tip articulates directly with the cartilage cap of the articulating process (Figure 3.15e,f). Scolecophidians are a special form of this condition and demonstrate the derived state of loss and degeneration within Type IV skulls where the ligament fails to calcify (Figure 3.15f)— the cartilage caps have been lost, the articulating process on the quadrate is lost, and the distal tip of the columella is linked to the quadrate by only a small ligament. The origin of these quadrate-derived cartilaginous nodules in Type II–IV skulls is problematic. Though it is tempting to consider them all to be splanchnocranial condensations (see Rieppel, 1983), some insight is gained by reading closely the observations of Kamal and Hammouda (1965a:187):

> In slightly younger embryos than the present stage, a heap of mesenchymatous cells is produced from the distal end of the shaft and is connected with the posterior projection of the quadrate.

And:

> Later in the fully formed chondrocranium, these cells release from the shaft of the columella and form a cartilaginous nodule which fuses completely with the posterior projection of the quadrate. The distal end of the shaft lies very close to the cartilaginous nodule.

This is a particularly insightful observation even though Kamal and Hammouda (1965a) go on to present three arguments against this cartilage nodule as an extracolumella, as they recognize the source of the nodule as arising from cell proliferation late in development, not early, and that those cells release from the cartilage on the distal tip of the columella. As I understand the embryology of the "stylohyal," it is similarly produced from the hyomandibula during hyoid arch formation.

### Ancient Snake Lizards

*Dinilysia patagonica* (MACN 1014) presents the one and only nearly complete middle ear osteology of any known fossil snake lizard (Figure 3.17a,b). The first description of that specimen by Caldwell and Albino (2002: fig. 2) identified a stapes/columella and an extracolumella, though in the text they remained more equivocal (Caldwell and Albino, 2002:864) based on the historical debate (e.g., Rieppel, 1980b) as outlined above, "That shaft contacts a small extracolumella/intercalary/stylohyal element that articulates with the suprastapedial process of the quadrate." In contrast, Zaher and Scanferla (2012:24) followed Rieppel (1980b):

> Following the criteria set out by Rieppel (1980a,b,c), it is not possible to consider this ossification as homologous to the extracolumella of non-ophidian squamates. We thus consider this structure as homologous to the stylohyal of other extant snakes (Rieppel, 1980a,b,c; Frazzetta, 1999).

However, this conceptualization of Rieppel (1980b:45) appears to be a misapprehension based on the following overarching homolog and character, as noted above:

> Assuming a constant course of the chorda tympani, the ophidian stylohyal can be homologized with the intercalary cartilage of lizards while the cartilaginous distal portion of the ophidian stapes represents the internal process.

What Rieppel was saying is that the stylohyal is not a stylohyal at all, but an intercalary element, even though it has been referred to as a stylohyal by numerous authors. While I disagree with Rieppel (1980a,b,c) (see above) and cannot conclude for any reason, including the course of the chorda tympani (see below for details), that the ophidian cartilage in Type I Anilioid skulls is an intercalary, it is clear that Zaher and Scanferla (2012) misread Rieppel (1980b). Thus for *Dinilysia patagonica*, contra Zaher and Scanferla (2012), the stylohyal is present and that element is homologous to the extracolumella of non-snake lizards. In every comparable sense of anatomy, the middle ear osteology (mandibular and hyoid arch derivatives and associated chondrocranial and dermatocranial elements) of *Dinilysia* (Figure 3.17a,b) is virtually indistinguishable from that of *Cylindrophis* (Figure 3.16c,d) and *Xenopeltis* (Figure 3.17e,f).

## Playing the Cartilages You Were Dealt: The Problem of Embryological Absolutism

A core thesis of this book is that historical burden disables our ability to critically review and overturn knowledge that began as, or has evolved into, absolute truth. It is also problematic when scholars attempt to blend multiple lines of evidence into the construction of absolutes, and still worse when several lines of evidence in that chaotic blend are given primacy as trump cards that can outweigh and overturn other lines of evidence. Without grossly overstating the case here, it has been my challenge in considering primary homolog concepts for the extracolumella and columella to untangle the complexities and wrong turns of both classical anatomy and embryology. Though these are certainly the observations of a naïve historian of science, it has become clear to me that nineteenth- and twentieth-century embryological scholars were struggling to produce absolute statements on homologies of elements, in this case, between vertebrates, that would empower arguments of evolutionary relationship. While there is nothing wrong with this in the least, the problem is that over time these historical statements and approaches evolve to become burden, not knowledge, in driving towards the goal of presenting absolute answers, not relative ones. For example, from Kamal and Hammouda (1965a:187) on the large number of opinions regarding the homology of the shaft and cartilaginous nodule of the snake lizard columella:

> ...three proofs weigh against regarding the distal part of the shaft as an extracolumella. The first proof is that the chorda tympani passes under it, **if it was an** extracolumella the nerve would pass dorsal to it as commonly found in Lacertilia.

And:

> The second one is that this distal part ossifies in the adult snake, **while it is generally accepted** that in reptiles the extracolumella remains cartilaginous.

And finally:

> The third proof is that the ill-defined tympanic diverticulum does not surround the distal part, **while it is well known** that the tympanic cavity surrounds completely the extracolumella. (boldface added for emphasis)

The three proofs provided by Kamal and Hammouda (1965a) are common to numerous authors (e.g., Parker, 1878b; Rieppel, 1980b; Versluys, 1898; de Beer, 1937), and in Kamal and Hammouda's (1965a) case outline the problem—they are appealing to authority in all three "proofs" regarding their conclusions of the presence or absence of an extracolumella and have completely ignored their observations of topology. It is true, there are anatomical differences between snake lizards and non-snake lizards, including the varied passage of nerves and blood vessels, the loss of a tympanum and the absence of a Eustachian tube, and ossifications versus chondrifications, but none of these preclude the observation of topological relations of elements consistent with their embryology in the hyoid arch. They are merely arguments appealing to what is "well known," "generally accepted," and "if it was" knowledge. Such atomistic and logical positivist logic and philosophy was common in the nineteenth and twentieth century, and remains so today as powerful, directing, and absolute. Embryological research atomized anatomical structures to their germ cell layers, cranial embryology, and so on, and argued that on such grounds, homologies would ultimately be traceable and proven between disparate species and higher groups around general conditions for form first, followed by more specific condition of form. As for the extracolumella, as nerves and vasculature develop early, they are useful as primary roadmaps for deducing primary homologies of other elements. However, this is just an assist, not a final arbiter. It cannot overwhelm connectivity and topology. The conclusion that if the chorda tympani passes under the extracolumella, that that element cannot be an extracolumella as the information on the position of the nerve is invariant and absolute, is, frankly, non-sensical. The passage of a nerve, and "common knowledge," are of little value in assessing the primary homologies of the cartilaginous/bony element distal to the columella of snake lizards, modern and fossil. Connectivity and topological relations are sufficient to inform the conclusion that in Type I Anilioid skulls and Type I *Dinilysia-Najash* skulls, the cartilage element(s) and preserved bony elements are extracolumellar elements, likely representing the pars superior or posterior process, though it is highly possible that the orientation of the columella is fundamentally different in these skulls and the element remains the internal process. There is no reason, either in terms of connectivity or embryology, to consider this element an intercalary derived from outside of the hyoid arch condensations. Likewise, the same is true for Type II–IV modern snake lizard skulls, but in the opposite sense—the cartilage element(s) and preserved bony elements are indeed extracolumellar elements, likely representing first the internal process and then other degenerate parts of the extracolumella; in the most highly derived snake lizards, such as some scolecophidians, the entire system is reduced to the columella and a ligament likely derived from the now lost condensations of the extracolumella (see de Beer, 1937 and the pars interhyalis as an example).

### The Ancestral Snake Lizard

The Late Mesozoic fossil record is clear—there is an extracolumella between the distal tip of the columella and the suprastapedial process in *Dinilysia patagonica* (Figure 3.17a,b). Contra Zaher and Scanferla (2012), this is not an intercalary element, and embryologically it is likely formed from the stylohyal condensation, but regardless, in non-snake lizards, it is an extracolumella and can only be homologized to that element in ancient snake lizards. Based on the criteria illuminated by Kamal and Hammouda (1965a) to exclude the possibility of the snake lizard element being an extracolumella, the element present in *Dinilysia patagonica* means that the modern snake lizard condition of a Eustachian tube being absent, tympanum absent, and chorda tympani passing below the element were all in place by the early Late Mesozoic. Predictions have value, and in using the Late Mesozoic record to predict the past, the picture that emerges of pharyngeal arch evolution in the skull of ancient snake lizards would hold that the Gondwanan snake radiation as exemplified by the unnamed Gondwanan snake from South Africa (Chapter 2), moves that inner ear/otic capsule/hyoid arch/middle ear condition for Type I Anilioid and Type I *Dinilysia-Najash* skulls to a point in geological time during at least the Late Jurassic–Early Cretaceous. If the Middle Triassic recognition of *Megachirella wachtleri* as the oldest known lizard (Simões et al., 2018a) is any indicator, then the condition of snake lizardness among lizards is very old indeed, likely Late Triassic to Early Jurassic, and revolves around a variety of features, including the degenerate extracolumella. Does this suggest

hearing loss and burrowing? Not at all—rather, the loss of the tympanum and modifications to the middle ear osteology are clearly deeply affected by aquatic habits as well, as in mosasaurians and *Lanthanotus*. As a final comment, Evans (2016:245) concludes in her review of the ear apparatus and osteology of lepidosaurs that, "Where known, the ears of early snakes more closely resemble those of burrowers than swimmers." Considering the review of anatomy presented here, I would conclude absolutely the opposite, and further that the ear of scolecophidians as a Type IV skull osteology is derived, not primitive—admittedly, there is nothing about the osteology of that system that resembles a pythonomorph nor ophidiomorph, but the same cannot be concluded of the ear of Type I skulls (e.g., *Dinilysia*, *Najash*, and *Anilius/Cylindrophis*).

## Tooth Attachment

Mike, it is your Interdental Vimy Ridge

AARON LEBLANC
June 27, 2018

With the exception of secondary loss, the presence of teeth in the mouth has been a feature of gnathostome vertebrates for more than 430 million years (Brazeau and Friedman, 2015) (Figures 3.18 through 3.21). The manner in which teeth attach and implant to the tooth-bearing elements has been examined in great detail for more than 160 years, both in gross anatomical studies and in histological studies (e.g., Owen, 1840; Tomes, 1875a,b, 1882; Peyer, 1968; Osborn, 1984; Gaengler, 2000; Bertin et al., 2018). It is no surprise that this has resulted in an excess of terms, observations, anatomies, and histologies that have provided ample fuel to the debate on dental evolution. Snake lizard tooth attachment—no surprise or I would not be writing about it here—has in recent years been at the center of a reinvigorated debate using both data from the fossil record and the data set of dental histology (Lee, 1997b; Zaher and Rieppel, 1999b; Caldwell et al., 2003; Rieppel and Kearney, 2005; Budney et al., 2006; Caldwell, 2007b; Luan et al., 2009; LeBlanc and Reisz, 2013; LeBlanc et al., 2016a,b, 2017a,b; Bertin et al., 2018). At issue are the classical terms of pleurodonty, acrodont, and thecodonty and the nature and presence of a variety of tooth attachment tissues. Whereas the

THE ORIGIN OF SNAKES

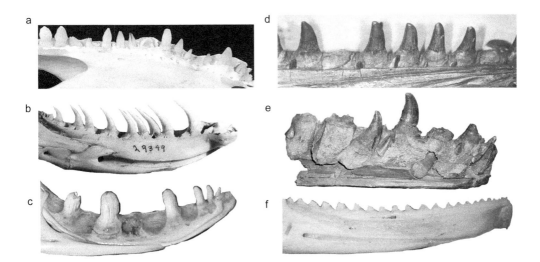

Figure 3.18. Classical tooth implantation categories for all lizards. (a) Crocodilian "thecodonty," *Caiman* sp. (UAMZ 799); (b) snake lizard "pleurodonty," *Eunectes murinus* (AMNH 29349), left mandible; (c) varanid lizard "pleurodonty," *Varanus exanthematicus* (unnumbered specimen, Eotvos University); (d) mosasaurian lizard "pleurodonty," *Mosasaurus hoffmannii*, right maxilla image reversed; (e) mosasaurian lizard "pleurodonty," close-up, *Mosasaurus hoffmannii*, right dentary, image reversed (note: numerous teeth in partial stages of development within alveoli in the dentary, see Caldwell [2007b]); (f) agamid lizard acrodonty, *Agama agama*, left dentary.

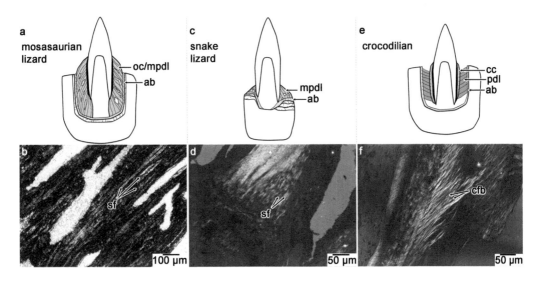

Figure 3.19. General conditions of tooth attachment in a mosasaurian lizard, snake lizard, and crocodilian with detailed micrographs of the alignment of mineralized and unmineralized periodontal ligament fibers. (a–b) Mosasaurian lizard. (a) Line drawing of tooth attachment tissues in cross-section; (b) cross-polarized light image, high magnification, of mineralized periodontal ligament. (c–d) Snake lizard. (c) Line drawing of tooth attachment tissues in cross-section; (d) cross-polarized light image, high magnification, of mineralized periodontal ligament. (e–f) Crocodilian. (e) Line drawing of tooth attachment tissues in cross-section; (f) cross-polarized light image, high magnification, of mineralized periodontal ligament. (Abbreviations: ab, alveolar bone; cc, cellular cementum; cfb, collagen fiber bundle; mpdl, mineralized periodontal ligament; oc, osteocementum; pdl, periodontal ligament; sf, Sharpey's fibers.) (Line drawings and photographs courtesy of Aaron LeBlanc.)

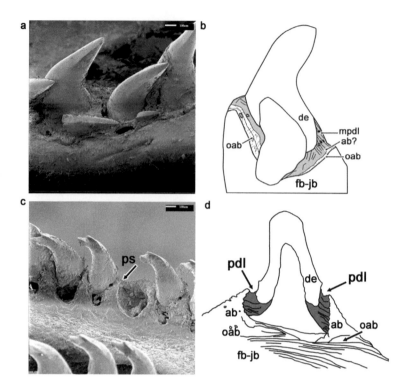

**Figure 3.20.** Periodontol ligaments and "hinged teeth" in Type I Anilioid modern snake lizards *Cylindrophis ruffus* (ankylosis) and *Xenopeltis unicolor* (gomphosis). (a–b) *Cylindrophois ruffus* (see also Figure 3.17a–c). (a) SEM photo; (b) line drawing of histology of attachment tissues. (c–d) *Xenopeltis unicolor*. (c) SEM photo; (d) line drawing of histology of attachment tissues (non-mineralized "hinge" ligament in red). (Abbreviations: ab, alveolar bone; de, dentine; mpdl, mineralized periodontal ligament; fb-jb, fibrolamellar bone of jawbone; oab, old generations of alveolar bone; pdl, periodontal ligament; ps, space for periodontal ligament.) (Line drawing (b) courtesy of Aaron LeBlanc.)

debate has primarily centered on the evolutionary relationships of snake lizards to other squamates, it has since had a ripple effect on the studies of dental evolution in amniotes more broadly.

### The Fallacy of Geometry

The classical approach to attachment and implantation has been focused on implantation geometries blended with tooth characteristics and histological attachment features ever since Owen (1840) and Tomes (1875a,b). The result has been the slow but steady construction of a common knowledge, archetype-driven three-part classification system (e.g., Owen, 1840; Leidy, 1865; Tomes, 1875a,b, 1882; Edmund, 1960; Osborne, 1984; Gaengler, 2000) (Figure 3.18a–f). Vertebrates broadly fall into three categories of tooth implantation: (1) thecodonty, considered the exclusive domain of mammals and crocodilians among modern amniotes, where teeth were set in alveoli made of alveolar bone have an obvious root covered in cementum and attached to the alveolar bone of the alveolus wall by a periodontal ligament, beginning where enamel ends and the root enters the alveolus (Figure 3.18a); (2) pleurodonty, characterizing most lizards, including many snake lizards, and not similarly detailed histologically, where no alveolus (or "true" alveolus) is present, the tooth is attached to the sidewall or pleura of the tooth-bearing element by "bone of attachment," and there is no obvious root (Figure 3.18b–e); and (3) acrodonty, characterizing some lizards, the Tuatara (*Sphenodon punctatus*) and some snake lizards, and, again, not detailed histologically, where no alveolus is present, the tooth is attached to the apex of the tooth-bearing element by "bone of attachment," and there is no obvious tooth root (Figure 3.18f).

From Owen (1840) forward, these numerous studies have recognized variations on the three categories in terms of geometry of implantation,

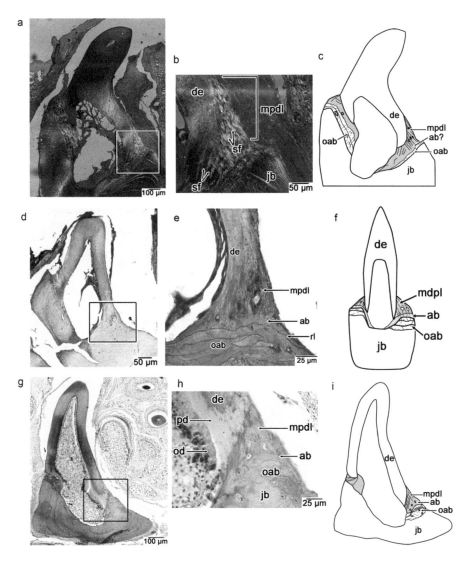

**Figure 3.21.** Histological sections of snake lizards with ankylosed teeth. (a–c) Type I Anilioid *Cylindrophis ruffus*. (a) Coronal section through an ankylosed tooth under cross-polarized light; (b) close-up of box in (a), in cross-polarized light (fibril organization more easily distinguished under cross-polarized light); (c) line drawing of histology of attachment tissues, mineralized ligament in gray. (d–f) Type II Booid-Pythonoid *Ungaliophis panamaensis*. (e) Close-up of box in (d); (f) line drawing of histology of attachment tissues. (g–i) Type IV Viperid-Elapid *Hydrophis annandalei*. (g) Coronal section through a nearly completely ankylosed dentary tooth; (h) close-up of box in (g), in plain light; (i) line drawing of histology of attachment tissues, mineralized ligament in gray. (Abbreviations: ab, alveolar bone; de, dentine; mpdl, mineralized periodontal ligament; mx, maxilla; ea, empty alveoli; jb, jawbone; oab, old generations of alveolar bone; od, odontoblasts; pd, predentine; pdl, periodontal ligament; pl, palatine; ps, space for periodontal ligament; pt, pterygoid; sf, Sharpey's fibers.) (Line drawings and photomicrographs courtesy of Aaron LeBlanc.)

resulting in additional terms to define such implantation architectures as "subthecodont," "ankylosed thecodont," "modified thecodont," "subpleurodont," "labial pleurodonty," "full pleurodont," and "subacrodont" (Gaengler, 2000) (Figure 3.18a–f). It is important to note here, however, that in no case were the actual histologies of the attachment tissues, apart from mammalian or crocodilian thecodonty, and the varying amounts of tissue they represented and their topological relations to each other reconsidered beyond the conventional wisdom of three distinct geometries of implantation and the absolutes they represented.

## Debating the Tooth

As mentioned, the newest form of the debate took shape when Lee (1997a) presented his interpretations of the dental anatomy of mosasaurian lizards, attempting to link them to snake lizards, noting that: (1) the developing teeth are recumbent and rotate into position similarly to the rotation observed in modern snake lizard developing dentitions; (2) the mosasaurian lizards were thecodont, that is, with sockets for implantation of the teeth and thus similar to snake lizards (which Lee [1997b] argued were not acrodont or pleurodont in classical terms, as outlined above) (cf. Figure 3.18b with 3.18d,e). Zaher and Rieppel (1999b) responded to Lee (1997b) and presented contrasting interpretations of tooth attachment and implantation in mosasaurs, other lizards, and snake lizards. Observing the same specimens as did Lee (1997b), Zaher and Rieppel (1999b) relied on conventional terms to describe tooth implantation and attachment, differing substantially in their conclusions as compared to Lee (1997b). Of significance, Zaher and Rieppel (1999b) differed on the definition of "thecodonty," such that mosasaurs were not sub-thecodont or thecodont as per Lee (1997b). Instead, Zaher and Rieppel (1999b) argued that mosasaurs were of the varanid-type (cf. Figure 3.18c with 3.18d,e); that is, they were "sub-pleurodont" and that these "sockets" were not real as compared to mammals and crocodiles, but only "apparent sockets" produced by a tissue long known as "bone of attachment" (see Tomes, 1882). Zaher and Rieppel (1999b:2) stated:

> However, the use of the term "thecodont" to define a synostotic mode of tooth attachment is misleading and must be avoided. This also applies to snakes and mosasaurs, which among squamates have been the only taxa erroneously considered as having a "thecodont" or "modified thecodont" mode of tooth implantation...

As per conventional wisdom, Zaher and Rieppel (1999b) argued that only mammals and crocodilians among extant amniotes were thecodont and possessed a "true socket," with a true socket having long been characterized by the possession of alveolar bone, cementum, and a periodontal ligament. As all dental histologists from Tomes (1875a,b) forward had not found alveolar bone, cementum, or a periodontal ligament in any

amniote except mammals and crocodiles, Zaher and Rieppel decided to followed Tomes's (1875a,b) thesis that any teeth that were ankylosed to the jaws were held in place by a standalone tissue called "bone of attachment"—importantly, and ironically, Tomes (1875a) identified this tissue in histological sections of snake lizards. Therefore, in Zaher and Rieppel's view, a "true socket" could not exist in a non-mammal or non-crocodile. Flawless logic as long as Tomes's (1875a,b) histological descriptions were not just correct, but sensible and logical—more to follow on this one.

Zaher and Rieppel (1999b) concluded that all lizards, non-snake and snake alike, were either acrodont, pleurodont, or exhibited a special form of pleurodonty—labial pleurodonty—after Lessmann (1952) (Figure 3.18b–f). For mosasaurs and snakes, Zaher and Rieppel (1999b:12) noted that:

> Although mosasaurs and snakes show teeth that appear to be set in sockets, they both have a pleurodont mode of tooth implantation, yet specialized in different respects.

Thus, it is clear that despite the shared presence of alveoli, and without investigation of the tissue types forming the alveoli other than lip service to the term "bone of attachment," they concluded the alveoli are a figment of the imagination and that both are derived from a plesiomorphic pleurodont condition Zaher and Rieppel (1999b:15) continued:

> Similarity in tooth implantation between higher snakes and mosasaurs is therefore derived independently from standard pleurodonty and is not evidence in support of a sister-group relationship of the two clades.

And in asserting the primitive nature of scolecophidian anatomies, and of scolecophidians as basal snake lizards phylogenetically, Zaher and Rieppel (1999b:15–16) reiterate that:

> By contrast, the plesiomorphic, pleurodont tooth implantation adds support to a basal position of Scolecophidia within snakes.

With the Zaher and Rieppel (1999b) versus Lee (1997b) positions firmly stated concerning traditional approaches to characterizing tooth implantation and attachment, in 1999 I began a project that took a completely different approach to the problem and questioned the reliability and validity of data gathered by simply looking at the geometry of tooth implantation. The result has been a paradigm shift from geometry back to the roots of dental histology (pun intended), cutting open the teeth and jaws of these animals and critically assessing tooth attachment tissue homology and evolution across amniote clades (Caldwell et al., 2003; Rieppel and Kearney, 2005; Budney et al., 2006; Caldwell, 2007b; Luan et al., 2009; Maxwell et al., 2010, 2011a,b,c; LeBlanc and Reisz, 2013; LeBlanc et al., 2016a,b 2017a,b).

After several years of data collection, writing, manuscript submissions, and rejections, Caldwell et al. (2003) shifted the questions away from gross morphology and artifacts of implantation to the data set of consequence—attachment tissues in a fossil mosasaurian lizard. Budney et al. (2006) followed that with the first-ever report of attachment tissues in a fossil snake, *Dinilysia patagonica*, and Caldwell (2007b) presented new data on the gross morphology of mosasaur tooth attachment. All three studies were a first step in addressing the issues raised by Lee (1997b) and Zaher and Rieppel (1999b), but also in tackling the long-standing problems of absolutes and essentials burdening the study of vertebrate tooth attachment.

Contra Lee (1997b) and Zaher and Rieppel (1999b), Caldwell et al. (2003) found that the geometry of the attachment site (whether it is thecodont, pleurodont, or acrodont) for a mosasaurian lizard tooth is uninformative on how the teeth are actually attached to the tooth-bearing element. Caldwell et al. (2003) examined the histological features of mosasaurian lizard tooth attachment and recognized that it is these attachment tissues that record the developmental and evolutionary history that is of interest to paleontologists. Moreover, Caldwell et al. (2003) concluded that tissue geometry is nothing more than an artifact of organization, not a driver of organization, and should be discarded in assessments of tooth attachment and evolution. While seemingly not profound at all, the results of tissue-level comparisons were in fact deeply impactful not only because of the report of stereotypically thecodont attachment tissues in a lizard, but because these new data challenged the identifications of tissues types such as Tomes's (1875a,b, 1882) "bone of attachment" and laid bare the traditional attachment categories of acrodonty, pleurodonty, and thecodonty. In mosasaurs, Caldwell et al.'s (2003) tissue histologies and tissue

types were consistent not just histologically, but also topologically, with the thecodont attachment tissues of mammals and crocodiles, and as recently clearly demonstrated by LeBlanc et al. (2017b) with those of snake lizards as well (Figure 3.19a–f). Similar to mammals and crocodilians, mosasaurian lizards possess: (1) a woven-fiber bone matrix that forms the tooth alveolus and an identifiable cribiform-plate (bundle bone, or a zone of compact alveolar bone lining the alveolus); (2) a periodontal ligament (i.e., preserved as the anchoring Sharpey's fiber scars in the bone of the alveolus); and (3) a dentine root around which is a thin layer of acellular cementum, which itself is surrounded by a massive collar of cellular cementum. Caldwell et al. (2003) also noted that the massive collar of cellular cementum was composed of two histologically distinct tissues: (1) a loosely organized ground matrix and (2) a laminar form surrounding the vascular tissues.

Budney et al. (2006) found similar tissues to those of mosasaurs (Figures 3.18b,d,e and 3.19a,b) (contra Zaher and Rieppel, 1999b) preserved in the Late Cretaceous snake *Dinilysia patagonica* (for a generalized snake lizard, see Figure 3.19c,d). They described the alveoli of *Dinilysia* as composed of alveolar bone, often retaining successive and unresorbed layers from previously attached tooth generations, which differed from mosasaurian lizards only in terms of the volume of cellular cementum forming the attachment zone for the insertion of the periodontal ligament and in the dimensions of the non-enamel covered root portion of the tooth (cf. Figure 3.19a,b with c,d). The identification of cementum in snake lizards, and acrodont lizards as well, was not new, as much earlier works by Bradford (1954) and Poole (1967) had identified "root cementum" on the tooth bases of both groups.

Importantly, Budney et al. (2006) also described for *Dinilysia patagonica* the absence of any mineralized periodontal ligament, noting teeth are seldom found with skull specimens and when they are, full-sized or not, they show no ankylosis to the rim of the alveolous. Budney et al. (2006) hypothesized, via a comparison to Type I Anilioid skulls, that *Dinilysia* did not show the *Cylindrophis*-like ankylosis and mineralization of the periodontal ligament (Figure 3.20a,b), but rather was similar to the hinged tooth condition and non-mineralized periodontal ligament of *Xenopeltis* (Savitsky, 1981) (Figure 3.20c,d). Considering how broadly known hinged teeth are for a variety of species of snake lizards (Savitsky, 1981) and pygopodid gekkotans (Patchell and Shine, 1986), it is perplexing that a primary homolog concept was never developed around the obvious connectivity and topological relation of this ligamentous hinge as a periodontal ligament—in all truth, what else could it be? A bridging ligament between the tooth base and the rim of the alveolus cannot be anything but a non-mineralized periodontal ligament. A brilliant and illustrated example is the "hinge ligament" of the Sunbeam Snake *Xenopeltis unicolor* as illustrated by Mahler and Kearney (2006:57; fig. 24), where the ligament is clearly attaching to an alveolar bone tissue forming the alveolus and to the dentinous root portion of the tooth, though the magnification is not high enough to determine the certain presence of a thin layer of cementum. This condition is not limited to just *Xenopeltis unicolor*, a Type I Anilioid snake lizard, but as described by Savitsky (1981) is present in higher colubroids as well, that is, *Liophidium rhodogaster, Scaphiodontophis annulatus* (refigured by LeBlanc et al., [2017b]), *Scaph. venustissimus, Sibynophis chinensis, Lycophidian capense,* and *Mehelya poensis*. Savitsky (1981) presents data indicating that the common theme connecting all of these snakes is not phylogeny, but diet choice and thus convergence—they all eat large, hard-bodied prey, and hinged teeth would flex inward under ingestion pressure and flex back as pressure was released, thus facilitating both prey capture and ingestion (note: neither Savitsky [1981] nor Patchell and Shine [1986] hypothesized a dental tissue homology for non-snake and snake lizard tooth ligaments).

For snake lizards, therefore, where root cementum had long been identified, but ignored, and ligaments hinging the teeth in the socket were long recognized but ignored for what they are, the only missing "thecodont" tissue remains alveolar bone. As mentioned previously, alveolar bone forms the tooth alveolus in crocodilians and mammals, and Budney et al. (2006) identified this as the alveolus-forming tissue in an ancient fossil snake. It is also obvious in the sections illustrated by Mahler and Kearney (2006). Not surprisingly, this bone tissue is located in exactly the same position (Figure 3.20c,d) as Tomes's (1875a,b, 1882) illustrated "bone of attachment."

FOLLIES OF THE FOLLICLE
Tooth development was an important component of the rationale presented by Caldwell et al.

(2003) and was a key component of Caldwell (2007b). In particular, those authors explored developmental work reported decades before by Ten Cate and Mills (1972), Freeman et al. (1975), and Ten Cate (1976). These studies reported on the observation that all tooth tissues in mammals, that is, thecodont implantation (enamel, dentine, cementum, periodontal ligament fibers, alveolar bone), develop from cell lineages present within the developing tooth germ bud; they are not external to the tooth germ bud/dental follicle, nor are they derived from any cells or tissues of the periosteum or cortex of the tooth bearing element. Ten Cate and Mills (1972), Freeman et al. (1975), and Ten Cate (1976) also reported the results of experiments that graphically illustrated this fact—tooth germ buds removed from the mouth of embryonic vertebrates were transplanted to skull and postcranial bones and developed little alveoli into which were anchored, by a periodontal ligament, teeth with roots covered by cementum. In its purest form, Ten Cate and Mills (1972), Freeman et al. (1975), and Ten Cate (1976), demonstrated that a jaw bone and its dental groove were unnecessary for the development of a tooth, making it easy to imagine how an epithelium of tooth germ buds might produce teeth not just on maxillae, premaxillae, and dentaries, but also on palatal bones, for example. Ten Cate (1989) reviewed several decades of research on this matter, including his own collaborative works, noting that the woven fiber bone of the alveolar process is derived from the basal portion of the dental follicle—in short, the tooth makes its own house. The continually emerging details of developmental pattern and process, as repeatedly shown in this chapter and throughout the history of anatomical and evolutionary research, illuminate problematic homologs and interpretations only if we are prepared to revise the dogmatic and historical burden of the past.

BITING OFF MORE THAN YOU CAN CHEW
Studies undertaken to provide a rebuttal to Caldwell et al.'s (2003) critique of the debate between Lee (1997b) and Zaher and Rieppel (1999b) (i.e., Rieppel and Kearney, 2005; Kearney and Rieppel, 2006; Mahler and Kearney, 2006) revealed themselves underinformed in their assessment of the basics of dental development, which is a necessary starting point for any assessment and

critique of tooth attachment and implantation. From Mahler and Kearney (2006:20):

> For example, is the bony alveolus seen in mammals and archosaurs homologous to the "socket" seen in snakes and mosasaurs, or is the latter a resorbed cavity within the attachment tissue that creates the superficial impression of a socket once the tooth is shed?

Even a superficial knowledge of the long-standing work of developmental odontology would have addressed this question. Alveoli are not an artifact of resorption and osteoclast activity. Alveoli are produced by the basal cell lineages of the dental follicle; the shape of an alveolus is the result of genetic and epigenetic factors, but the alveolar bone osteocytes are still present. Osteoclast activity can increase the size of an alveolus during resorption and shedding of a tooth, and, as has been demonstrated in the literature cited here (e.g., Caldwell et al., 2003), evidence of multiple generations of alveolar bone are present in histological section, demonstrating the fact that each tooth in a polyphyodont dentition builds its own alveolus, is shed, and the bone not always fully resorbed before attachment of the next tooth. What this means is that the alveolus and the alveolar process are not a resorption artifact. Further, if the research goal is to, at all costs, ensure there are no shared dental attachment characters between snake lizards and any kind of non-snake lizards by rescuing "bone of attachment" from its necessary grave, then the expectation is to diagnose the sources of the osteocytes that form the tissue and where they are sourced. This of course has never been done, and is in fact unnecessary, as "bone of attachment" can neither be histologically nor developmentally sourced from any part of the dental follicle except those that make avleolar bone and cementum.

Applying even the most simple extant phylogenetic bracket to the mammal and crocodile possession of alveolar bone would suggest that it should be present in all amniotes, including snake lizards, mosasaurian lizards, and in fact all lizards (Figure 3.19a–f). Alveoli are present in all amniotes because alveolar bone formed them, nothing more. Where alveoli are seemingly absent, the first hypothesis to test is whether the tissues are absent, not to conclude de facto that they are absent because of nothing more than taxonomy and geometry in gross morphological view. As an example of nearly 200 years of essentialism and archetypes, Mahler

and Kearney (2006) cling to the long-standing terminology of "bone of attachment," which lacks any connection to the known cell lineages of the dental follicle, rendering them unable to interpret the significance of illustrating a ligamentous tooth attachment system in an extant snake lizard (Mahler and Kearney, 2006: fig. 24).

At the same time as Mahler and Kearney (2006) challenged Caldwell et al. (2003) and Caldwell (2007b), Rieppel and Kearney (2005) undertook a histological study on tooth replacement in the mosasaurian lizard *Clidastes* sp. Rieppel and Kearney (2005: figs. 2 and 3) began the propagation of two significant scholarly and histological mistakes: (1) they confused Caldwell et al.'s (2003) identification of osteocementum with their "bone of attachment," which was not the "bone of attachment" of other authors (see Tomes, 1875a,b), and (2) they confused leftover alveolar bone/osteocementum from a previous tooth cycle with their "bone of attachment," which they had synomized with Caldwell et al.'s (2003) osteocementum, and then altogether ignored the tissue type they labeled as "interdental ridges" (Figure 3.19a,b). Not only were their figures mistaken, but Rieppel and Kearney (2005:690 and 691) continued by claiming that:

> ...mosasaur teeth are of rather uniform morphology (e.g., Russell, 1967), with a pointed and recurved tooth crown that sits on a distinct pedicel composed of bone of attachment ("cementum" of Caldwell et al., 2003).

And:

> Perhaps because of its extensive development in mosasaurs, the bone of attachment ("cementum" of Caldwell et al., 2003) is not fully resorbed and redeveloped during a tooth replacement cycle (see Zaher and Rieppel, 1999a,b: fig. 3), as is typical of other squamates (Osborn, 1984).

Kearney and Rieppel (2006) followed Rieppel and Kearney (2005) with an investigation of the occurrence of plicidentine in snake lizards and non-snake lizards, but continued with the propagation of earlier mistakes regarding tissues as identified by Caldwell et al. (2003), but now extended them to snake lizards:

> The nature of bone of attachment and its putative homology with mammalian cementum... remains poorly understood and deserves further attention.
>
> KEARNEY AND RIEPPEL, 2006:344

And:

> In snakes, bone of attachment is often extensively developed...covering the outside of the tooth and providing a bony sheath to its base...Tooth replacement in snakes occurs through resorption of this bone of attachment (Tomes, 1874), releasing the tooth from its sheath and from the tooth-bearing element altogether.
>
> KEARNEY AND RIEPPEL, 2006:345

Caldwell (2007b) responded to the mistakes and critiques presented by Rieppel and Kearney (2005), which no doubt (despite claims to the contrary) led to the most recent and more finely tuned odontological challenge to Caldwell et al. (2003) and Caldwell (2007b) as given by Luan et al. (2009). Luan et al. (2009:253–255) provided a point-by-point agreement or challenge to Caldwell et al.'s (2003) and Caldwell's (2007b) new data:

> It appears that our cellular cementum largely corresponds to the "osteocementum" of Caldwell et al. (2003), and to the "aligned cellular cementum" of Caldwell (2007).

And:

> Our "interdental ridge" is a reclassification of Caldwell's alveolar bone (Caldwell et al. 2003), and although we have identified an extensive mineralized fiber layer between interdental ridge and cellular cementum, the "mineralized periodontal ligament,"

And:

> Caldwell et al. (2003) report that the nonossified component of the periodontal ligament is unrecognizable. Instead, they find morphologies of a cribiform plate-like structure and remnants of Sharpey's fibers, which they believe to support the presence of a periodontal membrane.
>
> CALDWELL ET AL. 2003

Perplexingly, Luan et al. (2009) continued to refute the hypothesis that mosasaurian lizards could ever be thecodont. In contrast to Caldwell et al. (2003) and Caldwell (2007b), Luan et al. (2009) argued that the "interdental ridges" in mosasaurians and snake lizards were not alveolar bone, but "bone of attachment." Reading Luan et al. (2009:254–255) closely though, permits

an understanding of significant aspects of this intransigence, beginning with misrepresentation:

> ...Caldwell et al. (2003) thought of the mosasaurian sockets as real sockets that remain in place through multiple replacement events, similar to those found in crococilians.

What Caldwell et al. (2003:624) actually stated was exactly the opposite:

> The difference in osteon type between the inner and outer regions of the socket is interpreted as having been caused by multiple waves of tooth replacement and tooth attachment. The alveolar bone closest to the tooth is resorbed and replaced with every replacement event.

In addition to misrepresentation of the work they were criticizing, Luan et al. (2009:255) finally revealed their reasons for retaining the term "interdental ridge" with their version of "true thecodonty":

> We therefore retained Zaher and Rieppel's nomenclature of the interdental ridge in contrast to the mammalian alveolar bone... "true" thecodonty implies the presence of deep sockets which in depth exceed the length of the tooth crown, a lack of ankylosis, and the presence of genuine alveoli.

"True thecodonty" as per Romer (1956), or per any other author since Owen (1840), requires no such expectation of socket depth to crown height ratios—were this not so, then the saberlike canines of extinct cats, or the tusks of elephants or of walrus, or any one of the million mammalian variants on the theme of tooth crown height to root length, would not qualify as "true thecodont" (including those in the dentary of *Mosasaurus hoffmannii*; Figure 3.18d,e). Furthermore, regarding ankylosis versus gomphosis, in all mammals, including humans, ankylosis of the tooth into the alveolus is a common pathology. This means, biologically speaking, it is possible for osteoblast activity to ossify/calcify the periodontal ligament and produce a mass of bone locking the tooth into an ankylosis. Such an ankylosis is not just a mutation or a disease response representing dysfunction; rather, the presumed dysfunction is actually a function of an underlying and constrained biological process (see Rieppel [1988a,b] for a review of the history of

THE ORIGIN OF SNAKES

study of teratologies and their impact on thought in the eighteenth and nineteenth century by St. Hilaire, Serres, and Meckel). The fact that such an ankylosis can occur means there is a genetic and phenotypic capacity and plasticity that at some point in evolutionary history could well have been a typical tooth attachment type—the work of LeBlanc et al. (2016a) is exemplary evidence of such a condition in the evolutionary history of synapsids leading to mammals (e.g., Figure 3.19a–f). And, finally, it is demonstrably unclear as to what constitutes a genuine alveolus except to go back to their statement on page 254, "and Caldwell et al. (2003) thought of the mosasaurian sockets as real sockets that remain in place through multiple replacement events, similar to those found in crocodilians." Clearly, for Luan et al. (2009), a genuine alveolus survives in place through multiple replacement events. However, this is simply not true—they do not, not even in mammals. To use humans as an example, adult human teeth do not replace "baby teeth" in the exact socket, in fact, not even close. Were this not so, the practice of orthodontics would not be able to move teeth around on the jaw by steady manipulation of the alveolar process. In addition, when disease or old age effect the loss of teeth, the alveolar process, and the alveoli, disappear via resorption, leaving only the tooth-bearing element. Clearly, the "genuine alveoli" of Luan et al. (2009) are not real in any sense they were meaning, other than by appeal to essentialism and dogma.

Surprisingly, only two sentences later, Luan et al. (2009:256) contradict themselves and describe alveolar bone in squamates as a variable element under evolutionary pressure and continue with the following on mosasauroids specifically:

> Nevertheless, histological similarities between the interdental ridge and alveolar bone such as the organization into osteons (Caldwell et al., 2003) appear to suggest that the mosasaurian interdental ridge might be an early step in the evolution of thecodont modes of tooth attachment in vertebrates.

While mosasaurian lizards are not some early step in the evolution of thecondonty leading to mammals and crocodiles, it is clear that Luan et al. (2009) were internally confused and self-contradictory in both their observations and conclusions. Obviously, the need to retain "bone of attachment" provided consistency with Zaher

and Rieppel (1999b), Rieppel and Kearney (2005), Kearney and Rieppel (2006), Mahler and Kearney (2006), and the literature back to Tomes (1875a,b), and was thus consistent with the essentialist archetype that "true thecodonty" did not exist outside of crocodilians and mammals, and that the "interdental ridges" and alveoli in mosasaurs and snakes were not "true alveoli." Again, as noted earlier and in keeping with previous studies, Luan et al. (2009) disregarded completely the empirical expectation to identify the origin of the bone cell lineages that produce a woven fiber bone tissue attached to the jaw bone, to which the now acknowledged "trabecular cementum" adhering to the tooth is itself attached. However, one major deviation from party lines in Luan et al. (2009) is that those authors straightened out the confusion and conflation of Rieppel and Kearney (2005), Kearney and Rieppel (2006), and Mahler and Kearney (2006) regarding "bone of attachment" as a synonym of Caldwell et al.'s (2003) osteocementum, though they simply moved that term one tissue over and confused it with alveolar bone.

Interestingly, Luan et al. (2009) did not cite Budney et al. (2006) regarding tooth attachment tissues in the ancient snake lizard *Dinilysia patagonica*; did not refute the generalized snake lizard thecodont attachment tissues hypothesized by those authors for all snake lizards; and in fact never mentioned snake lizard attachment tissues at all, even though this issue was at the core of the debate between Zaher and Rieppel (1999b) and Lee (1997b). It is also curious, from the quote given above from Luan et al. (2009:257) regarding their discovery that mosasaur teeth were covered in "trabecular cementum," not "bone of attachment," that they could not tell the difference between bone of attachment on a tooth root and bone of attachment forming their "interdental ridge" and "false sockets"; these are neither the same tissues nor are they in the same position anatomically or histologically (Figures 3.18d,e and 3.19a–f). I would suggest, as Caldwell et al. (2003) did, that the reason for this histological confusion is because it is easy to confuse a tissue when it has never been characterized histologically and has no described or consistent connectivities nor topological relationships. As noted earlier, Tomes, in his fifth edition of his comparative odontology text (Tomes, 1898:229–230), was also confused in the sense that he likely conceived of mammals as fundamentally more advanced than snake lizards, did not characterize the tissue, and thus created this long-standing dogma around the notion that mammals and crocodiles were uniquely thecodont:

> ...the union takes place through the medium of a portion of bone...developed to give attachment to that one particular tooth, and after the fall of that tooth is itself removed.
>
> For this bone I have proposed the name of "bone of attachment", and it is well exemplified in the Ophidia... (p. 229)

And:

> But the "bone of attachment," is very coarse in texture, full of irregular spaces, very different from the regular lacunae, and its lamination is roughly parallel with the base of the tooth. (p. 230)

Caldwell et al. (2003) noted that histologically, there was no difference between bone of attachment and alveolar bone in mosasaurian lizards and stated:

> We can find no comparative histological description of Tomes' (1898) "bone of attachment" indicating how it differs from alveolar bone...that Osborn (1984) considered that bone of attachment might be the homolog of alveolar bone...We find that mosasaurid bone of attachment is ontogenetically, compositionally, and topologically indistinguishable from alveolar bone.

These debates continued most recently with LeBlanc et al.'s (2017b) challenge to Luan et al.'s (2009) interpretations for mosasaurian lizards, crocodilians, and their bearing on tooth attachment in modern snake lizards. LeBlanc et al. (2017b) presented overwhelmingly clear data on mosasaurian, crocodilian, and snake lizard attachment tissues (Figures 3.18a–f and 3.19a–f), refuting Luan et al. (2009) (in particular, the existence of "bone of attachment"). LeBlanc et al. (2017a,b) also modified the conclusions of Caldwell et al. (2003) with clear evidence that the "osteocementum" of Caldwell et al. (2003), and the "trabecular cementum" of Luan et al. (2009), is actually the mineralized portion of the periodontal ligament (considering they were willing to identify

cementum and a periodontal ligament, it remains confusing as to why they refused to identify the surrounding bone to which the ligament attached as alveolar bone). As mosasaur teeth develop, the increasing addition of calcifying collagen fibers in the periodontal ligament build up on the tooth base as the tooth erupts into position, the alveolar process forms, and the remaining collagen fibers of the ligament calcify and ankylose the tooth into the alveolus (see Caldwell, 2007b) (Figures 3.18d,e and 3.19a,b). Moreover, LeBlanc et al. (2017b) showed that the difference between ankylosed tooth and one suspended by a ligament is simply the amount of calcification, and even identified a ligamentous tooth attachment in some shell-crushing mosasaurian lizards.

For all amniotes, including snake lizards, this mosasaurian example makes perfect anatomical and embryological sense for a tissue such as the periodontal ligament—if the collagen fibers of the ligament calcify, they form Sharpey's fibers, and for a tooth, it is thus ankylosed to the alveolar bone (whether the latter forms an alveolus or not is of no consequence to the fact that it is alveolar bone); if the collagen fibers of the ligament do not calcify, they form a gomphosis where the tooth is now supported by a variably shaped ligament suspended between a four-sided, three-, two-, or even one-sided structure composed of alveolar bone (see Savitsky [1981] or Patchell and Shine [1986] as examples). Varying the amount of cementum and periodontal ligament (and its calcification or lack thereof) with the varied amount and placement of alveolar bone effectively describes the complete sweep of amniote "geometries" of implantation and attachment (Figure 3.19a–f). There is no such thing as acrodonty or pleurodonty, nor even thecodonty, for that matter. In every taxon so far examined, alveolar bone (attached to the tooth-bearing element), periodontal ligament tissue (calcified or uncalcified, and spanning the distance between the alveolar process and cementum), and cementum (surrounding the dentinous portion of the tooth, or even in some cases on the enamel) are easily identified even if the amounts of any or all are incredibly minute.

TOOTH ATTACHMENT IN SNAKE LIZARDS
While the debate since 1997 has been stimulated by Lee's (1997b) attempts to discover synapomorphies linking snake lizards to mosasaurian lizards, the

specifics of the debate have focused on mosasaurian lizards (Figures 3.20 and 3.21). In fact, prior to 1997, apart from Tomes's (1875a,b) original characterizations of ophidian bone of attachment in *Python* and the "Common Snake" (presumably *Natrix*), little descriptive histological work has ever reconsidered the dental attachment tissues of snake lizards. Budney's (2004) unpublished thesis work began the process, but left it incomplete until subsequent authors, including Maxwell et al. (2011a,b,c), and LeBlanc et al. (2017b), began to work with those histological data. Mahler and Kearney (2006) certainly provided detailed histological figures, but, as noted above, did not entertain any reanalysis of tissues of attachment. Budney et al. (2006) represents the first descriptive work devoted to a fossil snake lizard, but even then provided no real data on modern snake lizards.

The purpose of highlighting the last 20 years of the debate around mosasaurian lizards and snake lizard synapomorphies is to point out that in ancient and modern lizards, including snake lizards, tooth attachment tissues are now recognized as including alveolar bone, a periodontal ligament, and cementum adhering to the dentinous portion of the tooth root (LeBlanc et al., 2017b). The non-tissue of "bone of attachment" must now be ignored both as an anatomical reality, as a tissue of relevance to dental evolution of snake lizards, and as a possible primary homolog concept.

In all known snake lizards, including the oldest known snake lizards (Caldwell et al., 2015: fig. 2a), the tooth root, that is, that portion of the dentine exposed at the terminus of the enamel, is small (see Figures 2.1 and 2.2). This feature of snake lizard teeth—a very short "root" portion of the tooth—appears to have been a feature of snake lizard anatomy since before the Late Jurassic–Early Cretaceous (see *Parviraptor estesi*, in particular the dissociated tooth, likely a replacement tooth not ankylosed into position, see Figure 2.2a–d). It is no surprise that in the three modern snake lizards presented here (Figures 3.20a–d and 3.21a–i) (the red-tailed pipe snake, *Cylindrophis ruffus*; the Panamanian dwarf boa, *Ungaliophis panamanensis*; and the big-headed sea snake, *Hydrophis annandalei*), that that is a feature common to modern snake lizards.

Regarding ankylosis as opposed to a gomphosis or hingelike periodontal ligament, the oldest known snakes appear to present ankylosed teeth, though this is only a gross morphological assessment of that anatomy (Caldwell et al., 2015) (see Figures 2.1 and 2.2a–d), and the absence of teeth in *Eophis underwoodi* suggests an alternative condition. *Dinilysia patagonica* and *Najash rionegrina* also appear to present evidence of a non-mineralized periodontal ligament, though absence of teeth is not necessarily the best indicator; such a condition certainly describes the absence of teeth in skeletonized specimens of *Xenopeltis unicolor* (see LeBlanc et al., 2017a,b: fig. 6i). The modern Type I Anilioid *Cylindrophis ruffus*, on the other hand, presents excellent histological details, explored by LeBlanc et al. (2017b) (Figures 3.20a,b and 3.21a–c), indicating the presence of a mineralized periodontal ligament, current and past generations of alveolar bone, and a thin layer of cementum on the dentine portion of the tooth, ankylosed to the mineralized periodontal ligament. For Type II Booid-Pythonoids (Figure 3.21d–f) and Type IV Elapid-Viperids (Figure 3.21g–i), the same tissues were reported by LeBlanc et al. (2017b), demonstrating the same connectivities and topological relations described for *Cylindrophis ruffus*. The xenopeltid or colubroid ligament hinge as reported by Savitsky (1981) requires nothing more than for the periodontal ligament not to mineralize during the ontogeny of any one tooth, and thus represents a derived condition within Type I and Type IV skulls—convergent, but derived, and essentially pedomorphic, as the ligament must form before it can mineralize.

BUILDING THE ANCESTRAL SNAKE LIZARD FORM OF TOOTH ATTACHMENT

Zaher and Rieppel (1999b) were correct in one regard: the alveoli of snake lizards and mosasaurian lizards are not primary homologs and thus are not derived from a common ancestor that also possessed a "socketed tooth." The condition in the crown of each group is derived, in terms of ancestral implantation geometry, from a "pleurodont" geometry. This is very clearly observed for mosasaurian lizards via examination of the gross morphology of tooth attachment in basal pythonomorphs; the same is true for snake lizards by a simple comparison of even modern snake lizards such as *Xenopeltis* (a three-sided to two-sided alveolus at best) versus *Python* (a full sided, deep-set socket). However, for snake lizards and mosasaurian lizards, where full alveoli are observed, they are true alveoli produced by alveolar bone, not "apparent sockets," "false sockets," or artifacts of abnormal resorption. After all, the "true alveoli" of mammals and crocodilians share no common ancestry in the classical model either; why would snake lizard and mosasaurian lizards be less true because of close or distant phylogenetic relations?

Ancient snake lizards can thus be predicted to possess all three tooth attachment tissues of alveolar bone, periodontal ligament (mineralized or otherwise), and cementum. If *Eophis underwoodi*, as Middle Jurassic exemplary of snake lizard attachment, is "typical," then the alveoli were three-sided structures formed of alveolar bone and the teeth were loosely attached to that alveolar process by non-mineralized periodontal ligaments. I doubt, though, based on the attachment tissues in other coeval non-snake lizards, that *Eophis* represents the ancestral snake lizard form. My prediction is that ancient snake lizards resemble *Portugalophis* or *Parviraptor*, with substantive alveolar bone-produced alveoli and a mineralized periodontal ligament.

## Suspensorium: Anatomy and Morphology

While the two-element suspensorium (one element, quadrate only, in most scolecophidians and all uropeltids [Cundall and Irish, 2008]) of most modern snake lizards has been, as Cundall and Irish (2008:356) said, "a puzzle to anatomists," that puzzle has concerned the missing elements as compared to the three-element system of modern non-snake lizards (Figures 3.22 and 3.23). Surprisingly, the identities of the elements in the suspensorium have not been the subject of recent debate despite the divisive polarities of interpretation regarding other anatomies (see above for examples), though McDowell (2008) presented a contrary opinion to the arguments given by nearly all other workers (Romer, 1956) (i.e., that the lizard supratemporal is the ancient diapsid tabular bone, not the supratemporal of ancient diapsids). While there is little argument concerning the primary homologs of the suspensorial elements, and thus seemingly little reason to explore them in detail here, the suspension of the quadrate from elements of the braincase directly affects an important feature of snake lizard evolution, and that is oral gape size, which is directly linked to snake feeding, prey ingestion, and the general kineticism of the skull as pertains to ingesting large prey items. The morphology and function of the suspensorium, and its evolution from the non-snake lizard condition where gape size is smaller, informs our understanding of snake lizard evolution in important ways and links to discussions held previously in this chapter and throughout this book on the intramandibular hinge, general skull kinetics,

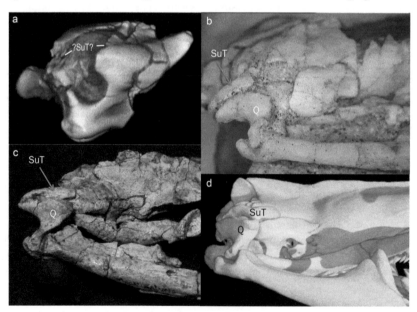

Figure 3.22. Suspensorium anatomy, in right lateral view, of Type I *Dinilysia-Najash* and Type I Anilioid skulls. (a) Unnamed Valanginian-aged snake lizard from South Africa; (b) *Najash rionegrina*, MPCA 500; (c) *Dinilysia patagonica*, MACN 1014; (d) *Cylindrophis ruffus* (Image from Digimorph.com). (Images not to scale.)

the presence/absence of an intermandibular joint, and palatal morphology and mechanics. Current broad-scale higher taxonomic groupings such as Macrostomata (originally, "*Ophidia macrostomata*" of Müller [1831] and elevated to clade status by Rieppel [1988a,b]) reflect supposed clade structures defined by synapomorphies recognizing presence of large gape size (Macrostomata = Booids + All Higher Snake Lizards) versus its absence (anilioids, scolecophidians).

## The Suspensorium

To define the suspensorium, it is that system of bony connections, articulations, and joints that connects the mandible to the braincase (Figures 3.22a–d and 3.23a–f). These include, proximally (next to the braincase) the supratemporal and distally the splanchnocranial mandibular arch element of

the quadrate. The quadrate then articulates with the mandible (and indirectly still, with Meckel's cartilage) via the articular portion of the compound bone. There is also a great degree of variation within snake lizards regarding the presence or absence of a paroccipital process arising from the exoccipital-opisthotic and projecting posteriorly—the extremes of the morphology are observed in Type I *Dinilysia-Najash* skulls (Figure 3.22a–c), and to a significantly lesser degree, in Type I Anilioid skulls (Figure 3.22d), even though there is no snake lizard paroccipital process matching the elongate structures common to most non-snake lizards (Estes et al., 1970; Cundall, 1995; Caldwell and Albino, 2002; Apesteguía and Zaher, 2006; Zaher and Scanferla, 2012). From Type II to Type IV skulls, the paroccipital trends from a small eminence to complete absence as an eminence of any kind (Figure 3.23a–f for Type II skulls, fossil and modern).

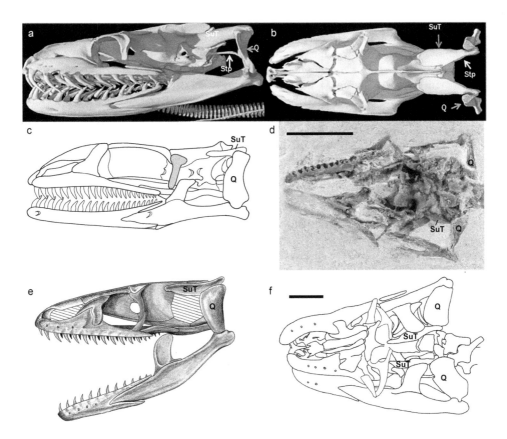

**Figure 3.23.** Suspensorium anatomy, in right lateral view, of Type II Booid-Pythonoid and Type II *Pachyrhachis-Haasiophis* skulls. (a) *Python molurus*, left lateral view (Image from Digimorph.com); (b) *Python molurus*, dorsal view (Image from Digimorph.com); (c) *Eupodophis descuensis*. (Reconstruction adapted from Rieppel and Head, 2004.) (d) *Eupodophis descuensis*, MSNM V 3661; (e) *Pachyrhachis problematicus*. (From Palci et al., 2013.); (f) *Pachyrhachis problematicus* holotype. (Line drawing adapted from Lee and Caldwell, 1998.) (Images not to scale.)

In all uropeltids (Type I Anilioid skulls) and most scolecophidians (Type IV), not only is the paroccipital non-existent, but the supratemporal is also absent, or if present, as in anomalepidids, it appears to float in the muscle tissue and neither articulates with the braincase nor with the quadrate, or if so, just barely (cf. *Anomalepsis* versus *Liotyphlops*) (Haas, 1962, 1964, 1968; Cundall and Irish, 2008; Rieppel et al., 2009). The quadrate is therefore "suspended" from the braincase, and in the case of uropeltids, from a fused braincase, the otooccipital complex (see Olori and Bell, 2012), in a position posterior to fenestra ovalis.

THE SUPRATEMPORAL BONE

Consideration of the supratemporal bone necessitates understanding its topological relations and connectivities among both non-snake and snake lizards. In typical non-snake lizards, an additional element articulates with the quadrate, lateral to the supratemporal, and that is the squamosal. The squamosal has both a suspensorium articulation with the quadrate and medially an articulation with the supratemporal. The squamosal is also part of the temporal series of dermatocranial bones, framing a portion of the posterior and lateral margins of the upper and lower temporal fenestra. If the squamosal retains an anterior ramus, it will form a fibrous joint and articulate with the posterior ramus of the postorbital or postorbitofrontal bone forming the upper temporal bar or arcade of the temporal region. While the squamosal is considered part of the suspensorium, the postorbital is not.

An important point of non-lizard context is the observed anatomy of *Sphenodon punctatus*, the modern Tuatara. Unlike lizards of all kinds, snake lizards included, the lower temporal bar is complete, along with the supratemporal bar, and connects the jugal to the quadratojugal, which is lateral to the quadrate, both of which articulate with the squamosal. While the suspensorium of *Sphenodon* is immobile (fixed quadrate and quadratojugal complex), the mandible, via the condyle of the articular bone, forms a mobile joint with the quadrate and quadratojugal bones. Of key importance to assessing the argument made by McDowell (2008) regarding the identity of the supratemporal of snake lizards (made much earlier in McDowell and Bogert [1954] regarding the varanoid lizards *Lanthanotus* and *Heloderma*) is that a supratemporal is absent in adult *Sphendon*

(Romer, 1956). However, the element was reported in a hatchling *Sphendon* by Rieppel (1992) in the expected position of the supratemporal in a lizard—between the squamosal and parietal ramus, extending anteriorly along the ramus.

McDowell's (2008:503) contrary perspective on snake lizard suspensorial homologies was as follows:

> I have no disagreement with identifying the bone in question of snakes with the bone identified…as the supratemporal of lizards. My disagreement with current usage comes from judgement of an hypothesis of homologies of the temporal dermal bones of diapsids in general…

McDowell (2008) argued that the supratemporal of all lizards, including snake lizards, is the same bone, but that that bone is not the supratemporal of ancient diapsids, but rather is the tabular bone, an element long considered to have been lost early on in amniote evolution (Romer, 1956; Carroll, 1988). McDowell (2008) also extended this argument, by logic, to all non-snake lizards, suggesting that if the snake lizard and non-snake lizard proximate cranial bone of the suspensorium is the tabular, then the more lateral element is not a squamosal in non-snake lizards that possess it, but a supratemporal, and that therefore the squamosal is lost.

Following Cundall and Irish (2008), I will not delve deeply into McDowell's (2008) rationale and arguments but will point out several problems of connectivity that do not support the latter author's position. First, I hold to connectivity and topological relations as key Tests of Similarity (*sensu* Patterson, 1982). Thus, if working in the opposite direction from McDowell (2008) and identifying the dermatocranial squamosal first, and starting with any non-lizard diapsid (e.g., *Sphenodon*) (Evans, 2008), the triradiate squamosal generally articulates with the quadratojugal, quadrate, postorbital, jugal, and supratemporal (sometimes the parietal when the supratemporal is missing as in adult *Sphenodon*). The supratemporal, in relation to such a squamosal, sutures to the parietal and squamosal, but most importantly to the exoccipital-opisthotic. The tabular, if present, is a small, "tabular-shaped" bone that articulates with the supratemporal (but is recognized to lose that suture as it is reducing through phylogeny

[Romer, 1956]), the parietal, postparietal (if still present), and supraoccipital. Importantly, in ancient diapsids, or in fact in modern *Sphenodon*, the supratemporal (if present) has never had, and does not have, an articulation to the quadrate, mobile or immobile (Evans, 2008).

I therefore take the position that the original and thus ancient suspensorial element is the squamosal, an articulation that has remained consistent right through to fossil and modern non-snake lizards still in possession of a supratemporal arcade. Through the evolution of diapsid dermatocranial elements, if a trend can be identified, it would be a reduction of the posterior and medial elements via loss (tabulars and postparietals), followed by dermatocranial-chondrocranial coevolution via fenestration of the temporal regions, leading ultimately to the loss of the quadratojugal, portions of the jugal (see above), and, in a number of non-snake lizards and all snake lizards, loss of the squamosal.

The argument for the suspensorial element of snake lizards being the supratemporal is linked to its position and articulations: (1) the tabular has never, in diapsid history as preserved in fossil diapsid skulls, been sutured to the paroccipital process articulating with the squamosal; that role belongs to a bony element identified as the supratemporal (I base this point on the squamosal as the start point for determining connectivity and topology, not on the postparietal in order to identify a tabular, not to mention the fact that these latter two elements are absent early on in diapsid evolution); (2) using the hatchling *Sphenodon* described by Rieppel (1992) as an example, McDowell's argument of tabular and supratemporal in all lizards would require the classically defined triradiate squamosal of *Sphenodon* to be a supratemporal and the splinter of bone identified by Rieppel (1992) to be the tabular— no supratemporal has ever played such a role in the temporal region of diapsids; (3) in Type I *Dinilysia-Najash* skulls, the "supratemporal" bone is strongly articulated in complex L-shaped fossa formed by the prootic, exoccipital-opisthotic, supraoccipital and parietal, dorsal and lateral to the sagittal plane; (4) that the supratemporal of all snake lizard skulls does not possess a cephalic cotyle as is observed in non-snake lizards with a squamosal, but rather, as in non-snake lizards, the supratemporal articulates with medial face of the quadrate, not the dorsal or cephalic condyle. For

all these reasons, I feel that McDowell's (2008) conjecture of primary homology fails the Test of Similarity and confidently hypothesize that the suspensorial element of snake lizards is the supratemporal. Uropeltids and scolecophidians bear out this conclusion as convergent solutions to quadrate suspension with increasing loss of the dermatocranial component of the temporal region to absolutely zero.

The fossil record of ancient snake lizard suspensorial morphology identifies two differing forms of supratemporal anatomy and connectivities: (1) Type I *Dinilysia-Najash* (Figure 3.22a–c); (2) Type II *Pachyrhachis-Haasiophis* (Figure 3.23c–f).

For Type I fossil skulls (Figure 3.22a–c), the supratemporal is short, thickened, and tightly sutured into an "L-shaped" fossa or facet on the dorsolateral surface of the skull roof, articulating with both chondrocranial (i.e., prootic, supraoccipital, and exoccipital-opisthotic) and dermatocranial bones (i.e., parietal) and extending a short distance past the superior margin of the foramen magnum as formed by the exoccipital-opisthotic. The latter element extends posteriorly, an equal length paroccipital process (nearly the full length of the supratemporal). This morphology is very well described for *Dinilysia patagonica* by Estes et al. (1970), Caldwell and Albino (2002), Caldwell and Calvo (2008), and Zaher and Scanferla (2012) (Figures 2.9; 3.4a; 3.9a,b; 3.22c)—the supratemporal is short, thickened, and tightly sutured into an "L-shaped" fossa or facet on the dorsolateral surface of the skull roof, articulating with both chondrocranial (i.e., prootic, supraoccipital, and exoccipital-opisthotic) and dermatocranial bones (i.e., parietal) and extending a short distance past the superior margin of the foramen magnum as formed by the exoccipital-opisthotic. The latter element extends posteriorly, an equal length paroccipital process (nearly the full length of the supratemporal). The medial face of the suprastapedial process of the quadrate articulates with the lateral facet of the supratemporal, stabilized medially the paroccipital process, thus suspending the quadrate bone (Figures 3.9a,b and 3.22c).

The modern snake lizards included within Type I Anilioid skulls (Figure 3.22d) (see also Cundall and Irish, 2008) show a virtually identical morphology of the supratemporal, and its complex anatomical connectivities as are observed in Type I *Dinilysia-Najash* skulls. Variation is observed around the size and posterior length of the paroccipital process, and whether the supratemporal extends past the posterior tip of the paroccipital of the exoccipital-opisthotic (e.g., *Cylindrophis* versus *Xenopeltis*). Type I Anilioid skulls also trend, in the more derived states, to extreme reduction and eventual loss of the supratemporal altogether (e.g., *Anomochilus* versus *Plecturus*) (see Cundall and Irish, 2008; Olori and Bell, 2012).

In contrast, the Type II *Pachyrhachis-Haasiophis* and Type II Booid-Pythonoid skulls (Figure 3.23a–f) present a very different supratemporal morphology, though the set of anatomical connectivities of the supratemporal are identical to those of Type I *Dinilysia-Najash* and Type I Anilioid skulls. Morphologically, the supratemporal is no longer set in an "L-shaped" facet formed by the parietal, prootic, supraoccipital, and exoccipital-opisthotic. Instead, it forms a fibrous connection to those elements, sitting like a broad spatulate-shaped element on top of or across them and extending a variable distance beyond the superior and posterior margin of the foramen magnum.

Common to the supratemporal morphology of snake lizards from Type II to IV, the element in Types II, III, and IV modern and fossil snake lizards can be extremely large both in length and width, and at its extremes, the anterior process of the supratemporal may extend past the anterior margin of the prootic to against a crest or fossa formed parasagittally on the parietal. Likewise, the posterior process articulating with the quadrate may extend some distance past the superior and posterior margin of the foramen magnum. The reverse of these several points is also observed among Type II–IV modern snake lizards, where the supratemporal is extremely reduced in size or altogether absent. It is reasonable to conclude, in the absence of the Test of Congruence, that these multiple examples of absence indicate convergence on supratemporal reduction and loss, and that these are derived conditions of form within each lineage and likely within snake lizards generally.

### The Quadrate Bone

ONTOGENY

A key aspect of the quadrate bone in snake lizards is its ontogeny. As was reported on by Kamal (1966) for a number of colubroid snake lizards, by Kamal and Hammouda (1965a,b) for *Psammophis sibilans*

(Colubroid, African Sand Snake), and reiterated by Bellairs and Kamal (1981), the quadrate develops from the only condensation remaining of the typical lizard pterygoquadrate cartilage (which is represented by four cartilages in typical lizards). The quadrate appears in Stage 2 (19-day-old) *P. sibilans* embryos and is in a horizontal position articulating with the Meckel's cartilage to form a "rudimentary retroarticular process" (Bellairs and Kamal, 1981:137). By Stage 3, (22 days old) the lower end of the quadrate begins to "rotate" ventrally and posteriorly toward a vertical position. By Stage 7 (44 days old), the mandibular condyle of the quadrate has rotated posteriorly past the vertical such that the entire element is posteriorly angled from the articulation of the quadrate cephalic condyle with the supratemporal, a position retained in full adults (Bellairs and Kamal, 1981: fig. 55). Similarly, Scanferla (2016) reported on a growth series of *Boa constrictor* and illustrated postnatal ontogeny where the orientation of the quadrate rotates from slightly anteriorly directed in smaller neonates (299 mm snout vent length) to strongly posteriorly directed in adults (snout vent length of 1620 mm). In contrast to alethinophidian snake lizards, scolecophidians present a horizontal quadrate in adults and presumably the same holds true through ontogeny upon first appearance of the element, consistent with the reported embryology of other snake lizards. Unfortunately, as Bellairs and Kamal (1981) lamented, "No developmental studies of the skull have been made on members of these families," and therefore ontogenetic data are sorely lacking (though recent studies such as Palci et al. [2016] are moving to rectify this problem). However, of great importance to both the polarity of such a character and its states in all lizards, snake lizards included, is that in non-snake lizards as the quadrate separates from the other elements of the pterygoquadrate condensation early in development, it is steeply angled anteriorly, though never matching the degree of horizontality observed in snakes except in amphisbaenians (Bellairs and Kamal, 1981). In most other groups of non-snake lizards, as ontogeny progresses, the quadrate rotates posteriorly, though it is seldom vertical or exactly perpendicular to the long axis of the skull. The question must then be asked, "In relationship to lizard phylogeny, is the scolecophidian condition primitive or pedomorphic?" My opinion needs to be made clear—pedomorphic—and I will elaborate further on this interpretation and its arguments later in this chapter.

Another important consideration in snake lizard suspensorial anatomy and evolution is both the morphology the quadrate and its orientation between the supratemporal-braincase and how these morphologies function and are characteristic of Type I–IV skulls, modern or ancient (see Caldwell and Lee [1997] and Caldwell [2000a], versus Zaher and Rieppel [1999a]; Polcyn et al., [2005]; and Apesteguía and Zaher [2006]). In Type I Anilioid skulls and Type I *Dinilysia-Najash* skulls, the quadrate is unique in its morphology and orientation. The quadrate is always oriented vertically, though the shaft is slightly sigmoidal, thus giving the appearance of a slight anterior "lean" of the proximal portion (Figure 3.22a–d). In terms of morphology, the quadrate is a "C"- or inverted "J"-shaped element with a deep medial conch and a large and robust shaft that is broad mediolaterally but thin anteroposteriorly. It has a large and often flat-topped cephalic condyle that does not articulate with any cranial element (i.e., squamosal is absent) and an extremely large and posteriorly directed suprastapedial process to which the stylohyal-extracolumellar element articulates. The medial face of the quadrate presents a broad, flattened facet that articulates with the supratemporal.

In contrast, for Type II–IV skulls, quadrate morphology is rather uniform (Figure 3.23a–f), though there are some exceptions. To generalize for more than 3000 species of modern snake lizards, the quadrate is an elongate to extremely elongate bone with little or no suprastapedial process development and a small cephalic condyle (Cundall and Irish, 2008). The stylohyal-extracolumella cartilage(s) articulate with or fuse to the quadrate shaft. As for orientation, the quadrate can be: (1) vertical or close to vertical; (2) horizontal, where the mandibular condyle is directed anteriorly; (3) angled posteriorly, at an extremely wide variety of angles, where the mandibular condyle is directed posteriorly. Complexity linked to function in Type II–IV skulls is not around quadrate morphology, but rather quadrate bone simplicity with an emphasis on quadrate orientation. As a functional system linked by developmental and anatomical outcomes, quadrate function as the cranial suspension element for the compound bone of the mandible means that when the latter elongates (and gape size increases), the posterior angle increases to meet that more posterior compound bone and correspondingly the quadrate must also become more elongate.

### The Ancestral Snake Lizard Suspensorium

Both Type I and II ancient snake lizard skulls are known from the Early Cenomanian (terrestrial Gondwana for *Najash* and marine aquatic for *Pachyrhachis*, *Haasiophis*, and *Eupodophis*). The most reliable data (i.e., most completely preserved) is Cenomanian in age and demonstrates unequivocally that the Type I Anilioid and Type II Booid-Pythonoid skull types had evolved much earlier, as these skull types are well developed in Type I *Dinilysia-Najash* and Type II *Pachyrhachis-Haasiophis* category skulls. In transformational terms, the fossil forms are clear antecedents to the categories defined by the modern forms and thus the modern would be subsumed within the fossil. If the unnamed South African braincase is accurately interpreted (see Chapter 2) (Figure 3.22a), then the Type I *Dinilysia-Najash* skull condition was in place even earlier (Valangian; Early Cretaceous), placing the origins of that skull category, and the clade it likely defines (though this is untested at this point), well back at some undefined point in the Jurassic.

While the absence of evidence is not evidence of anything at all—except its absence—the currently available data suggest that the ancestral snake condition for the suspensorium is significantly older than the Valanginian (Early Cretaceous), likely older than the Kimmeridgian (Late Jurassic). The known Middle to Late Jurassic snake lizards (Caldwell et al., 2015) do not yet provide any definitive data on the suspensorium. However, Type I *Dinilysia-Najash* and Type II *Pachyrhachis-Haasiophis* can be traced to the Valanginian (Early Cretaceous) and Cenomanian (Late Cretaceous) and have no anatomical traits that would suggest they share anything but a common ancestor that was not itself a snake lizard. That is to say, Type I *Dinilysia-Najash* is not demonstrably a series of progressive types evolving a suspensorium system "toward" the condition eventually demonstrated by Type II *Pachyrhachis-Haasiophis*. Such a hypothesized transformation series would suggest that the absence of evidence is indeed evidence and that snake lizard evolution proceeded along a simplistic scalae naturae of sorts. However, as I interpret it, the absence of evidence speaks to the opposite—snake lizard suspensorial evolution, while no doubt linked developmentally, functionally, and evolutionarily to all the other features so far discussed here in Chapter 3, proceeded independently in a number of early lineages of ancient snake lizards. Fossil evidence points to Type I *Dinilysia-Najash* as well established in the Valanginian, which makes the origin of this type even older, potentially Oxfordian or Kimmeridgian; Type II *Pachyrhachis-Haasiophis* is not yet excluded from the same or older geologic ages as there is no evidence yet beyond the Cenomanian. The balance of modern snake lizards possess Type III and Type IV skulls, which in morphological terms can be recognized as derived from Type II skulls (e.g., pedomorphosis and peramorphosis can sensibly explain derived states of III and IV as transformations of Type II), while Type I suspensorial anatomies are unique and ancient but still snake lizard and not typical lizard.

The extrapolation is that the ancestral snake lizard suspensorium is not characterized by Type I *Dinilysia-Najash* and Type II *Pachyrhachis-Haasiophis* skulls but remains more typically lizardlike. Potential candidates among typical lizards for modeling such a suspensorial form are found in basal pythonomorphs such as adriosaurs or among living forms such as *Lanthanotus borneensis*, as was argued by McDowell and Bogert (1954). In this cryptic varanoid lizard, the supratemporal is large (McDowell and Bogert's [1954] tabular [see above discussion]), the postfrontal is small, there is no postorbital bar, and the squamosal is present but reduced to a small element with no anterior ramus and joint with the postorbital. Importantly, though, the tiny squamosal remains the principal articulation for the quadrate bone, though the supratemporal does articulate with the medial face of the quadrate, as is the case in all snake lizards. This condition would be expected in the ancestral snake lizard prior to the evolution of the complete loss of the squamosal and postorbital. It is also important to state that while *Lanthanotus* displays a predicted condition in the suspensorium of the ancestral snake lizard, *Lanthanotus* itself is not such an ancestor, not a sistertaxon of snake lizards, but rather is a derived varanoid lizard presenting a reduced morphology of the suspensorium.

## Gape Size: Micro versus Macrostomate?

I will conclude my consideration of skull features in ancient and modern snake lizards with a summary of sorts focused on not one feature, but rather on the functional and evolutionary synergy of numerous features that affect and promote gape size. Many of these features have been discussed in detail in this chapter already, but have been considered in the context of their specific or even atomized function and not as a part of a characteristic feature of snake lizard evolution. Where birds are characterized by their evolution of feathers and the achievement of powered flight, most modern snake lizards are similarly characterized by their evolution of a remarkable number of kinetic joints in the skull and their achievement of the ability to swallow prey that are significantly larger than their heads and bodies (macrostomy). Limblessness and elongation are not unique to snake lizards, but variation around gape size most certainly is and comes in a variety of morphologies and functions supported by extremely unique anatomies (hard and soft tissues) that run the gamut from primitive big-mouthed non-macrostomy to macrostomy, extreme macrostomy, and likely the most derived form of macrostomy, microstomy, as displayed by maxillary raking and mandibular raking in scolecophidian snake lizards.

Consistent with the core theses of this book, both great and small, the modern science behind our understanding of gape size evolution is heavily burdened by the history of data, not-so-data, and the interpretations of both that have brought us to the present state of "knowledge"; that is, that small gape is primitive and large gape derived. And, as always, the human and scholarly predilection for essential features, idealized anatomy, and archetypes will swiftly bubble to the surface as I explore the history of thought on gape size in snake lizards and follow that with the data that have been used as evidence for the arguments rendered.

### History I: Developing the Archetype

As comparative anatomy was being born as a science in the mid-seventeenth century, these first anatomists had to start somewhere, and by happenstance, several classical studies of comparative vertebrate anatomy reported on various aspects of the anatomy of snake lizards, in particular, vipers. While venoms and their toxicology, and "magical powers" or not, were the primary focus of these early works (e.g., Redi, 1668; Charas, 1669), significant progress was still being made on the anatomy of snake lizards as compared to other vertebrates and to each other (as an aside, it is also a historical fact that these early students of herpetology were academic luminaries, including the likes of the renowned scholar Francesco Redi, who almost singlehandedly invented experimental biology, and in particular is recognized as a leading contributor to the demise, if you will, of the myth of spontaneous generation). A particularly insightful work by one of Redi's contemporaries, the English comparative anatomist Tyson (1683), focused not on the venoms and toxicology of vipers, but on the anatomy of a rattlesnake made available to him, which he dissected and compared to a common viper; a key innovation of this work was the detail of the comparative anatomy and the attempts made to link that anatomy to the function of the bony skeleton and its musculature. On these points, Tyson (1683:573) wrote, speaking of the upper jaw with its fang, his "upper maxilla":

This articulation seems advantageous…but its principal and most remarkable advantage for swallowing large bodies, is the curious articulation of the maxillae backwards to the cranium by 2 bones, which from their use (since we know no name to distinguish them by) we shall call maxillarum dilatores.

And:

Their shape, size, and aptness for this motion… For the lower jaw being not conjoined at the mentum…but parted at a good distance; upon the receiving a large body, as the membrane here to which they are fastened easily extends… it must needs considerably widen the rictus of the mouth…

His figure caption is also valuable text in understanding the anatomical nomenclature he applied to bones (Tyson, 1683:576) so as to understand his assessment of function and gape, or what he termed dilation.

Working from Tyson's text, figure and figure caption (1683: fig. 6), it is necessary to first decipher his anatomical nomenclature and identify modern corollaries in order to then understand his conception of gape size: (1) "kk," the "lower maxillae" references the dentary of the lower jaw (Tyson also recognizes the intramandibular

joint and a "mentum" or intermandibular joint, where the "lower maxillae" are "not conjoined" as in other vertebrates; (2) "ee," is the prefrontal, the small bone lying between the true maxilla (which Tyson thought was an ear bone) with the "poisonous fang" and the cranium; (3) his "upper maxilla" is the toothed pterygoid-palatine complex; (4) the "dilators maxillarum" or "maxillorum dilatores" are the quadrate bones, which articulate "with the upper and lower jaw," that is, the compound bone and the pterygoids-palatines (his upper maxillae); (5) "oo," the short bone joining the "dilators" to the cranium, or the supratemporals.

Anatomical nomenclature made clear, the key point here is that Tyson (1683) understood from his comparative anatomy the functional interactions of these numerous elements of the suspensorium and the inter- and intramandibular joints in controlling gape size and in protracting the fangs by movement of the maxilla (see the boldface text in the first quote) and in widening the gape by extension of the quadrates and spreading the tips of the dentaries ("most remarkable advantage for swallowing large bodies," and "it must needs considerably widen the rictus of the mouth").

As early as the mid- to late 1600s, the stage was being set by groundbreaking comparative anatomical works that recognized the relationship between osteology and function and could describe the achievement of great gape size and cranial kinesis in snake lizards. Some 120 years later, when Oppel (1811) characterized Squamata, he recognized two groups, "Saurii" and "Ophidii." While neither of these two groups is identical to the modern constitution now employed of "Lacertilia" or "Serpentes/Ophidia" (e.g., his "Ophidii" included amphisbaenians), Oppel (1811:14) wrote of "Ophidii" as differing from "Saurii" because:

Sternum nullum, pedes nulli, maxillae saepissime dilatabiles.

And translated:

No sternum, no feet, jaws again dilate.

Following on the comparative anatomical work of luminaries such as Tyson (1683), Oppel created classifications based on the fact that typical lizards could not "dilate" their jaws to widen their gape, while snake lizards most certainly could. For Oppel (1811), consistent with Tyson (1683), all of whom were naturally working in a pre-evolutionary paradigm, this was not only perfectly sensible, but frankly quite harmless because classification systems and comparative anatomy and its similarities had no linkage to either an underlying process or pattern, let alone one of evolution, except perhaps as revelation of the orderly creation of living things by God. Like all scholars of his time, Oppel, if challenged to identify his underlying theory set, would have professed to some form of essentialism and the recognition of bauplans and archetypes in nature. In stating this, I note that Oppel (1811:46–73) continued his treatise by recognizing seven families of snake lizards: Anguiformes, Constrictores, Hydri, Pseudo-viperae, Crotalini, Viperini, Colubrini. Of consequence here, he (Oppel, 1811: 53) distinguished his first grouping, "1. Familia Anguiformes" as follows:

Caput corpore fere minus, aut ab eo minime distinctum.

And translated:

The head of the body is small, and not distinct from it.

Important to the point of this citation of head size and its lack of distinction from the body is the generic composition of the "Anguiformes": Amphisbaenia sp., Typhlops sp., and Tortrix sp. (today, the included species in this genus are in the modern snake lizard genera Cylindrophis and Anilius). All of the snake lizards and non-snake lizards (Amphisbaenia) included in Oppel's "1. Familia Anguiformes" are, of course, either small-gape snake lizards or small gape non-snake lizards (Cundall, 1995; Kley and Brainerd, 1999; Kley, 2001). The remaining descriptions for the other six families note distinctively large heads that are well defined, and though none are distinctly linked back to the "maxillae saepissime dilatabiles," it is clear that large head size, and by extension gape size, is a feature of the six higher groups of "Ophidii" that are not present in his first and most basic family.

With head size and jaw dilation in place as diagnoses of the various groups of "Ophidii," subsequent authors proceeded to further refine species membership in various groups, usually by refining the diagnoses around more accurate morphological characterizations, and, of course, with the addition of literally hundreds of new species and genera. The value of the gape category, however, did not go away. Müller, continuing forward from Oppel (1811) wrote (1831:265), translated as follows:

The snakes can be conveniently brought into two large divisions, in those of the large-mouthed or Macrostomata, and in those of the small-mouthed ones or dim-sighted snakes, Microstomata.

Müller (1831:265) continues on this same page to then very accurately describe the anatomies and morphologies of his "Ophidia macrostomata" that are associated with supporting a large gape, such as the "längere oder kürzere bewegliche Ossa mastoidea = longer or shorter Ossa mastoidea" (=supratemporal bone), and the, "woran die Quadratbeine allein eingelenkt sind...= where the quadrate alone are deflected" And continues to both the palatal bones and the intermandibular joint (Müller, 1831:265), translated as follows:

Also their pterygoids and palatines move away from each other, and their lower jaw halves through tensile straps are fastened together and are capable of remarkable extension.

Müller (1831:266) continued following his characterization of his Ophidia Macrostomata with his second group, the Ophidia Microstomata, and note that the mastoideum is the supratemporal bone (translated as follows):

The second division of the serpents I call Ophidia microstomata, small-mouthed or narrow-mouthed, because they have a narrow non-expandable mouth and throat have and their quadrate on the skull itself and not suspended by a mobile mastoid.

Müller (1831:266–267) characterized their collective anatomy (genera Chirotes, Amphisbaena, Lepidosternon, Cephalopeltis, Alanus, Typhlops, Rhinophis, Uropeltis, Tortrix—a mixed bag of amphisbaenians, scolecophidians, anilioids and uropeltids), translated as follows:

The Ophidia microstomata are small mouthed, they can reach the pharynx and do not dilate the jaw apparatus for catching. In Tortrix, although

THE ORIGIN OF SNAKES

the lower jaw halves are loosely connected, for their remaining jaw apparatus no expansion is possible and their small quadrate is...hung from the skull itself.

And:

The small mouthed snakes have therefore an immovable mastoid fused with the skull, or none at all. All lack the orbital posterius; the eye pit therefore is completely open at the back.

Müller (1831:269–272) built on Oppel's (1811) classification, not just with his recognition of gape size, but by recognizing four families of "*Ophidia microstomata*" and seven families of "*Ophidia macrostomata.*" And, ever more assuredly, he created definable, diagnosable, and most certainly more sensible categories of snake lizards built around essential features of large versus small gape size. His work was detailed, anatomical, clear, concise, and very empirical—Müller (1831) recognized the solidly sutured supratemporal of Type I Anilioid skulls and the fact that Type IV skulls as presented by scolecophidians did not demonstrate the presence of a supratemporal. He correctly interpreted the tight intermandibular "joint" of *Cylindrophis* (see Cundall, 1995), and even considered the presence/absence of a jugal or postorbital as well. However, Müller (1831) also more firmly established a preconditioned stage for future works on snake lizards, where small head and gape size, regardless of differences anatomically that are not comparable (amphisbaenians are not morphologically similar to either scolecophidians, and the latter do not compare well at all to anilioids) were sufficient categories of similarity to require recognition of exclusion from a category reserved for all other snakes.

Müller (1831) also created essential gape size categories that would ultimately, and uncritically, be co-opted into scenarios of evolutionary transformation. The firmly established essential characteristic of macro- versus microstomate snake lizards could now transcend to the transformational with microstomate snake lizards as primitive. This obviously so as typical lizards are not macrostomate, but are microstomate, even though their skulls are extremely different from those of any modern microstomate snake lizard that Müller (1831) would have classified in his "*Ophidia microstomata.*"

In the nearly 150 years from Tyson (1683) to Müller (1831), the comparative anatomy of gape size had been established, along with the idea that you could be a snake but not have a large gape size. What was created by these scholars was the important foundations of classifications based on anatomy that would lead to a linear transformation set of small to large, that is, microstomatans to macrostomatans, looking for an overarching theory set (e.g., evolution) as the explanans. What would be hard to overturn, though, would be the synonymy of small gape and its association with snake lizards that were blind and degenerate and few in number, as compared to those snake lizards with a large gape and large heads that were clearly not blind, nor degenerate, but rather were advanced and specialized.

### History II: Fossils Bring the Potential to Reorganize the Modern "Essentials"

I wrote in Chapter 1 that Professor Edward Drinker Cope was the "Father of the Gilded Age of the Fossil." I made this claim based on his contributions, both empirical and theoretical, to the study of snake lizard evolution, origins, and relationships, and to the fact that he was the first scholar to consider these questions with the addition of data from fossils (Cope, 1869). But here I want to refine how Cope used fossils, combined with the concept of evolution and the anatomy of modern snake lizards, to reduce the essentialist categories of the archetype concept to zero (intentional or not). Only a few decades after Müller (1831), Cope (1869) introduced an entirely new data set, with a revolutionary paradigm backing it up that in no way supported a simple linear evolutionary progression of the first micro- then macrostomate condition for snake lizards. Wittingly or unwittingly, Cope (1864) had already begun this process when he cleaned up the lingering misclassification problems in the works of Müller (1831), Wagler (1830), and Duméril and Bibron (1844) by removing amphisbaenians from the Scolecophidia, thus limiting "Ophidia microstomata" to just scolecophidian snake lizards. Cope (1869) detailed the ten shared anatomical features of snake lizards and mosasaurs as he saw them and thus modified the diagnosis of Ophidia (note: Cope's opisthotic is the supratemporal in current parlance). On the shared anatomies facilitating a large gape in snake lizards and mosasaurian lizards, Cope (1869:253) wrote of the latter:

> 5. There is no symphysis mandibuli....7. The subarticular and splenial elements of the mandible are connected by articular faces.

And on page 257:

> That terrestrial representatives...inhabited the forests and swamps of the Mesozoic continents...their habit was to devour whole is evident...though the articulation of the lower jaw will not admit of as much extension as that of the Ophidia, it exceeds other reptiles in... the lateral motion of the splenial articulation.

Intentional or not, Cope's (1869) work on the comparative anatomy of modern and fossil reptiles found snake lizard characters, long considered essential and exclusive to modern snake lizards, in fossil non-snake lizards. Such characters of anatomy had long been considered and conceived of as exclusive and essential to the archetype of "snake" and "snake only" (e.g., the intermandibular joint and intramandibular joint [see Tyson, 1683]). In truth they were not and are not exclusive. For evolution as change over time to operate, they cannot be exclusive; logic compels this conclusion...just like feathers are not the exclusive domain of birds, though it took the fossil record a while to support this logical conclusion with data.

As I argued in Chapter 1, Cope's critics (Owen, 1877; Marsh, 1880; Baur, 1892, 1895, 1896; Boulenger, 1893; Nopcsa, 1903, 1908, 1923, 1925; Williston, 1904; Janensch, 1906; Bolkay, 1925) interpreted Cope's ideas to mean he was arguing that snakes evolved from mosasaurs (this malaise continues; see Rieppel [2012]). And, again, nothing could have been further from the truth, nor remains further from the truth now, but, as many of these critics were burdened by their essentialist archetype framework (e.g., Richard Owen), it is unsurprising that Cope was harangued for such a heretical approach to data and hypothesis construction. Unfortunately for Cope (1869), there were no fossil snake lizards available for him to compare to mosasaurian lizards in order to better understand gape evolution. Even the materials reported on by Nopcsa (1908, 1923) possessed no reliable cranial data revealing anything about gape size and cranial kinesis that would have assisted Cope's hypotheses. And, to find even more

support in the value of fossils for exacting severe constraints on essentialism and archetypes, Nopcsa (1908, 1923) also did not consider scolecophidians to be primitive snakes, but rather derived within his concept of snake lizardness from within the radiation of the fossil groups he was aware of, which themselves were not microstomate. Thus, for Nopcsa as for Cope, there was no linear transformation in gape size from small to large.

As should be abundantly clear, what is building in this essay so far is a clear need to define what gape is, what micro- and macro- are, and how these anatomies are linked to classifications, accurately and inaccurately. But there is more history yet to understand in the context of thought and burden on gape size and gape size evolution before I get to those points.

### History III: Essentials of an Archetype Re-Evolved

As discussed in Chapter 1, shortly after Nopcsa's death in 1933, the fossil record ceased to be of consequence to the discussion of snake lizard gape evolution, because there was no fossil record, and because everyone, including paleontologists, had ignored the data present in *Dinilysia patagonica* Smith-Woodward, 1901. As the fossil record was woefully unable to assist in providing insights, scholars dove back into the details provided by modern snake lizards. This necessity meant, quite frankly, that the arguments and data sets regressed back to "evolutionary essentialism," in this case a top–down essentialism, as dictated by a data set limited to the modern assemblage. Twenty-eight years after Nopcsa (1908, 1923), Bellairs and Underwood (1951:228) wrote:

> For a long while, perhaps for the greater part of the Jurassic, the ancestors of snakes, which may have been derived from the basal platynotid stock, pursued a subterranean existence along with other groups of more or less fossorial lizards which lacked the potentiality of becoming snakes.

And:

> The available data on the snakes of the Cretaceous contribute little towards an understanding of the early history of the suborder. It seems likely, however, that during this period the evolution of the group entered a new stage and the remaining essential features of ophidian structure were differentiated.

And finally:

> It is suggested that the resumption of epigean life and subsequent wide radiation of the snakes was made possible by the deployment of three evolutionary trends: re-elaboration of the eye, development of a wide gape, and increase in the efficiency of the mechanism of locomotion.

Without belaboring the meaning of "potentiality" and "essential features," the burden of a simple linearity of progressive evolution, and the essential features delimiting snake lizards, had clearly returned to the conversation, the evolutionary paradigm notwithstanding. The scenario envisioned by neontologists required the data of the modern assemblage to be interpreted against a "subterranean" habitus of the proto-snake lizard. The logical progression arising from this hypothesis/scenario/transformation, was, and has remained, that from among modern snake lizards, the burrowing, degenerate, blind, scolecophidians were the living remnants of this subterranean origin, and they are microstomate (see Chapter 5). Ergo, the microstomate condition is primitive for snake lizards, though this is fraught with logical and empirical complications.

For Bellairs and Underwood (1951), their notion of the essential "snake" held that the three key, essential features of snake lizards (re-elaborated eye, wide gape, and locomotory style) did not develop until the Cretaceous, and that the resumption of the snake lizard "epigean life" was not fully realized until these features evolved (see Chapter 5). This, of course, makes no sense whatsoever. No matter how fantastical one's imagination can be, no adaptive mechanism can possibly be envisaged in any empirical manner for a Jurassic-aged burrowing, wormlike, ancient proto-scolecophidian (note: the subterranean Jurassic forms were not snakes, but had the potential to become snake lizards) to evolve a large gape, re-elaborate the eye, and evolve alternative locomotory capabilities. Not to mention, why in the Cretaceous?

Taken to its logical expectation of outcomes, the arguments and data alluded to by Bellairs and Underwood (1951) do not even actually support the notion that a small-gaped animal like a modern scolecophidian should be considered a snake lizard, but rather they would be a protoform of stem snake lizard that never resumed its epigean habitus. Now,

I do not share this notion because I disagree with the logic they applied to their interpretations of the data, the scenario they developed as a result, and the approach taken by all subsequent and previous authors to such archetypes and their essentials—I see scolecophidians as a polyphyletic assemblage of highly derived, degenerate pedomorphic snake lizards, nothing more. They are regressed macrostomatans, not exemplars of the ancient archetype, and are incredibly distantly related to such basal snake lizard lineages such as anilioids—in fact, as I will argue in Chapter 6, garter snakes are more closely related to *Cylindrophis* than is *Afrotyphlops*. There are lineages of primitive snakes among the modern assemblage, but the scolecophidians are not among those lineages even though they are microstomous by comparison to other macrostomous snake lizards. However, I will get back to the balance of this chain of thought on snake origins in Chapter 5, but here, these ideas are explored around the key innovation of gape size as an essential feature of snake lizardness and for what subsequent scholars inherit from the era of thought represented by Bellairs and Underwood (1951) and their contemporaries (e.g., Gans, 1961; Haas, 1973).

### History IV: Big Heads, Big Mouths, Fossils, and the "Regressed Macrostomatan"

After Caldwell and Lee (1997) and the reinjection of fossil snake lizard data back into the debate, and with a host of new fossil data published in the decade after, Caldwell (2007a:291) wrote:

> In the final twist on this debate, at a point when it seemed that no additional new hypotheses might be forthcoming, Apesteguía and Zaher (2006) removed the Scolecophidia from their safe-haven as the basal-most snakes, and instead reconstructed the large-bodied, large-mouthed *Najash* as the basal snake plesion.

And:

> I would argue that their topology unequivocally supports an origin of snakes from a large-bodied, large mouthed snake.

I still see my 2007 review was prescient in a sense, as it rather closely predicted the counterarguments of Rieppel (2012) who posited instead that such a scenario required small-mouthed snake lizards to be "regressed macrostomatans," a term communicated to him by Garth Underwood. Instead of microstomate snakes being regressed macrostomatons, Rieppel (2012) preferred instead to call upon re-evolution and re-elaboration paradigms for a host of features (e.g., gape, eyes, limbs, etc.) to defend microstomate gapes as primitive. Rieppel's (2012) essay was formed as an argument against the increasing number (Apesteguía and Zaher [2006] among them) of then recent morphology-based phylogenies of snake lizard sistergroup relationships that had regularly been placing fossil taxa as basal to scolecophidian snake lizards + alethinophidians (note: nothing has changed in this regard, as more and more phylogenies have been proposed since that do not place scolecophidians at the base of the tree [e.g., Longrich et al., 2012; Caldwell et al., 2015; Martill et al., 2015]). Rieppel (2012:99) phrased his dissatisfaction with such phylogenies and their logic schemes for gape evolution as follows:

> Several recent morphology-based phylogenetic analyses of snake interrelationships...have placed fossil taxa of large body size and/ or with a macrostomatan skull structure basal to either all extant snakes, or basal to the Alethinophidia...This has led to the characterization of scolecophidians and/or anilioids as "regressed macrostomatans".

Rieppel (2012) explored the implications of this idea briefly by noting that scolecophidians as "regressed macrostomatans" means these snake lizards would have lost their macrostomatan feeding capacities as they adapted to a fossorial lifestyle, linked no doubt to miniaturization. However, such a scenario was incompatible with the data on the muscle morphology of the head skeleton. Rieppel (2012:100) began as follows:

> Such a conclusion is in conflict with earlier interpretations of the evolution of ophidian jaw mechanics...it renders scolecophidians and other non-macrostomatan snakes (*Anilius*, *Cylindrophis*, *Anomochilus*, and uropeltines) so-called "regressed macrostomatans" (Underwood, pers. comm.) that have lost macrostomatan features in adaptation to a fossorial or secretive mode of life.

As he developed his argument around muscle morphology and microstructure, his arguments "flex" by moving from taxic descriptions of anatomy to transformational arguments (Rieppel, 2012:100):

> However, as the snake jaws work themselves across the relatively large prey item, the jaw adductor muscles undergo significant passive stretching, which in turn requires changes in the jaw adductor muscle architecture.

And:

> But long- and parallel-fibered jaw adductor muscles that characterize macrostomatan snakes (Zaher, 1994) do not represent the plesiomorphic sauropsidan condition. It has long been recognized that the complex multipinnate jaw adductor muscle architecture is remarkably conservative across Sauropsida...

And the final clause linking muscle morphology + transformational scenario to taxa (pp. 100–101):

> Non-macrostomatan snakes (scolecophidians, *Anilius*, *Cylindrophis*, *Anomochilus*, and uropeltines) all retain a complex multipinnate jaw adductor musculature, a muscle architecture that is furthermore closely comparable with that of lizards...and hence represents the ancestral condition for snakes (given that snakes are deeply nested within squamates in all phylogenetic analyses reviewed above).

The final clause completes the equation and states unequivocally that all non-macrostomate snake lizards retain a complex multipinnate jaw adductor musculature, that this is "closely comparable" to typical lizards, and that this similarity means that it is the "ancestral condition for snakes." The remainder of this examination of gape in snake lizards will examine what big versus small gape, and big versus small mouth, actually is, and the empiricism and factual accuracy of claims made regarding bones and muscles in small and large gape snake lizards, as well as a discussion of what "closely comparable" means in relationship to the Test of Similarity, and the untested conclusion that "hence," small gape in all of its forms is the ancestral condition.

### Morphology I: Small Gape, Big Gape, Chicken or Egg?

With history somewhat lined up (see above), the core argument on morphology I will make is that

THE ORIGIN OF SNAKES

there is significantly more variation in the mouth of snake lizards than is captured by the simplistic dichotomy of micro- versus macrostomy, or small and large gape size. The linear transformation from small to large gape in no way reflects the fossil record and its linkage to phylogenies (see Chapter 6), but rather is seen to be "true" or "accurate" for the simple reason that it is the very same weak data that were used to create the phylogeny in the first place (as an example, see character transformidans in Gauthier et al., [2012]). A more comprehensive understanding of the existing literature on the distribution of mouth size, gape size, the assignations of micro- and macrostomy, and where that literature is constrained by archetypes and essentialist categories is critical to evaluating the character state constructions that have created the self-prophecies of the linear transformation series and what I see as the continuing confusion on gape size evolution.

In this examination of morphology and function, I will reference Cundall (1995) quite heavily, as well as Cundall's subsequent work on the details of gape-facilitating anatomy in a variety of snake lizards (Cundall and Greene, 2000; Close and Cundall, 2014; Close et al., 2014; Cundall et al., 2014). This is because I see Cundall's (1995) study of *Cylindrophis* and his subsequent work on other snake lizards as having revealed a greater diversity of morphologies and function than the usual dichotomy of micro- versus macrostomy implies. Directly and indirectly inspired by Cundall's work are a small number of studies by other research groups that have investigated the complexities of gape size evolution in other snake lizards, almost all restricted to Type III and Type IV snake lizards (Kley and Brainerd, 1996, 1999; Queral-Regil and King, 1998; Kley, 2001; Vincent et al., 2007, 2009; Herrel et al., 2008; Hampton and Moon, 2013) with the exception of Segall et al. (2016), who also examined Type I and II snake lizard heads.

Conventional approaches to defining macrostomy among modern snake lizards mean that a large number of species, nearly 99%, in fact, are macrostomate—this includes all species in my Type II, III, and IV skull categories, with the exclusion of scolecophidians (in an interesting twist of proposed phylogeny and function, which I will discuss below, the Type I Anilioid skull types of *Xenopeltis* and *Loxocemus* are considered "macrostomatans" [e.g., Frazzetta, 1999; Gauthier

et al., 2012; Reeder et al., 2015]). However, macrostomy and microstomy, along with big gape, small gape and big mouth, small mouth, are also very poorly defined, incorrectly assigned, and confusingly synonymized such that gape reduces down to a simplistic and unrealistic dichotomy.

Macrostomy/big gape is usually characterized, as it has been since Tyson (1683), as the functional capacity to eat prey of a size (dimension) greater than that of the snake lizard's head; some modifications to this kind of description follow the lines of that characterization as given by Hampton and Moon (2013:194):

> Although many cranial elements probably contribute to gape, it is typically estimated from jaw length or jaw width, or occasionally from a combination of these two measures.

While this seems simple enough, and in fact is sufficient for most studies focusing on ecological aspects of feeding, function, and behavior, it is not sufficient as means of characterizing anatomy and its morphology for phylogenetic and evolutionary studies. As an example, the meaning of jaw length is unclear. Does this mean toothed portion and thus dentary only or dentary plus compound bone? And, if it is the latter, then how do you account for non-macrostomy in one taxon versus macrostomy in the second, when two "jaws" of absolute equal length are examined? For example, how do we resolve this question when comparing *Cylindrophis* and *Naja*, where the dentary is the same length as the compound bone in the first, and in the second taxon, the compound bone is 2.5× the length of the dentary?

From my perspective, the required level of detail is exemplified by the anatomical and functional studies of Cundall (1995) for *Cylindrophis*, and Close et al. (2014) and Close and Cundall (2014) for higher snake lizards. There are surprisingly few studies that present the appropriate detail permitting a diagnosis of macrostomy (Scanferla, 2016; Segall et al., 2016), microstomy (Kley and Brainerd, 1999; Kley, 2001), or the third condition of non-macrostomy as displayed by Type I Anilioids (e.g., *Cylindrophis* [Cundall, 1995]) if they do not describe the anatomical features that must act in synergy to support gape size. In terms of considering fossil data in such analyses Frazzetta (1970) still remains the only focused study on the functional morphology of an ancient snake lizard

(*Dinilysia patagonica*). The only recent study on the evolution of macrostomy that involved fossil snake lizards was that of Scanferla (2016) who added to the discussion insights on cranial ontogeny from both extant and fossil snake lizards (i.e., *Dinilysia patagonica*), as well as general data and observations from ancient snake lizard taxa.

Typical lizards have big mouths and use a variety of feeding mechanisms—among carnivorous lizards, this includes inertial feeding among those that ingest larger prey item (e.g., *Varanus* can ingest very large prey and have a very large mouth). In complete contrast, scolecophidian snake lizards have a very tiny mouth even though their skull osteology is Type IV, and their musculature is a somewhat simplified form of the caenophidian condition (see also Johnston, 2014) resulting in highly specialized and diverse feeding mechanics within the three principal groups (see Kley and Brainerd, 1999; Kley, 2001).

In no certain order, the anatomies, morphologies, and behaviors linked to gape size, either promoting large size or in restricting it, include: (1) supratemporal extension posterior to basioccipital process to its quadrate joint; (2) quadrate vertical to inclined posteriorly extending its joint with mandible even more posteriorly than contact with tsupratemporal; (3) suspenorial joint between supratemporal and quadrate (common to all snake lizards with a supratemporal) on medial face of quadrate; (4) compound bone portion of mandible elongate (i.e., extends past craniomandibular joint; (5) dentigerous portion of dentary extends posterior to orbit; (6) bony crest development for CID muscles on basicranial elements; (7) maxillae "mobile" with respect to cranium, and functionally linked to dermal palate via ectopterygoid articulation; (8) palatines toothed and independently "mobile" within dermal palate; (9) maxillae lack sutural contact with small and toothless premaxilla; (10) intramandibular joint permits flexure of mandible at midpoint; (11) intermandibular symphysis absent (degree of jaw tip separation variable, but extreme in henophidian snake lizards); (12) anterior and ventral displacement of tracheal opening to facilitate breathing during ingestion; (13) soft tissues throughout buccal and pharyngeal area capable of extreme stretching and elasticity (Close and Cundall, 2014; Close et al., 2014); (14) rearrangement of gut tube to anterior position in buccal region (see Cundall

et al., 2014); (15) constriction as a behavior and constriction functionally as a coordination of postcranial anatomy; (16) a wide variety of behaviors associated with prey stabilization for ingestion (coiling, neck bending, etc.); (17) strike position (head on, lateral striking, etc.).

The number of anatomical features and their resultant composite morphologies and behaviors that must act synergistically to facilitate macrostomy or microstomy in snake lizards is astonishing. In fact, it hints at great complexity and specialization and in no way can be modeled to have evolved easily, just once, nor in a linear transformation pattern from small to large throughout snake lizard evolutionary history (e.g., see arguments given by Shine and Wall [2008] and Caldwell et al. [2015]). It is also, as a result, unlikely to have been a specialization of snake lizards early in their evolutionary history, but rather is a late-stage specialization taking full advantage of numerous other evolutionary innovations, as well as sidesteps and missteps in that process, resulting in a specialized functional morphology of the most recent radiation of snake lizards, the caenophidians, where it reaches its acme (note: I am referring here to my reading of Cundall et al. [2014] and will elaborate).

To address the question of what I think gape is in an evolutionary sense, I will conduct a conceptual exercise. Typical lizards all have a "gape"; that is, they can all open their mouths to varying degrees and do so many reasons, an important one of which is to ingest food; they do so whether they are insectivores, piscivores, carnivores, frugivores, or herbivores of some kind. For all typical lizards, they possess large mouths of varying gape size, but they are not macrostomate in the snake lizard sense, cf., *Iguana iguana* with *Crotalus horridus*. The differences between the gape of an iguana and that of a rattlesnake vary around the numerous anatomies, morphologies, and behaviors identified above. Accepting the lizard origins of snake lizards as I do, and recognizing the anatomies present in basal lizards such as *Megachirella wachtleri* (Simões et al., 2018a), the big mouth non-macrostomate condition would be the primitive condition for all lizards, snake lizards included. This also means that microstomy among typical lizards is also derived. The only question arising is whether it is derived once where microstomate non-snake lizards such as amphisbaenians are the sistergroup to snake lizards, or convergently, with snake lizard microstomy being derived within snake lizards, independently of amphisbaenian lizard microstomy. Here, my conceptual exercise finds no anatomical similarities between amphisbaenians and snake lizards except the gross morphology, not anatomy, of burrowing adaptations (similar to the gross morphology of wings between birds and bats). Because all snake lizards cannot be derived from a common ancestor reconstruction where amphisbaenians are the sistergroup to scolecophidian snake lizards and then all other snake lizards, that hypothesis is abandoned a priori. That this phylogenetic scheme appears repeatedly in phylogenetic analyses of both morphology and molecules will be examined in Chapter 6. To continue, because scolecophidian snake lizard anatomies are "snake lizard" anatomies, not typical lizard anatomies, and cannot be derived from a big mouth, non-macrostomate condition of typical lizards without first being snake lizard big mouths, at the very least, they are only understandable as degenerate conditions highlighted by pedomorphic states of those anatomies present among ontogenetic stages of non-macrostomate or macrostomate snake lizards.

My working hypothesis, arising from this conceptual exercise, is that the ancestral snake lizard gape condition was an even more primitive form of Type I Anilioid as observed in Type I *Dinilysia-Najash* fossils, and more particularly in the available maxillary and mandibular elements of *Parviraptor*, *Portugalophis*, *Eophis*, and *Diablophis*. Cundall (1995) noted for *Cylindrophis* that the two constraining factors on macrostomy were the linkage between the dermal palate and the snout, and the non-elastic separation of the dentary tips, despite the absence of a bony symphysis. The fossil record of ancient snake lizards of Type I present an even more rigid dermal palate anatomy (see Chapter 2) coupled with the absence of a bony symphysis at the dentary tips. It appears from such a comparison that osteological anatomies for permitting enlarged gape evolved long before the super elastic properties of ligaments, skins, nerves and vasculature did (Cundall et al., 2014).

## Morphology II: "Microstomy," Heterochrony, and Heterodoxy

Scolecophidians, monophyletic or not (Wiens et al., 2008; Harrington and Reeder, 2017), truly have a small gape, but anatomically, it is neither a typical lizard nor Type I Anilioid anatomy that restricts gape size in these snake lizards, and it comes in at least two forms in terms of underlying anatomies. I therefore view scolecophidian microstomy as a modification of derived snake lizard anatomies (Type IV) that restricts gape, and it is these same modifications that support the unique oral functional morphology of modern scolecophidians. Surprisingly, these statements are heterodoxy in the extreme as they violate more than a century of "wisdom" arising from what I will argue here are mischaracterizations of the idealized anatomy of scolecophidians. Not too surprisingly, the data presented here rattle the "comfort zone" of snake lizard origins that view it as an epigean re-evolution (see Chapter 5) of scolecophidianlike features for all other ancient and modern alethinophidian snake lizards (i.e., scolecophidians are basal to *Parviraptor* and all other early snake lizards, and thus to all snake lizards).

While I am hardly the first to posit such heterodoxical interpretations and hypotheses, I suspect mine are the most serious, not delivered as one of several options, and the least favorable at that to the orthodox view (see Bellairs and Underwood [1951]; Rieppel [1988a,b]; Kley [2001]; Rieppel [2012]). The interpretations given here are also based on morphology, not molecules (the latter are the current reigning counterpoints to small gape as primitive for snake lizards [e.g., Vidal and Hedges, 2004, 2009]). Conventionally, morphology, not molecules, is seen as the bastion of defense for the perspective that the scolecophidian small gape is primitive among not just modern snake lizards, but all snake lizards (e.g., the unpalatable "regressed macrostomatans" of Rieppel [2012]).

The scolecophidian small gape breaks down into two different anatomical types specialized around feeding and prey acquisition using either the maxilla (typhlophids and the unstudied anomalepidids [see Kley, 2001]) or the dentary/mandible (leptotyphlopids). These two small gape anatomies are highly derived snake lizard osteologies, not primitive lizard anatomies, as their unique function relies on derived snake lizard synapomorphies, not non-snake lizard symplesiomorphies. In leptotyphlopids, this unique anatomy results in a mandibular raking system (Kley and Brainerd, 1996, 1999) dependent on synchronized right, left, abduction of the mandibles and a widening of the gape by deep outward flexion at the intramandibular hinge (see above). In complete contrast, the typhlopid condition requires the extremely reduced mandibles to abduct and open the mouth, but for feeding to proceed by right and left asynchronous erection and retraction of the mobile maxillae; erection or extension of the maxillae is accomplished by the highly advanced snake lizard anatomies (Type IV Viperid-Elapid-Scolecophidian skulls) of the CID musculature of the m. *levator* and m. *protractor pterygoideus* acting on the dermal palate bone chain of the pterygoids and palatines as described earlier in this chapter (Boltt and Ewer, 1964; Cundall, 1983; Greene, 1983; Kley, 2001; Cundall and Irish, 2008).

Thus, even though the received wisdom of snake lizard essentialism holds that the scolecophidian gape is the primitive gape for snake lizards, such a perspective fails completely under any Test of Similarity for the observable anatomies of scolecophdians. While I have reviewed a great deal of scolecophidian snake lizard anatomy so far in this book, I will introduce new data from old literature to support the statements given above, and point the reader back to various sections and details of this book rather than explore it in detail a second time.

Of particular relevance to my interpretations and conclusions on scolecophidians has been the work of Kley and Brainerd (1996, 1999) and Kley (2001, 2006), though their hypotheses differ from what I propose here. Added to the older literature on scolecophidian anatomy and embryology (Haas, 1930, 1959, 1962, 1964, 1968, 1973; Kamal, 1966; List, 1966; Bellairs and Kamal, 1981), the more recent studies by Kley and Brainerd have served to highlight the core archetype constructs that view scolecophidian anatomy as primitive. Kley (2001) reviewed a long history of anatomical works on scolecophidians and summarized the characteristics of microstomy as demarcated by the two principle types of feeding mechanisms observed in scolecophidians. Regarding both maxillary rakers such as typhlopids (*Typhlops* and *Rhinotyphlops*), and mandibular rakers such as *Leptotyphlops* Kley (2001:1331) wrote:

> ...movements of the lower jaw in *Leptotyphlops* are morphologically constrained to be bilaterally synchronous. As in typhlopids...

the left and right halves of Meckel's cartilage in *Leptotyphlops* extend beyond the distal tips of the mandibular rami and fuse together to form a robust interramal nodule of cartilage.

This bizarre condition presented by scolecophidians, where there is no bony sympyhsis, nor an intermandibular joint formed by ligamentous connections of variable elasticity, is unique to scolecophidians as compared to all other snake and non-snake lizards (Bellairs and Kamal, 1981). This extension of the Meckel's cartilage, anterior to the dermal dentary bone, was first noted by Smit (1949) and confirmed by Bellairs and Kamal (1981) for *Typhlops*. However, the above quote from Kley (2001) corrected Bellairs and Kamal's (1981:184) statement regarding *Leptotyphlops*:

> *Typhlops* is of special interest in possessing a relatively rigid mandibular symphysis, in contrast to the condition in snake generally. The tips of the dentaries are not fused, but the anterior ends of the two Meckel's cartilages are connected by a mass of cartilage (65C).

And:

> In *Leptotyphlops*...The dentigerous lower jaw has a well developed hinge between the dentary and the more posterior bones (McDowell and Bogert, 1954), and has no firm connection anteriorly with its fellow.

And regarding the mandibular raking of *Leptotyphlops* specifically, he reiterated (Kley, 2001:1331):

> ...Meckel's cartilage in Leptotyphlops extend beyond the distal tips of the mandibular rami and fuse together to form a robust interramal nodule of cartilage. This cartilaginous link between the tips of the mandibular rami permits hingelike rotations between the dentaries, but allows almost no...separation of the mandibular tips.

On maxillary raking in *Typhlops* and *Rhinotyphlops* specifically, Kley (2001:1331) wrote:

> In contrast to mandibular raking...maxillary raking bears at least some similarity to the feeding mechanisms of alethinophidian snakes. In particular, prey is transported into and through the mouth via independent ratcheting movements of the upper jaws...

more differences than similarities can be found between the feeding mechanisms of typhlopids and alethinophidians.

And further (Kley, 2001:1332):

Due to the relatively wide separation between the pterygoid and quadrate and the loss of the pterygo-quadrate ligament (Iordansky, 1997), the upper and lower jaws are functionally decoupled in Typhlopidae.

The final blow to the scolecopidian gape as primitive for snake lizards is delivered by ontogeny, and in this case what appears to be a pattern of extreme pedomorphosis. I discussed quadrate ontogeny earlier in this chapter while examining the suspensorium as an isolated anatomical system, and will only reiterate here that the quadrate of scolecophidians is horizontal with the articulation with the compound bone located far anteriorly; there is variation among the various genera of scolecophidians on the angle of quadrate orientation, but it is slight (Cundall and Irish, 2008). However, as this is the position and angle of development of the quadrate as a cartilage condensation and remnant of the hyoid arch (pterygoquadrate condensation; see Kamal [1966] and Bellairs and Kamal [1981]), this is clearly a pedomorphic retention of the juvenile condition of its ancestor—a heterochronic pattern of pedomorphosis. Via reference to Scanferla's (2016) profile of *Boa constrictor* quadrate rotation, it is clear that higher snake lizards are peramorphic through ontogeny on the juvenile condition (i.e., quadrate rotation proceeds from horizontal in embryos to vertical in late development and neonates, to extremely posteriorly directed in older adults). As has been noted repeatedly (Bellairs and Kamal, 1981; Kley, 2001), the posterior rotation of the mandibular condyle of the quadrate results from extreme elongation of the Meckel's cartilage during development, thus lengthening the lower jaw. The more prolonged the growth of the Meckelian cartilage, the more posterior the rotation of the quadrate. During later developmental stages, bony dermal dentary and compound bone growth, along with the Meckelian cartilage, would continue the posterior exaggeration of the quadrate.

It is unclear to me why the scolecophidian anteriorly inclined to virtually horizontal quadrate, with a short compound bone and dentary, has never been viewed as a pedomorphic condition, yet I can find no such reference even in Bellairs and Kamal (1981), where the feature was explicitly treated in the discussion of scolecophidian embryology. The intransigence to accepting such an obvious pattern of heterochrony is not merely puzzling, but rather seems entrenched beyond even essentials and archetypes to the defense of dogma regarding the primitive condition of all snake lizards as exemplified by scolecophidians (e.g., Rieppel, 2012).

The same is true of the second pedomorphic feature associated with the mandible/dentary and gape—the retention of an anterior extension of Meckel's cartilage beyond the terminus of the dermal dentary bone and its cartilaginous fusion to the Meckel's cartilage of the opposite mandible. This is not just a highly unusual feature in vertebrate development, but is demonstrative of extreme pedomorphosis—exposure of the Meckel's cartilage anterior to the dentary—coupled with the anterior fusion of the right and left Meckel's cartilage to form a rigid symphysis. In all typical lizards, the right and left Meckel's cartilages are separated from each other (Bellairs and Kamal, 1981; Cundall and Irish, 2008; Evans, 2008) and contained either completely or partially within the bones of the mandible (i.e., usually some exposure along the medial face of the dentary). With the exception of scolecophidians, the Meckel's cartilages terminate within the dentary bone, usually a short distance from the anterior tip of the dentary and, where present, posterior to the intermandibular symphysis. The scolecophidian retention of an anterior portion of Mecke's cartilage being uncovered by the dermal dentary bone can only be viewed as a pedomorphic pattern of development and thus derived, not primitive. The anterior fusion of the right and left Meckel's cartilages is by no means unusual, as this condition is observed in modern cartilaginous fishes, that is, sharks, that lack a bony skull, and have stabilized jaws as feeding structures by fusing the anterior tips of the cartilaginous palatoquadrate and Meckel's cartilages.

Characterizing only two bony features of the mandible of microstomous scolecophidians, and there are more, presents an obvious heterochronic pattern of development that renders viewing at least these two anatomies and a primitive condition of morphology impossible. They can only be considered in terms of primary homology, as likely derived. The same is true of the usual conditions of the toothless pterygoid and palatine, and reduced dentition of the dentary—these are pedomorphic states in a miniaturized snake lizard who shares a large number of anatomies and morphologies with higher snake lizards. My working hypothesis going forward is that scolecophidians are highly derived snake lizards where microstomy of the kind displayed by these animals is a derived condition of macrostomy, and likely extreme macrostomy (see below).

## Morphology III: Extreme Macrostomy

Basal macrostomatans, such as boiines and pythoniines, as macrostomate as they are, do not exhibit what I will term here the "extreme macrostomy" displayed by Type III and Type IV skulls and the snake lizards that present it (Close and Cundall, 2014; Close et al., 2014; Cundall et al., 2014). Despite their capacity to engulf enormous prey, much larger in fact by proportion than caenophidians (e.g., giant pythons ingesting entire wallabies, deer, pigs, goats, crocodiles, humans, etc. [see Greene, 1997]), they accomplish their ingestion via different modifications of their anatomy, what I will refer to here as "normal macrostomy." Whether anacondas can ingest proportionately larger prey than cobras is of no consequence to the variations in anatomy of hard and soft tissues under my respective categories of "normal" versus "extreme" macrostomy.

In modern snake lizards, with the exception of Type I Anilioids and Type IV scolecophidians, this ratio of prey size to head size is amazing (large adult pythons or anacondas eating humans or other big vertebrate prey), though at the same time, most snakes most of the time eat rather smaller prey simply because the biomass of the latter is usually much higher and availability and capture success scale with biomass and overall size (mice, voles, rats, other snakes, birds, fish, numerous invertebrates, etc.).

Extreme macrostomy, as I define it here and diagnose it anatomically, is observed in Type III and Type IV skulls (with the exception of scolecophidians) and goes beyond the macrostomy of Type II skulls by extreme posterior relocation of the suspensorial joint between the quadrate and compound bone. This is effected by the elongation of the supratemporal to move the joint it forms

with the medial face of the cephalic condyle of the quadrate posteriorly; an elongation, often extreme, of the quadrate bone itself, and an elongation of the Meckel's cartilage and the compound bone and dentary in order to move the inferior articulation of the mandibular condyle of the quadrate posteriorly (see Cundall and Irish, 2008). Coupled with these osteological modifications, Type III and IV snake lizards present extreme modifications of all soft tissues associated with trunk and skulls (i.e., nerves, vasculature, musculature, and endodermal structures of the mouth, pharynx, and anterior portions of the trachea and esophagus). The posterior relocation of the suspensorium means that anterior portions of the pharynx and gut, normally demarcating the end of the head and the beginning of the neck and trunk, become effectively "captured" into the head (Queral-Regil and King, 1998; Cundall and Greene, 2000; Herrel et al., 2008; Vincent et al., 2007, 2009; Hampton and Moon, 2013; Close and Cundall, 2014; Close et al., 2014; Cundall et al., 2014; Segall et al., 2016). The complete package results in a highly modified feeding system associated with extremes of gape that are noted to characterize Type III and IV snake lizards, thus my terminology recognizing extreme gape (see Palci et al., 2016; Catie Strong and Tiago Simões, personal communication). Such skulls can also be seen to display an intriguing mix of pedomorphic and peramorphic anatomies, not unlike some conditions already examined for microstomy in scolecophidians.

## Morphology IV: Non-Macrostomate "Macrostomatans"

As reviewed by Cundall (1995), *Cylindrophis ruffus* (Type I exemplar taxon) often ingest comparatively large prey (equivalent diameter to head/neck diameter); I need not point out that true microstomate snake lizards, the scoleclophidians, do not do this and cannot (Kley and Brainerd, 1999; Kley, 2001). However, the same is true of *Xenopeltis and Loxocemus* regarding prey choice; both ingest similarly large-bodied prey but are conventionally considered to be macrostomatans. Importantly, Frazzetta (1970, 1999) noted that for *Xenopeltis*, the strong sutural connection between the premaxilla and maxilla, the large number of teeth on the premaxilla, and the solid connection between the anterior portion of the toothed palatine and the septomaxilla and vomer were very similar features to those of anilioids. I would agree, and to this I would add, based on observation of feeding in *Xenopeltis unicolor* (Caldwell, pers. observ.), that similar to Cundall's (1995) observations of *Cylindrophis ruffus*, the dentary tips of *Xenopeltis* do not separate during prey ingestion—in fact, the mandibles form a broad v-shape even during ingestion of a large-bodied mouse meal, but appear to "flex" significantly at the intramandibular joint, as noted by Cundall (1995) for *Cylindrophis*. In other words, from both a functional morphology point of view, and based on anatomical data, *Xenopeltis* is not at all macrostomate, but clearly can ingest large prey because of convergent anatomies of the dermal palate kinesis and elasticity in buccal and pharyngeal tissues and musculature similar to *Cylindrophis*. I would predict exactly the same anatomies and functional morphology for the "gape" in *Loxocemus*—the phylogenetic implications of Type I skull categories for both *Loxocemus* and *Xenopeltis* are examined in Chapter 6.

The question is: Why are *Xenopeltis* and *Loxocemus* considered to be macrostomatan alethinophidians to the exclusion of uropeltids, *Anilius*, *Cylindrophis*, and *Anomochilus* (Cundall and Rossman, 1993; Cundall et al., 1993; Rieppel and Maisano, 2007), especially when marked similarities between these snake lizards have been noted for a long time (e.g., Haas, 1955)? Or, why are they not recovered in a clade with typical anilioids? Apart from the obvious answer, which is based in received wisdom and thus unconscious and conscious bias in character constructions and coding, there is likely a citation or two where data of some kind or another were provided in support of such a hypothesis. From Gauthier et al. (2012:43), there are twenty-four supposed "unambiguous" synapomorphies, two of which were reported as "unique and unreversed," that support Macrostomata, inclusive of *Xenopeltis* and *Loxocemus*. I will not examine all twenty-four unambiguous synapomorphies here, but will state that most of them fail to support *Xenopeltis* and *Loxocemus* as macrostomatans to the exclusion of anilioids, as do the two unique and unreversed synapomorphies, which I examine in detail here. From Gauthier et al. (2012:43):

224(2)*, vomer septum (vertical lamina) nearly completely separating olfactory chambers along with septomaxilla and nasal

And:

274(1)*, anterior end of ectopterygoid located dorsal to maxilla, invading the dorsal surface of the maxilla to a variable degree.

As always, the character data and their accuracy, or not, and thus their susceptibility to simple falsification if inaccurate or miscoded, is where the problems begin. The problems inherent in Character 224 were noted by Simões et al. (2017), and the character was recoded (for details see Simões et al., 2017: Supplementary Information). But, as published by Gauthier et al. (2012), and to reiterate here, the state description for Character 224 as excerpted above is actually a new character inserted into the actual character, so that it defines macrostomatans and excludes anilioids and scolecophidians (it includes assessing the septomaxilla in its state, which deviates from a simple assessment of the height of the medial and vertical septum of the vomer). For the latter taxa, they are coded against a different character. More importantly, this character also includes two state assignments for other macrostomatans presumably assigned state 2 as well, which means state 2 has substates of 2–3 and 2–4. This character is unambiguous because it is a conflation of two characters combined into one, with some states including conditional features not included in the original character description (see Simões et al., 2017). In addition, from the figures provided by Gauthier et al. (2012:160) to illustrate this character, both *Cylindrophis* and *Epicrates* have a tall vertical vomerine septum that, with the septomaxilla, separates fully the right and left olfactory chambers (the difference is in the relative contribution of the septomaxillae; also, I would argue that *Epicrates* possesses state 4, as does *Aparallactus*, and that state 3 simply does not exist). Here is how the full character number 224 reads (Gauthier et al., 2012:160):

Vomer septum (vertical lamina) height: (0) low, not forming septum; (1) partly separating olfactory chambers; (2) nearly completely separating olfactory chambers along with septomaxilla and nasal; (3) only ventral edge of septum remains; (4) V-shaped notch separates dorsal and ventral rami of vomer septum.

Accessing whole specimens of *Xenopeltis*, as well as online data available at Digimorph.com for *Xenopeltis* (sagittal cutaway, Section 223), it is observed that in the vomerine septum there is an equal contribution to separating the olfactory chambers similar to that observed for *Cylindrophis*. As recoded by Simões et al. (2017), all alethinophidians snake lizards were coded state "1" with states 2–4 being discarded. Simply put, this putative unambiguous synapomorphy of Crown Macrostomata is neither unambiguous nor a synapomorphy of anything but the concept of Alethinophidia. The remaining question is to decide whether or not scolecophidians, which were coded as "0" by both Gauthier et al. (2012) and Simões et al. (2017) do indeed share the same state. I would suggest in fact, they do not and should be coded "?" But that discussion is outside of the issue here of gape and Macrostomata and the synapomorphies of that purported clade.

The final putative unambiguous character is Character 274 (Gauthier et al., 2012:175):

> 274. Anterior end of ectopterygoid: (0) restricted to posteromedial edge of maxilla (*Anilius scytale*, dorsal close-up of anterior skull); (1) located dorsal to maxilla, invading the dorsal surface of the maxilla to variable degrees (*Python molurus*, dorsal close-up of anterior skull).

For this character, Gauthier et al. (2012) coded all typical lizards and all scolecophidians as "?" and coded *Cylindrophis*, *Anilius*, *Anomochilus*, and *Dinilysia patagonica* as "0." However, as the character description is written, *Cylindrophis* and *Dinilysia* should both be state "1," because *Cylindrophis* has two processes, one of which onlaps the dorsal surface of the maxilla, and in *Dinilysia* the entire ectopterygoid overlaps the dorsal surface of the maxilla. Oddly, *Anilius* as illustrated (Gauthier et al., 2012:175), and as confirmed by observation, does not actually possess an ectopterygoid that makes complete contact with the maxilla, a condition shared with *Anomochilus*, and so both taxa should be scored "?" not the "0" as scored by Gauthier et al. (2012:293). Based on what I assess as incorrect scorings and poorly constructed characters Gauthier et al. (2012:42) wrote:

> All morphology-based analyses of snake interrelationships (except Caldwell and Palci, 2010) have concluded that "anilioids" are located outside Macrostomata…This contrasts with molecular studies that found *Anomochilus*, *Cylindrophis* and uropeltines variably nested within Macrostomata…

And with specific regard to their purported characters, states, and the recovered synapomorphies regarding *Xenopeltis* and *Loxocemus*, which as I have indicated above share both purported unambiguous synapomorphies with *Cylindrophis* and *Dinilysia*, and *Anilius* and *Anomochilus* on one trait, Gauthier et al. (2012:43) wrote:

> In most morphology-based analyses, *Xenopeltis unicolor* and *Loxocemus bicolor* are successive sisters to other crown macrostomatans… however, New World *Loxocemus* and Old World *Xenopeltis* form a clade albeit with very poor support…That they are closer to other extant macrostomatans than to "anilioids," scolecophidians and stem snakes seems secure…

So, other than received wisdom, the current most comprehensive efforts to separate alethinophidians into basal non-macrostomatans and crown macrostomatans are fraught with empirical problems of the most fundamental kind—essential morphologies coded and characterized incorrectly (see Simões et al., 2017). What conclusions, or at least, new hypotheses open for testing can be taken away from this discussion so far? It is that *Xenopeltis*, *Tropidophis*, *Ungaliophis*, and *Loxocemus* require careful reconsideration in relationship to the Type I Anilioid characteristics of their skulls, and the lack of evidence supporting them as macrostomate, let alone their purported relationships within Macrostomata.

### Morphology V: Ancient "Non-macrostomatans"

Though it was completely ignored by paleontologists and neontologists alike from its description by Smith-Woodward (1901) until Estes et al. (1970) and Frazzetta (1970), and then ignored again until Caldwell and Albino (2002) (Rieppel [1988a,b] discussed the specimen at length but had not seen it), *Dinilysia patagonica* (see Chapter 2) has returned to share center stage with *Najash rionegrina* in the ongoing theatrical revue of fossil snake lizard comparative anatomy. Both of these ancient snake lizards are big-bodied animals with big heads and big mouths, but they are not macrostomate, and I would not argue so either. They are Type I Anilioid snake lizards, through and through, and all anatomical features presented suggest big mouths with big heads but not the cranial kinesis of the palate and suspensorium necessary to judge them macrostomatan. I also cannot support consideration of them as microstomate, either, and find it is a categorical misrepresentation of Type I Anilioid and Type I Dinilysia-Najash anatomy, both osteological and muscular, to lump them together with scolecophidians as small-gaped snake lizards (contra Rieppel, 2012). Both Type I Anilioid and Type I Dinilysia-Najash are not scolecophidian-like in any feature of anatomy, particularly as concerns gape and feeding mechanics, and share a virtually identical cranium-otic capsule morphology, a comparatively rigid and akinetic palate, hinged teeth in several taxa, an identical suspensorium system, long jaws, long maxilla, free premaxilla, and so on. As Type I Anilioid snakes are not macrostomate, it is no surprise to find, as did Haas (1973), Rieppel (1980c), and Cundall (1983, 1995), that their jaw adductor musculature is more lizardlike than any other modern snake lizards; doubtless Type I Dinilysia-Najash possessed a similar musculature, if not more primitive still. However, for both of these groups, it must be a guarded "lizardlike," as it is still a snake lizard skull with snake lizard musculature. Considering the ancient Type I Dinilysia-Najash affinities and similarities, modern Type I Anilioid skull musculature should retain an ancient configuration not seen in any other modern snake lizards, including Type IV scolecophidians. It is unclear why Rieppel (2012) elected to summarize both gape size and anatomy so coarsely as to present a case where non-macrostomy was a single category inclusive of scolecophidians and Type I Anilioids.

Adults of Dinilysia are estimated to be upward of 2 meters in length (Caldwell and Albino, 2002), while adult Najash were at least 1 meter long, likely significantly longer. This is indirect contrast to what represents "small" among the modern snake lizard assemblage, where scolecophidians typical present a size range between 5–10 cm in length (e.g., Indotyphlops braminus at 5–10 cm in length). The surprising exception is a typhlopid, Aphrotyphlops schlegeli, Schlegel's Giant Blind Snake, that can attain sizes of ~95 cm, and likely more. While there is no reason to suspect a correlation between average size and microstomy (arguments of miniaturization as in [Rieppel, 1996]), there appears to be one in this case. Type I Anilioids may not be macrostomate, but they are assuredly not gape limited as are the tiny-bodied, tiny-mouthed, highly modified gapes of scolecophidians—I would argue there is no justification for considering these bizarre snake lizards as anything but gape specialists, not gape primitives.

A second group of ancient snake lizards that do not appear to show all of the necessary features associated with functional macrostomy, though they may be basalmost Macrostomata phylogenetically, are the simoliophiids/ pachyophiids inclusive of Pachyrhachis problematicus, Haasiophis terrasanctus, and Eupodophis descouensis, though their macrostomatan relationships have been forcefully and passionately argued for in numerous studies (Zaher, 1998; Zaher and Rieppel, 1999a; Rieppel and Zaher, 2000a,b, 2001; Tchernov et al., 2000; Rieppel and Kearney, 2001; Rieppel et al., 2003a; Rieppel and Head, 2004; Apesteguía and Zaher, 2006; Zaher and Scanferla, 2012; Rieppel, 2012). However, despite these appeals to anatomy and morphology, the actual issue of gape size was never appropriately considered and it is important to examine it in the context of the characters and anatomies used to hypothesize such a relationship.

The four synapomorphies shared by Pachyrhachis and modern macrostomatans from Rieppel and Zaher (2000a:61) are:

Character 47 (1) supratemporal at least half of maximum skull width

Character 233 (2) posterior dentigerous process of the dentary enlarged

Character 238 (1) tooth-bearing anterior process of the palatine present

Character 239 (1) suprastapedial process of the quadrate absent

These four supposed key synapomorphies suffer numerous problems. For Character 47, it is absolutely unclear how the skull width was measured in Pachyrhachis with respect to the width of the supratemporal, let alone what the underlying homolog concept of this character actually is. It is also not clear why or how this character relates to gape size. Character 233 reflects, I believe, the posterior elongation of the dentigerous process of the dentary as it onlaps the coronoid (see Figures 3.11 and 3.12); however, enlarged is not only subjective, it is impossible to demarcate as a character state concept, and, again, is not linked to being "Macrostomatan" to the exclusion of Anilius, Cylindrophis, Dinilysia, and Najash, all of whom present a dentigerous process of the dentary (note: the only modern snake lizards for which such a feature is absent are scolecophidians). Character

238, the tooth-bearing portion of the palatine, is a character present in *Cylindrophis* and *Anilius* and is hardly exclusive to *Pachyrhachis* + Macrostomata. So, in truth, though Rieppel and Zaher (2000a) felt that in the context of their analysis that these four characters were exclusive support for *Pachyrhachis* as a macrostomatan in the sistergroup position to Macrostomata, they are not. Additionally, with the exception of Character 238, teeth on the palatine, none of these characterizations of anatomy are even functionally linked to macrostomy.

As noted above, Gauthier et al. (2012:43) identified twenty-four "Crown Macrostomata" synapomorphies that were among the many characters strongly criticized by Simões et al. (2017) in their analysis of gigantic data sets and quality of character constructions. Without reviewing Simões et al. (2017) here, it is enough to say that these twenty-four characters include problematic ratio characters and character constructions including non-comparable states (e.g., What does nasal anterior width have to do with frontal anterior width?). In addition, many of them are incorrectly coded for non-macrostomatans such as *Dinilysia*, *Anilius*, *Anomochilus*, *Cylindrophis*, *Najash*, *Xenopeltis*, and uropeltids. More importantly, many of them cannot be coded for putative fossil macrostomatans such as *Pachyrhachis* and *Haasiophis*, or are coded incorrectly. I will review briefly those that cannot be coded for in the fossil "macrostomatans" and which ones are even linked to gape size (Gauthier et al., 2012:43).

*Character 18(2), nasal anterior width less than frontal anterior width.* This character is equivalent in *Pachyrhachis* and *Haasiophis* and cannot be assessed in *Eupodophis*. Also, as described it depends on "where" you measure the nasals in *Anilius* and *Cylindrophis*— if you measure the "front" of the nasals, they are narrower than the middle and back, and narrower than the frontals.

*Character 115(2), loss of maxilla facial process. Pachyrhachis,* *Haasiophis*, and *Eupodophis* possess actual ascending facial processes, so this does not diagnose these three within Macrostomata at all, and, based on the requirement for unilateral maxillary-palatal kinesis, would suggest non-macrostomy; the "processes" of *Anilius* and *Cylindrophis* are nothing more than tiny abutments for an anterior prefrontal process and *Anomochilus* has no process. The scorings of this character in now way supports the conclusions.

*Character 170(3), supratemporal lies dorsally on parietal (or braincase alone).* This is true for *Haasiophis* and

*Eupodophis* and unclear for *Pachyrhachis* but is present in all snake lizards with a supratemporal except for *Liotyphlops*.

*Character 174(1), supratemporal tip extends freely posterior to oto-occipital.* This feature is present in all three simoliophiids, but is also present in *Dinilysia*.

*Character 185(2), quadrate 70%–74% of braincase depth.* The character annot be scored for *Pachyrhachis*, *Haasiophis*, and *Eupodophis*, and it is also unclear what the homolog concept is in the intention of this character construction.

*Character 194(3), fenestra ovalis opens posterolaterally.* See the section on the crista circumfenestralis above, and I note here that this character cannot be scored for *Pachyrhachis*, *Haasiophis*, and *Eupodophis*.

*Character 203(1), septomaxilla with long posterodorsally directed, blade-like process (medial flange) extending nearly to frontal.* This character cannot be scored for *Pachyrhachis*, *Haasiophis*, and *Eupodophis*.

*Character 208(0), nervus ethmoidalis medialis passes dorsal to septomaxilla.* This character cannot be scored for *Pachyrhachis*, *Haasiophis* and *Eupodophis*.

*Character 210(1), cupola for vomeronasal organ closed medially (even if only narrowly).* This character cannot be scored for *Pachyrhachis*, *Haasiophis*, and *Eupodophis*.

*Character 224(2), vomer septum (vertical lamina) nearly completely separating olfactory chambers along with septomaxilla and nasal.* Though this character is a problem (see above), it also cannot be scored for *Pachyrhachis*, *Haasiophis*, and *Eupodophis*.

*Character 247(1), choanal process of palatine touches or abuts the vomer without articulation.* This character cannot be scored for *Pachyrhachis*, *Haasiophis*, and *Eupodophis*, but also not present in Type IV skulls.

*Character 248(1), choanal process of palatine forms a short vertical or horizontal lamina.* This character cannot be scored for *Pachyrhachis*, *Haasiophis*, and *Eupodophis*.

*Character 256(2), palatine teeth enlarged, similar in size to marginal teeth.* Such a character makes no sense when characterizing higher caenophidian snakes in Type IV.

*Character 260(1), pterygoid articulates with palatine in a tongue-in-groove joint.* This feature is present in non-macrostomatan snake lizards.

*Character 265(1), quadrate ramus of pterygoid blade-like and with distinct longitudinal groove for insertion of the protractor pterygoidei muscle.* This feature is present in non-macrostomatan snake lizards, as in all snake lizards, the insertion of the m. protractor pterygoidei is on the quadrate ramus of the pterygoid.

*Character 268(2), pterygoid teeth enlarged, similar in size to marginal teeth.* Such a character makes no sense

THE ORIGIN OF SNAKES

when characterizing higher caenophidian snakes in Type IV where extreme macrostomy sees a reduction to zero of pterygoid teeth.

Character 274(1), *anterior end of ectopterygoid located dorsal to maxilla, invading the dorsal surface of the maxilla to a variable degree.* This character is a problem and has been dealt with (see above), but is present in *Pachyrhachis, Haasiophis,* and *Eupodophis.*

Character 299(2), *temporal muscles spread onto supraoccipital to form a Y-shaped crest.* Not present in *Haasiophis* and *Eupodophis,* and difficult to score in *Pachyrhachis*; crests are present in *Dinilysia* and *Najash.*

Character 300(2), *supraoccipital nuchal crest spreads laterally onto oto-occipital.* This character cannot be scored for *Pachyrhachis* and *Eupodophis.*

Character 324(2), *dorsum sellae enclosed in a distinct fossa, a cup-like depression walled laterally and ventrally by the basisphenoid and anteriorly by the parasphenoid rostrum.* This character cannot be scored for *Pachyrhachis* and *Eupodophis.*

Character 337(0), *posterior opening of Vidian canal within the basisphenoid.* This character cannot be scored for *Pachyrhachis* and *Eupodophis.*

Character 351(0), *perilymphatic foramen faces ventrally.* This character cannot be scored for *Pachyrhachis, Haasiophis,* and *Eupodophis.*

Character 387(2), *coronoid eminence formed mainly by surangular.* The opposite condition is observed for *Pachyrhachis, Haasiophis,* and *Eupodophis,* where all three have a tall, prominent coronoid sitting atop a low eminence on the compound bone. In addition, detecting the surangular in snake lizards is impossible due to compound bone formation.

Character 420(3), *maxillary tooth count 16–27.* While this range of tooth counts is nonsense, it also excludes most Type IV skulls from the Macrostomata.

Regardless of the quality of the character and mistakes in coding (e.g., Type I skulls, modern and ancient, the inapplicability to *Haasiophis, Pachyrhachis,* and *Eupodophis,* and the exclusion of crown macrostomatans, particularly Type IV skulls), there are still only five characters that actually reflect the issue of concern here—gape size and macrostomy. Indeed, looking specifically at *Haasiophis, Pachyrhachis,* and *Eupodophis* and whether they were truly macrostomate, regardless of phylogenetic position in or out of Crown Macrostomata, they appear not to be macrostomate, but rather fall into a category of function mechanics somewhere between *Cylindrophis* and *Tropidophis* or *Loxocemus.*

The quadrates in all three fossil snake lizards are vertical, the supratemporal extension beyond the otoocciput is short, the maxillae all have strong ascending processes and robust articulations with the prefrontals that do no suggest kinesis at all, and the palates appear to be fixed in place, not too differently from that of modern anilioids. The intramandibular hinge appears flexible like that of a pythonomorph or of *Cylindrophis,* but yet the intermandibular hinge development, judged by the straightness of the dentaries through to the tips, points to a v-shaped jaw and limited flexibility. In short, these putative macrostomatans do not appear to have been macrostomate at all, rendering the evolutionary history of macrostomy even more complex, occurring when it did most likely in parallel in distantly related lineages of snake lizards.

### Ancestral Snake Lizards and Gape

I will not belabor the point—the Type I Anilioid skull and its associated soft tissue anatomies, and the gape that they collectively facilitate, are derived from the much more primitive snake lizard system exemplified by Type I *Dinilysia-Najash* skulls and their anatomies. The scolecophidian skull and its soft tissues are pedomorphic derivatives of the Type IV Viperid-Elapid skull, not primitive for all snake lizards, and share no discernible features as primitive primary homologs with non-snake lizards that do not in fact characterize all snakes. Therefore, the ancestral snake lizard skull type should present "more" non-snake lizard skull features regarding gape than do *Dinilysia, Najash,* and so on. Potentially some of these features are present in the Jurassic and Early Cretaceous forms, but as of yet, most of the necessary anatomies to support such a characterization are not yet known from the fossil record (see Caldwell et al., 2015). What is clear, though, in fact stridently clear, is that microstomy in any guise is not the ancestral condition for snake lizards, but rather the ancestral condition is to be big-mouthed and non-macrostomate.

### THE POSTCRANIAL SKELETON

#### Basics of the Postcranium and "Short Tails"

While it is a commonly recognized fact that modern snake lizards are legless, having lost the fore- and rear limbs, what is less widely known is that limblessness, and in particular limb reduction to complete limbloss, is quite common in modern and ancient lizards of all kinds, in fact, in all tetrapods (Caldwell, 2003), not just snake lizards. According to Greer (1991), limb reduction and limb loss have occurred in modern lizards (snake lizards included) at least sixty-two times in fifty-three different lizard lineages, with the exceptions of the non-snake lizards Varanidae, Lacertidae, Gerrhonotidae, and Iguanidae. The number of clades presenting completely limbless forms includes the dibamids, amphisbaenians, pygopodid geckos, annielline anguids, ophisaurine anguids, gerrhosaurids, cordylids, acontine scincines, feylinid scincines, numerous genera of scincids (twenty-five different scincid lineages [Greer, 1987, 1990, 1991]), gymnophthalmids, and of course nearly 4000 species of modern snake lizards (see also Caldwell, 2003).

Interestingly though, while there is absolutely no vestige of the pectoral girdle and thus no forelimb, there are a rather large number of modern species that have retained recognizable vestiges of the pelvic girdle, and a few even possess a femoral spur proximate to the cloaca inside of which is a tiny "femur" articulating with a highly modified and reduced pelvic girdle. However, the balance of modern snake lizards, that is, the several thousand species of modern caeonophidians, possess no remnants of the pelvic girdle elements as evidenced in more basal modern snakes (i.e., booids, anilioids, etc.). Of particular interest, though, is the presence of pelvic girdle remnants in scolecophidians, especially considering the arguments rendered above concerning scolecophidians as derived caenophidian snake lizards.

An associated, and likely evolutionarily and developmentally linked result of this process of limbloss to limb reduction is that modern snake lizard bodies are also extremely elongate. In the absence of girdle anatomies to assist in demarcating body regions, generations of scholars (e.g., Bellairs and Underwood [1951]; Hoffstetter and Gasc [1969]) considered the neck lost in snake lizards and homogenized the axial skeletal regions of snake lizards from the usual cervical, thoracic, lumbar, sacral, and caudal to only three barely distinguishable regions, that is, precloacal, cloacal, and postcloacal/caudal. Not too surprisingly, elongation has not occurred uniformly throughout the postcranial skeleton

of snake lizards (Rochebrunne, 1881: Table 1), and as with most axially elongate vertebrates is accomplished developmentally by increasing the number of somites (i.e., length increases by increased segmentation [e.g., Gómez et al., 2008]) and not by increasing the size of any one somite (though there are a couple of vertebrate exceptions, for example, the necks of giraffes or tanystropheids [Carroll, 1988], where the neck vertebrae are grossly elongated by 3–4 times, the length of a typical thoracic or dorsal vertebra).

As was reviewed first by Rochebrune (1881) (note: this nineteenth-century French zoologist is also the namesake of the first-ever described fossil snake, *Simoliophis rochebrunei* [Sauvage, 1880]—see Chapter 2), followed by Sood (1946) and of course, the twentieth-century classic by Hoffstetter and Gasc (1969), the number of pre- and postcloacal vertebrae in modern snake lizards is surprisingly variable (Rochebrunne, 1881:Table 1) and, from a fossil snake lizard from the Eocene (*Archaeophis proavus*), surprisingly high. Rochebrune (1881) noted the highest count among modern snake lizards in *Python molurus* at 435 individual vertebrae, and the lowest in *Leptotyphlops* (=*Stenostoma*) *nigricans* at 141. As was reviewed by Sood (1946) and originally described by Massalongo (1859), the fossil snake lizard *A. proavus* possesses an astonishing 565 vertebrae. From Rochebrune's (1881) table, "*Tableau des especes et du nombre de leurs vertebres,*" he ascribes two cervical vertebrae to *Typhlops lumbricalis*, presumably based on the presence of an atlas-axis pair of vertebrae and then presents ditto marks for all other snake lizards he observed and provided data on, obviously considering that these two vertebrae were the only vertebrae still demarcating the neck or cervical region in modern snake lizards.

While Hoffstetter and Gasc (1969) discounted the presence of a neck and cervical region altogether (see below), and recognized only the precloacal, cloacal, and postcloacal/caudal regions, they also could not confirm Rochebrune's (1881) observed range number of vertebral numbers. They reported a maximum of 400 vertebrae for *Python molurus*, not 435, and a lower bound in scolecophidians of 160 as opposed to 141. Of consequence, though, as was reported by Rochebrune (1881), Hoffstetter and Gasc (1969) noted that in *Python* there were 320 precloacal vertebrae, and for *Archaeophis* 454— in other words, elongation via increased number of somites affects vertebral number between the occipital condyle and the cloaca or sacral region.

Interestingly, the postcloacal region of snake lizards, or the tail, never gets longer, either by proportion or by increases in the number of vertebrae or somites, even though the body, or precloacal region can increase in somite/vertebral number by tenfold in snake lizards (30–40 in a typical lizard to >300–400 in many snake lizards). However, in the tail, the reverse is true—through snake lizard evolution it decreases in somite number in derived burrowing forms in all lineages and also decreases in relative proportion even if the somite number remains high, so that in all snake lizards the tail appears "short." A "short" tail with a large number of vertebrae is proportionally short because the size of each vertebra is significantly reduced compared to the size of precloacal vertebrae. However, somitogenesis still proceeds at a rapid pace compared to other tetrapods, but allometric growth later controlled the size of each somite/segment/vertebra (Gómez et al., 2008).

The number of caudal vertebrae in snake lizards is highly variable, that is, from ~98–100 down to 7–9 and seems to decrease to its smallest number in its most derived snake lizard forms, that is, scolecophidians. For example, in *Python molurus*, Rochebrune (1881) reported sixty-one caudal vertebrae and in the scolecophidian *Indotyphlops* (=*Typhlops*) *braminus*, seven vertebrae. The average of the reported sixty-two species of snake lizards in Rochebrune's (1881) table is 49.4 caudal vertebrae. Typical non-snake lizards, excluding amphisbaenians where the highest number is around twenty-eight or so and the tail is also proportionately short (Hoffstetter and Gasc, 1969), have significantly more caudal vertebrae than presacrals, or oftentimes a similar number, and the tail is also proportionately larger even if the number is less than that observed in snake lizards (a tail of 80 vertebrae in a snake lizard can be a small proportion of total body length but still have a large segment count). This contrasts remarkably with a non-snake lizard where 80 caudal vertebrae would mean that the tail is long compared to the body both in number of vertebrae (an average number of presacrals in anguid or an iguanid is around 40–42) and in the size of each element and the proportion of the tail length to body length) (Hoffstetter and Gasc, 1969:267).

Importantly, in the scolecophidians, the tail is exceptionally short (less than ten in *Typhlops* [Rochebrune, 1881; Hoffstetter and Gasc, 1969]) and therefore is demonstrative of a reduced number of segments or somites. Such a number should not be, prior to the Test of Similarity and the Test of Congruence, the primitive condition of non-snake lizards, as tail length in snake lizards is actually, on average, equal to or higher than the average in non-snake lizards. Relative proportions of shortness compared to body length are meaningless. It is only accurate to examine number of somites as demonstrated by number of vertebrae when discussing tail length. An example of the problems created by such mischaracterizations is that of Martill et al. [2015:416], who stated *Tetrapodophis* possessed a "short tail" even though it has 118 postsacral/caudal vertebrae. This is certainly not a short tail; in fact, by comparison to any modern snake lizard, it is exceptional, as it would be the longest if indeed *Tetrapodophis* were a snake lizard. Among modern and ancient snake and non-snake lizards, caudal vertebral counts range from as many as 163 in the aquatically adapted pontosaur, *Pontosaurus kornhuberi* (Caldwell, 2006); the elongate, limbless anguimorph lizard *Ophisaurus apodus* presents 90–100 caudals (Hoffstetter and Gasc, 1969), while, as noted above, scolecophidian snake lizards present the smallest number of caudal vertebrae among all lizards at 10. Type I Anilioid snake lizards (presumably modern members of more basal snake lizard clades inclusive of fossil snake lizards such as Type I *Dinilysia* and *Najash*) such as *Cylindrophis* present twenty-four caudal vertebrae (cloacals + postcloacals), while *Anilius* has thirty-seven and *Xenopeltis* twenty-seven. The problem, of course, is to decide if these caudal counts are primitive for all snake lizards, or are derived within the clades inclusive of ancient snake lizards—unfortunately for *Dinilysia* and *Najash*, caudal vertebral counts remains unknown, though new unpublished data for *Najash* suggests it possessed more than thirty-seven caudals.

Regardless of the vector of transformation for Type I Anilioids + *Dinilysia* and *Najash*, it is difficult to model just how the scolecophidian snake lizard vertebral count can be considered symplesiomorphic for all crown ophidians. Considering the high vertebral count anterior to the cloaca, and in concert with the other features of morphology discussed thus far in this chapter, scolecophidians are clearly derived snake lizards in terms of somitogenesis as compared to typical lizards (Gómez et al., 2008). The short tail, proportion, and somite number are obviously a pedomorphic truncation of rapid snake lizard somitogenesis in the precloacal region and not a primitive condition, especially when compared to the long tails (proportion and somite number) of typical lizards. It is, in evolutionary terms, and not in the cladistic sense, more parsimonious to consider the short scolecophidian tail as a truncation of somitogenesis via pedomorphosis, and not a primitive condition. A tail truncated by limited somitogenesis cannot be viewed as primitive for any clade of lizards, but rather must be seen as a derived condition within the group, just as extreme tail elongation in *Pontosaurus* (163 caudals) is also derived within its lineage and arises due to a prolongation of caudal somitogenesis.

## Cervicals and Intercentra: The "Lost Neck" of Snake Lizards

Going back to the beginning, that is, leaving the tail behind and restarting this discussion rostrally or anteriorly along the axial skeleton, I will examine here one of the most intriguing problems in snake postcranial evolution, and it is not the loss of the limbs, but rather the evolution of the "neck." Hoffstetter and Gasc (1969), in their review of snake lizard vertebrae and ribs, followed a long tradition of considering the neck of snake lizards to be lost, and that the absence of a pectoral girdle made it impossible to identify the boundary between neck and thorax—so that is problem one that I will consider here. Problem two concerns osteological details of that "neck. In this case, I am referring to the ancient vertebrate vertebral component of the centrum body, the intercentrum, which in most lizards is restricted to the vertebrae of the cervical and caudal regions (though in gekkotans they are continuous throughout the column)—for snake lizards, these questions are: where are they, how many do they have, and what insights do intercentra provide on the size and length of the snake neck? Problem three concerns recent developmental and statistical approaches to assessing postcranial evolution and regionalization and the problems arising when important data such as fossils have no impact on data collection, approaches taken to an analysis, and, of course, to the final act of hypothesis construction (e.g., Hoffstetter and Gasc, 1969; Head and Polly, 2015). I recognize, of course, that few fossil taxa are in any way complete enough to be compared throughout the length of the vertebral column to modern snake lizards. However, I would argue most forcefully that the questions being asked, and the analyses thus employed, should be reworked in order to accommodate fossil data. If not, the analysis is inherently incomplete, and asks, as always, the essentialist, top–down question about only the "organization" of the modern assemblage and cannot ask any meaningful questions nor provide meaningful answers to the question of snake lizard evolution—this latter question is a question of time, as evolution is, in my opinion, most importantly change over deep time, not just change within lineages recognized within the modern assemblage.

## Genes, Processes, and Linkages to Patterns: A Neck Found

With the development of technologies allowing visualization of domains of expression of various gene products, in particular Hox gene regulatory networks, efforts have been renewed in the search to understand snake lizard body and limb development and thus evolution. Research has focused on gene expression domains early in development in an attempt to link those patterns of expression to the developmentally distant phenotype of adult forms and the various semaphoronts throughout ontogeny. Specifically, there has been a small but growing body of research on snake lizard axial elongation and limb loss that began with Cohn and Tickle (1999) and more recently has been followed by the empirical studies of Woltering et al. (2009), Di-Poi et al. (2010), Infante et al. (2015), Kvon et al. (2016), Leal and Cohn (2016, 2017) and reviews by Woltering (2012), Kaltcheva and Lewandoski (2015), and Villar and Odom (2015). These studies and reviews have all provided new insights and perspectives on gene activity and action in snake embryogenesis and development but have produced two widely differing hypotheses on the "how" of Hox gene expression shifts leading to limbloss and elongation. Woltering et al. (2009:82) summarized the original Cohn and Tickle (1999) hypothesis as explaining how the "…peculiar deregionalized limbless anatomy results from a corresponding homogenization of mesodermal Hox expression domains along the primary axis." In contrast, they proposed a second hypothesis not as homogenization, but rather as a retention of subtle regionalization but with downstream regulation being the key player (Woltering et al.,

2009:82), "Our results suggest that the evolution of a deregionalized, snake-like body involved not only alterations in Hox gene cis-regulation but also a different downstream interpretation of the Hox code." Head and Polly (2015) confirmed Woltering et al.'s (2009) findings by adding postaxial morphometric data indicating anterior and posterior precloacal regions were in fact present, though again subtle, in snake lizards.

I completely agree with recent recognition that axial skeleton regionalization still exists in snake lizards. In fact, I have agreed with this notable regionalization for a very long time, beginning with Lee and Caldwell (1998:1538), who in their detailed redescription of *Pachyrhachis problematicus* Haas, 1979, presented anatomical evidence for that ancient snake lizard that demarcated twenty cervical vertebrae that themselves could be distinguished to present two distinct vertebral forms within the "cervical" region (Figure 3.24):

> ...the putative cervicals are differentiated from the putative dorsals by a marked difference in size, the shape of the neural spine, and the length, robustness and ossification of the associated ribs...cervical-dorsal boundary (vertebrae 20–22) presumably represented the approximate original position of the shoulder girdle and forelimb.

Caldwell (2000b:731–732) followed Lee and Caldwell (1998) with a redescription of the forelimb and pectoral girdle-bearing ophidiomorph dolichosaur, *Dolichosaurus longicollis* Owen, 1850, which possessed at least seventeen and possibly nineteen cervical vertebrae, and made the following argument:

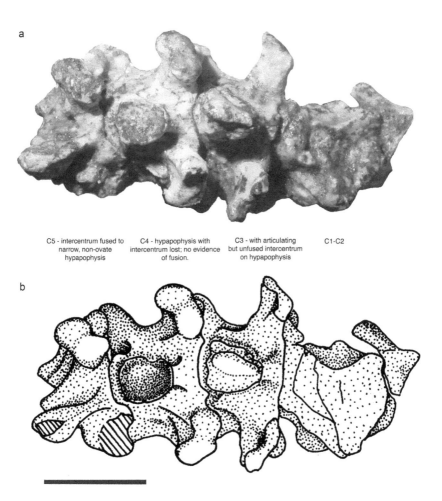

C5 - intercentrum fused to narrow, non-ovate hypapophysis

C4 - hypapophysis with intercentrum lost; no evidence of fusion.

C3 - with articulating but unfused intercentrum on hypapophysis

C1-C2

Figure 3.24. Ventral view of the "cervical" series (anteriormost precloacals) of *Dinilysia patagonica* (MPCA-PV 527). (a) Photograph; (b) stipple drawing of same. Scale equals 1 cm. (Modified from Caldwell and Calvo, 2008.)

A neck is a body region that contains a variety of organ systems…hyoid apparatus…esophagus… trachea, a discrete series of arteries, veins, motor nerves…As snakes still possess all of these structures, it can only be concluded that snakes have not lost their necks, only their pectoral girdles and forelimbs.

And:

Redefining the neck in terms of features possessed by the constituent vertebra resolves this problem, and allows snake evolution to be better understood in the context of neck elongation and forelimb loss.

Further, Caldwell (2000b:732) presented data on the "cervical" vertebrae of *Pachyrhachis* derived from Caldwell and Lee (1997) that noted distinct morphologies not requiring morphometric analyses nor molecular genetic techniques to identify and provide insight on the neck in snake lizards:

…the primitive snake *Pachyrhachis* has at least 18 (see Lee and Caldwell, 1998) vertebrae with hypapophyses, similar to the number in Dolichosaurus (contra Zaher and Rieppel's (1999a,b) recent dismissal of any notable cervical zonation in *Pachyrhachis*).

Caldwell (2000b:732) went further, calling for the Test of Congruence:

"For snakes, the unique evolutionary feature is the cervicalization of a significant proportion of the presacral column thus creating a very long neck. The next step is the test of congruence as outlined by Patterson (1982).

These data and conclusions on the distinctiveness of the snake lizard cervical series, first raised by Lee and Caldwell (1998) for *Pachyrhachis problematicus* Haas, 1979, and examined comparatively by Caldwell (2000b), were found to be consistent with the cervical anatomy of *Dinilysia patagonica*, which showed distinct intercentra articulating with the anterior cervical hypapophyses (Caldwell and Albino, 2002) (Figure 3.24) and later with that of *Haasiophis terrasanctus* Tchernov et al., 2000 (Palci et al., 2013) (Figure 3.25a–d). Thus, in the parlance of this treatise, Type I and Type II ancient snake lizards retained, likely independently, much more distinctive regionalization of the axial skeleton than is retained in the continually evolving modern

snake lizard fauna. These results were reiterated in Caldwell (2003, 2007a), who also noted that Cohn and Tickle's (1999) original insights were only "miscolored" by a reading of thoracic character printed forward because they mistakenly viewed ribs on the "cervicals" as primitive throwbacks, if you will, not normal for most lizards. Caldwell (2003, 2007a) reviewed the then state of knowledge, and identified one consistent theme— the fossil record is essential to understanding the pattern and process of evolution of all anatomies in snake lizards, including the axial skeleton. While modern molecular genetic techniques are of inestimable value in illuminating process, in truth, they are strictly limited to providing insights from the modern fauna only, and thus, as with all studies of living forms, are limited in their temporal reach to current pattern and process, but not ancient pattern. Fossils become the only tools for engaging deep time in addressing evolutionary questions and thus are essential data points to deconstructing essentialist archetypes.

The classic approach to understanding the evolution of the cervical region in snake lizards, or the neck, has been through the study of anatomy and its morphology, inclusive of hard and soft tissues and is most perfectly summed up by the review study given by Hoffstetter and Gasc (1969:281), which swiftly became the gold standard on rules defining the presence or absence of a neck in lizards, inclusive of snake lizards:

In snake-like organisms, the presacral region is quite uniform. The absence of a pectoral girdle in all snakes makes it impossible to refer to the site of this organ in defining the boundary between the cervical and the trunk regions.

And from Hoffstetter and Gasc (1969:293):

…cannot be interpreted as the simple result of an extreme lengthening of the cervical region; such a region, in the literal sense, no longer exists, its diverse components having been scattered even before the girdle disappeared… (variable extension of the hypapophyses) has therefore a different meaning.

In other words, leading up to 1969, the general conclusion was that snake lizards had no neck, that the presence of hypapophyses meant something else other than as a demarcation of the cervical region, and that the absence of a pectoral girdle

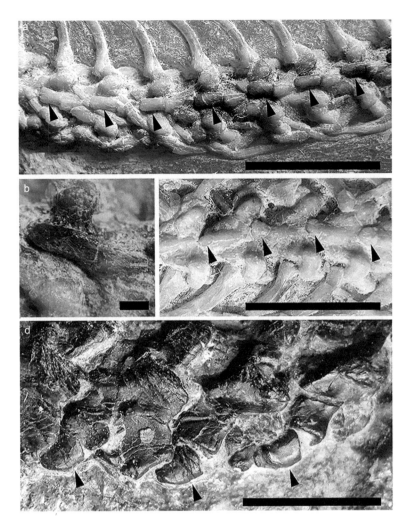

Figure 3.25. Comparison between anterior presacral (cervical) vertebrae of *Haasiophis terrasanctus* (HUJ-Pal EJ 695) and cervical vertebrae of the ophidiomorph lizard *Adriosaurus skrbinensis* (SMNH 2158). (a) Anterior cervical vertebrae of *Haasiophis terrasanctus* (HUJ-Pal EJ 695) in ventral view; (b) close-up on one of the intercentra of *Haasiophis terrasanctus* (HUJ-Pal EJ 695) in ventrolateral view; (c) posterior cervical vertebrae of *Hassiophis terrasanctus* (HUJ-Pal EJ 695) in ventral view; (d) anterior cervical vertebrae of *Adriosaurus skrbinensis* (SMNH 2158) in lateral view; black arrows point at free intercentra. Scale bars equal 10 mm in (a), (c), and (d), 1 mm in (b). (Modified from Palci et al., 2013.)

meant the neck could not be identified. The presence and number of hypapophyses was a critical anatomy for Hoffstetter and Gasc (1969: 293), as they interpreted their presence to not indicate cervical vertebral form:

> Was, then, the acquisition of the apodous state by snakes accompanied by a considerable lengthening of the neck, as Abel (1924) supposed? Here again we find contradictions; in *Typhlops*, the hypapophyses do not appear beyond the fifth vertebra, but the heart lies at the level of the fifty-fifth vertebra.

It is now possible, however, to move beyond the character of absence of a neck because of the absence of the pectoral girdle, and this is because fossil snake lizards possess distinctly regionalized vertebral columns despite the absence of pectoral girdle. They are snake lizards even if they are missing essential categories observed in the crown group or displaying anatomies lost in the crown group, but in the case of axial skeletal characters, they possess unfused intercentra articulating with distinct hypapophyses. What appears to be lost in modern snake lizards are the intercentra, fused or unfused, at least in the anterior/cervical portion of the axial skeleton, and likely posteriorly as well. Large numbers of unfused intercentra are present most noticeably in Type II *Pachyrhachis-Haasiophis* snake lizards (~20) (Figure 3.25a–d), and fewer in Type I *Dinilysia-Najash* forms (~2–3) (Figure 3.24). Much has been said about hypapophyses in the literature (e.g., Hoffstetter and Gasc, 1969; Tsuhiji et al., 2006, 2012), but very little has been written and thus considered concerning intercentra, which are a more distinct indicator of cervicalness, neckness in all lizards including snake lizards, and whether, like *Dolichosaurus longicollis*, the snake lizard neck was elongate before the pectoral girdle and forelimb were lost.

## Cervical Intercentra: Check the Lost and Found

In all non-snake and snake lizards, with the exception of gekkotans, intercentra are restricted to the neck and tail (cervical and caudal vertebrae), while in gekkotans they begin on the first cervical and extend along the length of the vertebral column to the last caudal (Figures 3.24 and 3.25). Gekkotan intercentra follow the primitive amniote pattern of being located between the centra at the intervertebral joint. In non-gekkotan lizards, cervical intercentra are single elements articulating with a process arising from the base of the centrum, of varying height and width, referred to as hypapophyses (a bony pedicle or articulating surface). The hypapophyses are located either anteriorly, centrally, or posteriorly on the centrum. The intercentrum element may fuse to its hypapophysis, but never fuses to the centrum if bridging the joint between centra. It is also extremely common for hypapophyses to be present for which there is no evidence of an intercentrum element anymore; that is, the intercentrum is lost, not fused to the hypapophysis.

In modern snake lizards, free, or sutured, cervical intercentra (three are generally recognized) are present only on the centra of the atlas-axis complex; throughout the rest of the precloacal column and caudal/postcloacal, they are absent (Hoffstetter and Gasc, 1969; Scanlon, 2004; Zaher et al., 2009). Hoffstetter and Gasc (1969) considered the non-atlas-axis intercentra to be fused to the hypapophyses in modern snakes, but recent embryological work finds no evidence of separate chondrifications or ossifications to support fusion of the intercentra to the hypaophysis in modern snakes (Gauthier et al., 2012). The

hypapophyses arise during development from the base of the pleurocentrum and thus the intercentra are considered lost.

The first description of non-atlas-axis cervical intercentra in a snake lizard, ancient or modern, was given by Caldwell and Calvo (2008) based on a new specimen of *Dinilysia patagonica* (MPCA–PV 527) (Figure 3.24); that specimen included a skull and the first five cervical vertebrae with free intercentra to the fourth vertebra and possibly a fused intercentrum on the fifth cervical. The second report of free intercentra was given by Palci et al. (2013) who found free cervical intercentra on the type specimen of *Haasiophis terrasanctus*, extending posteriorly on no less than forty-five precloacal vertebrae. It is critical to note that these are not just forty-five hypapophyseal outgrowths of the pleurocentrum body, but forty-five hypapophysis-bearing vertebrae with true and free intercentra. While Hoffstetter and Gasc (1969) may have had difficulties tracking cervical identity using only hypapophyses, this ambiguity is significantly reduced when considered by assessing the additional presence of intercentra.

Hypapophyses articulating with unfused cervical intercentra are only observed among non-snake lizards in mosasaurs (Russell, 1967), dolichosaurs (Caldwell, 1999b, 2000b), pontosaurs (Caldwell, 2006), and adriosaurs (Palci and Caldwell, 2007; Caldwell and Palci, 2010) and now ancient snake lizards. The only non-snake lizard with a truly long neck (using the usual criterion of presence of pectoral girdle) with articulating intercentra and hypapophyses is *Dolichosaurus longicollis*. Sistergroup relationships aside, the model system for identifying the neck in snake lizards, in the absence of a pectoral girdle, is the hypapophyses and associated intercentra. Modern snake lizards, which have clearly lost the intercentra in the precloacal column, have certainly cervicalized the column in developing the extreme condition of hypapophyses on almost all precloacal centra. If *Haasiophis* is an indicator of original neck length in snake lizards, at forty-five vertebrae (Figure 3.25a–d), which is not inconsistent with the position of the heart and other organs in many snake lizards, then the neck of ancient snake lizards was much longer than that of even *Dolichosaurus longicollis*, and neck elongation could be predicted to have occurred before the forelimb and pectoral girdle were lost.

The myological studies of Tsuhiji et al. (2006, 2012) lend support to this prediction, as they identified in a broad survey of snake lizards and non-snake lizards that "neck" muscles were still present in modern snake lizards and the posteriormost origin of a variety of "neck," in particular the *Musculus longissimus capitis, pars transversalis cervicis* in the Type I Anilioid snake lizard *Loxocemus bicolor*, was at the eleventh vertebra. Other alethinophidians presented cranioverterbral muscles, such as the M. *rectus capitis* anterior extending posteriorly to about the twentieth vertebra. However, using these same criteria, Tsuihiji et al. (2012) certainly found that other modern snake lizards presented evidence of "shorter necks."

Tsuihiji et al. (2012:1008) carefully hedged their bets however with regards to long versus short necks and the origins of snake lizards, and the phylogenetic utility of osteological and myological characters when they wrote (boldface added):

> Optimization of the observed lengths of cranioverterbral muscles on an ophidian cladogram suggests that the most recent common ancestor of **extant** snakes would have had M. rectus capitis anterior elongated only by several segments…the trunk, not the neck, would have contributed most to the elongation of their precloacal region.

It is not clear, without carefully reading the text as written, that Tsuihiji et al. (2012) were not speaking about the origin of snake lizards, but rather only the most recent common ancestor of extant snake lizards—these are two very different "ancestors" and should not be confused. As I have said many times, the data available from modern snake lizards cannot address questions on the origin of snakes, and in fact may not actually provide an accurate picture of the most recent common ancestor considering the scrambled mess that is snake lizard phylogeny, particularly if only examining relationships using data from extant/modern snake lizards.

From my reading of the literature and my synthesis of my own work and that of other authors, I can only predict that in the early phase of their origin from within non-snake lizards, snake lizards evolved long necks, likely long before they lost the forelimb and pectoral girdle. Some 70 million years later, where the body-plan of the "modern snake lizard" appears multiple times, and likely independently, snake lizard evolution had convergently undergone modifications of the anatomy of the neck and thorax. The modern fauna is so distant in time from the origin of snake lizards, and even from the Late Mesozoic origin of the modern snake body plan, that in many cases evolution through the last 100 million years has overprinted the neck identity onto the thorax and reduced the obviousness of the regionalized anatomy that normally characterizes the tetrapod axial skeleton. It is not gone, as Head and Polly (2015) correctly note, but it is subtle and reduced.

### Caudal Intercentra: Tales of the Tail Chasing Itself

While the modern snake lizard tail cannot be accurately characterized as "short," it does not mean that snake lizard tails are not modified in some manner as compared to all other lizards (Figure 3.26). An example of an evolved and modified caudal morphology is the anatomy of the caudal intercentrum system and its corresponding apophyseal structures.

In modern snake lizards, the caudal intercentrum system presents a very different osteology compared to the cervical intercentrum system. Free intercentra are absent in both systems but the cervical series appears to have a single apophysis and intercentrum, while the caudal series appears to be paired in some manner (see Gauthier et al., 2012; Palci et al., 2013; Garberoglio et al., 2019a). The received wisdom holds that the caudal intercentra of modern snake lizards are paired, fused to the haemapophyses, but are not fused at their tips as they are in non-ophidian lizards (Hoffstetter and Gasc, 1969). For either the condition of "intercentrum absent" or "intercentrum fused and paired" to make sense, it is necessary to first review the anatomy of non-snake lizard caudal intercentra and their apophyseal articulations with the centra.

In a typical non-snake lizard, and most are typical except for gekkotans (e.g., Holder, 1960), the remnant morphology of the caudal intercentrum resembles a tuning fork or "y"-shape (chevron or haemal arch and spine), with the paired arms of the "y" contacting right and left apophyses on the base of the vertebral centrum (in some lineages, the haemal arch is fused to the apophyseal processes). In gekkonid lizards, as per Holder (1960), the chevrons or haemal arches

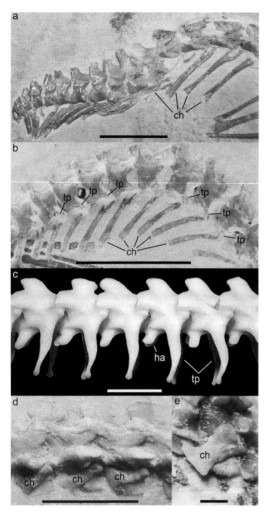

**Figure 3.26.** Caudal vertebrae of the snakes *Eupodophis*, *Morelia*, and *Haasiophis* in lateral view, anterior to the right. (a) *Eupodophis* sp. (MSNM V 4014); (b) *Eupodophis descouensi* (Rh-E.F. 9001-3); (c) the extant snake *Morelia viridis* (ZFMK 53538); (d) *Haasiophis terrasanctus* (HUJ-Pal EJ 695); (e) close-up of a chevron bone of *Haasiophis terrasanctus* (HUJ-Pal EJ 695). Scale bars equal 5 mm in (a)–(d), and 1 mm in (e). (Modified from Palci et al., 2013.) (Abbreviations: ch, chevron bones; ha, hemapophyses; tp, transverse processes.)

hypothesize that the most modern lizard clades (including snake lizards) have lost the base of the intercentrum, but not the developmental anlage, which is retained as the chevron/haemal arch element. As an endochondral bone tissue preformed in cartilage, the haemal arches (my preferred term) must have an embryological origin in the axial skeleton. And, considering that the ancient condition of the haemal arch is as a distal component of the intercentrum body in ancient early tetrapods, for example, the anthracosaurid *Archeria* (Holmes, 1989: fig. 27), it is reasonable to conclude by comparing *Archeria* to the Gekkonidae that even as the intercentrum body is reduced in size over time, the haemal arch component of the intercentrum is retained.

For the remainder of this discussion, I will refer to the lizard portion of the intercentrum as the haemal arch, though, as noted, it is often referred to as the chevron bone, and the apophyses on the centrum as the haemapophyses. The "haem-" prefix makes reference to the arterial system (caudal artery, nerves, and branching aterioles) running ventral to the caudal vertebral centra, passing through and protected by the collective haemal arch system of the complete caudal intercentra. The more distal portion of the haemal arch forms an elongate spine, usually as long as, but maybe even longer than, the dorsally positioned neural spine positioned above the neural tube and dorsal hollow nerve cord. The ventrally directed haemal arch spines serve as muscle and ligament attachment sites for the hypaxial muscles. As with the hypapophyses of the cervical intercentrum system, the caudal haemapophyses arise from the pleurocentrum, but in contrast to the cervical system are not single apophyses, but rather are paired, left and right. Similar to the cervical system, the haemapophyses are positioned either posteriorly on the centrum or at the intercentral joint; the haemal arches may also fuse to the haemapophyses, though this is not common among lizards, with the most common condition of form being a non-sutural articulation.

As discussed above, modern snake lizards do not possess an articulating haemal arch, but instead possess paired haemapophyses that display varying shapes and sizes and arise from the centra. And, as I stated earlier, while the terminology has been to refer to these structures as haemapophyses, it was accepted that these apophyses were fusion

develop as cartilage processes from the remnants of an intercentrum element that in the caudal region of gekkonids is a small platelike wedge of bone that articulates at the intercentral boundary between anterior and posterior vertebral centra with apophyseal facets on each of those two centra. Assuming that the gekkonid condition of intercentra being present throughout the column is representative of the primitive condition for all lizards, snake lizards included, I am left to structures formed as a composite of the right and left limbs of the haemal arch intercentrum (i.e., the haemal spine was lost). While developmental data from studies of snake embryogenesis could easily identify a two-part chondrification pattern for modern snake lizard "haemapophyses," the fossil record, once again, provides the kind of data (morphology bounded by time) that are sufficient to understand the highly derived, and likely independently derived, condition of modern snake lizards. That is, that the right and left structures in modern snake lizards are not fused haemals lacking a chevron, but rather are tall haemapophyses lacking the haemal element altogether (see Palci et al. [2013] contra Hoffstetter and Gasc [1969], and Rieppel and Head [2004]).

Palci et al. (2013) reported new observations of previously described ancient snake lizards that presented distinct haemal arches (intercentra) and haemapophyses for both *Eupodophis descouensis* and *Haasiophis terrasanctus* (elongate haemal arches in *Eupodophis* [Figure 3.26a–c] and short, blunt, spatulate-shaped haemal arches in *Haasiophis* [Figure 3.26d,e]). Their re-analysis supported the previous conclusions of Rage and Escuillié (2000) for *Eupodophis descouensis* that had been oddly discounted as neomorphic elements by Rieppel and Head (2004) in their description of three new specimens of the taxon, and supported the identifications give by Scanlon and Lee (2000) for the caudal skeleton of *Wonambi naracoortensis* contra Rieppel et al. (2003a). Garberoglio et al. (2019a) recognized articulating haemal arches and haemapophyses in new specimens of *Dinilysia patagonica* and reinterpreted the caudal structures of *Najash rionegrina* (previously interpreted as "non-articulating blunt haemapophyses" by Apesteguía and Zaher [2006] and Zaher et al. [2009]).

Consistent with my long argument on the derived, not primitive, condition of scolecophidian anatomy, these snake lizards can be considered to have lost intercentra throughout the axial skeleton, and to have retained hypapophyses on a small number of cervical/precloacal vertebrae, and along with their shortened tails (<20 vertebrae) have also lost their caudal haemapophyses. Type I through to Type IV modern snake lizards, excluding scolecophidians from the latter category, present the same morphology as observed in *Anilius scytale* (Type I)—cervical and caudal intercentra absent, single cervical hypaphophyses and paired caudal haemapophyses present.

The linkage for the modern snake lizard condition to that of the ancestral non-snake lizard anatomy is easily found among the recently identified anatomies of the cervical and caudal skeletons of *Dinilysia*, *Najash*, and other "madtsoiids," as well as *Eupodophis* and *Haasiophis*. Here we find ancient snake lizards presenting the non-snake lizard condition of unfused cervical and caudal intercentra in articulation with their respective hypapophyses and haemapophyses. A key point arising from these new fossil data is that snake lizardness cannot be defined around the loss of cervical and caudal intercentra. Rather, the loss of intercentra from the axial skeleton in modern snake lizards is best considered as convergent loss, particularly as concerns potential phylogenetic relationships nesting various lineages of Type I to IV snake lizards within the clade structure implied by Type I and II ancient snake lizards (i.e., that Type I modern snake lizards are derived from within the radiation of Type I ancient snake lizards, etc. [for elaboration, see Chapter 6]). Common to all the organ systems discussed in this book, a working hypothesis for the evolution of the caudal intercentrum system of snake lizards is that there have been multiple independent losses, not one single event suggesting inheritance of this condition from a single common ancestor. Once again, "simple" finds no support as a real pattern of change.

## Limbloss and Limb Reduction

Aristotle's prescience on snakes (Aristotle, *History of the Animals*, Book I, Chapter VI) and their state of limblessness is arresting, as he most certainly was thinking beyond the box of essential characters to imagine his notion of the condition of snake lizardness as a result of a variance of morphology, in this case, limblessness: "The serpent genus is similar and in almost all respects furnished similarly to the saurians among land animals, if one could only imagine these saurians to be increased in length and to be devoid of legs." It is certainly not possible to interpret Aristotle as having considered that organisms might change over time as a process, that is, to evolve, in order to explain the workings of his imagination, but it is clear the he recognized that limblessness and elongation were the only real differences between snake lizards and all other lizards. This is a non-trivial point, because, as I have made clear in this treatise, snakes are lizards, and while elongation and limblessness are common among lizards, snake lizards have gone their own way on at least elongation. Though a core thesis of this book is that skull evolution and innovation have driven snake lizard evolution, it is absolutely true that limbloss and limblessness have hardly been a limitation to snake lizards through at least the last 100 million years (using *Najash rionegrina* or *Pachyrhachis problematicus* as the temporal indices for full modern snake lizard postcranial anatomy, that is, elongate and limb-reduced to functionally limbless).

The overarching question inhibiting further understanding on the origins and evolution of limbloss in snake lizards is the absence so far of the great "missing link" fossil, the "four-legged snake." *Tetrapodophis amplectus* Martill et al. (2015) was certainly heralded as that four-legged snake lizard fossil, but in fact has come up short on snake lizard anatomical characters in all systems except for the number of presacral/precloacal vertebrae (a not insignificant deficiency [see Caldwell et al., 2016, in prep.]), even though it is elongate and does possess four small legs (forelimbs decidedly smaller than the rear limbs). But to be considered in the discussion of snake lizard evolution, you must first be a snake lizard, not something else. *Tetrapodophis amplectus* is most certainly something else altogether (see overview in Chapter 2).

In exploring the morphological attributes of the oldest known snakes, that is *Parviraptor* and kin, Caldwell et al. (2015) noted for the most part the absence of substantive postcranial fossil material and that the Middle to Late Jurassic and Early Cretaceous forms showed numerous snake lizard characters of their skulls. Considering the substantive addition of some 70 million years to the known fossil record of snake lizards, and by comparison to the patterns of skull and postcranial anatomy displayed in other non-snake lizard clades (diagnostic features of the skull of skink, anguid, or lacertid, but the body has become limbless and elongate), the prediction would be that 170 million-year-old snake lizards such as *Parviraptor* and *Portugalophis* could possibly still retain all four limbs yet have snake lizard skulls. One hundred million-year-old snakes such as *Pachyrhachis* and *Najash* still possess obvious though reduced hindlimbs, and thus it is reasonable to predict that 170 million-year-old snake lizards might still have the forelimbs as well. Caldwell (2003) presented pattern data

from a wide variety of elongate and limb-reduced tetrapods that indicated that elongation only becomes extreme when the pectoral girdle finally disappears, along with the forelimb. This pattern remained a robust observation among vertebrates until the discovery of *Tetrapodophis amplectus*, whatever it might be phylogenetically speaking. The one unique aspect of its morphology is its extreme elongation between the pectoral girdle and pelvic girdle. It possesses a relatively long neck (~10 cervical vertebrae), but the number of dorsal vertebrae (~150) exceeds by more than two times (see Chapter 2, Figure 2.13) the number observed in all other elongate tetrapods that have not lost the pectoral girdle even if the complete forelimb is lost (Caldwell, 2003). *Tetrapodophis* does have a few more cervical somites (10) than does a typical lizard (7–8), but by no means has it increased somitogenesis at a level displayed by *Dolichosaurus longicollis* (~19), nor is it close to the number observed in *Pachyrhachis* or *Haasiophis* (~20). What is unique, and thus hard to explain, is the extremely large number of dorsal vertebrae in *Tetrapodophis*, a number unheard of in tetrapod evolution in taxa still possessing forelimbs, with the exception of the amphisbaenian *Bipes biporus*, which still has fewer dorsal vertebrae (4 cervicals, 106–128 dorsals, and 23–24 caudals). Like *Bipes biporus*, *Tetrapodophis amplectus* is unusual and unique with regards to numbers of presacral vertebrae, but like *B. biporus*, *T. amplectus* is also not a snake lizard just because it is elongate and limb reduced; for both of these lizards, skull anatomy testifies to affinities, and neither of them display snake lizard skulls, neither ancient or modern.

There are a number of questions about limbloss that remain unanswered and in fact, for the most part, can be addressed by modern techniques applied in studies of evolutionary developmental biology. For example, why are the forelimb and pectoral girdle lost first in all limbless to limb-reduced tetrapods, with the exception of course of the amphisbaenian, *Bipes biporus*? And arising from elucidating the answer to that question is: Why does *Bipes* retain the forelimb while the rear limb is lost? A third question? Why is that limbloss and elongation are linked to the distal reduction of digits in the hand and foot, and these losses begin preaxially and proceed postaxially in the limb before proceeding to proximal losses? Is this just a simple reduction via reverse recapitulation of the patterns of limb formation? Is it that easy?

But, if so, why is it stepwise within a limb but yet out of sync between the forelimbs and rear limbs (see the scincid lizard *Lerista* where literally every pattern imaginable is present [Greer, 1987, 1990]). Why, as Caldwell (2003) noted, does elongation of the presacral column seem to be limited until the pectoral girdle is lost, with, of course, the recent exception being *Tetrapodophis?* Why are there exceptions to these seemingly constraining rules in the first place, or better yet, why does the basic pattern of almost all tetrapods exist in the first place? In essence, why tetrapods only? Where are the hexapods or octapods among vertebrates? And finally, because these questions could go on forever, what patchwork quilt of loss and reduction of limbs and girdles is the real and true pattern of snake lizard evolution through more than 170 million years? If one thing is clear, it is that the real history of limb evolution and loss in snake lizards is not a simple linear transformation from limbs to no limbs, but rather a complex nexus of numerous lineages of ancient snake lizards losing limbs and becoming elongate, likely within lineages long after they became distinct from their closest snake lizard sistertaxa. I think it is safe to conclude that limblessness does not characterize anything but the outmoded and outdated archetype of snake lizards, and that in truth limblessness occurred not once but many times within ancient snake lizard lineages. While I think we know a great deal, I also think we know enough to realize that we know very little, and that each answer only rattles the next questions even harder regarding the patterns and processes of limbloss and limb reduction.

## BUILDING THE ANCESTRAL SNAKE LIZARD: "ESSENTIALS OF AN ARCHETYPE"

Despite the fact that I have railed against essentialism and archetypes for their negative impact and resulting historical burden, I have also noted it is impossible to avoid them—thus we simply must be aware of them and not allow them to be elevated to the level of absolutes and to be dogma that resists change in the face of new evidence. I cannot, nor can we, define an object, nor outline an idea or concept, in the absence of language that does not by its very nature and how we use it, create boundaries between the object or idea of interest and other ideas or objects. If I want to eat an orange, I refuse to be fooled by a potato into thinking it is an orange, no matter how charming the spud might be, because I know what an orange looks like and the flavors I will be rewarded with if I follow the rules of identifying it by its essential characteristics. I therefore cannot build a prediction, based on anatomical details, of an ancient snake lizard without resorting to categories that delimit an ancient snake lizard around its probable and very essential characteristics. Ultimately, the value of that essentialist prediction is real because when that fossil is found, it will be recognizable for what it is—the fossilized remains of an individual of an unknown species of ancient snake lizard. It will not be the ancestor of all snake lizards, but it will be a member of a species demonstrating that during the early evolution of snake lizards, their evolution led at least one lineage to experiment with some or all of the morphologies predicted here by reference to the anatomies of known ancient Mesozoic snake lizards and their contemporaneous closest relatives. My hope is that the details of this chapter have outlined those features that are expected to be observed in their most ancient form; this is based on the predictive power of combining the anatomies of the modern and ancient snake lizards we currently know something about. Predicting the morphology of an ancestral snake lizard as remains to be found in the fossil record is no easy task. However, I think I have presented sufficient data to propose, or at least to suggest, what we should expect to find: a four-legged, long-necked snake lizard with well-developed cervical hypopophyses and unfused intercentra, along with a longish tail and articulating haemal arch intercentra; attached firmly to the "neck," we should predict the presence in an ancestral snake lizard of a big head presenting anatomical features consistent with a big mouth but not demonstrating macrostomy, and certainly not extreme macrostomy; specific features would present as an absence of a sutural joint between the dentary tips; a developing intramandibular joint and probable absence of a compound bone; a vertically oriented quadrate with a large suprastapedial process; immobile palatopterygoid complex; the presence of "thecodont" tooth attachment tissues but with a more "pleurodont" geometry of implantation; presence of a prominent jugal (no quadratojugal); likely remnants of the squamosal and postorbital (completeness of the bar is difficult to predict); possible epipterygoid and membranous braincase only if the anterior shift of the CID musculature, in particular from the crista prootica, has not evolved yet; an enlarged otic capsule but no juxtastapedial recess; the presence of a "lizardlike" fenestra rotunda and a short columellar shaft, with most likely two extracolumellar elements. The ecology and evolutionary relationships of this ancestral snake lizard, and the subsequent radiation of fossil and modern snake lizards, are the subjects of the remaining chapters.

# CHAPTER FOUR

# Ancient Snake Lizard Paleoecology

## READING THE ROCKS FOR
## HABITS AND HABITATS

I have often found, as I have read through the paleontological literature, that I hoped beyond hope that no sedimentologist was trying to make sense of the interpretive paleoecological leaps made by eager paleontologists from the rocks in which their favorite fossils were found. The bulk of the vertebrate paleontological literature presents a five- to ten-sentence "Geological Setting" section usually wedged uncomfortably into the Methods and Materials section with a few words on the lithology of the rock, for example, "sandstone," and a rough stratigraphic position for the unit containing the fossils, all tied up with a bow by a paleoenvironmental statement such as, "interpreted as a proximal shallow marine to lagoonal depositional environment," or, "channel sandstones in the area represent anastomosing channels on a floodplain." While the first few pieces of this logic string are harmless enough, there is no depth or detail to such descriptions that would legitimize the paleoecological conclusions for the animal in question. If the manuscript avoids the latter, which they seldom do, then a detailed facies analysis based on detailed sedimentological and stratigraphical analyses is unnecessary and a light and fluffy "Geological Settings" paragraph is sufficient. But if paleoecologies are proposed, then light and fluffy is completely unacceptable. Facies analysis is complex work requiring real interest in, and a dedication to, rocks for the sake of rocks and their genesis in specific depositional

environments under defined and diagnosed processes. Sedimentary rocks are not simple crystal balls that can be held up to the light, palantír-like, to reveal to the all-seeing paleontologist with but a passing glance the paleoecology and life habits of their favored ancient animal.

While I cannot claim a career of scholarly work on facies associations and their genetic interpretations, I can, however, claim some reasonable experience with such studies, coauthored with my students and postdoctoral fellows and dedicated sedimentologist and ichnologist collaborators (e.g., Caldwell and Albino, 2001; Zorn et al., 2005; Abdel-Fattah et al., 2009, 2016; Diedrich et al., 2011). So on behalf of rocks, sediments, lithology, stratigraphy, and facies analysis, I present a thesis statement for this chapter—the depositional environment of the sediments preserving the fossil in question generally provide very little direct evidence of the paleoecology of a fossilized organism(s).

For free-living organisms such as vertebrate animals, 99.999% of the depositional environments in which they are preserved are not where they lived. For example, of the hundreds of thousands of fossil fishes in thousands of museums around the world, spanning some 400 million years of earth history, few if any of them lived in the muds, detritus, and chemical precipitates in which we find them. These sediments were deposited under very specific circumstances in very specific environments; even if they were deposited in a lake or shallow marine setting and were not formed in the water column, they were formed at the bottom of the lake or marine lagoon. By and large, ancient fishes, like modern fishes, were living, moving, breeding, and feeding somewhere in the column of water above the accumulating sediments on the bottom. Within that water column, segregated by physical barriers or niche specializations, are an astonishing variety of habitats, and thus paleoecologies, that cannot be directly interpreted from the now lithified sediments defining that lacustrine, fluvial, or coastal marine depositional system.

Again, for fossil organisms, no matter where they lived in that water column, or in lateral equivalents to the place where they were fossilized, due to processes linked to, but not necessarily driven by, the depositional environment, upon their death they sank to the bottom and were caught up in the sedimentary processes of that depositional environment inclusive of diagenesis and lithification. We can obviously generalize and make a statement about "fish," and say, "They lived in a "shallow lake environment," but that really says little about that lacustrine system in the absence of a detailed presentation of facies associations and lateral equivalents (i.e., following Walther's Law). Why is this essential to an accurate assessment of paleoecologies? Well, what if the fish assemblage being studied was not lacustrine at all, but rather simply ended up in that lake as a death assemblage washed downstream from a large and proximate fluvial system? The rocks can tell you such a story if you look long enough or care to discover what might still be preserved, but what is certain is that you must look beyond the unit preserving your fish death assemblage. This problem is even more profound when you find associated with the fish assemblage the occasional pterosaur, bird, mammal, or even snake lizard. Just because they are in lacustrine-deposited sediments, does this mean that they too, like the fish assemblage, lived in the lake? Unlikely. What ancient paleoecology can be proposed for such animals from the depositional environment of lacustrine systems, or even fluvial systems? Answer—nothing. We know that most birds fly, but not all. We also know that some birds swim, dive, and feed in aquatic environments. There are a few mammals that fly, but not likely a Cretaceous mammal, and pterosaurs are certainly considered to have flown, though direct observational evidence will be forever lacking. Snake lizards? Well, modern forms swim in ponds, lakes, rivers, seas, and broad oceans. They locomote in their varied snake lizard manners in terrestrial environments expressing a wide number of ecologies and are not considered to have flown under powered flight, though some modern forms are uncannily good at gliding through tropical forests.

Our understanding of these various vertebrate habits, ecologies, and functions have one thing in common and it is not depositional environment, it is observed functional morphology coupled with behavior and the correlation of both to observed habits in modern members of their respective groups, with the exception, of course, of pterosaurs. For fossil forms, their paleoecology is inference only, made possible by modeling and comparison to modern forms of a similar kind.

To use the most simple language possible, sedimentary environments, paleoenvironments and paleoecology are very different things and should not be confused, one with the other. Such mistakes of simplistic overinterpretation are common in the paleontological literature on snake lizards (e.g., Longrich et al., 2012; Martill et al., 2015) and, though seemingly harmless, create long-lasting concepts of paleoecology that have implications for functional morphology as interpreted from preserved fossil anatomies and generalizations made around the ecologies of modern snake lizards (see Chapter 5). The remainder of this chapter will review the basics of facies analysis and propose expectations on hypotheses of paleoecology, particularly in reference to paleoenvironmental interpretations. Case studies of a kind are given for several important and overinterpreted ancient snake lizards, and, finally, I will review the facies associations at La Buitrera where I have conducted fieldwork in search of the fossil snake lizard *Najash* in the Argentine Patagonia.

## Sedimentary Rocks: Depositional Environments versus Paleoenvironments

Beginning at the beginning is always a good thing, so to first principles we go. To begin, though there are rare exceptions, fossils are found almost exclusively in sedimentary rocks. It is possible for fossils to be preserved in rocks derived from volcanogenic rocks (e.g., pyroclastic debris flows, tuffs, and so on, which are often treated as a form of sedimentary rock), but not primary igneous rocks. It is also possible for some aspect of fossils remains to preserved in certain kinds of metamorphic rocks that are themselves derived from sedimentary rocks (e.g., certain kinds of marble, some low-grade quartzites, etc.), but neither of these are common nor is preservation very good if something is preserved in some fashion or another.

The primary sources of fossils, as I said, are sedimentary rocks, which come in two broad forms, clastic and biochemical-chemical sedimentary rocks, the varieties of which are deposited in very specific and very different kinds of depositional environments. If we wish to understand those depositional environs and what they inform us on regarding an ancient fossil organism, in modern geological sciences we proceed to collect data and develop interpretations by conducting a "facies analysis." The study of sedimentary facies, that is, facies analysis, is the study and interpretation of the textures, sedimentary structures, fossils (biofacies), trace fossils (ichnofacies), lithological associations (lithofacies), and even hydraulic properties (hydrofacies, i.e., how water moves through sediments and rocks) of sedimentary rock units and their collective organization or strata. The tools of facies analysis are intended to develop three-dimensional models and interpretations of sedimentary strata between or within outcrops, core, and well-log sections, and across and potentially even between, basins. The facies concept of sedimentary facies analysis can be extended beyond a mere descriptive statement to one that is more genetic, such as a depositional process or depositional environment. However, the genetic definition of facies is not the primary goal of facies analysis, which is by intention, purely descriptive; that is, a fine-grained quartz sandstone is a lithofacies descriptor of a sedimentary unit, while a fine-grained "aeolian" quartz sandstone adds the genetic component to the description by positing a broad, though not specific, depositional environment and process. The genetic component identifies recurring groups of facies referred to as a facies associations that are interpretative models using all the data available from all levels of the facies analysis.

The temptation for someone looking for a quick "paleoenvironmental" diagnosis or to posit a meaningful "paleoecology" for an origins hypothesis is to use a few lines of data from a facies analysis without having first conducted a

complete facies analysis and developed sound facies associations. The simplistic approach, all too common in the literature is, "My ancient snake lizard was living where sand grains can be moved by wind, which of course is in a hot, dry, desert!" Unfortunately, this is an incorrect co-option of a genetic term that only suggests a potential and very specific depositional environment for the sediments, not a "home" for an organism. Such genetic descriptors are also little more than lithofacies fragments of a fully developed facies analysis incorporating a great deal more data than just a genetic lithofacies descriptor. Extrapolating to the broad scale of a paleoenvironment only proves my stereotyped excited scholar does not know that depositional environments and paleoenvironments are not the same thing. Anderton (1985:31) noted:

> Short cuts can lead to fallacious interpretations. Reliable facies models cannot be produced without careful facies analysis.

To best extrapolate to a paleoenvironment from a section of outcrop containing a fossil or fossil assemblage of interest, extreme effort must be expended on collecting all the data available in order to first conduct a facies analysis. These data will, assuming that a stratigraphic section includes multiple lithofacies, biofacies, ichnofacies, and so on, result in a suite of facies assemblages reflecting the lateral association of these assemblages (Walther's Law [Middleton, 1973]) and thus the lateral association of specific depositional environments. With all these data and associations in place, it is possible to posit specific depositional environments and then extrapolate from these lateral associations of depositional environments to a bigger picture of the "paleoenvironment."

I use as an example of how to empirically justify paleoenvironmental reconstructions based on facies analysis and the construction of facies assemblage the study presented by Diedrich et al. (2011), "High-resolution stratigraphy and paleoenvironments of the intertidal flats to lagoons of the Cenomanian (Upper Cretaceous) of Hvar Island, Croatia, on the Adriatic Carbonate Platform." This paper presented the results of work conducted on Hvar Island in 2004 and 2005 as part of a larger fieldwork project prospecting for pythonomorph lizards (Cajus Diedrich was a postdoctoral fellow working in my lab in 2004–2005). Diedrich et al. (2011) began by providing a geological overview of the Dalmatian Coast, including the Island of Hvar, the location of the studied outcrops and their stratigraphy and sedimentology. They then detailed the stratigraphic framework connecting their measured sections (217 m in total) by presenting the lithostratigraphic and biostratigraphic details of the complete section and placed all of this information into a sequence stratigraphic framework. They then presented facies associations/assemblages based on (Diedrich et al., 2011:387):

> The main carbonate rock lithologies of the Cenomanian of Hvar Island include biolaminates, stromatolitic layers, tempestite-generated fossiliferous layers, platy limestones, platy limestones with scattered fossil debris, chert beds, massive rudist packstones/floatstones, intraclast beds, and bedded anhydrites. These lithologies are discussed below in the context of recurring facies associations.

Six facies associations were identified in 217 meters of measured section that represented, as per Walther's Law, the lateral association of numerous depositional environments in the context of a larger paleonenvironmental setting—in this case the carbonate sediments and facies associations presented sufficient data to allow interpretation as intertidal flats, sabkhas, and shallow carbonate island lagoons. The sequence stratigraphic context of the six facies associations presented data indicating repeated emergence and subsidence, that is, regressive and transgressive events, within the measured sections on Hvar Island. It is a similar level of detail that is needed to empirically justify any and all claims for a proposed paleoenvironment for a fossil organism.

A paleoenvironment, like a modern environment, is a complex concept that includes living (plants, animals, microbes) and nonliving (soil, chemistry, air, water, rocks, etc.) things in all their variety and seemingly chaotic interactions (this is why a depositional environment for a sedimentary rock can only be accurately interpreted from a complete facies analysis, as such an approach combines the available biotic and abiotic data). For example, if we are talking about a "river environment," we are talking about the living and non-living things, both on land and in the water and in the air and below the ground that interact in a "river environment." In terms of organisms in this complex environment, we would want to

consider bugs, mammals, fishes, birds, reptiles, amphibians, crustaceans, microbes (prokaryotic and eukaryotic), fungi, and plants of all kinds (vascular, nonvascular, and cryptogamic). We have organisms living on land, in the water, near the river's edge, and up into the surrounding shrubs and trees, along with organisms underground and in the air. For paleontologists, paleoenvironments are indeed complex entities that, in truth, the rock record provides virtually no insights on, as nearly 99% of the needed data are simply not preserved. Interpreting paleoenvironments and paleoecologies for fossilized organisms is limited to the data—it may be surprisingly complete in one case and shockingly incomplete in another. Only a fully developed facies analysis permits extrapolation beyond the specifics of a depositional environment as read from the immediate rocks containing the fossilized remains.

As indicated, depositional environments (sometimes referred to as sedimentary environments) are extremely specific environs nested within the larger concept category of a paleoenvironment, or even a modern one. A depositional environment defines and diagnoses the biological, physical, and chemical processes that influence the deposition of a particular kind of sediment in a particular place or circumstance. For example, the lee side of Barchan dune (a crescent-shaped sand dune structure where the "horns" of the crescent face downwind and the arc-side of the dune is a steep face) is a depositional environment where sand grains moving from the windward side fall down the slipface or lee-side and are deposited along that face at the bottom of the dune structure. That slipface on the lee-side is a depositional environment, but the entire desert is not a depositional environment, but rather is "desert environment." Within that desert, there are most probably a large number of depositional environments, that is, other dunes, seasonal rivers, creeks, lakes, and so on, all of which would present different facies characteristics upon entering the rock record. It is these specific facies-level features that diagnose depositional environments, and from there, paleoenvironments, not the other way around.

## Facies and Paleoenvironments Do Not Equal Paleoecology

Another common mistake is to confuse "environment" with "ecology" and thus "paleoenvironment"

THE ORIGIN OF SNAKES

with "paleoecology." As discussed, environments can scale up or down in size, but invariably the term is meant to combine the biotic and abiotic and not just focus on the biotic interactions, which are the domain of "ecology." When we speak of an organism's ecology, or an ancient organism's paleoecology, we are speaking of its ecological relationships within its ecosystem, that is, animal, plant and microbial interactions. Depending on the scale of the question being asked or the data being studied, a study can scale up ecologically to an ecosystem (of which numerous definitions abound, but few are agreed upon [see Schulze et al., 2005]), or down to mesocosms, habitats, and niches. At the ecosystem level, in some form, the questions and data focus on interactions between all organisms, while at the niche level, the focus is small and specific, and relates to a single species and how it has specialized or fits into its particular habitat and interacts with its specific predators, prey, members of its own species, and so on.

An organism's paleoecology is also not the same thing as its functional morphology, though there is a direct relationship of outcomes that connect the two in a somewhat linear cause-and-effect fashion. For example, Icarus dreamed of flying, but lacked wings, and so when with his invention of feathers and wax, he flew too close to the sun and the wax melted...well, his true functional morphology of arms, not wings, led to his failed attempt to escape from Crete, which we can view as a disastrous attempt at claiming the ecology of a bird. Thus, he fell into the sea and drowned. Which also means he clearly lacked the functioning morphological systems of a fish and thus failed a second time at yet another ecological boundary breakout.

The anatomy of a fossil organism, which is at the root of its functional morphology, can provide strong clues as to the function that those composite morphologies would have permitted or perhaps even excelled at, but, again, sometimes they do not. Flipperlike appendages in mosasaurian lizards strongly suggest they would, by analogy, have functioned like the flippers of modern whales—as steering and sometimes propulsive structures for an animal living in water. However, sometimes anatomies provide no such insights on life habits; for instance, the limbs of the Galapagos marine iguana are virtually identical to the limbs of the Galapagos land iguana or to the tree-climbing iguanas of the continents of South and North America, yet the Galapagos marine iguana

locomotes and feeds in the shallow marine waters around the Galapagos Islands. If it were found fossilized in a marine nearshore sandstone, or a lagoonal limestone, the conclusion would be simple: a terrestrial lizard preserved in marine deposited sedimentary rocks—died, swept out to sea, and fossilized in a depositional environment that was not its habitat.

To extrapolate a function for the preserved anatomies of fossil organisms, we often combine the data of anatomy with the outcomes of facies analysis, that is, paleoenvironment interpretations, in order to understand aspects of an organism's paleoecology—habitat selection, niche specializations, and so on. It is in fact extremely difficult to conduct true paleoecological studies because you need to understand the more complex organization of the ecosystem as a whole, not just extrapolate from anatomy to function and then link function to environmental interpretations.

Where such premature extrapolations go off the rails is when the functional morphology of an organism and an incomplete hypothesis of paleoenvironment, or paleoecology, are also used to support a particular origins hypothesis. The latter has most certainly been the case in the game of snake origins hypotheses—a snake fossil is found in "terrestrial sediments," said snake fossil shows morphologies that look like they belong to a subterranean animal; and so subterranean + terrestrial equals a burrowing origin for snakes. That the reasoning applied here is circular—well, that is obvious, but there is also an egregious abuse, mostly by simplification to the point of being wrong, of sedimentological data, lithofacies analysis, the interpretation of depositional environments and of taphonomy.

## ANCIENT SNAKE LIZARD DEPOSITIONAL ENVIRONMENTS

The following provides overviews of reported depositional environments, though not always derived from detailed facies analyses, for the seventeen ancient snake lizards presented in Chapter 2. They are organized under the broad headings of geologic period and ancient supercontinent, followed by the geologic stage (all in rock time units) relevant to the more inclusive category of geologic period, and the modern place name where the fossils are found today.

*Eophis woodwardii: Bathonian (Modern England)*

The balance of the sediments preserved at the Kirtington Cement Works quarry are packestone to grainstone fossiliferous shelley limestones, grading to sandy and muddy marls (clastic muds and silts mixed with carbonate sediments) (McKerrow et al., 1969; Wyatt, 1996). In terms of sea level rise these litho- and fossil facies represent mainly transgressive sequences of marine deposition, with some regressive deposition as channel splays and unconsolidated marly mudstones lenses with lignites; on smaller scales almost every depositional environment represents a regression of some kind if it is preserved (M. Gingras, personal communication). Within the Forest Marble Formation, considered a "regressive unit" in the sequence stratigraphic sense (Wyatt, 1996), one of these mudstone lenses, Bed 3p of McKerrow et al. (1969), immediately overlying the "Coral *Epithyris* Limestone," has produced most of the microvertebrate remains, including those of *Eophis underwoodi* (see Figure 2.1a). From the litho and biofacies assemblages, the specific depositional environment has been interpreted as estuarine mud flats, intermingled with back-beach, back estuary lacustrine environments marginal to the estuary or beaches.

**Evolving Interpretations**—Evans (1991), in a study describing the lepidosauromorph *Marmoretta oxoniensis* from the same beds as *Eophis underwoodi*, noted that "The paleoenvironment at Kirtlington has been reconstructed…as a freshwater lagoonal habitat like the present Everglades of Florida." The lithological and sedimentological reality is decidedly more complex than a simple "everglades" environment (McKerrow et al., 1969; Wyatt, 1996) and so such a characterization is essentially meaningless. All specific paleoenvironments (e.g., everglades) existed in their time, as they do today, as a series of continuous, lateral successions of environments. In other words, the Everglades of modern Florida are connected to distinct river systems on one end that pass through swamp and everglade forests, while on the other end, the everglades grade into distinct deltas and delta fronts, tidal estuaries, shorefronts, and open seas.

Evans and Milner (1994) refined the paleoenvironment description by describing the Kirtlington locality as "on or near the shore of an island barrier some 30 km northwest of the London-Ardennes landmass," citing Palmer and Jenkyns (1975). The barrier-island paleoenvironmental scenario for the period of time during which the Forest Marble Formation was being deposited is a very interesting one when considered in terms of the onlap-offlap cycles of marine transgression and terrestrial sediment regressions, particularly with proximity to the larger Ardennes-London Island land mass. The known vertebrate remains are certainly intermingled with numerous marine and brackish water vertebrate and invertebrate remains in Bed 3p. Most vertebrate fossils are worn, abraded, completely dissociated, and broken, suggesting a great degree of transport and even possibly reworking. Though not impossible that the source was from the large island land mass to the east, it seems more sensible to conceive of the swiftly transgressive-regressive cycling as the source of island environmental shifts from above water to below water and back again.

What is important to keep in mind, however, is that even the large Ardennes-London land mass was still an island chain in the epicontinental sea covering most of Europe at the time. It is intriguing that on these series of islands, we find not just evidence of the oldest known snakes, but also an early anguimorph such as *Dorsetisaurus*, early scincomorphans such as *Paramacellodus*, poorly understood choristoderes, lepidosauromorphs, frogs, salamanders, mammals, and so on (Evans and Milner, 1994).

*Portugalophis lignites: Kimmeridgian (Modern Portugal)*

The locality at Guimarota, Leiria, Portugal, and its contained snake lizard materials (see Figure 2.1f–i) is astonishing—all the material, and it is extensive, diverse, and beautifully preserved, comes from subterranean excavations of a relatively thin and limited Kimmeridgian-aged, Upper Jurassic coal seam set that had been part of a larger set of strata that had long been mined for industrial brown coal. The recognition of a rich vertebrate paleontological resource in the coals began in the late 1950s and early 1960s, and between 1972 and 1982, the site was excavated only for its fossil resources (for details, see Martin and Krebs, 2000).

In very broad terms, the depositional environment at Guimarota is a coal swamp (Myers et al., 2012), though the detailed geological setting is much more complex. The coal deposits of

THE ORIGIN OF SNAKES

Guimarota are positioned along the eastern rim of the Lusitanian Basin, which developed as a north-south trough in the rift system that developed between "Iberia," "North America," and northern Gondwana during the rifting of Pangaea. By the time the Guimarota coals were deposited, the rift valley system was now fully marine and the coal swamps were positioned along the coastline of an "Iberian" island system. The resulting "Guimarota strata" are composed of upper and lower lignitic coal layers separated by marly limestones in a lithofacies stack of marly limestone, shaley marls, lignitic coal, repeat. The unit is referred to the Alcobaça Formation, which is considered a lateral equivalent to Kimmeridgian-aged Abadia Formation (Schudack, 2000). The coal swamp has been interpreted as a brackish water lagoonal environment with freshwater input, similar to a wooded swamp environment not unlike a modern tropical mangrove forest (Van Erve and Mohr, 1988).

## Diablophis gilmorei: Kimmeridgian (Modern Colorado)

Diablophis gilmorei (see Figure 2.1b–e) is found in rocks outcropping at the Fruita Paleontological Area, from the Morrison Formation (Turner and Peterson, 2004), and are Upper Jurassic (Kimmeridgian) in age. In contrast to the island/coastline, estuarine lake, and coal swamp paleoenvironments that preserve Eophis and Portugalophis, the lithofacies and depositional environments of the Morrison Formation at the Fruita Locality are all decidedly channel sands and overbank muds deposited in a large-scale continental basin, bordering the northern margin of ancient Lake T'oo'dichi'. At the Fruita Paleontological Area, the snake lizard fossils, along with the well-preserved and dense record of other vertebrates, are to be found in what has been interpreted as lithofacies deposited in a set of three or so small lakes or pools associated with a meandering upland river system (Galli, 2014:104)—"Major accumulations of bones are restricted to the pools at three bends of meandering-river channel sandstone 2 at FP where preservation potential was enhanced." The larger-scale paleoenvironment is best conceived of as an extensive floodplain environment with seasonal precipitation moving east from western uplands in and around the margins of a large inland lacustrine system (Turner and Peterson, 2004; Galli, 2014).

## Lower Cretaceous of Laurasia

### Parviraptor estesi: Berriasian (Modern England)

Parviraptor estesi Evans, 1994 (see Figure 2.2), is now restricted to the holotype specimen NHMUK R48388; the second specimen, NHMUK R8551, shows no overlapping characters with P. estesi and is assigned to aff. Parviraptor. The specific locality data for both specimens was never recorded in detail, as they were collected a very long time ago (likely late nineteenth century), and contains no information suggesting they are from the same horizon or layer, though the specimen tags for both describe them as Middle Purbeck Formation, Durlstone Bay, near Swanage. From personal observation, I can confirm that the two slabs of rock (NHMUK R48388 and R8551) do not share the same lithology, and thus are likely from different beds at Durlstone Bay. This is further supported by Evans's (1994) reported discovery of the specimens in independent collections at the Natural History Museum. She found NHMUK 48388 among a number of undescribed specimens, and presumably not among the Museum's squamate materials, though this is possible; NHMUK R8551, on the other hand, was found mixed in with specimens considered to be pterosaur remains.

The rocks of the Purbeck Limestone Formation exposed along the shoreline at Durlston Bay, Swanage, are roughly 110 meters in thickness and in places quite significantly deformed, and are subdivided into the lower Lulworth and upper Durlston Beds (West, 1975; Riboulleau et al., 2007); the Lulworth is mainly Early Berriasian (earliest Lower Cretaceous), while the Durlston Beds are Middle to Late Berriasian. The exposures at Durlstone Bay can be broken down into approximate thirds where the lower third is considered to represent lagoonal evaporites deposited in a sabhkalike setting; the middle third is an arid climate with freshwater to brackish waters, with locally hypersaline littoral facies; and the upper third is more a humid climate with similar freshwater to brackish water littoral facies (West, 1975; Clements, 1993; Samankassou et al., 2005; Riboulleau et al., 2007). The biofacies assemblages (molluscan, ostracod, charophyte, and palynomorphs) suggest strong salinity fluctuations from freshwater to hypersaline (Riboulleau et al., 2007). Dinosaur footprints and mudcracks occur throughout the sediments of the Purbeck

Limestone Group, indicating repeated sequences of transgression-regression. Most importantly, the middle part of the Durlstone Bay outcrop exposes the regional marker bed, the Cinder Bed, and the locally important and extremely fossiliferous Mammal Bed, just below the Cinder Bed. The balance of the Purbeck vertebrate assemblage was collected during the nineteenth century excavation of the Mammal Bed and likely includes the materials ascribed to *Parviraptor* from Durlstone Bay. This unit of rock is in the Lower Durlston Beds, considered Middle Berriasian in age, and identifies the paleonenvironments as freshwater to brackish waters, periodically hypersaline, and bordering littoral zones on island chains in the boreal portion of the Tethys Sea.

## Lower Cretaceous of Gondwanan Africa

### Unnamed Snake Lizard Cranium: Valanginian (Modern South Africa)

The cranial remains of AM 6001 (see Figure 2.3) were found during preparation of the skeleton of a small, basal coelurosaurian dinosaur, *Nqwebasaurus thwazi* de Klerk et al., 2000, from the Kirkwoods Cliffs Quarry, Kirkwood Formation (de Klerk et al., 2000). The Kirkwood Formation is approximately 2000 meters thick (Muir et al., 2015, 2017), conformably overlies the Enon Formation (alluvial fan deposits), and is overlain conformably by the Sundays River Formation. The Kirkwood Formation is subdivided into three parts, the basal Swartskop Beds, Colchester Beds, and an unnamed unit recognized in outcrop that is the source of the fossil vertebrate materials, including AM 6001. The age of the Kirkwood Formation remains problematic, as it contains no specific fossil assemblages that delimit ages within the formation (Muir et al., 2017); the Enon Formation provides a lower bound age of Upper Tithonian (~145 ma), while the Sundays River Formation, a marine unit containing diagnostic foraminifera (McMillan, 2003), is Upper Valanginian (~134 ma) at its contact with the Kirkwood. The fossiliferous redbeds in the unnamed upper unit of the Kirkwood are likely still within the Valanginian; however, the borehole most used by McMillan (2003) is 7 kilometers to the southeast of the Kirkwood Cliffs Quarry, leaving the exact age within the Valanginian in question.

Despite the uncertainty of geologic stage in the Upper Jurassic to Lower Cretaceous, the Kirkwood

Cliffs exposures present interesting lithologies: grayish to yellowish, medium/coarse-grained sandstones, with red to gray-green mudstone and siltstone interbeds (Muir et al., 2017). Quite importantly, there are also highly bioturbated paleosols, showing important similarities to arid environment dune paleosols from the localities in Argentina that are the source of the numerous specimens of *Najash rionegrina* Candia Halupczok et al., 2018. In addition to these probable aeolian deposited sediments, there are large silicified logs and woody fragments throughout the Kirkwood Formation, along with remains of ferns, cycads, and conifers, and, most interestingly, significant charcoal deposits as laminae within bedforms (Muir et al., 2017). The charcoal has been interpreted to represent fluvial transport of burned woody plant materials from significant forest fires (Muir et al., 2017).

From Muir et al.'s (2017) field observations of these varied sedimentary facies and biofacies, they identified a number of depositional environments within the unnamed upper beds of the Kirkwood Formation, varying laterally from a meandering fluvial environment with mature, vegetated floodplains to arid aeolian-deposited paleosols and likely dune systems, proximate to alluvial fans descending from the highlands (e.g., Enon Formation), and distally toward open marine systems (e.g., Sundays River Formation).

## Upper Cretaceous of the Afro-Arabian Platform

### Rear-Limbed Snake Lizards: Cenomanian (Modern West Bank and Lebanon)

*Pachyrhachis problematicus* (see Figure 2.4) and *Haasiophis terrasanctus* (see Figure 2.5) were collected from Lower Cenomanian–aged platy limestone beds quarried at Ein Jabrud, near Ramallah in the West Bank, approximately 20 kilometers north of Jerusalem, Israel. The exact formational unit remains unknown, though Chalifa and Tchernov (1982) considered the fossil-bearing unit within the Bet-Meir Formation (basal Lower Cenomanian), while Chalifa (1985) revised this conclusion and instead suggested they were to be found in the overlying Amminidav Formation (upper Lower Cenomanian). In either case, the age of the sediments is bounded within the Lower Cenomanian.

The five known specimens of *Eupodophis descouensi* (see Figure 2.6) were collected from platy

THE ORIGIN OF SNAKES

limestones collected from two different quarries in Lebanon: (1) Holotype locality, at Al Namoura in the valley of Nahr Ibrahim in Northern Lebanon, and (2) Ouadi Haqel in Northern Lebanon. In contrast to the Ein Jabrud snake lizards from the West Bank, the snake lizards of Lebanon are from geologically younger sediments placed at or slightly above the boundary between the Lower and Middle Cenomanian (Hückel, 1970, 1974a,b). Though *Eupodophis* is the geologically younger snake lizard, the entire Cenomanian Stage, broken into three substages, is only about 7 million years in duration. Relatively speaking, *Eupodophis* and *Pachyrhachis/Haasiophis* are separated by only a few hundred kilometers of distance, even in the Cenomanian, and if the Ein Jabrud beds are uppermost Lower Cenomanian (Chalifa, 1985), then the three taxa are not separated by very much time, if any at all, considering the uncertainties of the West Bank locality and its stratigraphic and temporal position. What is very clear is that all three snake lizards share virtually the same depositional environment.

**Conflicting Interpretations**—In very broad terms, from the Albian to the Turonian, the Tethyan Realm and Boreal Realm of the Neo-Tethys were submerged during a major transgression that reached its eustatic sea level peak in the Cenomanian. The result was the formation of an extensive carbonate platform across the Afro-Arabian Plate linked to the entire Tethyan and Boreal Realms across what is now central and western Europe (Follmi, 1989). The paleoshoreline of northeast Africa was approximately 900 km to the southwest of modern Al Namoura and Ouadi Haqel, and around 450 km to the southwest of modern Ein Jabrud (Scanlon et al., 1999; Rage and Escuillié, 2003). Along the western edge of the shelf margin, there was an extensive carbonate platform patch reef and intraplatform basin system, similar complexes dominated by rudist bivalves and scleractinian corals as described by Jurkovsek et al. (1996) along the north and south shelf margins of the Tethys Seaway in Europe.

As was reviewed by Caldwell (2007a), Gayet (1980) characterized thin sections made from the fossil-bearing limestones at Ein Jabrud, which are virtually identical to those from Al Namoura and Ouadi Haqel, and from the Cenomanian deposits on the Island of Hvar, Croatia (Diedrich et al., 2011): (1) microcrystalline limestone with microsparite-sparite; (2) absence of organics;

(3) microsparite and sparite between laminations; (4) "sheetcracks" filled with calcite (microsparite and sparite); (5) deposited in a low-energy environment without currents; (6) no description of clastics, that is, clays. Gayet's (1980) thin section description of the lithology of the Ein Jabrud chemical precipitate platy limestones is at odds with Chalifa's (1985) interpretation of the depositional environment at Ein Jabrud, that is, a shallow marine bay close to land and receiving terrestrial alluvium. As such, Scanlon et al. (1999:135–136) revised the depositional environment as follows:

> The fossil bearing horizons at Ein Jabrud are interpreted as having been deposited in low energy, quiet water, inter-reef basins. The sediments are fine grained to crystalline limestones that precipitated as carbonate muds to form thinly bedded laminae.

I retain this interpretation of the depositional environments at Ein Jaburd, Al Namoura, and Ouadi Haqel, as it fits everything we currently understand about the plate tectonic framework of the Afro-Arabian Platform and the Boreal and Tethyan Realms in the Cenomanian, that is, gigantic carbonate platform populated by millions of reef biostromes and lagoons, no doubt with thousands of small exposed reef mound islands, hundreds of kilometers north of the Afro-Gondwanan shoreline. This depositional environmental interpretation is also consistent with the observed lithofacies, biofacies, and ichnofacies (see also Diedrich et al., 2011).

## Upper Cretaceous of the Adriatic-Dinaric Platform

### *Pachyophis woodwardi* and *Mesophis nopcsai*: Cenomanian (Modern Bosnia-Herzegovina)

The age of the rocks at Selisca–Bilek, Bosnia-Herzegovina, has been debated for some time, ranging from the "Neocomian" (Lower Cretaceous) to Cenomanian–Turonian (Langer, 1961), but now the consensus seems to be that they are Upper Cenomanian in age (Sliskovic, 1970). The type specimen of *Pachyophis* (see Figure 2.7a) is preserved in finely bedded, carbonate mudstones showing some degree of dolomitization and no evidence of bioturbation (personal observation and testing). The exact location and horizon at Selisca-Bilek is unknown. The specimen of *Mesophis* (see Figure 2.7b) shows exactly the same kind of lithology as has been described in hand sample for *Pachyophis*; again, the exact horizon and location of collection of the specimen, and its position stratigraphically with *Pachyophis*, is unknown.

Though the limestone blocks preserving both specimens are less platy then those of the Middle Eastern snake lizards described previously, the surrounding tectonic framework, bedforms, biostromes, biofacies, and lithofacies support a similar paleoenvironmental interpretation, and thus depositional environment, as has been given for the localities at Ein Jabrud and in Lebanon— patch reef lagoonal environment on the Adriatic Platform portion of the massive Tethyan Carbonate Platform and reef system. Throughout modern Slovenia, Croatia, and Bosnia-Herzegovina, there are substantial carbonate reef systems of rudist bivalves and corals forming massive biostromes within which are numerous shallow water lagoonal systems serving as microbasins for chemical carbonate precipitation and the formation of platy limestones. During the Cenomanian, the land masses of the Adriatic and Dinaric microplates were largely submerged during a major marine transgressive event. These shallow water submerged margins and extensive platforms supported the development of extensive reef complexes (Follmi, 1989) that were great distances off shore of exposed continental rocks and their environments, but were biologically extremely rich communities of reef organisms associated with all manner of benthic and planktic organisms.

## Upper Cretaceous of Gondwanan Africa

### *Menarana nosymena*: Maastrichtian (Modern Madagascar)

To date, *Menarana nosymena* is known only from the Upper Cretaceous (Maastrichtian) Maevarano Formation, an outcropping near Berivotra in the Mahajanga Basin of northwestern Madagascar. The fossiliferous Maevarano Formation sediments (Rogers et al., 2000) were deposited on an alluvial plain in the Mahajanga Basin (Maastrichtian, Cretaceous) with mountains to the southeast and marine waters to the northwest. Rogers et al. (2000) subdivided the Maevarano Formation

into three subunits: Masorobe, Anembalemba, and Miadana. The basal Masorobe Member is characterized by redbed and paleosol facies that are highly bioturbated by root traces associated with caliche, indicting stable paleosol surfaces and biotic communities developed under arid conditions. Complex sandstone lithofacies and a variety of bedforms indicate that the Anembalemba Member sediments accumulated in a fluvial system of braided shallow channels under seasonal flow control, with evidence of seasonal or flash-flooding channel scours; such systems are consistent as lateral equivalents in an arid environment such as a vegetation-stabilized aeolian system with seasonal fluvial flows on the margins of dune systems and dune field interbasins. Rogers et al. (2000) noted that the sediments of the Miadana Member are fine-grained sandstones that were deposited in a "lower coastal plain setting" relatively close to marine environments. The overlying sediments to the Maevarano belong to the Berivotra Formation, which marks a transgressive phase and is a marine unit whose biofacies assemblages include planktonic foraminifera, ammonites, and other marine invertebrates that indicate a Late Maastrichtian age for the Berivotra.

An important discussion, first undertaken by Rogers et al. (2000), was to compare and consider the similarities between Maevarano Formation in terms of climate, lithofacies, biofacies, and stratigraphic position, with the Maastrichtian-aged sediments of western India in the area of the Deccan Traps. This was done without any knowledge whatsoever of the future linkages that would be made between *Menarana nosymena* as a madtsoiid snake lizard and the depositional environments of the Maervarano Formation (Laduke et al., 2010) to the Maastrichtian madtsoiid from Gondwanan India, *Sanajeh indicus* Wilson et al., 2010, nor of the depositional environmental interpretations associated with the Gondwanan madtsoiid-like snake lizards, *Najash* (Candia Halupczok et al., 2018) and *Dinilysia* (Caldwell and Albino, 2001).

## Upper Cretaceous of Gondwanan South America

### *Lunaophis aquaticus*: Cenomanian (Modern Venezuela)

The block containing the vertebrae was collected from black shales exposed in a cement quarry mining carbonate rocks of the La Luna Formation, a major petroleum-producing formation within Venezuela's Maracaibo Basin. The quarry is east of Lake Maracaibo, 10 km to the northeast of Monay, in Trujillo State, Venezuela. The fossiliferous horizon, containing mollusks, fishes, snakes, and large calcareous concretions, is in a black shale layer within the La Luna Formation, 28 meters above the base of the La Aguada Member (Upper Cenomanian) (Albino et al., 2016). The entire sequence of rocks represents a transgressive highstand in the region during the Upper Cenomanian. Sediments within the La Luna include black calcareous shales interbedded with cryptocrystalline limestones and calcareous cherts; the rocks give off a strong petroleum odor on fresh break.

These lithologies represent two important features of the La Luna: deep water depositional environment and anoxic-euxinic bottom conditions preserving organic matter. While a deep water environment with anoxic bottom sediments on the outer edge of a continental shelf seems not to be the sort of habitat suitable for an aquatically adapted snake lizard, it is also true that among modern sea snake lizards, the yellow bellied sea snake, *Pelamis platurus*, is Pan-Pacific and Pan-Indian Ocean, ranging from the coasts of Central America and South America, across the Pacific to all island archipelagos, Japan, China, Korea, all of southeast Asia, Australia, Africa, India, Madagascar, and all points in between. This is an amazing distribution for such a small-bodied snake lizard and clearly indicates it lives in, travels through, feeds in, and breeds in deep, deep water environments as well as shallow-water coastlines, reefs, and lagoons. Like *Lunaophis*, *Pelamis* skeletal remains will no doubt be found far into the future in deep water sediments in unexpected places.

## Najash rionegrina: Cenomanian (Modern Argentina)

To date, I have enjoyed three field seasons (2015, 2016, and 2017) with the crews (students, technicians, volunteers, and geologists) of Sebastian Apesteguía and Guillermo Rougier, hunting fossil snakes in the sediments of the La Buitrera Paleontological Area in Rio Negro Province, Argentina (see below for details). Not only are these rocks rich beyond belief with the remains of numerous three-dimensionally preserved skulls and articulated skeletons of *Najash rionegrina* Apesteguía and Zaher, 2006 (see Figure 2.8a–f), they also preserve a fossil Lagerstätten

of numerous other small-bodied vertebrates, including mammals, sphenodontians, and dinosaurs (Apesteguía et al., 2001; Apesteguía and Zaher, 2006; Rougier et al., 2011; Gianechini et al., 2017). The La Buitrera Paleontological Area includes a large number of localities (Cerro Policía, El Loro, La Escondida, and El Pueblito) and sites within those localities, in the northern part of the Patagonian Desert, just to the northwest of the small town of Cerro Policía. The sediments outcropping within the La Buitrera Paleontological Area belong to the upper portions of the Cenomanian-aged Candeleros Formation (Neuquén Group [Leanza et al., 2004]), which has recently been recognized as a paleoerg desert and dune/interdune system (~826 km²), dubbed the Korkorkom Desert (Candia Halupczok et al., 2018).

A summary statement of the depositional environment is that articulated skulls and skeletons of most of the tetrapod specimens are preserved on the slip faces or within the sands deposited by large, migrating aeolian dunes; many of these dune faces were stabilized for long periods of time by plant roots and possibly burrowing insects, and so formed paleosols. As described by Candia Halupczok et al. (2018), these same slipface surfaces also preserve numerous dinosaur trackways in the Candeleros, likely because they were stabilized as paleosols and paleosurfaces, and because they were seasonally wet. Fossil remains are certainly found in the interdune basin depositional environments (fluvial, lacustrine, ponds, dune fringes), but by and large these fossils are disarticulated to heavily weathered, reflecting exposure and transport, and taphonomically, therefore, are not predisposed to preserve small, delicate skeletons such as those of snake lizards. A detailed examination of the La Buitrera Paleontological Area is given later in this chapter.

## Dinilysia patagonica: Upper Santonian/ Lower Campanian (Modern Argentina)

The sediments at all the known *Dinilysia patagonica* (see Figure 2.9) localities, with the exception of the possible vertebrae described by Scanferla and Canale (2007), are red- to white-weathering, coarse-grained sandstones currently assigned to the Bajo de la Carpa Formation, Río Colorado subgroup, Neuquén Group (Santonian–Campanian, Upper Cretaceous; Leanza et al., 2004; Sánchez et al., 2006). Sediments of the Neuquén Group range in

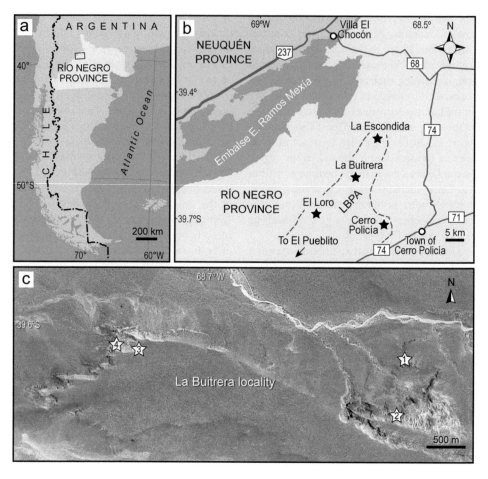

**Figure 4.1.** Locality map and profile. (a) Area of study in Río Negro province, northern Patagonia, Argentina; (b) detail of La Buitrera Paleontological Area and different paleontological localities within (black stars); (c) detail of the La Buitrera locality and sites of provenance (white stars) of the holotype and referred specimens of *Najash rionegrina*: (1) Tefa (MPCA 418); (2) Hoyada de Muñoz (MPCA 385); (3) Med4 (Holotype, MPCA 399), (4) Vifa (MPCA 386, MPCA 388). (Illustration by Copyright Fernando Garberoglio.)

age from the Cenomanian to the Campanian. These sediments are part of the Argentine portion of the Andean Basin (Figure 4.1), one of the Pacific basins of southern South America. The Río Colorado subgroup is the last continental formation and is overlain unconformably by the Allen Formation of the Malargüe Group (the Allen Formation was laid down during a major marine transgressive phase occurring in the Late Cretaceous). The Bajo de la Carpa Formation is currently considered Santonian–Campanian in age (Leanza et al., 2004; Sánchez et al., 2006). The balance of the known *Dinilysia* specimens come from the exposures at the Tripailao Farm site, Paso Cordoba, worked in 1994 by Bonaparte's field teams, and again in 2001, by

me, Jorge Calvo, and Adriana Albino on a National Geographic Fieldwork and Exploration Grant; I was first in the field at Paso Cordoba in 1998, with staff from the Universidad del Comahue, Neuquén (Mr. Pablo Possé) and collected the data used in Caldwell and Albino (2001). Subsequent firsthand sedimentological analysis is used here to build on the data presented in Caldwell and Albino (2001).

There are two broad lithological groups represented in the studied sections of the Bajo de la Carpa Formation exposed at the Tripailao Farm site (see Caldwell and Albino, 2001). The first is a poorly sorted, coarse-grained near conglomeratic sandstone that weathers red to bright pink, and on fresh surfaces is light pink. The second lithology is a fine-grained sandy siltstone that on fresh and weathered surfaces is bright iron-red. The finer-grained, non-conglomeratic units are at the base of all measured sections at the Tripailao Farm. As was true in 1998 and 2001, these lower beds at the Tripailao Farm have not produced any snake lizard fossils. In stark contrast, the beds characterized by the poorly sorted, coarse-grained, nearly conglomeratic lithofacies are rich in very well-preserved vertebrate fossils. Two broad sedimentological-facies groups are recognized in these upper beds: (1) a massive, structureless sandstone with a distinct and very well preserved ichnofabric; the tops of these beds are clearly paleosurfaces and thus the bioturbated units are also paleosols; and (2) a massive, structureless sandstone that shows neither primary bedforms, nor any obvious traces, and varies from 20 to 200 cm in thickness.

These two facies alternate, that is, bioturbated unit/non-bioturbated unit, throughout the sections exposed at the Tripailao Farm Site. Fossils of Dinilysia are found at the contact between the top of burrowed beds and the bottom of non-burrowed beds; at this same interface, there is often a caliche conglomerate of variable thickness. The caliche conglomerates are interpreted here as deflation lags preserved by subsequent burials.

**Conflicting Interpretations**—Taken together, the litho- and biofacies assemblages, and all other sedimentological, stratigraphic, and broad-picture paleoenvironmental data strongly indicate that the depositional environments present in the Bajo de al Carpa Formation are a suite of lateral facies variants consistent with an aeolian-dominated dune-interdune basin flanked most probably by seasonal fluvial and lacustrine systems. The units producing Dinilysia specimens at Paso Cordoba are all found in the aeolian sediments linked to dune slip faces and root stabilized paleosurfaces; the disarticulated remains occur in interdune depositional environments. As was noted by Caldwell and Albino (2001), these observations are consistent with those made by Eberth (1993) for the paleoenvironments of a dune-interdune basin exposed today as redbeds near Bayan Mandahu, Inner Mongolia, Peoples' Republic of China. Eberth (1993) subdivided the Bayan Mandahu deposits into three sedimentation zones. "Zone 3" at Bayan Mandahu is dominated by aeolian dune and structureless aeolian deposits with lacustrine

and interdune fluvial deposits, intrabasin streams and ponds, extremely common caliche nodules forming caliche conglomerates, and dense trace fossil accumulations radiating downward from the base of the caliche conglomerates. The Bayan Mandahu facies are virtually identical to lateral and vertical facies successions observed for the Bajo de al Carpa Formation at the Tripailao Farm (Caldwell and Albino, 2001).

### The Non-Snake Lizard—Tetrapodophis amplectus: Albian (Modern Brazil)

The Araripe Basin of northeastern Brazil is divided into two principal sub-basins: the Feitoria in the West and Feira Nova in the east (Viana and Neumann, 2002; Martill et al., 2007). Stratigraphically, these sub-basins are composed of the Santana Group, where in outcrop and stratigraphic order, top to bottom, it is possible to find the Arajara, Romualdo, Ipubi, Crato, and Rio da Batateira Formations (Neumann and Cabrera, 1999; Valença et al., 2003). Though the exact locality remains hidden, the one and only specimen of Tetrapodophis amplectus (see Figure 2.13) reputedly comes from the Crato Formation, which is highly fossiliferous; has produced thousands of specimens of fishes, plants, arthropods, crustaceans, chelicerates, turtles, crocodilians, pterosaurs, fossil feathers, and several lizards; and is considered a global Lagerstätten (Maisey, 1993; Kellner and Campos, 1999; Simões, 2012; Simões et al., 2014) (from Martill et al. [2015: Supp. Mat. Pg. 2]):

> Comparisons with other specimens confirm that it is from the Crato Formation, however, and specifically from the Nova Olinda Member of the formation. The specimen is most likely to have come from exposures between Santana do Cariri, Nova Olinda and Tatajuba…

The Crato Formation consists mostly of thinly laminated limestones, and according to Martill et al. (2015: Supp. Mat.), the lithology of the Tetrapodophis slab is a fine-grained micritic limestone displaying finely bedded planar laminations. Numerous small coprolites are present in the laminae and are thought to be from small fishes, which would introduce a significant phosphate component into the sediment. In addition, there is no evidence of burrowing on either a macro or micro scale, indicating likely hypersalinity or hypoxia. There is a small clastic component present as clay particles within the micritic limestone (<1%), which is common for chemical precipitates in proximity to coastlines, islands, or other terrestrial environments where most of this clay is introduced as aerial deposition and not as water-borne clastic sedimentation.

The standard depositional environmental interpretation is that the sediments of the Crato Formation were deposited in an carbonate rift lake setting, developed during a period of significant aridity and rifting during the opening of the South Atlantic Ocean in the late Aptian to Aptian/Albian (Maisey, 1993; Valença et al., 2003).

## Upper Cretaceous: Gondwanan Asia

### Xiaophis myanmarensis: Cenomanian (Modern Myanmar)

The amber clasts containing the snake lizard remains of Xiaophis (see Figure 2.11) were found at what is referred to as the "Angbamo Site" in Tanai Township, Myitkyina District, Kachin Province, Myanmar. The age of the clastic sediments in which the amber nodules are preserved as clasts of widely varying sizes has been dated as earliest Cenomanian (98.8 ± 0.6 ma) (Cruickshank and Ko, 2003). The mine exposes a variety of clastic sedimentary rocks, abundant carbonaceous materials (i.e., carbonized plant remains), with thin limestone interbeds. The amber is found in a fine-grained sandstone facies as generally flattened, disk-shaped clasts that are oriented parallel to the bedding planes (Cruickshank and Ko, 2003). The amber clasts are considered to have been formed some distance away from where they were found, and because amber floats quite well, were only subsequently captured as clasts in nearshore, like upper-shoreface or even beach, accumulations in transitional marine environments, such as bays or estuaries.

What is really exciting about these amber clasts is that they were transported from where they formed to a place where they were deposited, the latter of which is a remarkably different environment than the one in which they were formed. Why is this exciting? Because of the manner in which amber forms from tree sap droplets, both small and sometimes extremely large, that once they fall from the tree, will usually collect evidence of where they landed. The inclusions within the amber clast include much more data pointing toward the ancient

environment than just the remains of a snake lizard; they in fact provide taphonomic support for identifying a forested ecosystem, as both DIP-S-0907 and DIP-V-15104 contain abundant insects, carbonized insect feces, and fragmentary plant materials, which are usually associated with "litter amber," or resin produced near the forest floor. As a result, both specimens present exciting new data on a previously unknown ancient snake lizard ecology—a terrestrial (possibly arboreal) ecosystem marginal to inland and coastal fluvial environments.

### Sanajeh indicus: Maastrichtian (India)

The Lameta Formation (Maastrichtian) in the Kheda District, Gujarat State, west central India, is, by comparison to many formations discussed in this text, rather thin (∼8 meters), and is overlaid by the flood basalts of the famous Deccan Traps. Lithologically, the sediments are characterized by coarse-grained arkose sandstones derived from the underlying bedrock, with some conglomerates and occasional marl interbeds. Though the formation is very thin, the sediments at the Sanajeh locality near Dholi Dungri show extensive paleosol development (Wilson et al., 2010) that is clearly expressed in the presence of calcrete nodules and other pedogenic structures, and likely root bioturbation that has obliterated all primary sedimentary structures and bedforms.

Wilson et al. (2010) identified a lithofacies-biofacies succession at the snake lizard-bearing interval at Dholi Dungri that has important implications for depositional environment interpretations. Within a lag concentration of boulders (up to 60 cm), cobble and large- to medium-grained sands were located the articulated and associated vertebrate remains, including the eggs and bones of Sanajeh (see Figure 2.10). The spatial distribution of the boulder lag and fossil was quite small (∼100 meters). They interpreted the clastic lag as having been deposited in an intermittently flowing fluvial channel. Wilson et al. (2010) continued to reason from their facies succession observations toward a depositional environment interpretation at the site that involved a storm-induced debris flow. I would suggest, based on personal experience at the localities producing Dinilysia, the unnamed South African snake lizard, and Najash, that Wilson et al. (2010)

were much closer to a more accurate answer when they stated:

> Debris flows, subsequently modified by bioturbation (e.g., root formation, nesting sites) and pedogenesis, would lack clearly defined primary sedimentary structures, would be limited in areal extent, and would be expected given the inferred paleotopography present in the immediately adjacent area.

A common theme arising for all Gondwanan snakes of the "madtsoiid-kind," is an ecology linking them to arid/semi-arid environments located within dune-interdune basins, seasonally heavy fluvial activity, maybe seasonal lakes, important paleosoil development, dune environments stabilized by long periods of plant and insect bioturbation, and a diverse vertebrate fauna. This pretty much describes the depositional environment reconstructed by Wilson et al. (2010) and is consistent with everything I have seen for the sediments and tectonic settings that preserved other contemporaneous Gondwanan snake lizards in South Africa and South America (see above on *Najash*, *Dinilysia*, *Menarana*, and the unnamed South African snake lizard).

## Form Taxon I: Upper Cretaceous of European Eastern North Atlantic

### *Simoliophis rochebrunei:* Cenomanian (Modern France)

The neotype locality is in the Les Renardieres Quarry at Tonnay–Charente, the sediments of which are Lower Cenomanian (99.7–94.3 ma) in age. The known specimens of *Simoliophis rochebrunei* (see Figure 2.12a,b) have been found at a wide variety of localities between Ile d' Aix in the west to quarries to the north east of Angouleme in the east (Néraudeau et al., 2009; Rage et al., 2016). The Cenomanian rocks in the Charentes region, outcrop on the northern margin of Aquitaine Basin of south central France. These Albian–Cenomanian deposits (Beds A1sl to A2sm) sit unconformably on Upper Jurassic rocks in the Basin (Néraudeau et al., 2002, 2009; Vullo et al., 2002) and represent a gradual eustatic sea level rise leading into the Cenomanian, as well as a significant rise in ambient water and air temperatures; in sequence stratigraphic terms, the lower Cenomanian marks a significant transgressive phase throughout the Aquitaine Basin. Lithofacies in the lower units are clastic dominated by fine-grained sands, muds, and sandy marls, and in higher units include "falun" beds of unconsolidated echinoids, rudists, bivalves, and marly carbonates.

These sediments were deposited in a wide range of environments from tide-dominated estuaries and delta fronts to upper and lower shorefaces and paralic zones. Biofacies are mixed assemblages dominated by marine invertebrates along with significant ichnofacies assemblages of both burrowing organisms and rooted horizons from plant communities similar to those of modern Mangrove systems (Népraudeau et al., 2002). The vertebrate remains, including those of *Simoliophis*, occur in stratigraphically higher beds, that is, B1, in the "Departmentes" of Charentes Maritime and Charentes, and then disappear toward the end of the Lower Cenomanian (i.e., are not found beyond bed B2).

## Form Taxon II: Upper Cretaceous of North America

### *Coniophis precedens: Maastrichtian (Modern Wyoming/Montana)*

The Hell Creek Formation is a fine-grained, siliciclastic unit that is approximately 175 meters thick and occurs in the Powder River Basin, eastern Wyoming and Montana, United States. The sediments forming the rocks of the formation are considered to have been deposited during the last 1 to 2 million years of the Upper Cretaceous (Upper Maastrichtian, ~63 ma). In general terms, the paleoenvironments are recognized as a complex suite of lateral equivalents including meandering/aggrading river systems, marginal lakes and swamps, and toward the south, a subtropical coastal lowland floodplain, estuary, and delta. Depositional environments are numerous and complex, as are the lithofacies preserved: cross-bedded channel sands, crevasse splays, overbank muds, lacustrine silts and muds, deltaic sands, silts and muds, and lignitic horizons suggesting plant rich swamps (Fastovsky and Bercovici, 2016), creating a wide variety of habitats for all kinds of organisms, including, of course, the enigmatic snake lizard, *Coniophis precedens*.

**Conflicting Interpretations**—A recent example from the fossil snake literature comes from Longrich et al. (2012:3): "*Coniophis* occurs in continental floodplain sediments as do other basal snakes" (numeric citations removed). This is an important example for several reasons: (1) in part because *Coniophis precedens* is such a problematic snake taxon, (2) because the paleoecological claim as justified against lithofacies analysis and depositional environment inferences, and (3) because this interpretation was used to claim further support for a particular "origins of snakes" hypothesis, which naturally is of intimate relevance to the subject matter of this book.

Longrich et al. (2012:205):

> *Coniophis* occurs in a continental floodplain environment, consistent with a terrestrial rather than a marine origin; furthermore, its small size and reduced neural spines indicate fossorial habits, suggesting that snakes evolved from burrowing lizards.

To be honest, I have no idea how the conclusion "continental floodplain" links unequivocally to "terrestrial," which links unequivocally to "reduced neural spines," which leave us with nothing less than a "burrowing lizard" ancestor. Miraculous. What is unfortunate, however, is that the reality of the complex facies associations and the critical details of lithologies, sedimentology, and stratigraphy are glossed over and ignored. Referring back to Chapter 2, I briefly detailed lithofacies, and so on, with an interpretation of the numerous depositional environments of the Hell Creek Formation in which are found the completely disarticulated remains of the snake lizard be referred to as *Coniophis precedens* Marsh, 1892.

I reiterate: At any one point in time, the geographical area that would become the Hell Creek Formation was not a one-trick pony of a depositional environment—that is, a uniform "continental floodplain environment," but rather was a complex collection of numerous depositional environments in lateral equivalence to each other. In other words, a river, straight, braided, anastomosing, or meandering, it does not matter, always has banks, and over those banks are swamps and ponds, lakes and oxbows, that during floods become vastly different depositional environments than the river channel proper. Banks and channels disappear in floods, crevasse splays open up in those banks, and new channels are created that erode the old as the fluvial system moves dynamically across its landscape. Should the fluvial system be proximate to the coastline, which

the Hell Creek environment in the Late Cretaceous most certainly was (Scholz and Hartman, 2007), then the number of depositional environment possibilities for sediments increases significantly as tidal, delta, and estuarine processes will modify both the quality and quantity of possible depositional environments.

## Summary

The Mesozoic record of depositional environments in which are found the remains of ancient snake lizards are nearly as varied as are the number of snake lizards fossils. If I were to loosely extrapolate to ancient environments from these data, I could only conclude that snake lizards had occupied as many habitats and niches in as diverse an array of environments and ecosystems as they have today—marine environments of all kinds, forests, islands and island chains, coastal back beach lakes and inland lakes, deserts of many varieties including ergs or sand seas, and deltas, estuaries, and their associated rivers moving great distances inland. While depositional environments provide little or no information on upland environments, or terrestrial environments some distance lateral to the extent of floodplain systems, there is no reason at all to think that ancient snake lizards were not present in these paleoenvironments as key members of their vibrant ecosystems. All we lack are the depositional systems to empirically support the conclusion, but the probability exists regardless.

The current oldest known snake lizards (Caldwell et al., 2015) are found in four different depositional environment settings, three of which are linked to each other as lateral equivalents—silt-mud lenses in coal swamp deposits (*Portugalophis*), marginal marine lacustrine settings (*Parviraptor*), and a backbeach lacustrine to estuarine setting (*Eophis*), all associated with island systems located along the margins of the widening Atlantic Ocean at its intersections with waters of the Tethys and Boreal Seas. The fourth snake lizard, *Diablophis*, is the depositional and paleoenvironmental outlier by comparison to these other three snake lizards; in this case, the remains of this ancient North American snake lizard (Kimmeridgian, Upper Jurassic) are found in mudstones and claystones deposited in small lakes or pools associated with a meandering upland river system (Galli, 2014:104) on the east slopes which were some distance from the nearest coastal margin. In the early Kimmeridgian, this was most certainly a receding sea, the Sundance Sea; by the middle Kimmeridgian the general environment was high, arid uplands with seasonal lakes and streams draining to the east and north; and by the Upper Kimmeridgian, the period of time in which Diablophis occurs, the southwest corner of the "Morisson" inclusive of the sediments of the Brushy Basin and Salt Wash Members, was part of the northern margin of an saline/alkaline rich lacustrine complex referred to as Lake T'oo'dichi' (Turner and Peterson, 2004).

Xiaophis myanmarensis is preserved in amber, which provides a direct sampling of the ancient ecosystem and environment in which the amber formed and the baby snake lizard was preserved—a forest full of insects, fossil poop, and plant debris. The amber assemblage preserves evidence of a particular clastic sedimentary process along a coastal margin in nearshore beach sand deposits. The lithofacies details from the quarries do not yet reveal the delta front and deltaic litho-, ichno-, and biofacies, but they are coming. What is clear, however, is that the entire suite of facies is associated with a sedimentary system occurring on an island system that was itself an allothonous terrane captured into large scale early Mesozoic plate tectonics that moved a piece of Austral Gondwana north over time, and across the incipient Indian Ocean to accrete to southeast Asia in the Cenozoic. In short, we have a Gondwanan snake lizard fauna evolving in isolation in an island forest—presumably—from the mid-Mesozoic onward.

At the same time as Xiaophis was happily dangling from large sappy trees on its drifting island ark across the sea, Pachyrhachis, Haasiophis, Eupodophis, Lunaophis, Pachyophis, Mesophis, and Simoliophis were swimming along merrily in those same oceans and seas in which Xiaophis was drifting along, blissfully unaware of the committed radiation of its marine snake lizard cousins. The Cenomanian explosion of aquatically adapted marine snake lizards is no mere "toe-dipping" by a quaint lineage of Mesozoic snake lizards, but rather represents a major secondary invasion of the marine realm by a diverse assemblage of snake lizards. Their remains have been found now in marginal marine to carbonate platform deposits in modern South America, Moroccan Africa, Southern France, Israel-Westbank, Lebanon, and the northern carbonate platform deposits of Croatia and Bosnia-Herzegovina. And these are not snake lizards desperately clinging to the cliffsides of continents wondering how they fell off. Rather these animals are hundreds of kilometers offshore of any continental landmass. Here they were living, feeding, and no doubt breeding, in, on, and around tiny carbonate reef islands, in a vast patchreef-lagoon complex of unbelievable dimensions. It stretched from what today is the North Sea across all of Western Europe to the north coast of Africa, and east across what would become Persia, spilling into the developing Indian ocean—this is the Tethys Seaway. The opening Atlantic Ocean on the European and African western margin was all that kept this great platform of shallow-water and reef-building organisms from creating a carbonate platform connected to North and South America. However, that growing Atlantic Ocean did not stop the flow of life from moving across it, and so the Super Tethys Seaway flowed to the west and brought its reef-forming organisms with it (e.g., rudists) and also its vertebrates, and, more importantly, its snake lizards (e.g., Lunaophis). The Tethys Sea carbonate platform was the largest of its kind of reef system to have ever developed in the last 100 million years—the Tethyan-Boreal Realm Platform began in the Cenomanian and persisted to the Eocene. It was only finally choked of its resources when the northward movement of the African plate finally buckled the plates skyward, lifting the Persian sediments above sea level along with all of Europe west to Spain.

Perhaps the most interesting and broadly distributed paleohabitat is associated with the Gondwanan Type I Dinilysia-Najash snake lizards, Najash, Dinilysia, Menarana, and Sanajeh, and potentially the currently unnamed snake lizard from South Africa (Ross et al., 1999)—dry, upland, desert systems, in some cases marked most notably by sand seas or ergs, for example, Najash. It is intriguing that lithofacies, biofacies, and ichnofacies indicative of facies models and depositional systems representative of a wide variety of paleoenvironments associated with vast arid landscapes (deserts and paleoergs) on at least three Gondwanan fragments—South America, Madagascar, and India—would also all preserve evidence of large bodied Type I Dinilysia-Najash snake lizards. These animals range through some 30 million years of geologic time (Cenomanian to Maastrichtian [Najash to Menrana/Sanajeh]) yet maintain a common thread of likely common descent and common ancient environments and probably paleoecologies. Coincident with the Gondwanan heritage of these southern land masses, their snake lizard fauna, and their shared common ancestry was a southern hemisphere climate trend that led to major aridity through the Mid to Late Cretaceous and the development of largescale aeolian systems (Rodríguez-López et al., 2014).

In summing up the summary, the only paleoecology that cannot be hypothesized from the known materials of ancient snake lizards is the extreme specialization of fossoriality leading to a burrowing, subterranean life habit as exemplified by modern scolecophidian snake lizards. Considering how specialized and diverse ancient snake lizards were, from sand seas to the ancient world's oceans, it would seem unlikely that some form of subterranean adaptation would not have occurred. However, the problem is the environment of deposition and the rarity of entire soil profiles entering the rock record from terrestrial systems short and long distances away from depositional basins and water or wind influence. I cannot for a moment imagine that 100 million years ago, if snake lizards were plying the oceans of a mid-Mesozoic earth, dangling from the branches of amber forests, and coursing the wind-swept dunes of massive sand seas, that they had also not already colonized fossorial realms as well. However, and here I foreshadow the next chapter, if they were underground, burrowing merrily away and chasing whatever small creatures they were in search of, then 100 million years ago this was a derived state for ancient snake lizards, not a primitive one.

## HUNTING FOR ANCIENT SNAKES

The pursuit of knowledge is, in the best metaphoric sense I can conjure, a never-ending intellectual high of discovering new things. The discovery highs of paleontology are made manifest in several ways—sometimes it is in the opening of dusty and musty old museum drawers to find a slab of rock overlooked or misinterpreted for years (for me, that was *Pachyrhachis problematicus*), but most often it is the pirate treasure moment when you brush away ancient sediments and are the first to see the remains of a mythical beast never seen by anyone before you. It is the fine, fine drink that brings you back, time and again, to search for more myth and magic in the ancient past of this very ancient planet. For a few of the pages in this treatise on ancient snake lizards, I want to bring you to a place on this planet where I have "gone a-hunting" for ancient snake lizards, the Upper Cretaceous fossil Lagerstätten in northern Patagonia known as "La Buitrera," and to build a picture of paleontological fieldwork at its finest—a rich fossil locality nestled within a complex sedimentological and stratigraphic wonderland, coupled with an exotic and magical place in the world where I get to work with great friends and wonderful students on an ancient assemblage of legged snakes. It does not get much better than this.

## "LA BUITRERA" AND *NAJASH*

I first met Dr. Sebastian Apesteguía in 2005 at the 2nd Congreso Latinoamericano de Paleontologia de Vertebrados in Rio de Janiero, Brazil. At the congress, I gave a couple of talks, one of which was on the Argentinian Cretaceous snake lizard *Dinilysia patagonica*, and Sebastian presented data on the sphenodontians from his relatively new locality in Rio Negro Province, Argentina, "La Buitrera"—I remember very clearly he made a point of mentioning he had new snake lizards fossils, though he provided no details. In 2006, he published in *Nature* on a new taxon of very important rear-limbed Upper Cretaceous snake lizard from "La Buitrera" that he and coauthor Hussam Zaher named *Najash rionegrina* Apesteguía and Zaher, 2006. Seven years later, I finally got to see the type and referred materials at the museum in Cipoletti, Rio Negro Province, following which I published a critique of Apesteguía and Zaher (2006), Zaher et al. (2009) with my Ph.D. student Alessandro Palci and collaborator Adriana Albino (Palci et al., 2013). Shortly after that publication, Sebastian invited me to join his field team at La Buitrera for the 2015 field season. At the same time, I was invited by the Argentinian national science foundation, CONICET, to co-supervise Fernando Garberoglio's Ph.D. dissertation (with Raul Gomez and Sebastian) on new materials of *Najash*. To date, I have spent three field seasons with Sebastian and Guillermo Rougier and their large group of students, prospecting for vertebrates in the sediments of the Candeleros Formation (Cenomanian–Turonian) (Figure 4.1). The outcrops being prospected are exposed within a 25–30 kilometer radius of the tiny village of Cerro Policia, Rio Negro Province, Argentina. This part of the world is arid, sparsely populated by human beings, but full of guinea pigs, armadillos, rheas, colorful burrowing parrots, feral dogs, and a lot of goats and sheep.

Prior to this opportunity, I had spent 8 weeks over two field seasons (1998 and 2001) prospecting for materials of *Dinilysia patagonica* in the Rio Colorado Formation (Santonian), again in Rio Negro Province, but about 100 kilometers to the north of Cerro Policia near the village of Paso Cordoba, just across the Rio Negro River from the city of General Roca. There are definite similarities between the sedimentary systems of these two temporally and spatially distinct localities, and, of course, both snake lizards common to these two systems are Type I *Dinilysia-Najash* snake lizards. But the primary difference is a mixture of age and what I will term "clarity," which is itself a blend of quality of preservation and quality of the sedimentary rock record and its exposure. While the *Dinilysia* material and the rocks at Paso Cordoba are well preserved and exposed, La Buitrera and *Najash* are a marvel of excellent "rocks" and easily represent the world's finest lägerstatten of spectacularly preserved ancient snake lizards, not to mention other small-bodied dinosaurs, mammals, and sphenodontians.

Because this is a treatise on fossil snake lizards, I will detail here, as a case study of sorts, a brief facies analysis of La Buitrera, an overview of the depositional environments proposed for the various facies highlighted (lithofacies, ichnofacies, biofacies). I also want to pass on a sense of the rigors and pleasures of fieldwork in this part of the world.

## Geological Overview

The holotype of *Najash rionegrina* (see Chapter 2), along with numerous other articulated and disarticulated snake lizard remains, has been found at the "La Buitrera Locality" (LBL) within what is now known as the La Buitrera Paleontological Area (LBPA), a large geographical area in Río Negro province (northern Patagonia, Argentina), mostly to the northwest of the village of Cerro Policía (Figure 4.1a,b). A number of other proximate localities within have also produced fossil snake lizard remains, including "Cerro Policía" (which is a large series of outcrops at the actual "hill" of Cerro Policia, not the village), "El Loro," "La Escondida," and "El Pueblito," the latter of which

is to the south west of Cerro Policia (Figure 4.1b). Exposed outcrop in the LBPA is recognized as upper Candeleros Formation, Neuquén Group, and considered Upper Cenomanian in age (Leanza et al., 2004; Candia Halupczok et al., 2018). The overlying deposits of the Huincul Formation have been dated from a preserved tuff unit at ∼88 ma (Corbella et al., 2004). The Candeleros Formation is considered to preserve fluvial and aeolian sediments accumulating along the fringe of a large erg referred to as the Kokorkom Desert by Candia Halupczok et al. (2018). Numerous depositional environments are thus preserved in the Candeleros Formation sediments, with specimens collected representing the diverse taphonomies such environments effect on preservation and preservation styles. For example, some specimens have been collected from thick, cross-bedded sandstone beds that are considered to represent dune slip faces of large aeolian dunes in the ancient Kokorkom Desert, which obviously bears some consideration here, as it is a unique paleoenvironment whose signature lithofacies, ichnofacies, and biofacies seem to be common to Type I *Dinilysia-Najash* snake lizards regardless of geologic age and continent (see above).

## The Ancient Kokorkom Desert, a Sand Sea or Paleoerg

Finding evidence of aeolian deposited sediments in the rock record is relatively common. However, finding evidence of a true ancient dune field, a sand sea, or "erg"—well, this is much more rare. Kocurek (1999:239–240) characterized the rarity of sand sea preservation as follows:

> ...route from geomorphic expression to rock record might be viewed as occurring in three phases...The first phase must be the construction of sand sea...the second phase is the accumulation of a body of strata...the third phase is the incorporation of this strata into the rock record.

Kocurek (1999) was expressly considering ergs or sand seas, but at the same time acknowledged his statements here echoed the long-standing geological questions of what the rock record really reflects as an expression, via preservation as "rock," of the scale and kind of ancient processes and dynamic environments. In every sense possible, Kocurek's (1999) comments reflect my opinion of the caution that must be employed when interpreting ancient paleoecologies and habits from what is preserved in the rock record—like all data, fossils and the rocks they are found in come with real constraints and these should be respected.

Still, limitations recognized, there is a great deal to be learned from the rock record and what it contains. In the case of the Kokorkom Paleoerg, we can call it an "erg" because of its size (>100 square kilometers in extent and with a sand cover greater than 20%). As estimated and characterized by Candia Halupczok et al. (2018), the Cenomanian-aged Kokorkom Paleoerg may well have been a small sand sea, at approximately 826 square kilometers (covering the modern map area loosely defined in Figure 4.1c), but its dimensions are well beyond the required minimum, and its true size remains a problem of preservation, not actual size. Aeolian-deposited sediments occupying less geographical area squared, and with less sand cover, are still wind-deposited sediments, but are considered only a collection of dunes, not a sand sea, and would not dominate the sedimentological profile of the environments they occupy (Rodríguez-López et al., 2014). On a modern global scale, the largest dune sea/erg on earth is greater than 650,000 sq/km—the Rub' al Khali, which today is found in Saudi Arabia, Oman, Yemen, and the United Arab Emirates.

The sedimentary rocks preserving evidence of Kokorkom Paleoerg are approximately 100 meters in thickness in the northwest corner of the paleoerg (northwest Rio Negro Province) and thin to the south and east, where we have been prospecting for *Najash* and other vertebrates south of Cerro Policia (Candia Halupczok et al., 2018). These aeolian deposits are overlain by fluvial sediments, while the lower boundary of the erg sediments are, not too surprisingly, an erosive surface. The aeolian succession set preserves numerous fluvial-aeolian stacks representing discontinuities or bounding surfaces suggesting the sand sea was mobile, preserving evidence in the vertical stacking of units of the lateral variation of environments. As identified in detail by Candia Halupczok et al. (2018), the vertical succession preserved in the Candeleros Formation Aeolian set begins with a sandy facies deposited in seasonal, shallow fluvial system in large floodplain, that is, associated

with cross-bedded, small- to medium-grained aeolian-deposited sandstones. It is easy to imagine that as the sand sea moved across this landscape, the interdune spaces were seasonally wet, with ephemeral streams and floodplains developing between dune sets. Small ponds and lakes would also be expected, which also means there would be semipermanent water sources and vegetation stabilizing interdune and perhaps even some dune structures. The variation of annual precipitation, coupled with long-term precipitation or cessation, and water table stability and fluctuations, would explain quite clearly the stability when wet, or instability when dry, of sand sea sediments and the subsequent erosion and movement of the sediments, or their stability and the development of paleosols and paleosurfaces or hardgrounds. These stabilized surfaces are often found on what are easily recognized as dune slipfaces, but also in wet interdune spaces or small interdune "basins" as wet, vegetated, and thus bioturbated and rooted sandsheets (low-angle Aeolian deposits in interdune spaces where clasts are still moved by wind currents but dune structures are not formed). These same three surfaces were also recognized by Candia Halupczok et al. (2018) as preserving dinosaur footprints.

Within the Candeleros Formation, particularly at Cerro Policia, there are at least three units or bounded sequences ("D1-D3" of Candia Halupczok et al., 2018) representing alternating dry and wet periods in the development of the basin and its accumulation of sediments. Dry periods are associated with dune growth and movement, while wet periods are linked to dune stability, the development of soil profiles, vegetation, and fluvial and lacustrine deposits on the paleoerg margins, but more importantly in interdune regions and in the case of vegetation, on the dune faces and between dunes (see Candia Halupczok et al., 2018). Above the final aeolian zone, unit "D3" of Candia Halupczok et al. (2018), these authors described a fluvial unit of mixed clastics that demarcates the top of the Candeleros Formation and the beginning of the Huincul Formation. In broad terms (Neuquen Basin in the Upper Cretaceous), the dry periods are considered linked to lowstand phases and major regression altering the paleoenvironment significantly, with wet phases linked to transgressions and a decrease in aridity (Rodríguez-López et al., 2014).

## Facies Analysis and Facies Association

*Lithologies and Lithofacies* (Figure 4.2a–c)—Without sounding trite, the dominant lithology is a fine-grained sandstone, but while sand is everywhere, packages of strata are never so homogeneous, and on closer inspection, the Candeleros Formation within the LBPA is riddled with coarse gravel and medium-grained sandstone units, medium-to fine-grained sandstone units, fine-grained to silty sands, and in some place silty muds to clays.

As noted above in discussing the Kokorkom Paleoerg, the lithofacies range widely, from lacustrine clays, silts, and sandstone to channel sands and gravels, aeolian-deposited sand sheets, and large-scale climbing dune structures composed of medium-grained sandstones. The lateral associations of the various facies are mixed; that is, on the margins of the Kokorkom Paleoerg, there would have been fluvial and lacustrine facies associations proximal to the

Figure 4.2. La Buitrera Lithofacies. (a) Photograph, section at "El Pueblito Locality" (see Figure 4.1b), of stacked aeolian and fluvial sands and gravels. Massive sand sheets on the bottom are heavily burrowed, and are then eroded by gravel sands from fluvial system, with planar crossed bedded sands above, capped by aeolian fine-grained sands at top. (b) Photograph, section at El Pueblito Locality (see Figure 4.1c) showing massive and heavily burrowed sand sheet on the bottom, thin fluvial channel sandy gravels at middle, and stacked fluvial possibly lacustrine silts and sands at top. (c) Photograph, section at La Buitrera Locality, Med 4 (see Figure 4.1b), large dune foresets in two lower units, bounded by discontinuities, and capped by lacustrine deposited lithofacies. (F. Garberoglio for scale; all photos by the author.)

sand sea itself. Within the sand sea and its dune fields, as dune sets shifted around the sand sea, creating interdune basins, and with activity and movement of fluvial and lacustrine systems in the interdune regions both seasonally and over time, the lateral facies associations would result in a stacking of the various facies one on top of the other. When dune sets were stabilized during wet periods by vegetation, interdune fluvial and a lacustrine environments would also have time to stabilize and cut into former aeolian sediments, developing channel and lake systems in the interdune.

*Ichnofossils and ichnofacies* (Figure 4.3a–c)—The ichnofossils, or trace fossils, vary widely in terms of kinds, extent, and variety. As reported by Candia Halupczok et al. (2018), there are numerous dinosaur footprints throughout the various units of the Candeleros Formation within the LBPA. Paleosol horizons are dominated by root traces or rhizoconcretions that are laterally extensive in their distribution across a stratum and extensive in their penetration into a paleosol and the underlying sediment (Figure 4.3a). The most pervasive ichnofossils, dominating the ichnology in the LBPA, though sometimes, if not frequently, hidden

**Figure 4.3.** La Buitrera Ichnofacies. (a) La Buitrera Locality, around "Hoyada de Munoz" site showing rhizoconcretions (root trace fossils) on paleosol surface of stabilized dune slipface; (b) detail of "massive" sandstone units displaying complete animal-produced bioturbation and loss of primary bedform structures (note: weathered surfaces present a "coating" of sandgrain detritus that masks complexity of burrow structures; removal of that patina reveals complexity of bioturbation); (c) detail of animal burrow structure showing internal burrow morphology. (a, La Buitrera; b,c, Cerro Policia.) (All photos by the author.)

THE ORIGIN OF SNAKES

by hardened weathered surfaces (Figure 4.3b), are animal burrow networks (Figure 4.3b,c) which have completely deformed and obliterated any primary bedform structures within numerous "massive" strata within the Formation at the LBPA. These burrow networks appear to belong to the ichnotaxon *Taenidium barretti*, tentatively identified as such from the stacks of unwalled, grooved, sinous, lobate half-ovals or half-discs (~1.5 cm diameter), backfilled by the same sandy matrix as the ground matrix of the dune/sand sheet structures (meaning it is homogeneous, not heterogeneous), forming what are referred to as adhesive meniscate burrows. The burrows are thought, in non-marine environments, to be produced by insect larvae or other arthropods with an exoskeleton; in non-marine systems is a typical component of the Scoyenia Ichnofacies (Keighley and Pickerell, 1995; Buatois et al., 2002). Fernandes and Carvalho (2006) reported *T. barretti* from aeolian Cretaceous deposits in Brazil, and at the generic level it is broadly recognized throughout the Mesozoic of Brazil. The presence of both *Taenidium* and the dense rhizoconcretion paleosols is consistent with stabilized dune and interdune settings during prolonged wet periods with intense colonization of sediments by burrowing invertebrates.

*Fossils and Biofacies* (Figure 4.4a–c)—The vertebrate fauna from the LBPA are extremely diverse and

**Figure 4.4.** La Buitrera Biofacies—fossil snake lizards *in situ* and under recovery. (a) Surface discovery of two articulated *Najash* vertebrae; (b) Leonardo J. "Harry" Pazo preparing field jacket around articulated skeleton discovered as sediment was removed from around the two tiny vertebrae in (a) (locality south and east of Cerro Policia); (c) articulated *Najash* postcranium as exposed on "fossil dune" slipface (locality at Cerro Policia). (All photos by the author.)

well-preserved. It is also, by itself, indicative of a complex series of paleoenvironments existing in lateral proximity to each other with specific kinds of vertebrates being habitat generalists throughout the erg system, while some were no doubt specialists. The known assemblage includes dipnoans (Apesteguía et al., 2007), fishes, anurans, non-snake lizards (Apesteguía et al., 2005), snake lizards (Apesteguía and Zaher, 2006), chelid turtles, eilenodontine sphenodontians (Apesteguía and Novas, 2003), araripesuchid crocodyliforms (Pol and Apesteguía, 2005), carcharodontosaurid and coelurosaurian theropods (Makovicky et al., 2005, 2012), diplodocimorph and titanosauriform sauropods, and dryolestid mammals (Rougier et al., 2011).

Biofacies associations within Candeleros Formation are critical for interpreting the snake lizard materials and providing empirical support for proposing paleoecologies. A detailed analysis of these facies associations is in progress so I will only briefly examine them here. My observations thus far note that articulated snake lizard materials of *Najash rionegrina* are found exclusively in aeolian deposited medium-grained sandstones, deposited as sandsheets, and often on the bedding planes between giant trough cross-stratified beds on dune slip faces (Figure 4.4c). Isolated snake materials occur throughout the fluvial and in some lacustrine units as isolated vertebrae, with some dissociated remains being preserved at contacts/discontinuities between fluvial and aeolian units. Rooted and bioturbated paleosols also contain snake lizard remains, but, again, they are usually dissociated and disarticulated.

## Fossil Localities in the La Buitrera Paleontological Area

As reported by Garberoglio et al. (2019b), the La Buitrera Paleontological Area (Figure 4.1b) sits in the southern "center" of Candia Halupczok et al.'s (2018) Korokorum Paleoerg. Within the LBPA there a number of productive localities for fossil snake lizards, though in particular, there are three larger localities and a number of sites or "sublocalities" that have produced particularly excellent articulated skulls and skeletons (Figure 4.1c). The prospected localities include the La Buitrera Locality, La Escondida Locality, El Loro Locality, El Pueblito Locality (EPL), and the Cerro Policia Locality (CPL). The best specimens

and most productive outcrops have come from the LBL, EPL, and CPL. Productive sublocalities or sites from the LBL include: (1) Tefa Site, (2) Hoyada de Muñoz Site, (3) Med4 Site, (4) Vifa. At the EPL and CPL localities, snake lizard remains are surprisingly common and easily found, but again many of these are isolated finds not associated with bone beds or microsites, though they are most certainly beautifully preserved specimens (see Chapter 2, Figure 2.8a–f).

### Fieldwork

I am anticipating that some people who may read this book are themselves not professional paleontologists (Figures 4.5a–c and 4.6a,b). Therefore, as a bit of an indulgence, mostly for me, but in part for those who will never experience the delicious adventure that is the act of discovery of ancient animals never before seen nor even dreamed of, I pass on here a little bit of that great joy of discovery and outright fun that is camping in the middle of nowhere…I hope.

The arid landscape that is northern Patagonia around Cerro Policia is beautiful. It is steep, gullied, flat, rocky, dry, dusty, hot, cold, and covered in mean, prickly scrub brush without green leaves in many cases, and if greenness exists, it is oftentimes because the wood and spines are green (Figure 4.5a). Bright lemon-yellow lizards are everywhere (see Figure 1.1b), as are their limbless cousins, though the latter are harder to find, though once we got lucky on a very cold morning and found a very tiny baby cat's eye snake, *Pseudotomodon trigonatus* (see Figures 1.1b and 4.5b). Cranky ants and giant tarantula wasps are everywhere, and the latter are gigantic, as are the big, hairy, salad-plate-sized tarantulas they parasitize. Kicking up dust as they move six-leggedly across the landscape are big roving beetles known as crazy beetles, or bitcu loco—huge beetles, big mandibles, and mad. Scorpions abound, both big and small, and hide under the very rocks and in cracks and crevices that we fossil hounds like to turn over or dig up… these descendents of ancient chelicerates also like to sleep under your tent, which means the zipper stays tight both day and night. While it does not rain very much during our fieldwork adventures, when it does, it rains in torrents, and our routes in and out of La Buitrera along dry river and streams beds become dangerous flood channels choked

a

b

c

Figure 4.5. (a) Prospecting in the Hoyada de Munoz on what I call Mount Cronopio, a rich site for small dryolestid mammals skulls, *Cronopio dentiacutus* (Rougier, Apesteguía, and Gaetano, 2011), otherwise known as "Scrat" (from left to right, A. Lindoe, G. Rougier, and "Willi"). (b) Photographing a modern snake, a very small baby of *Pseudotomodon trigonatus* (see Figure 1.1 for a close-up) the cat's eye snake, while hunting for *Najash rionegrina* in the rocks of the Candeleros Formation at the La Buitrera Locality, near Cerro Policia, Rio Negro Province, Argentina. From left to right, Fernando Garberoglio, me with camera, Sebastían Apesteguía, and Guillermo Rougier. (c) G. Rougier at the wheel of his 4 × 4 adventure wagon, about to drop down into the dry river bed of the Rio Limay upstream of the reservoir at El Chocon, Rio Negro Province. (Photo courtesy of L. Allan Lindoe.)

with deep mud—you and your 4 × 4 go nowhere for a day or two after the rain stops (Figure 4.5c).

Water? It is nowhere to be found. It must all be carried in from Grandma Irma's artesian well some 15 km away (Figure 4.5c). It is a major weight, a necessary weight, and a most precious treasure. Who is Grandma Irma, you ask? She is a fourth-generation sheep rancher on the edge of a seasonally dry river of some width but is only a few kilometers long, as it drains the mesa above her ranch running east to evaporate into the desert. Grandma Irma's farm is a port in the storm—a place to make base camp when not north at La Buitrera, have a bone-chilling cold shower, eat her fresh homemade breads, and enjoy a feast

with her family—and they feast! Her garden is a cornucopia of garlic and tomatoes, potatoes, and root vegetables. The asado? Sheep, goat, and pig, slowly roasted and marinated with chimichurri over an open scrub wood fire (Figure 4.6a,b)— this is the stuff of culinary legend, dining al fresco under a star-bright night sky on the finest meats, baking, and garden-fresh vegetables.

As the crow flies, La Buitrera is some 15 kilometers northwest of Grandma Irma's ranch, into the arid scrubland of northern Patagonia. However, this is a long 15 kilometers, more like 25 kilometers with twists and turns and ups and downs, that is truly only accessible via a slow, careful 2-hour-long caravan of rugged 4 × 4 vehicles, heavy laden

Figure 4.6. Camp. (a) Great food, great wine, rustic and real. (b) Far right, top corner, Sebastian Apesteguía, Guillermo Rougier, and a wonderful group of students, collaborators, Grandma Irma, and her family.

with water, gear, plaster, burlap, food, and a host of Argentine students and professors, and, lately, me (Figure 4.5c). Field camp rests against the foot of giant cliff carved out of the paleoerg deposits of the Upper Candeleros Formation. By day it is a splendid bright yellow-tan arc of stone rising up against the blue sky. By night, in the campfire light, it is a haunting amphitheatre from humankind's ancient past when we sought shelter in cliffs and caves and left evidence of our art and ideas on their walls. Shadows flicker and wave up the cliff wall, blending into the absolute black of the night sky, pierced gloriously by ancient starlight from an infinity of stars billions of light-years away. It is impossible to stare at that sky and not be lost in the fact that many of them were dead or dying when the ancient winds of Kokorkom Desert swept across this part of Gondwana some 100 million years ago—but yet here they are today, their fossil light still making its way across the universe to my eyes. This is truly mind boggling.

Sitting round that campfire, we drink mate, wine, and an Argentine favorite, cola with Fernet Branca, a bittersweet digestif from Italy. We eat like kings—Guillermo Rougier is a master in his

kitchen of iron cookware and tools, and is the chef de cuisine with fresh meats and vegetables, sauces, and his infectious joie de vie. The moon rises above this haunting landscape to the guitar of the master troubadour Sebastian Apesteguía, accompanied by the drunken talk of life, evolution, fossils, who thinks what and about just about everything, and the reason of the universe. Surrounded by these "Seers of Deep Time," with a fine glass of Argentine malbec in my hand, I am at peace with my thoughts. I can see a moment in deep time, a polaroid of a sort that collages me, my life, ancient animals and worlds without number all gone by captured into a single snapshot, then another and another.... I float in it, revel in it, scratch at the sunburn on the back of my neck and legs…and then pour another cup of wine…a pleasant nightcap.

Morning comes quickly if there's too much drink before going to bed…which happens often, as there is always much to talk about. We rise early; Fernando is usually up first, though Guillermo is not far behind, and a fire is lit in the morning chill to boil water, make mate, and scrounge up breakfast and wake the camp. Colorful tents dot the scrubland, their colors returning in the morning light, and their nylon walls shake and grow bulges as people rise, roll around to get dressed, shake out the sleep and the hangovers, throw their gear together, pee, and get set to leave for a day in the field.

By 7 a.m., we are gone, usually by foot, as most sites within the La Buitrera Locality are a comfortable walk from camp—2 or 3 kilometers one way, and some are closer (see Figure 4.5a). We leave early to beat the heat and the unrelenting sunshine. Lunch is carried along by the grad student army and usually is a plastic tub with cans of vegetables, beans, tuna, and sauce all mixed together and passed round the crew, closely followed by the communal mate cup. Sebastian and Guillermo have been coming back to these sites for almost 20 years and yet, miraculously, the fossils still keep being found (2018 is the 20-year anniversary). Weather and erosion do their work every year, and more and more material is exposed. Water turns over more small pebbles, each with a perfect little ancient mammal or snake lizard skull preserved in it, or knocks just enough sand off a cliff face to expose a tooth of a complete beautiful theropod dinosaur still in its sandy tomb.

Prospecting is quiet, slow, close-to-the-ground kind of work. Each person finds a spot and begins crawling along, or walking stooped over like a rice farmer planting rice grains one by one, slowly but surely moving forward (Figure 4.5a). You find your zen place—hum a tune, or drift off quietly on your own to some mind place while you look closely for bone sign. A white fleck in the brown-tan sandscape all around you, or a purple discoloration where the light of the sun has UV-bleached a bone and discolored it. In the sandstones of La Buitrera, unweathered bone is creamy white, but when the sun gets to it for a while, it turns to bluish-brown. You look for both. You also look for regularity in a landscape of weathered rock where sandstones are regular in their composition, but irregular and unsculpted in their weathered and eroded form as fragments and concretions. Bone is not irregular; it has predictability and that is what you look for, predictability. Teeth repeat; they are serial homologs of each other. Ribs are elegant, long, and curved, and vertebrae, blocky as they are, have regularity, like all tissues, prescribed by phenotype and its underlying genetic code. When you see something, you stop moving. You place the object where it lies, avoiding anxiously grasping at it while you scan it again to be sure you have not been fooled by bird droppings or a dried-up and bleached beetle carapace. You scan around your new precious gem for more bone sign in case the object came apart and you will need to find its bits and pieces in order to make it whole again. Then slowly and carefully, using a pin vice or dental pick if need be, you extract your fragment from the earth. Cupping it in your hand, you lift hand lens to eye and survey your new find to see if it is a gemstone of preservation, or just a chunk of bone crud that broadly fits into the category of "shoulder bone"…that is, thrown over your shoulder and back to the earth whence it came. This is how Fernando Garberoglio found MPCA 500—in my opinion the best-preserved Mesozoic snake lizard skull currently known (see Figure 2.8f). He was at the Cerro Policia Locality, crawling along on his hands and knees, scanning each and every pebble, picking each one up, turning it over, double-checking, discarding, repeating, discarding, repeating, and…then… there it was, the full right side of its breathtakingly beautiful face staring back at him. Perfect. Ninety-five million years after it died, in his hand, the Rosetta Stone of Type I *Dinilysia-Najash* snake lizards, jugal and all.

Twenty-plus years later, La Buitrera continues to produce, not just snake lizards, but countless other never-before-seen animals. The diversity ranges from lungfishes that likely lived in seasonally wet lakes, ponds, and streams, and then spent the dry season, as they do now, estivating in dry mud tombs, to a diverse assemblage of snake lizards and a breathtaking fauna of small-bodied mammals, including the very young and the very old among the currently known specimens, and there are hundreds to date. The bias at La Buitrera is toward the small-bodied animals, yet there is plenty of evidence of large-bodied animals as well—splintered in millions of bits and pieces and spread across the landscape like glittering rubble are large sauropod limb bones and ribs. The big animals are there, but it seems as though they passed through the Kokorkom less often then the smaller-bodied animals, who no doubt fitted better into the shifting landscapes and ecosystems such environments would have defined. Between wind, water, and shifting climate patterns, the Kokorkom was an unstable place for a long time, and adaptations to such transience would have been critical to a species' long-term success in the ecosystems at play in the environments of the sand sea.

With these scenarios swimming round in my head, I sit back in the sandy dust, hold up the tiny squished skull of a baby *Cronopio*, cupped carefully in my hand, hand lens to my eye, and marvel at the beauty of its ancient mammalian dentition. Somewhere around me, hidden for a moment, resting in their long-held silence, are my snake lizard quarry. I wonder where they are, hoping my colleagues will find them if I do not, but am pleased for the moment that a young *Cronopio*, with its odd sabertooth canine, has been saved from destruction by wind and rain and radiation to tell its story to someone who has come to understand it best. Guillermo will be thrilled. That makes me happy, too. I am also thrilled that "Gee-jay" will offer to "buy" me a drink back at camp tonight. For a minute yet though, I hold onto to my newest little mammal skull and enjoy the landscape of its little skull through the hand lens.

Science and life, all rolled into a single experience. I love it.

# CHAPTER FIVE

# Origin Myths as Opposed to Scientific Hypotheses

Not only are facts and theories in constant disharmony, they are
never as neatly separated as everyone makes them out to be.

FEYERABEND, 1975:66

Facts are constituted by older ideologies, and a clash between
facts and theories may be proof of progress.

FEYERABEND, 1975:33

Sacred Cows make the best burgers.

ANONYMOUS

I am introducing this chapter on the subject of the
science of positing origins and ancestor scenarios
versus the non-science of origins myths, and how
to tell the difference, with an observation statement
and three thesis statements. These four "points"
will be expanded upon in detail throughout the
chapter, but deserve to be highlighted in brief
at the beginning in order to set the tone for the
ensuing discussion. The basic premise of this
chapter, apart from the obvious point of exploring
origins hypotheses, is that there is a correct way to
do science, and in the case of postulating origins
scenarios, that correct method requires a phylogeny
that includes your group of choice, your ingroup,
placed within the context of the larger group to
which it belongs, that is, a sistergroup hypothesis.
This not-too-surprising expectation of methodology
linked to outcomes was also correctly identified

by Cope (1869) in his proposal of a snake lizard/
pythonomorph phylogenetic relationship (though
Cope, as I will show here, never explicitly proposed
a marine origin for snake lizards). Surprisingly, this
requirement of a lizard "ancestor" hypothesis for
understanding ecological origins of a groups was
also explicitly expressed by the first herpetologist
credited with proposing a burrowing origin for
snake lizards, Mahendra (1938:34) who wrote:

> ...the question naturally arises whether the
> "scale-surrounded" type of eye should be
> regarded as primitive in comparison to the other
> type, or vice versa. The answer to this question
> depends on our location of the saurian type
> most closely allied to the ancestors of Snakes.

Mahendra's (1938) insight on the methods and
outcomes necessary to posit both polarity for a

transformation and thus ancestors and origins hypotheses was brilliant, accurate, and insightful, but then he never applied his recommendation on method to the remainder of his study on the phylogeny of the Ophidia with a proposal for such an ally to the "saurian type." Unfortunately, this oversight, as I will show, despite his call to method and a necessary phylogeny as the foundation for understanding the vector of transformation of a character, and thus the shared origin hypothesis of the ingroup and sistergroup, was mimicked by those who supported the burrowing origins hypothesis he championed and as it was championed subsequently. For example, the principal proponents of the mid-twentieth century growth and development of the burrowing origins hypothesis, Bellairs and Underwood (1951:223), found the ancestor hypothesis search "*unprofitable*," gave up, and settled instead for an ad hoc, inductively derived ingroup hypothesis with scolecophidian snake lizards at the base of their phylogeny as justification for the burrowing origin of all other snake lizards. To be blunt, my criticism of this effort is simple: the answer was found before the question had ever been tested. What is impressive is that it was not until Conrad (2008), followed by Gauthier et al. (2012), that there was finally a global lizard phylogeny that found a burrowing lizard group in the sistergroup position to all snake lizards (see Chapter 6). Not even the current deluge of molecular phylogenies (e.g., Reeder et al., 2015; Streicher and Wiens, 2017), derived from nucleotide sequences, have found a burrowing lizard group in the sistertaxon position to snake lizards (see Chapter 6). I will continue to argue throughout this chapter that the foundation of phylogeny is essential before a researcher can proceed to propose a higher-order hypothesis (or scenario if you wish), or more than one for that fact, of just about anything, origins included. Anything less than the rigor imposed by a foundational phylogeny and an origins scenario is nothing more than arm waving, at best. At their worst, these empirically unconstrained stories rise to mythical proportions and become easily digested dogmatic fairy tales based on nothing at all except the unstoppable inertia of the political and social framework that supported them in the first place and are now invested in keeping them real at all costs. This chapter on origins overlaps in no small part with the following chapter on phylogeny, and while the issues discussed here rely on phylogenies as their foundations and empirical support, the

questions addressed in the two chapters are quite different, though in no small part, the citations are the same. I have attempted to minimize overlap as much as possible, but recognize it is not possible to completely eliminate it.

## AN OBSERVATION

Human beings are obsessed with origins and origins myths—for science, this is a problem.

This obsession is a psychosocial hangover of the first order that looms over the scholarship of evolutionary biology like a grim reaper of expectation demanding that we, and thus everything, must have a beginning. For the scholarly pursuits of science, it is the worst kind of historical burden possible because it is a non-science concept of origins linked very neatly to Aristotelian causality and thus a design and a designer, arising from religious dogma, pre-evolutionary thinking, and our passion for creation myths. Today, one thread of this burden demonstrates itself in our expectation that a continuously varying "entity," a lineage, can have a beginning. We find ourselves compelled to identify the essential characteristics of observable "groups" as putatively non-arbitrary boundary points along the trajectory of a probably continuously varying collection of entities. That is, we look to create break points in the evolution of a constantly evolving lineage of organisms through time and space by creating the concept of essential characters of a "species." If we are correct, and evolution is real, then a species has no origin point in time and space. Instead, it arises out of a four-dimensional cloud of continuously varying individuals that a posteriori present fossil or living members of that clade that, due to neo-Darwinian mechanisms of evolution, appear different from other fossils or extant members of similar form. But there is no origin moment for a clade nor any of its contained species/terminal taxa. There is no Adam or Eve snake lizard individual nor ancestral clade, no stem snake lizard taxon that is the ancestor at the origin point, yet we keep looking for one and talking about origins as though they can be discovered. This is clearly both a logical and empirical mistake, as there is nothing to discover; there is only a higher-order hypothesis arising as a logical consequence of a foundational phylogenetic hypothesis.

THE ORIGIN OF SNAKES

## STATEMENT ONE

> The burrowing origins myth describing the origin of snake lizards must be discarded as the inductive, phylogenetically unsupported, antiquated, and dogmatic story paradigm that it is.

A companion to this first statement is that the burrowing origins myth exists in the absence of any phylogenetic support whatsoever and is based completely on the illogical assumption that if scolecophidians are burrowers, and scolecophidians are basal snake lizards, that burrowing is primitive for all snake lizards. This is both ridiculous and illogical. If scolecophidians were indeed basal snake lizards in a sistergroup hypothesis of all snake lizards, it does not mean that their ancestors were burrowers, only that as extant members of the clade or clades to which they belong, they are burrowing animals—the higher-order hypothesis of burrowing habits for all snake lizards requires the closest sistergroup of snake lizards to be a non-snake lizard burrower. A burrowing origin scenario cannot be hypothesized in the absence of a sistergroup hypothesis where the sistergroup is itself a burrowing lizard. The burrowing origins myth was proposed nearly 50 years (Walls, 1942) before the first hypothesis of a sistergroup relationship between snake lizards and a burrowing lizard group was ever formalized (e.g., amphisbaenians and dibamids [Estes et al., 1988; Rieppel and Zaher, 2000a,b]) (see Chapter 6). Prior to Walls (1942), snake lizard origins were not interpreted via a burrowing scenario, but rather via a "grass swimming" scenario (e.g., Camp, 1923) nested within a phylogenetic scheme with an anguimorph ancestor—for Camp (1923), burrowing was a specialization of later snake lizards, not an ancestral condition. The full-blown burrowing myth begins with Walls (1942) and is further developed in the absence of a supporting phylogenetic scheme by Bellairs and Underwood (1951) and supported by later authors (e.g., Haas, 1973).

Numerous authors have stated in the literature (e.g., Haas, 1973; Kley, 2001, 2006) that snake lizards such as leptotyphlopid scolecophidians (Haas, 1973:434) are, "highly specialized and have many degenerate characteristics; however, they also preserve a series of primitive characters." Logically it should follow that burrowing is an extreme specialization derived from a non-burrowing state of anatomy, functional morphology, and ecology, not a primitive condition from which snake lizards, or any organisms for that matter, then secondarily reradiate back into non-burrowing habits and environments—the evolutionary ratchet leading to the burrowing specialization simply does not turn the other way. Such specializations are unidirectional due to the extreme canalization of morphology and habits and the always-associated degeneration of morphology—such morphologies as those linked to the extremes of burrowing and subterranean life are not primitive, they are simplified or degenerate, and simplified and degenerate are derived states of essentially incomplete or unusually truncated development. A linked statement is that while a few modern snake lizards are true obligate, ecological burrowers (e.g., scolecophidians, atractaspidids), they are also highly derived snake lizards, not primitive ones sitting at the base of the modern snake lizard radiation, let alone representing the base of *all* snake lizard evolution.

Embedded within these two observations/statements are a number of conceptual problems and fascinating facts. In no certain order, these revolve around an annoying conceptual problem related to the manner in which we use words—in ecological terms, and you will note that above I used the word string "obligate, ecological burrowers," (niche selection, adaptations, habitat, habits, reproduction, feeding, and environments), our concepts and characterization of what it means to be "fossorial," no matter what kind of animal we are talking about, snake lizard or ground squirrel, are in need of refinement and recharacterization. A problematic and thoroughly surprising fact is that a number of so-called "fossorial/burrowing" modern snake lizards forage, mate, and hunt above ground, usually at night. These are the purported "fossorial/semi-fossorial/burrowing" snake lizards (see Yi and Norell [2015] versus Palci et al., [2017, 2018]), Type I Anilioids such as *Anilius*, *Cylindrophis*, and *Xenopeltis*, all of which are cryptic by day, emerge from underneath logs, rocks, out of burrows, or from the leaf litter to feed at night by attacking large-bodied prey, usually in the water, or to forage widely across the surface of the ground in search of prey and mates. Another surprising fact is that numerous species of modern snake lizards, from *Cylindrophis* to large-bodied pythons, dig holes with their heads, which leads me to ask the question: Does this make them fossorial, semifossorial, burrowing, semiburrowing? Which is it? I think not. I will argue in this chapter that true fossoriality should be only ascribed to a group of organisms when essentially all of an animal's entire life cycle is completed below the surface of the ground. Such an ecology is of course an extremely derived condition in every group of organisms we know of, not a primitive one, and in snake lizards it is only demonstrated by modern scolecophidians.

## STATEMENT TWO

> There never has been an aquatic/marine origins scenario for snake lizards, only a sistergroup hypothesis proposing pythonomorph/ophidomorph lizards as the sister to all snake lizards.

The rise to dogma of the aquatic/marine origin scenario as the antithesis of the burrowing origins hypothesis was created by the proponents of the burrowing origins myth, not by the systematists working on the sistergroup relationships of snake lizards. What I will identify here is just how that happened and the authors who birthed that myth of polarized dichotomy in the complete absence of a supporting phylogeny (see also Chapter 6). Surprising fact number one is that no one, or at least not the scholars cited as having said so, ever said that mosasaurs were the ancestors of snake lizards—not one, not once. Caldwell and Lee (1997) never said so, nor did Scanlon et al. (1999), nor did Caldwell (2007a), and neither did Nopcsa (1908, 1923). Most importantly, Cope (1869, 1872, 1878) never made such a claim, even though it has been claimed to be so for all these authors by proponents of the burrowing origins hypothesis (Zaher and Rieppel, 1999a, 2000; Tchernov et al., 2000; Longrich et al., 2012; Rieppel, 2012). From Cope (1869) forward, scholars supporting Cope's hypothesis of phylogeny have unequivocally proposed phylogenetic hypotheses reconstructing a relationship between Mesozoic aquatic lizards and snake lizards, and then concluded that this is strong evidence of a shared common ancestor that may have been aquatic—but this is completely different from the ad hoc burrowing origins scenarios that have been proposed since Walls (1942), for example, Bellairs and Underwood (1951), in the complete absence of a phylogenetic hypothesis as the necessary supportive foundation for the origins scenario. In fact, as I will show, the first published

cladistic hypothesis of a close relationship between snake lizards and a burrowing lizard group (i.e., amphisbaenians or dibamids), where the authors considered that relationship "real" was that of Rieppel and Zaher (2000a), but that analysis was not a complete test of global lizard relationships and fell short as a result (see Chapter 6). Estes et al. (1988: fig. 5) reconstructed a similar relationship (Amphisbaenia, Dibamidae) (Serpentes, Helodermatidae [Lanthanotus, Varanidae]), but did not consider it accurate but rather an artifact of limblessness and elongation characters and how they had constructed the same. Importantly, though, and this is a critical point of logic, both the Estes et al. (1988) result and that of Rieppel and Zaher (2000a) reconstruct not just snake lizards as having a burrowing origin, but also all, and I mean all, anguimorphs as sharing that common burrowing ancestor. Meaning everything about an anguimorph is secondarily re-evolved from a blind, limbless, elongate burrowing ancestor. This point is importantly ignored by the supporters of the burrowing origins myth/hypothesis, likely for the same reasons it was ignored by Rieppel and Zaher (2000a)—more to follow on this point. Setting Rieppel and Zaher (2000a) aside, it was not until the global analyses of Conrad (2008) and then Gauthier et al. (2012) that a burrowing group was indeed reconstructed in the sistergroup position to snake lizards.

## STATEMENT THREE

> Alternative scenarios for snake lizard origins have been completely ignored as the dogma of the burrowing origins myth trudged forward to reassert its historical position as the hypothesis by creating from scratch its polarized antithesis and then repeatedly pushing that Straw Man down, over and over again.

Even a mildly scholarly reading of the literature shows that terrestrial origins hypotheses based on phylogenetic hypotheses have been on the table since Cope (1869) (see Chapter 6). Janensch (1906) proposed the second one, though it was done in the absence of a phylogeny and in the light of a significant misread of Cope (1869, 1879), where he claimed that Cope had claimed that pythonomorphs were the ancestral stock from which snake lizards arose. Camp (1923) based his terrestrial origins hypothesis around his phylogenetic scheme of an ancestral grass-living

THE ORIGIN OF SNAKES

anguimorph lizard. And numerous other authors since, for example, Scanlon et al. (1999) and Caldwell et al. (2015), have presented alternative origins scenarios based on varied "reads" of their presented phylogenetic hypotheses that were supportive of a terrestrial origin of snake lizards, not a marine origin. Despite this long history of presenting well-reasoned and logical alternatives, supporters of the burrowing origins myth have elected to ignore the literature as written and to create their own myth version, with pseudosupporting citations, of a marine origins hypothesis as the polar opposite of the burrowing origins myth (e.g., Tchernov et al., 2000; Vidal and Hedges, 2004; Apesteguía and Zaher, 2006; Longrich et al., 2012; Martill et al., 2015; Da Silva et al., 2018).

In a long argument that hopefully does not ramble too badly, I hope to examine, in relation to Statement Three, how such burdened theoretical and empirical problems collect like logjams in a river and literally stop up the flow of new data and hypotheses to make sure they are drowned in favor of the well-verified and received wisdom.

## THE PROBLEM WITH ORIGIN MYTHS

These famous words, "In the Beginning..." and their impact on and burden on Western science and thus modern scholarship cannot be overstated—the ethos or character of that Old Testament phrase frames our notion that everything must have a beginning, or in the parlance of this chapter, an "origin." The only difference today is that we are hunting for the origin of snake lizards via natural causes, not ultimate causes such as a god or deity of some kind. Still, conceptually and methodologically, origins quests are loaded with expectations and limitations of causality, causal paradigms often expressed in adaptational-teleological language, and notions of progress in evolution, even if it is not intended to reflect design and a designer. As J.B.S. Haldane is most famously quoted, "Teleology is like a mistress to a biologist: he cannot live without her but he's unwilling to be seen with her in public." While we are much more guarded about how we speak and write about origins scenarios today, that was certainly not the case during earlier decades of thought on snake lizard evolution and origins (e.g., Walls, 1942), the outcomes of which still burden progress if progress is measured in proposing new

paradigms and scenarios for snake lizard evolution (e.g., Da Silva et al., 2018).

In truth, the origin of a something—a clade, species, object, or even me—is so much easier to "understand" when it has an identifiable beginning, date of birth, moment of conception, act of creation, and so on, and so on, and so on. I have, for a very long time, believed my mother and father regarding the date of my birth—had they been misrepresenting the date of my birth to me, or to the government, then such important things as the day I turned 16 and obtained my first driver's license would have been rather a problem. Origins and the date on which they occur have meaning to us because they make sense and have impact that can be measured. This is also one of the reasons why the notion of deity, of a God or Gods, is and are so appealing—they make understanding beginnings and endings very simple, for example, "In the beginning...," and then God created everything in 7 days. This simple version of a possibility neatly explains your beginnings; that is, you were created by an adoring deity.

The same virtue to beginnings, to origins, that is—simple and with a beginning—also remains even if you consider the universe in the absence of the supernatural and thus constrain your intellectual and scholarly pursuits to the divination of natural phenomena via natural causalities. As should be obvious by now, the beginnings or origins problem takes us back to the ancestor problem, because it is the same problem, just with a different name, that I discussed briefly in the introduction and Chapter 1 of this book. If it is true, and I will argue until I am blue in the face that it is, that real ancestors are unknowable and unfindable, then so are the origins of a group of things.

As an example of both unfindable ancestors and origins, I present the following thought exercise. As a member of a hominid group called Caldwells, I know my birthdate and the parent pair of which I am an offspring—I know my beginnings as a real event. I know the groupings from within which my parents were in turn offspring, and even know the names and faces of a large number of my old and distant family members, and the dates on which they too were born. However, after only a couple of generations, I can confirm that they really do not look like me. I expect it is safe to conclude that going back in time some 300 generations that my distant ancestors did not likely look very much at all like me or my dad or grandfather. If 1000 generations equals about 20,000–25,000 years, then if we slip back in time by 100,000 generations, or 2.0–2.5 million years ago, we are now in the geologic time range of "Lucy" or *Australopithecus afarensis*, just before *Homo habilis* appears in the fossil record. And while jokes may abound regarding the number of shared similarities between me and Lucy, modern humans do not resemble these ancient hominids beyond being "hominid," and in fact, on either end of that timeline with Lucy and the Queen of England juxtaposed as generation 100,000 versus generation 1, the differences are profound. The two ends of this generational/phylogenetic spectrum can easily be counted, but the evolutionary change has been breathtaking and the time, well, it has been vast, and that is only ~2.5 million years! My point here is simple: go back far enough in time and eventually the individuals of the past stop looking like individuals of the present, but in terms of continuity, the lineage has remained exactly that, continuous. Were it not so, there would be no individuals of the present connected to the lineages of the past. If a lineage survives to the present, then there have been no break points, just gradual changes, from one reproductive event to the next, that connect individuals along this chain of being, in any direction, as but a point in a long and continuous variable. However, fossil species can be easily identified from break points in this lineage, that is, fossils of individuals from 2.5 million years ago, 1.63 million years ago, 25,000 years ago, 5000 years ago, or even a distinctive species from the skeleton of someone who died in WWI—each is distinct and forms a nicely defined terminal taxon in a phylogenetic analysis, but the origin of their clade is not distinct at all, as every new form we could identify would simply, by operational requirements, be a new terminal, and not an ancestor. Finding ancestors and thus identifying origins is both ontically and operationally impossible.

This stuff is so interesting as an intellectual exercise that I could genuinely, and with great passion, linger on this point for chapters of text—but I will not. Rather, to remain germane to the topic of this book, that is, the evolution of snake lizards through time, and to the objectives of this chapter, that is, the problems of origins myths, it suffices for me to say that if I disagree with the archetype concept and its idealized conception of the essentials of snake lizardness, then, in fact, the

modern assemblage of millions of individuals of living snake lizards did not have an "origin" at a definable point in time and space—we can ballpark it, but we can never pinpoint it. As with the example given above, these earliest snake lizards are but points on a continuously varying lineage of individuals through enormous tracts of time— if a generation of a modern snake lizard occurs about 2 to 3 years after birth or hatching, then beginning some 170 million years ago (the oldest known snake lizards [Caldwell et al., 2015]), we are talking about a lineage history of approximately 56–85 million generations if snake lizards had only evolved anagenetically within a single lineage. As this is not the case, but rather there have obviously been numerous lineages over time, the number of generations across all lineages spirals out of control. Recognizing that snake lizardness evolved well before 170 million years ago, which is what this book is all about, we are now likely well past 90–100 million generations into the past for a single anagenetic lineage. Finding the snake lizard ancestor and its origins is impossible. What we can do, though, which is unique to the fossil record, is to approximate that ancestor—as a sistergroup only—from the remains of an ancient snake lizard from even older rocks than the currently oldest known snake lizards (i.e., >170 million-year-old rocks).

Origins quests, like grail quests, should only be engaged in if you are aware that the truth is impossible and you thus begin the quest knowing you are looking for facsimiles of origins and ancestors—engage knowing your search is really for the closest sistergroup among the taxa in your data set. While boring by comparison to the "truth" grail you seek, it really is all that you can hope to "know"—a relative truth of a sort derived from the best phylogeny possible. It is for this reason that I borrow from the language of cladistics, which frankly is the only language to borrow from because its expected analytical outcomes result in a found object with a testable hypothesis supporting the conclusion—taxon A1 is the closest relative to my ingroup A. To be fair to the method of cladistics and the extrapolation of phylogeny from the output of cladograms, origins quests should therefore be explicitly phrased as an outcome of the search for sistergroups—for example, a phylogenetic origins hypothesis. I also accept that along the stem below the node shared by my newly discovered sistertaxon to my study

THE ORIGIN OF SNAKES

group, there is an ancestor concept and thus an origin concept implied along that stem. In the conceptual world in which that common ancestor and its origins exist, there are also concepts of an ancient paleoecology to be inferred, a collection of anatomies, and an evolutionary history that tells a story of clade "origins." But it must remain a story because that is all it is; it cannot really be discovered. I do accept, however, that these concepts arising along the stem are also empirically predictive as individual and collective search tools, but only as search tools and not absolute answers suggesting epistemic and ontic "problems solved." For example, if *Najash* and *Dinilysia* are found to be sistertaxa, the stem branch origins concepts arising would predict that they share a common ancestor that itself could be found in modern Argentina in rocks older than the Cenomanian and that an older snake lizard taxon would display suites of characters found in both taxa, but be potentially more primitive. There is power in such predictions, as they provide a place for continued fieldwork, suggested ages of sediments to begin scouring, and a search image for anatomy.

To reiterate, we must remember that individuals come and go through exceedingly short periods of time that are not measurable by geologic time metrics, and that their species boundaries are defined by us for our convenience, not by them as individuals or as members of lineages. Those boundaries reflect our desire to communicate what we think we know and what we think is real—no more and no less. We have come to agree that things and kinds of things change through time; that is, they evolve. So if they do, and there are some rules of how this process works, then if individuals are selected for, and species are the agreed-upon unit that demarcates the summation of evolution between a timeline of individuals of a previous species, then it follows that in the long history of the progression from that ancient first species of snake lizard, you have quadrillions of individuals being born, living, reproducing, and dying, through hundreds of millions of years, and changing slowly and insensibly, as Darwin (1859) put it, from generation to generation. The first would in no way resemble the last, and the first would grade almost insensibly out of the swirl of individuals of the closest non-snake lizard sistergroup, making the origins of its kind impossible to divine. Even worse, this means that there are vast anatomical differences between the most ancient species of snake lizard, the so-called "ancestors," and the modern forms. However, as I have said, asking these questions in the context of testable sistergroup hypotheses remains of value because we create predictive power for what we expect to find in the fossil record, where to look, and so on, which, as stated at the beginning of this book, is precisely the point of such research programs—predicting the ancestral snake lizard anatomy and how it arose from within the ancient radiation of what we like to call typical non-snake lizards.

To consider a bit further the limitations of origins hypotheses, I will step forward 10 million years into the future. While snake lizards might still look like snake lizards, they will not be what they are today, as evolution will not have stopped at all—50 million years from now, those differences will be even more profound. Will future scholars and thinkers (highly evolved cockroaches, that is) try to determine the origin of their "modern assemblage" by looking only at their modern assemblage? Fifty million years is a long time, and I expect extinctions will have winnowed the future fauna by 99%, leaving but a shadow assemblage from the Anthropocene, not to mention the impact on morphology and molecules of 50 million years of evolution. Will those future scholars thus try to posit the origins of their assemblage by looking back to the fossil record of today's assemblage? If so, what will survive? Perhaps a boiine or two will be preserved in some degree of disarticulation from Amazon River delta sediments, or maybe some marine elapids will be preserved in flattened but perfect articulation from Great Barrier Reef patch reef lagoon deposits? Fifty million-year-old sediments preserving some record of ancient snake lizards are far and few between today—Messel Formation (articulated booids), Green River Formation (articulated booids), and a variety of coastal deposits around the world with disarticulated snake vertebrae. How will scholars 50 million years from now look to one or two places in their future world, using a mere handful of taxa from today's nearly 4000 species, and even come close to an absolute answer on the origins of their future assemblage as it arose from within the one we know? I would argue it will be very, very difficult indeed—in fact, impossible. The real answers, as we are aware today, to the origins of their future assemblage are in fact old even now, and will be 50 million years older in their future. Like scholars of today, they will need to content themselves with testable sistergroup hypotheses first, linking their assemblage of the future to the assemblage of today and from our collective pasts.

## DISCARDING THE BURROWING ORIGINS MYTH

The history of a science and its creation of hypotheses is not, and I repeat not, the mere chronological accumulation of data and hypotheses blended in an orderly fashion with brilliant and collaborative scientists all lined up in temporal and spatial rows, altruistically contributing to a succession of discoveries and ideas. Quite the contrary, in fact—the real history of a science, such as the one focused on snake lizard evolution, is instead a chaotic jumble of arguments around data, pseudodata, non-data, metadata, made-up stuff, competing hypotheses generated by competitive personalities, contrasting falsifications, quasiverifications, supposed absolute truths, and frequent and virulent interpersonal disputes where leading personalities, not the data nor testable hypotheses, become the focal point of a reigning paradigm—wading through this morass to find the plums of knowledge is difficult, and overturning the paradigm that emerges as supreme is nearly impossible. This jumbled mess, particularly around the weight of persona, dogmatically becomes absolute truth, with data and current analyses having "solved" or "resolved" or "proved" or "found" the final or most complete answers. To paraphrase Max Planck on the method of overcoming the inertia of such received wisdom and the associated cult of personality is—"Science progresses funeral by funeral" (Samuelson, 1975). What I still find amazing, though, is that in the distillation of this chaos, and under the weight of absolutes and cults of personality, that ultimately, funerals or not, questions eventually arise that bring "knowledge" and myth structures back to the chopping block of falsification and, upon suffering the weight of that axe, are found wanting, discarded in part with the good bits saved, and a new paradigm rises from the ash, awaiting its own future testing somewhere down the road.

The problem, of course, is that within the ruling paradigm, it is relatively easy to keep adding to the received wisdom. In comparison, those working toward a paradigm shift face an impossible task

of scholarly opposition. Should a researcher dare to attempt revision of this monolith of wisdom, their heresy is decried and excoriated from the pulpit as bad science, and their work misquoted and misinterpreted against the truth myths. Their efforts to investigate the "state of the art" become a gruesome Sisyphean task—the work required to overcome the inertia of the old and revise it into something new (and hopefully better) requires, in my experience, literally 1000 percent more effort than it takes for the High Priests of Dogma to merely sit back, blade of grass between their teeth, and recite scripture and verse of received wisdom in defense of that same wisdom. Historical burden (Caldwell, 2007a, 2012) is real and pervasive and indignantly resistant to change as it is maintained by the political, social, and historical networks of science (Feyerabend, 1975).

To the point, though, I am arguing here that the burrowing origins hypothesis is a non-science myth story that needs to be permanently deposited in the wastebasket of the history of science. Even though it is charming in its narrative and in its simplicity, it is antiquated in its view of evolution and biodiversity and is driven by fanciful and whimsical imagery, not data and realistic hypotheses. Even using modern molecular phylogenetic hypotheses to test the burrowing hypothesis, this old and tired myth fails, as there are no phylogenies that find burrowing non-snake lizards such as amphisbaenians or dibamids to be the closest sistergroup to Serpentes to the exclusion of anguimorphan, mosasaurian, or iguanian lizards (see Chapter 6).

## KEEPING A CLOSE "EYE" ON THE BIRTH OF A MYTH

In my reading of the literature, the earliest published arguments countering an aquatic/marine origin of snake lizards were those given by Janensch (1906). However, as was discussed earlier (see Chapter 1), Janensch (1906:27) proposed a decidedly terrestrial origin for snake lizards, perhaps with a burrowing phase, but not one where scolecophidians represented the primitive snake lizard condition, nor where burrowing was the primitive habit of the earliest snake lizards:

> If we thus see that the living snake-like lizards are exclusively land animals, we must assume

THE ORIGIN OF SNAKES

that their predecessors gained their body form on the land, but not in water.

And (Janensch, 1906:27–28):

Of the forerunners of snakes, it may…be assumed that the acquisition of the snake type is to be attributed to adaptation to residence on ground covered with thick vegetation, perhaps also a burrowing way of life…

And:

…we certainly do not have to think of such marked fossorial forms as today's typhlopids and glauconiids…The possibility may also not be ignored that the these recent fossorial groups could have developed secondarily from snakes living above ground.

And from the conclusions (Janensch, 1906:32):

11. The snakes could not descend from the pythonomorphs. It is further improbable that they could be derived from the dolichosaurids and aigialosaurids. Probably they developed from an unknown terrestrial lizard, not adapted to aquatic life.

Janensch (1906) most clearly initiated and articulated the terrestrial origin for snake lizards from an unknown terrestrial lizard, not an aquatic one, and could in no way see fit to derive snake lizards from either pythonomorphs *sensu* Cope (1869) (which he mischaracterized as having claimed pythonomorphs as ancestors of snake lizards) nor from dolichosaurs as per Kornhuber (1901). Seventeen years later, his terrestrial origins model fitted with Camp's (1923) sense of phylogenetic relationships for both snake lizards and pythonomorphs/dolichosaurs, such that Janensch's terrestrial lizard became Camp's (1923) (see Chapter 6) more specific "grass-living" anguimorphan lizard. But, most importantly, neither Janensch (1906) nor Camp (1923: fig. 1) considered other burrowing lizards such as dibamids and amphisbaenians to be even remotely closely related to snake lizards (the

necessary sistergroup hypothesis for proposing origins), but instead derived snake lizards in their origins scenarios from a four-legged terrestrial lizard that became limbless or limb-reduced and elongate through adaptations to above-ground environments. Both burrowing and aquatic habits were secondary adaptations, and neither viewed scolecophidians as primitive snake lizards. The rise of scolecophidians to the position of most primitive snake lizards and the physical archetype of the ancient snake lizard is due, at least in the published literature, to the work of another and somewhat later scholar, Gordon Lynn Walls, and his herpetological supporters, Angus d'Bellairs and Garth Underwood.

Walls wrote a massive tome entitled *The Vertebrate Eye and Its Adaptive Radiation* (Walls, 1942), derived in its snake lizard details from a prelude paper on the snake lizard eye written 2 years before (Walls, 1940). The book remains comprehensive and valuable in its coverage of the basics and complexities of the anatomy and histology of the eye for every group of vertebrates and must have consumed Walls' attention for several decades (he says nothing of the time frame required to write the book anywhere in the Preface, nor in his bibliographic section, so I am merely guessing here, but the scope of the work is enormous). It is intentionally evolutionary in its discussions and descriptions, but decidedly thin on a guiding and thus foundational phylogeny or classification other than the six pages presented in Chapter 6 as "Elements of Vertebrate Phylogeny." Walls' (1942: fig. 60) figure on "Inter-relations of the major groups of vertebrates" where snake lizards arise from lizards is actually surprisingly progressive, a trend that also appears in a number of his other supposed ancestor-descendent relationships (e.g., he derives birds directly from dinosaurs long before such ideas were popular in the twentieth century, echoing older nineteenth-century ideas [Huxley, 1868a,b, 1870]). But just when it appears things are going well from his diagram, in his text Walls (1942:138) writes on the phylogeny and evolution of all lizards:

The lizards, however, came into existence only recently as an offshoot of the extinct mosasaurs. The snake originated as legless lizards, so very recently (as geological intervals go) that the most primitive of them, the boas and pythons, still have vestiges of the hind legs.

Without question, Walls (1942) was utterly unfamiliar with even the most basic paleontological and herpetological literature of his day, where in his bibliography he does not cite even Camp (1923), for if he had, he would not have considered all lizards to be descended from mosasaurs, which of course begs the question: Where did mosasaurs come from, and how did the great variety of non-snake and snake lizards find their ancestral source in the giant, secondarily adapted mosasaurs? Likewise, if he was aware of the work on scolecophidian snake lizards by his contemporaries, such as Mahendra (1936, 1938), or Haas (1930), he does not list their work in his bibliography, nor are any of these works, let alone Camp (1923), cited in Walls (1940).

Considering Walls's (1940, 1942) complete lack of knowledge of lizard evolution and phylogeny, it is in no small part rather alarming for me to profess that Walls (1940, 1942) is without doubt the great-grandfather of the remarkably teleological and adaptationist, if not neo-Lamarckian, scenario of the burrowing origins of the snake lizard "eye." He is single-handedly responsible for creating the fundamental view of the eye evolution myth in snake lizards that led to the "burrowing origins scenario." However, the development and propagation of the full-blown burrowing origins scenario has nothing to do with Walls (1940, 1942), who after his book ceased to make any notable contributions to the question of snake lizard evolution.

Before attending to the beyond-Walls birth of the burrowing scenario, an actual reading of Walls (1940, 1942) is necessary in order to understand the depth of my criticism of his prosaic writing as a teleological origins myth, and to understand the inspiration created for the subsequent conclusions of Bellairs and Boyd (1947, 1950), Bellairs and Underwood (1951), Underwood (1967, 1970) and Bellairs (1972). Walls (1940) is a brief manuscript at eight pages including figures, but it differs from Walls (1942) by explicitly linking the amphisbaenian eye and its evolution to that of snake lizards (Walls, 1940:6—boldface reflects Walls's original use of italics in his paragraph):

But all of the differences, including these two, are entirely consistent with the hypothesis that **the early snakes were subterranean and their eyes underwent wholesale degeneration into a condition fairly well represented by the modern Amphisbaenidae, from which they have recovered by a remarkable evolutionary "come-back."**

And from Walls (1942:627–640), though the prose is different than the 1940 study, the teleological and progressive adaptationist paradigms remain, and are written with an "eye" to explaining each essentialist detail of the evolution of the anatomy of the eye:

...above-ground nocturnality would not...have called for any greater changes in the ancestral lizard eye than have occurred in the night-lizards, snake-lizards, and geckoes. The pattern of the whole snake eye is consistent... with the hypothesis that the first snakes lived underground or originated there from lizards which had become fossorial.

And:

...The long persistence of the light-shunning habit would permit the degeneration of the whole apparatus of accommodation....As the eye shrank, then, it also became spherical.

And:

...the eye finally "touched bottom" in a condition not much if any better than that of a modern *Typhlops*. Indeed, the organization of the *Typhlops* eye is such that this worm-like form could well have been the "first" snake...

And:

Then, as the race became better able to stand the light, the retina became duplex. The eye enlarged, but in the absence of stiffening structures in the sclera it was forced to remain forever spherical.

And:

And these losses and defects were so numerous that the snakes had almost to invent the vertebrate eye all over again.

And finally:

We can perhaps understand now why a legless lizard is not a snake simply because it is legless. The snake-shaped lizards such as *Ophisaurus* and

THE ORIGIN OF SNAKES

*Pygopus* originated above-ground, and escaped the painful period of near-extinction which the true snakes experienced and which they have so gloriously survived.

Walls (1940, 1942) is most clearly the source of the Adam and Eve–style origins myth at the very root of the idea that scolecophidians are the most primitive snake lizards. He assumed degenerate and primitive were the same thing (see Underwood, 1976) and, empowered by his narrative, took that state of "primitive degeneration" of the non-snake lizard eye and rebuilt it to its "glorious" phoenixlike rebirth and reincarnation as a newly built vertebrate eye.

The honor of building and developing what becomes the modern version of the burrowing origins scenario belongs to Bellairs and Boyd (1947, 1950), Bellairs and Underwood (1951), Underwood (1967, 1970), and Bellairs (1972). All of these studies relied heavily on Walls's (1940, 1942) detailed myth story; in fact, they outright praised it (e.g., Bellairs [1972:165]: "so brilliantly argued by Walls [1942]"). Yet Walls's arguments were little more than teleological prose for the evolution of each anatomical detail of the snake lizard eye designed to place scolecophidians in the position of most primitive snake lizard via a comparison to amphisbaenians, dibamids, and acontine and feyliniid skink lizards. It is of scholarly relevance to see just how this myth was built into an apparent hypothesis by Bellairs and Boyd (1947, 1950) and Bellairs and Underwood (1951) as they found support from one or two previous herpetological studies such as those of Mahendra (1936, 1938) or Brock (1941) on which to build from Walls (1940, 1942). From my reading of the literature, the most certain statement of a burrowing origins hypothesis, going beyond the "eye myth" of Walls (1940, 1942), is that of Bellairs and Boyd (1947:107), who wrote what I believe is the first, albeit tentative, statement on snake lizard origins from a burrowing ancestor, which was based on their critical assessment of anatomical features of the eye and its adnexae:

7. These findings are discussed in the light of current theories of ophidian phylogeny; the hypothesis of the origin of snakes from burrowing ancestors in particular is considered.

I can only conclude that Bellairs and Boyd (1947, 1950) and Bellairs and Underwood (1951) cited Mahendra (1938) as the first study to propose scolecophidians as the most primitive snake lizards and to construct the framework of what would become the burrowing origins hypothesis. While accurate, it is also clear it was necessary in order to provide valid herpetological support for Walls's (1940, 1942) comparatively weak "eye myth" in herpetological terms, even though Bellairs and Underwood (1951) held in high regard his anatomical and histological descriptions (1940, 1942). Walls's (1942) slides and data become central to the detailed work of Underwood (1967, 1970) on the eye and in particular the retina and its tissue histology, thus lending even more support to the burrowing origins hypothesis.

Bellairs and Boyd (1947), Bellairs and Underwood (1951), Underwood (1967, 1970), and finally Bellairs (1972) developed the burrowing origins myth well beyond Walls (1940, 1942) and gave it substance that was both empirical and decidedly more formal—it gained weight and merit on its path to becoming what I see as dogma. Whether this was the intention of these authors cannot be helped, and in truth I suspect it was not their goal, as Bellairs (1972:170–171) wrote as follows:

The theory that the snakes have evolved from burrowing ancestors still seems to provide the most satisfactory explanation for many of their peculiarities. **It rests on circumstantial evidence however, and should not become accepted as dogma.** (boldface mine)

And, from Bellairs (1972:171):

As Underwood (1957) has emphasized, there is still a need for workers in various fields to investigate every line of evidence bearing on the origin of snakes, and they should not be hampered by preconceived ideas.

However, dogma it became, preconceived ideas and all as historical burden, and as evidence all that needs to be presented (see later in this chapter) is the ferocity of challenges given to opposing hypotheses, the absence of the use of an appropriate method to hypothesize a lizard sistergroup relationship for snake lizards, and the continual effort to misrepresent the "marine origins hypothesis" ever since Cope (1869), but in particular after Bellairs (1972).

To return though to the point of this section of this chapter, a salient criticism for me of the empiricism of the burrowing origins scenario, quite apart from Walls's simplistic myth story, is that the full-blown scenario was created in the complete absence of a supporting phylogeny placing any kind of burrowing lizards in a close relationship to snake lizards. Simply concluding that scolecophidians were the most primitive snake lizards by comparison to booids and versus Boulenger (1893) is little more than a "he said, she said" kind of argument. The same is true of all the rhetoric and detail expended on the hyperdetailed anatomical and histological description written in the context of "discovery by observation" of homology by pairwise comparisons. In the end, such detailed homology descriptions written to discover *a priori* notions of phylogeny test nothing at all with regard to either homologs or phylogeny. In the absence of a defined sistergroup hypothesis, primitive scolecophidian snake lizards mean nothing—it is even more telling if the sistergroup is in fact anguimorphans, as was hypothesized by Camp (1923). Bellairs and Boyd (1947) were aware of this fact and so they attempted strenuously to find data, similar to that of Brock (1941), upon which to refute Camp (1923) (thus also refuting Cope [1869] and Nopcsa [1923] without trying very hard, as Camp had already accomplished that feat) and establish some lineage or another of burrowing lizards as the ancestral stock from which snake lizards arose (see Bellairs's [1949] study on *Varanus* and his attempts to refute *Varanus* and varanoids as closely related to snake lizards). In fact, the inductive reasoning applied to this burrowing hypothesis around the scenario of the "eye" was so powerful and such a focus of Bellairs and Boyd (1947, 1950), Bellairs and Underwood (1951), and Underwood (1967, 1970), that I would describe Walls's "eye evolution" myth as having risen to the status of an "Ocular Backbone" that served as the overarching phylogenetic constraint on all subsequent data and hypotheses on snake lizard evolution. For example, the "Ocular Backbone" was so constraining that Haas (1980a:100), who early on in his career had specialized on the anatomy of scolecophidians (e.g., Haas, 1930), wrote in the text of his manuscript describing *Pachyrhachis problematicus* that it was an aquatic varanoid, not a snake lizard, because:

The fact that also other reptiles of at least ophidian relationship, like *Pachyophis* and *Simoliophis*, are marine or estuarine, recalls Nopcsa's old ideas about marine snake origins (in which I do not believe, in view of many characters of the most primitive recent snakes).

Those many characters could easily be enumerated around the multitude of features highlighted by Walls (1940, 1942) as reiterated by Bellairs and Boyd (1947, 1950), Bellairs and Underwood (1951), and Underwood (1967, 1970), not to mention Haas's own works on other features (Haas, 1930, 1959, 1962, 1964, 1968, 1973). Once the paradigm is in place, all observations are fitted into that existing paradigm, with *Pachyrhachis* serving as an example of what happens to new data that do not fit the old or existing model.

The history of hypotheses, data, and conclusions, informing us on snake lizard origins, evolution, and phylogeny, as happens in all sciences, is buried beneath the weight of its chaotic and accretionary history. It is as burdened by its myth structures as it is by its subsequent forcing of empirical observations, hypotheses, and the theories and secondary and tertiary concepts arising to support those myth structures. As I have said before, overturning such an establishment of myths and forced supporting data and hypotheses is extremely difficult if not nearly impossible. Under the burden of Walls (1940, 1942), not only were Cope (1869) and Nopcsa (1903, 1908, 1923) unacceptable competing hypotheses, but so were the far more robust hypotheses of Janensch (1906) and Camp (1923) arguing for a more straightforward pattern of snake lizard evolution where burrowing and degenerate morphologies were derived states for snake lizards, not primitive ones. The fate of such serpentian heresies was firmly sealed by the iron fist of the burrowing myth when Caldwell and Lee (1997) revised *Pachyrhachis* (following closely on ideas earlier expressed by McDowell [1987]) first as a snake lizard and second when they found snake lizards to be the sistergroup of pythonomorphs— if Cope was wrong the first time, he was of course wrong the second time.

With regard to the scholarly purpose for which I write, it is of paramount importance to understand the complexities, not simplifications, of the historical burden of ideas bearing weight on the science of snake lizard origins, evolution, and phylogeny. Philosophers have expended a great deal of energy on how sciences progresses, and I am afraid my sympathies extend to the Kuhns (1962) and Feyerabends (1975) for their rejection of method as a prescription and the recognition that new ideas in science require

THE ORIGIN OF SNAKES

seeming anarchistic-scale effort in overcoming old paradigms (note: dogma as a synonym) to create new and hopefully better replacement hypotheses. For all of the reasons discussed so far, I continue recommend that the burrowing origins scenario be abandoned for the myth that it is, as much as for the science that has been argued to support it.

## INVERTING THE PHYLOGENY AND REVISITING THE EYE

Bellairs and Underwood (1951), Underwood (1967, 1970) reviewed the conclusions and scenarios of Walls (1940, 1942) based on their own detailed study of Wall's actual histological slides and also on the detailed and critical reviews and conclusions of a burrowing origin and ancestry for snake lizards as presented by Bellairs and Boyd (1947, 1950) and the work of Mahendra (1936, 1938), Haas (1930), and a host of contemporary neontological studies (Figure 5.1). As they were

convinced the fossil record was too threadbare to be informative, and because their search for a lizard ancestor had proven "unprofitable" (Bellairs and Underwood, 1951:223; see also the discussion in Chapter 6), these authors highlighted instead their preferred ingroup phylogeny of snake lizards with leptotyphlophids and typhlopids at its base. The summation of observations on the anatomy and histology of the eye of snake lizards, as cited here, was critical to the thesis that Bellairs and Underwood (1951) developed. Therefore, it is imperative to briefly highlight what I see as alternatives to Bellairs and Underwood (1951), to Walls (1940, 1942), and to subsequent authors conclusions with respect to some of those data. There is no room in this treatise to fully examine all the minute details, but an overview of the critical features of retinal cell anatomy and histology is warranted.

Walls's origins scenario may be nothing more than a myth story, but it does not mean that his

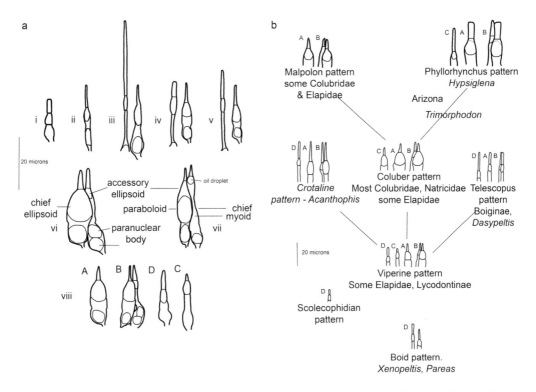

**Figure 5.1.** Snake lizard rod and cone cells redrawn from Underwood (1967:50 and 52). (a) (i) Single cell-type (rod) of *Leptotyphlops humilis*. (ii) Single-cell type (rod) of *Typhlops jamaicensis*. (iii) Rod and cone of boid, *Epicrates subflavils*. (iv) Rod and cone of *Xenopeltis unicolor* (suboptimum fixation). (v) Rod and cone of *Pareas margaritopliorous*. (vi) Double cone of *Dromiclis callilaemus*, showing features of a snake double-cell. (vii) Double cone of *Anolis lineatopus*, showing features of a lizard double-cell. (viii) Visual cell types of *Vipera berus* to show "viperine pattern." (b) Suggested evolutionary relationships/phylogeny of visual cell patterns of modern snake lizards from Underwood (1967). Labels "A, B, C, D" for the cone and rod cells are identified in "(a) viii" as visual cell types of *Vipera berus* by Underwood (1967).

observations of anatomy and histology were incorrect, only that perhaps his conclusions were deeply constrained by his notions of *primitive* as a synonym of *degenerate*. Likewise, it does not mean that Bellairs and Underwood (1951), or any of the permutations they authored before or after that work, nor any authors who have since labored under their paradigm, were wrong either, or that their observations were necessarily skewed by their preconceptions at time of analysis. But what is clear is that the absence of a clear answer on the identity of a lizard ancestor was for Bellairs and Underwood (1951) extremely problematic. In attempting a workaround, they ended up creating an ingroup hypothesis of relationships upon which to derive their burrowing origins hypothesis by finding scolecophidians as the most primitive snake lizards (Bellairs and Underwood, 1951:229):

> The Leptotyphlopidae and Typhlopidae, while possessing many primitive characters and still remaining in the fossorial stage of evolution, have deviated considerably from the main line of ophidian descent. It is likely that these aberrant families have been derived independently from very ancient ophidian types.

Regardless of the constraint on their data and that of Walls (1940, 1942) as imposed by an ingroup phylogeny rather than a sistergroup phylogeny, the data and conclusions used to support the burrowing origins scenario bear scrutiny in order to either verify them, falsify them, discover if they are correct but for the wrong reasons, or reinterpret their data and conclusions in a different manner and falsify their hypothesis. As I will shortly present, my conclusion is the last one—a complete rereading of their data and thus an inversion of the resulting hypothesis. Walls (1940, 1942) did not organize his study and presentation of snake lizard retinal cells in any kind of phylogenetic scheme. However, Bellairs and Underwood (1951) began the process of doing so, and Underwood (1967, 1970) continued that trend and produced a suite of phylograms of a sort for the evolution of retinal cell types in modern snake lizards (Underwood, 1967: figs. 11 and 12). Reproduced here as Figure 5.1a,b, these diagrams present the variety of visual cells in the retina of snake lizards in two formats. Figure 5.1a shows Underwood's (1967) detailed observations on the rod and cone cells of snake lizards, predictably organized from the scolecophidian visual cell of a single type of rod

(Figure 5.1a[i–ii]) to the more complex mixed cell types (rods and cones) of the "viperine pattern" as displayed by *Vipera berus*. From Underwood (1970: fig. 18), there is a very similar diagram, arranged scala naturae style, with the scolecophidian rod at the bottom and the viperine mixed cell set at the top, with various less complex cell combinations filling the orthogon between. The possession of a simplex retina (a term coined by Underwood [1967:48]), the condition in scolecophidians, indicates that there is only a single retinal cell type present, and these are identified as rod cells (Figure 5.1a[i–ii]). The duplex retina, on the other hand, is a term used to describe the presence of two retinal cell types, that is, rods and cones, and is observed in all other snake lizard retinae to the exclusion of scolecophidians (Figure 5.1a[iii–vi, viii]). In very broad terms, rod cells function best under low light conditions, while cone cells function best in bright light conditions and also are sensitive to color wavelengths reaching the retinal cells.

From Underwood (1967, 1970), and of course from Walls (1942), it is fair to say that it has long been recognized that colubroid snake lizards have complex duplex retinas with a wide variety of "cone" cells, at least three varieties, as illustrated by Underwood (1967) (Figure 5.1a[viii]), and one rod type. From Underwood (1967: fig. 12) (Figure 5.1b), it is very clear that there is even more complexity in the duplex retinal cone cells of higher colubroids, a fact also made very clear in Underwood's (1970) linear representation of the same. Underwood (1967) also created three groupings around his recognition of varying degrees of complexity, two of which have been mentioned so far, the Scolecophidian Visual Cell Pattern and the Viperine Visual Cell Pattern, as extreme ends of the spectrum of simplicity versus complexity; the third type, which Underwood (1967:43) considered to have retained a "conservative condition" as the simplest duplex retina found in snake lizards, was present in boids, pythons, *Xenopeltis unicolor*, the sunbeam snake, and the colubrid *Pareas margaritophorous*, the Asian spotted slug snake.

In discussing the evolution of these three patterns or conditions of the visual cell system, Underwood (1967:53) said, in particular with respect to the scolecophidian condition that (boldface mine for emphasis):

> The boid type retina may well be general in all of the Henophidia and...Pareas suggests

that it may persist into the Caenophidia... **the Scolecophidia would be derivable from the boid pattern by loss of the cones in association with reduced visual function in highly modified burrowing animals**.

The impact of this last sentence cannot be overstated, as it is counter to the foundational hypothesis that the visual cells of scolecophididan snake lizards are the remnant of that first primitive snake lizard's degenerate eye and visual cells as inherited from the supposed burrowing ancestral lizard form (see Walls, 1940, 1942; Bellairs and Underwood, 1951; versus Caprette et al., 2004; Simões et al., 2015). By Underwood's (1967) own admission, the visual cell pattern of scolecophidian snake lizards can be derived quite simply by the loss of the boid-type cones as a result of burrowing specializations; this is also exactly what Underwood illustrated in his figure 12 (Figure 5.1b). Profound, and I agree completely: the scolecophidian retina is derived, not primitive, not primitive in the least. Its eye is also unique, uniquely snake lizard, that is, and snake lizards are clearly unique among all lizards, but like all lizards, they primitively have rods and cones, not just rods.

Based on my assessment of skull types in Type II, III, and IV snake lizards, as outlined in Chapter 1 and explored in detail in Chapter 3, I would go even further to suggest that the Scolecophidian Visual Cell Pattern can easily be derived from a Viperine Visual Cell Pattern as well, and this is consistent with the numerous features scolecophidians share with Type IV and Type III snake lizards. All that is required, as Underwood (1967) noted, is the loss of any and all cone cells. I agree that the snake lizard visual cell system is complex, and were I completely informed on the comparative complexity across Vertebrata, I might agree with Walls (1940, 1942) that it is uniquely complex, a sentiment shared by Underwood (1970) in his treatise on the eyes of reptiles.

In the absence, though, of such knowledge, what I can speak to harkens back to matters of the vector of transformation and polarity of character change of complex systems—in my observation of the evolution of morphology, the morphology of complex systems changes in one of two ways. A system either continues to add complexity (e.g., from a non-snake lizard visual cell system to that of a viperine pattern), or it moves toward simplicity and potentially toward a degenerate condition that

is even more derived than mere simplification (e.g., from a boid visual cell system or that of a viperine visual cell system, to that of scolecophidian visual cell system) (see an excellent discussion by Underwood [1976] on the distinction of simplification of the eye as in scolecophidian snake lizards versus degeneration as in amphisbaenids). As the visual cell system of a scolecophidian presents with a rodlike retinal cell virtually indistinguishable from that of a boid or a colubroid, the vector of transformation is obvious. It began as a more complex snake lizard eye, and simplified to its current state (Underwood, 1976). It did not begin as a non-snake lizard eye, simplify to its current state, and then spontaneously re-elaborate to become a boid and then progressively a viperine cell system. Some time ago, Caprette et al. (2004) examined the phylogenetic distribution of eye characters in a large-scale "vertebrate phylogeny" and found that the snake lizard eye characters were shared with many "clades" of aquatic vertebrates that included, not too surprisingly, aquatic snake lizards. Whether this is evidence of a marine/aquatic origin or not is of little consequence here. My interpretation of Caprette et al. (2004) is that the snake lizard eye is very different from the eye of other burrowing vertebrates. A similar conclusion was drawn by Simões et al. (2015) who considered it unlikely that the ancestral snake was "fossorial" as are extant scolecophidians, regardless of the problem of their monophyly (e.g., Miralles et al., 2018).

Other authors may argue that there remain other character complexes supportive of a basal position of scolecophidians within Serpentes, but I do not know what anatomies and morphologies those might be. In Chapter 3, I examined in detail a large number of the "big classics" and found them wanting. Here I think it is made plainly obvious that the retinal cell system of snake lizards presents no conclusive *a priori* homologies in support of the position of scolecophidians at the base of a snake lizard phylogeny. Boulenger's (1893) notions of boid-type snake lizards (inclusive of Type I and Type II snake lizards) as basal are revived, consistent with the phylogenetic relationships espoused by McDowell (1987), and of course, by Caldwell and Lee (1997).

## "FOSSORIAL": ECOLOGY VERSUS EVOLUTIONARY TRANSFORMATION

While the Burrowing Origins Hypothesis is easily exposed as an inductive construction born in the absence of a phylogenetic hypothesis, with its roots in an adaptationist myth story, there remain other components of such an hypothesis that still require consideration. One of these is most certainly the manner in which we examine and understand ecologies, habits, and habitats, and the niches that organisms are pigeon-holed into as occupants of (see also Chapter 4); one of these is most certainly how we have created and used the term "fossorial," quite apart from the Burrowing Origins Hypothesis and its obvious linkage to fossoriality. The original Old Latin "fossorius" was a term applied to gravediggers, a so-called "fossor," a trade that has historically been reserved for the lowest castes of most societies. The modern biological application of the Anglicized derivative, "fossorial," is no longer used to described the profession of gravedigger, but is now applied broadly, and, I will argue here, problematically, to many kinds of living things across a broad spectrum of morphologies and ecologies and habitat adaptations (for a review of vertebrate adaptations/morphologies, see Kley and Kearney, 2008).

Among invertebrate animals, we can begin with arthropods, where the term fossorial has been applied for a long time to numerous groups of insects (e.g., Westwood, 1843; Ashmead, 1896) and has been used as an adjective to modify "limbs"; that is, a fossorial insect has "fossorial limbs" used for digging, or there are fossorial insects, that is, species or clades that dig into the ground/burrow for some part of their life cycle and/or ecology but may or may not possess fossorial limbs (Ashmead, 1896). It is also true among insects that many fossorial insects, that is, insects that can burrow or dig, have no obvious morphological digging specializations for digging, yet are amazingly good diggers (e.g., tarantula hawk wasps). Others, such as the mole cricket, have highly specialized forelimbs and heads for digging and for living underground through almost all of their life cycle. The list of "fossorial" or "burrowing" insects is vast, and, as with the two highlighted here, the tarantula hawk and the mole cricket, the range of burrowing abilities, morphological specializations, or the lack thereof, and ecologies, illustrates the problematic use of the term "fossorial." Both can dig a hole, but only one is truly fossorial in the ecological sense. This is the problem I will begin to explore here with more examples—the ecological sense of the term.

Continuing on the invertebrate theme, and linking modern faunas to ancient faunas and to ichnofacies assemblages, in this case in the marine and freshwater realms, it is clear that numerous invertebrate groups throughout all of earth history have burrowed into the substrate, whether it is soil or marine sediment, in order find safety and food, and reproduce (see the edited book *Trace Fossils* for an extensive overview [Miller, 2011]). The list is virtually endless and includes the tens of thousands of varieties of cephalochordates, echinoderms, mollusks, brachiopods, crustaceans, insects (freshwater), and annelids, to name but a few. What is interesting in reading through this vast literature is that all of these animals are described in terms of their modern ecologies, or in terms of their tracks and traces through the sediment, as burrows or bioturbators, but never as "fossorial." If the broad-spectrum environment occupied ecologically by an animal is an aquatic one, then "fossorial" is never applied to describe or modify the aquatic (freshwater/marine) ecology; this is done by using the terms "infaunal" or "benthic." Functionally, they simply burrow and are never "fossorial" unless they are land-dwelling invertebrates, such as the fossorial land crab *Gecarcinus quadratus* (e.g., Sherman, 2003). Here we find a qualifier in the use of the term "fossorial" that is not obvious in the common definitions—burrowing on land qualifies an organism as fossorial, but burrowing into the substrate while underwater does not. Here you are simply a burrower. So now what we have with respect to "fossoriality" is a ecological sense of the application of the word, and a habitat exclusion in using the word where terrestrial ecologies are modified as "terrestrial fossorial," but not "aquatic fossorial."

Across the vertebrate zoological spectrum, "fossorial" has been applied to all vertebrates except for fishes, even though many "burrow" in sediments in and out of aquatic environments. Among the modern assemblage of amphibians and reptiles, almost all recognized clades of lizards, turtles, crocodilians, frogs, salamanders, and caecilians "burrow" in some fashion or another, or use as refuge the burrows of other animals. This applies to both large-bodied (e.g., alligators dig burrows beneath the water surface into the banks of ponds and fluvial systems) and small-bodied (caecilians and amphisbaenians live underground for most of their lives) forms. Some living reptilians and amphibians do not dig

THE ORIGIN OF SNAKES

holes, burrow, or bury themselves in the leaf litter or soil, such as modern sea turtles (though they dig holes for their eggs), many kinds of frogs, and of course large-bodied tortoises, varanoid lizards, and many large-bodied arboreal, aquatic, and terrestrial snake lizards. Quite surprisingly, though, for many large-bodied snake lizards, such as some Australian pythons, they do in fact use their heads to dig holes in which to seek refuge, lay eggs, and so on.

A number of birds also "burrow"; the usual purpose is for nesting, and digging is done with their feet—the term "fossorial" is less frequently applied, with a preference for describing bird fossoriality as "nesting in burrows." "Fossorial" has been applied to mammals for some time (Shimer, 1903), where it is considered that fossoriality or semifossoriality is present in monotremes, many marsupials, and among numerous clades of placentals, that is, rodents, moles, mustelids, canids, edentates, mustelids, and insectivores. The mere act of digging extends to a much larger group of mammals, for example, those that dig a hole to find food, but do not "live" in the hole they dig, or occupy a hole in the ground dug by another animal. Likewise, among mammals that are considered to occupy an ecological/ habitat of complete fossoriality, such as moles (e.g., Shinohara et al., 2003), species of the larger group to which a truly fossorial mole (e.g., *Scalopus aquaticus*, the eastern mole of North America) belongs, the Talpidae, occupy a wide range of habitats, including mixed habitats/ecologies characterized as terrestrial, semiaquatic, aquatic/ fossorial, semifossorial, and fully fossorial.

By comparison, the term "subterranean" is less frequently applied to diggers and burrowers, but is usually used to describe cave-dwelling animals, though this is inconsistently applied in the literature, particularly among amphibian-focused herpetologists (e.g., Wollenberg and Measey, 2009) where "subterranean" is a term often applied to caecilians and burrowing salamanders and frogs. It is accurate to summarize the use of the term "burrowing," or "fossorial," as applied by biologists, as characterizing any animal that digs or can dig a hole in the ground, with any part of its body. As Shimer (1903:819) wrote, speaking of fossoriality:

The highest specialization in this direction is found…in those forms which secure not only safety but also their food within the earth… procuring of food above ground and the use of the burrow merely as a safe place…may be called semi-fossorial…between fossorial and semi-fossorial no fixed line can be drawn.

If Shimer (1903) is correct, and no "fixed line" can be drawn, then we are at an impasse regarding our understanding of the ecology of burrowing vertebrates and the evolution of adaptations that leads to this condition. However, I must admit that I disagree with Shimer (1903) in the sense that his "line" is not biological, but rather one of terminology and the problem of empirically identifying discontinuities in a yet another continuously varying trait, the condition of being "fossorial."

There is also a serious problem to deal with in determining the vector of transformation for fossoriality; that is, is fossoriality always derived from a non-fossorial state, and, once the evolutionary ratchet has turned, can a fossoriality and its extreme adaptations and modifications to morphology be reversed? I would argue from logic alone that the answer to the first question is "always"; that is, a non-fossorial condition of both form and ecology precedes evolution and adaptation toward varied forms of fossoriality, with the extreme condition of form and ecology being the final state along the transformation vector. With respect to the second question, I believe the answer to this is "no," as extremes of any morphology and adaptation never appear to turn round to re-evolve the preceding state, in this case the non-fossorial state. The proof of such a claim, however, cannot be in the inductive analysis of a variety of morphs, but rather in the empirical result arising from a phylogenetic analysis. Only in a phylogeny produced independently of the question being asked around evolutionary ratchets and the "one-way" street that is extreme specializations (e.g., it is unlikely a lineage of flying cetaceans will ever evolve from a future descendent clade of blue whales) can a question such as this be settled through reference to that phylogeny. Reflecting back on the burrowing origins myth we find that snake lizards are not the only group suffering this fascination with fossorial as primitive (e.g., turtle origins have also been proposed as fossorial, though this scenario links function and ribcage stability to limb driven digging, not head driven digging as in snake lizards

[Lyson et al., 2016]). It is perplexing to realize that fossoriality is indeed treated as primitive for snake lizards and that subsequent re-elaborations and re-evolution events are blithely invoked to explain the morphological state of the modern assemblage (see Walls, [1940, 1942] for the birth of such myth statements).

As a point of comparison, there is no such pattern of fossoriality as primitive for mammals such as rodents, where rodent systematists have concluded that the Bathyergidae, the clade including the blind burrowing mole rat, *Heterocephalus glaber*, among many kinds of African mole rats, are the most primitive of rodents, fossil and modern. Rather, the state of rodent systematics is quite the opposite—the Bathyergidae are one of the most highly derived members of their clade, the hystricomorphs (Fabre et al., 2012). This of course makes perfect sense, as you must first be an hystricomorph before you can become, regardless of the scenario, a blind, burrowing hystricomorph. Likewise, a scolecophidian must first have been a "snake" before it became a blind, burrowing fossorial snake lizard—why all snake lizards must be derived first from a blind, burrowing non-snake lizard, that slowly became a blind, burrowing snake lizard, that then evolved to become all other snake lizards, is beyond me; the perspective presented by Simoés et al. (2015) lends some support to my point of view on this matter. Except, of course, in this model, snake lizards are not derived from a burrowing ancestor, but from a non-burrowing terrestrial one, with modern scolecophidians derived as a specialization of whatever their most recent common ancestor was within Serpentes. It is well past due to reject this 80-year-old hypothesis of scolecophidians as basalmost snake lizards and start over where Mahendra (1938) began, or, at very worst, let us follow Bellairs (1972) or Underwood (1957) and let the burrowing origins idea stand again when there is evidence to support it, rather than letting it go the way of the geocentric solar system model and be dogma despite evidence to the contrary. A key empirical element is a robust snake lizard sistergroup hypothesis finding a burrowing lizard clade as sister, and second, for scolecophidians, at least in term of the ingroup argument, to actually be robustly reconstructed as something other than highly derived pedomorphic snake lizards with Type IV skulls.

## TO BE MARINE, OR NOT TO BE MARINE, THAT IS THE QUESTION

Great is the power of steady misrepresentation; but the history of science shows that fortunately this power does not long endure.

CHARLES DARWIN
*Origin of Species* (6th ed., 1872:421)

This section of this chapter is as easy to write as it is controversial in its content and tone, as it corrects a number of long-held misconceptions due to the misrepresentation of data, hypotheses, conclusions, and pretty much everything that has had anything to do with the debate on snake lizard phylogeny and evolution since Cope (1869) and his supposed marine origins hypothesis. I consider it insufficient to merely sweep all of this misrepresentative discourse under the carpet of time and proceed merrily along as though it never happened as it did, but that of course solves nothing and adds to the impediment to progress known as historical burden and dogma. So, in plain terms, I am doing what I must do, or as Martin Luther said when asked if he wished to revoke his position on the Protestant Reformation he was leading, said in a speech at the Diet of Würms in 1521, "*Hier stehe ich. Ich kann nicht anders. Gott helf mir. Amen.*" While I am not appealing to a god by any means, and the debate on snake lizard origins is hardly on the scale of the Reformation, laying bare misrepresentation and misunderstanding remains the right thing to do in scholarship. Importantly, it addresses one of the goals of this treatise, which is to reveal the real absence of a formulaic marine origins scenario and in so doing do away with the polarity of the supposed dichotomy of competing marine versus burrowing origins hypotheses (in other words, the debate on snake lizard origins should exist around data, not phantoms of ideas, one of which is a made-up straw man created in support of the ruling paradigm, not by the "antiburrowing" proponents). In identifying misrepresentation, purposeful or not, it also allows the entire community of scholars to begin anew the discussion of realistic alternatives to the burrowing origins scenario, and marine one for that matter, and move forward with new data, and new and existing hypotheses, toward a sensible and less constrained origins hypothesis based on current phylogenies.

THE ORIGIN OF SNAKES

## E.D. COPE: MOSASAURS AND SNAKE LIZARD ANCESTRY AND ORIGINS

To deal with misrepresentation, it is imperative that you cite the source of both the original statement followed by the misrepresentation of that statement. Following that requirement, here is what Cope (1869:257–258) wrote:

> …we may now look upon the mosasauroids and their allies as a race of gigantic, marine, serpent-like reptiles, with powers of swimming and running, like the modern Ophidia….That terrestrial representatives now unknown to us, inhabited the forests and swamps of the Mesozoic…is probable….

And:

> On account of the ophidian part of their affinities, I have called this order the Pythonomorpha.

Though there are more recent examples (e.g., Longrich et al., 2012; Martill et al., 2015), Rieppel (2012), in his essay on why scolecophidians cannot be "regressed macrostomatans," cited Cope (1872, 1878) as stating that snakes evolved from mosasaurs. Nothing could be further from the truth. Not only is it impossible to find any such text anywhere in either of the cited E.D. Cope papers, it is also impossible to find such statements in anything Cope wrote before or after the claimed citations made by Rieppel (2012). If I were to analyze the roots of this misrepresentation, I would conclude that Rieppel (2012) was not likely even quoting Cope (1872, 1879), but most likely reciting scripture and verse as derived from much older quotations of Cope by Bellairs and Underwood (1951:221) who misquoted Cope (1869, 1875) when they wrote the following:

> Cope (1869, 1875) originally suggested that the mosasaurs, which he termed the Pythonomorpha, were directly ancestral to snakes.

Working through this from start to finish, the problem begins to resemble the "Telephone Game," where the first player whispers a phrase into the second player's ear, and so on, and so on, around the circle until the last player is left to recite the phrase as it has been passed round by word of mouth—invariably, the final version is different from what the first player whispered into the second player's ear.

Thus, if this misrepresentation is a form of the Telephone Game, to be fair to Bellairs and Underwood (1951) in the absence of a quote for their statement, they were likely referencing their reading of Camp (1923), and correspondence held with him (Bellairs and Underwood, 1951:223 and 226) in composing their misquote of Cope (1869, 1875). Likewise, Camp (1923) can be read as referring his impressions of what Cope had written to his readings of Baur (1892, 1895, 1896) versus Cope (1895a,b, 1896a), Boulenger (1891, 1893, 1894, 1896) versus Cope (1896b), or Owen (1877) versus Cope (1878), or Janensch (1906), and not the original works of Cope. Nopcsa (1923) tried to correct those misquotes and misrepresentations of Cope, but obviously to no avail; Nopcsa's contemporary Camp (1923) would likely have surmised that these summaries from Owen and others were in total correct, as they were respected scholars all opposed to only one author, Cope. In either case, to correct for this historical trend that has continued to the present, I refer first to the quote from Cope (1869:257–258) as given at the beginning of this discussion. Following that, I quote Cope (1878:309) again in responding to Owen (1877):

> The demonstration of my second assertion, i.e., that the Pythonomorphous order presents more points of affinity to the serpents than does any other order…Professor Owen doubtless believes with me that the Lacertilia are more nearly allied to the Ophidia than is any other order…

And:

> …I only need to show that the Pythonomorpha are nearer to the Ophidia than are the Lacertilia to establish the truth of my position. Five of the seven characters enumerated above are so clearly of this nature that my statement is abundantly justified.

There is nothing here that suggests Cope, in any of his works, considered mosasaurs directly ancestral to snake lizards—to state otherwise is a gross misrepresentation of the facts and of Cope's science and his hypotheses, and is here corrected, hopefully for the last time. And regarding

Cope's (1878:310) defense of his use of the term Pythonomorpha, at which Richard Owen was so offended (Owen, 1877), and again making it abundantly clear to any reader that he did not think snake lizards had evolved from mosasauroids, he said:

> As to the use of the term sea-serpent, since I have not referred these reptiles to the Ophidia, the term involves no error.

So, it is only possible to conclude that from Owen (1877) to Janensch (1906), and Camp (1923) to Bellairs and Underwood (1951) and Rieppel (2012), that misrepresenting Cope (1869, 1878) has incorrectly proven valuable, as it makes for a convenient "straw man" that can easily be toppled over at will. While the end result of the "Telephone Game" is not malicious but rather is often downright funny, it is demonstrative of the problem of accurate communication, which is that it must be done accurately if information is to be relayed with a goal to accuracy. In the academy of scholars, however, the metaphor of the "Telephone Game" is serious stuff indeed, and is not funny and not trivial even if it happened innocently. It remains unscholarly misrepresentation and must be corrected. As Cope cannot defend himself, it is left to other scholars to do so. So here it goes.

Cope (1869) considered snake lizards and mosasauroids closely related—certainly a point of contention with other authors around data and observation—but at the level of distinctiveness at which he considered "Lacertilia" to be distinct from both snake lizards and mosasauroids, in no way was he suggesting anywhere in his work that snake lizards were derived from within a mosasaurian radiation. If anything, following his statements from Cope (1869:257–258) he considered they might well have shared an ancient and unknown terrestrial ancestor that was neither a snake lizard nor a mosasauroid! He never once in anything he wrote used the words "marine origins" in a single sentence, though he did explicitly talk about a terrestrial representative for pythonomorphs (see quote above) that leads me to conclude that a terrestrial ancestor was implicitly shared with snake lizards as well.

One hundred and forty-three years later, scholars are still misquoting Cope's work and hypotheses (see below), and, as noted earlier in this chapter and in Chapter 6, consciously or not, Cope correctly avoided the pitfalls of positing origins scenarios and instead worked to find closest relatives among fossil and living taxa based on the possession of shared characters. Not too surprisingly, as misquoting Cope has been so popular in terms of this debate, the same has been true regarding the revisiting of a snake lizard—mosasauroid sistergroup hypothesis by Caldwell and Lee (1997) and the misquotes and misappropriations of an exclusive championing of the marine origins hypothesis to that study. The same even applies to the misquoted study of Nopcsa (1923); though he did discuss snake lizard origins quite explicitly, he clearly avoided the pitfalls of deriving snake lizards from some kind of ancestral mosasaur by choosing instead to express his phylogenetic hypothesis of their close relationship (see Figure 1.4) as follows:

> The lineages of dolichosaurians…did not follow a one-sided, mechanically determined direction of development, and thus they have the form of a bush; but two groups of forms shoot out of this bush, which each shows a quite specific direction of adaptation, namely mosasaurs and snakes.
>
> TRANSLATION, J.S. SCANLON

Some groups of scholars have been more fair to Cope than have others, and should be recognized as such for their consilience and scholarship (e.g., Rieppel and Kearney, 2001), even if they also stood on the side of the received wisdom and dogma of misrepresenting Cope in other works (e.g., Zaher and Rieppel, 1999a; Tchernov et al., 2000; Longrich et al., 2012).

## CALDWELL AND LEE (1997): CORRECTIONS TO TWENTY YEARS OF MISREPRESENTATION

The following discourse may read like I am defending myself and my colleagues, and I suppose it will indeed read as such in several places. However, that is not my intention in the least. Rather, I am simply trying to set the record straight for the same reason that Darwin defended himself against such misrepresentations (see the quote at the beginning of this section), and for the same reasons that Cope's works need to be understood for what was actually written, and not from what others have said he wrote in the absence of quotations. Scholarship cannot proceed behind a veil of criticism that has no merit—problems will exist in any study and its epistemic and ontic claims, as humans and their powers of observation, reasoning, and logic are fallible (see also Simoes et al. [2018b] for a recent example of the power of misrepresentation). There is nothing wrong with being criticized for making mistakes, as that is how science presumes to operate around its model of test, retest, falsify or verify, improve on a hypothesis, and move on. However, when the criticisms are simply wrong, and a replacement hypothesis is proposed so as to fill a void that was not there, the challenge against misrepresentation is necessary. In making that challenge, progress is possible.

There is always a beginning to scholarly debates, and this one, that is, the debate over what Caldwell and Lee (1997) did and did not say, begins with a series of papers by Zaher and Rieppel (1999a, 2002), Rieppel and Zaher (2000a,b), Tchernov et al. (2000:2010), and Rieppel et al. (2003a). The earliest misrepresentation of Caldwell and Lee (1997) on snake lizard origins is surprisingly not Zaher (1998),which instead was critical of little more than a few isolated and highly selective characters (see rebuttal by Caldwell [2000a]), but rather comes from Zaher and Rieppel (1999a:832). The second, and not very heavily modified, misrepresentation is found in Tchernov et al. (2000:2010):

> …a series of recent publications…have interpreted *Pachyrhachis* to be basal to all other snakes, indeed to represent "an excellent example of a transitional taxon"…On the basis of this pattern of phylogenetic relationships, it was claimed that snakes had a marine origin (8)…

While Chapter 6 discusses the phylogenetic hypothesis of Caldwell and Lee (1997), I will provide here a brief but important rebuttal statement to both Zaher and Rieppel (1999a) and Tchernov et al. (2000). Caldwell and Lee (1997) never proposed that *Pachyrhachis* was basal to "all other snakes," only basal to scolecophidians and alethinophidians as characterized in the taxon-character matrix they tested and from the resultant cladograms and from there to their proposed phylogeny. This is a non-trivial distinction because, as has been stated by Caldwell (2007a) and repeated in Chapter 6, the real issue at stake was not the position of *Pachyrhachis*, but rather the supposed basalmost position of scolecophidians

(see above arguments arising from Bellairs and Underwood [1951] on the identity of the most primitive or basal snake lizards). Making this distinction in light of the remaining companion statements from Zaher and Rieppel (1999a) and Tchernov et al. (2000) changes each of them importantly: (1) Lee et al. (1999a,b) never said that the mosasauroid jaws were a start point for the evolution of the ophidian feeding system; using their cladistically derived phylogenetic hypothesis finding mosasauroids in the sistergroup position to snake lizards, they accurately inferred that mosasauroids and snake lizards shared features indicative of inheritance from a common ancestor, not mosasauroids as ancestral to snake lizards. This is exactly what is supposed to be done with higher-order inferences of function, origins, and so on, as read from a phylogeny derived from a cladistic analysis. The word intermediate is not a poison suggestive of ancestry, it is a descriptor of morphology connecting two terminal taxa in an analysis to an inferred common ancestor at the branch below a shared node, nothing more. It is also interesting to note that Zaher and Rieppel (2002:108) used the term "intermediate" as well, but at their own convenience, when describing the condition and relationships of *Xenopeltis*:

> Although *Xenopeltis* is a macrostomatan, the adult retains a very similar quadrate morphology and stapes–quadrate articulation to that found in anilioids, and it may represent an intermediate stage between the plesiomorphic scolecophidian–anilioid, and the more advanced, macrostomatan, auditory systems.

I recognize what Zaher and Rieppel (2002) were attempting to describe, yet see no fundamental difference between their use of the word "intermediate" compared to the term "transitional." However, not to put too fine a point on it, such criticism is a necessary construct in misrepresentation. (2) Most importantly, however, it is necessary to correct the complete mischaracterization, bordering on fallacious, in fact, as perpetuated by both Zaher and Rieppel (1999a) and Tchernov et al. (2000) of Scanlon et al. (1999) as having claimed that snake lizards "had a marine origin" (note: it is important to note that the Telephone Game analysis would find Zaher and Rieppel [1999a] as stating "implies" while Tchernov et al. [2000] state "claimed," and thus the misrepresentation is building on its own myth). Yet, in truth, absolutely nothing could be further from the truth, as Scanlon et al. (1999) argued for absolutely the opposite conclusions on origins hypotheses, let alone the suggestion they implied or claimed anything on one origins hypothesis over another. Instead, Scanlon et al. (1999:146) wrote the following:

> At present, there are two equally parsimonious scenarios… (1) A marine habitat, primitive for pythonomorphs (the mosasaur-snake clade), is retained in mosasauroids and *Pachyrhachis*. More derived snakes reverted to life on land. (2) Pythonomorphs were primitively terrestrial, with convergent marine adaptations evolving separately in mosasauroids and *Pachyrhachis*.

Scanlon et al. (1999) continued by elaborating on the necessary data needed to support scenarios one versus two and gave no preference for one over the other, and in fact noted that a marine hypothesis might represent convergence between mosasauroids and pachyophiid snake lizards with respect to aquatic adaptations. It could not be more clear that the misrepresentations by Tchernov et al. (2000) were nothing more than that—misrepresentations, and they served their purpose whether they were intentional or not. But, as I have noted multiple times throughout this book, correcting such statements is markedly more difficult than it is to write them and thus insert them into the defense system of the reigning paradigm, in this case, the burrowing origins scenario. For example, Lee and Scanlon (2002a:140) tried to correct these misrepresentations when they wrote the following in their paper describing new materials of *Mesoleptos zendrinii*:

> The aquatic hypothesis is often ascribed to Cope (1869), but Cope never suggested that the aquatic mosasaurs were ancestral to snakes: rather, he suggested that both had a close common ancestor, which might even have been terrestrial. However, critics subsequently misquoted Cope…

And:

> In describing the second specimen of *Pachyrhachis* (=*Ophiomorphus*), Haas (1980a:191) stated that the fossil "points to the fact that the snakelike body and loss of limbs did develop in a marine surrounding."

And finally:

> Thus, the position of these poorly known taxa close to snakes might reflect a false signal caused by marine adaptation and body elongation… More complete fossil finds…are required before their phylogenetic relationships can be conclusively ascertained and the early evolution of snakes clearly understood.

Quite clearly, Lee and Scanlon (2002a) wrote some 16 years ago, consistent with Caldwell and Lee (1997), Scanlon et al. (1999), Caldwell (2007a), and the text I have written here, where and how misrepresentation had ignored what had been written and what conclusions had been drawn. The other fascinating and extremely complicating factor is that, once written, misrepresentations self-propagate like parasitic paramecia and infest the literature for great periods of time. For example, similar statements to those of Tchernov et al. (2000) are to be found in Longrich et al. (2012:205):

> In particular, there is a longstanding controversy over whether the elongate body and reduced limbs of snakes evolved in a terrestrial setting5,6, perhaps as an adaptation for burrowing, or in a marine environment2–4, as an adaptation for swimming.

Similar phrases to Longrich et al. (2012) are to be found in Martill et al., 2015:417):

> As the only known four-legged snake, *Tetrapodophis* sheds light on the evolution of snakes from lizards. *Tetrapodophis* lacks aquatic adaptations…and instead exhibits features of fossorial snakes and lizards…*Tetrapodophis* therefore supports the hypothesis that snakes evolved from burrowing (2, 5, 6) rather than marine (19) ancestors.

In the case of the Longrich et al. (2012) study references "2–4" are Caldwell and Lee (1997), Cope (1869), and Lee (2005a), while in the Martill et al. (2015) study, reference "19" is Caldwell and Lee (1997). While it is nice to be cited, it is also better science and scholarship to be cited accurately for what you have written, the science conducted, and the conclusions you have actually drawn. In nearly all studies claiming support for a burrowing origins scenario and

THE ORIGIN OF SNAKES

calling for the defeat of the straw man of "marine origins," the culprit cited as the proponent of the latter, initiated no doubt under the inspiration of Tchernov et al. (2000), is Caldwell and Lee (1997). To properly address this false dichotomy and miscitation, I quote myself and Mike Lee yet again:

> The anatomy of *Pachyrhachis* provides strong new evidence for the hypothesis of mosasauroid-snake affinities. (pg. 705)

This is an affinity statement of sistergroup relationships using the character data of *Pachyrhachis*, scolecophidians, and alethinophidians, not an ancestor-descendent statement for the origins of *Pachyrhachis* from among mosasauroids.

> The recognition that the most primitive snake is marine does not support the fossorial hypothesis, and suggests that an aquatic ancestry for snakes merits serious reconsideration. (pg. 707)

This is a phylogenetic outcome statement of sistergroup relationships, to which there has never been a companion fossorial sistergroup hypothesis until Gauthier et al. (2012), where a fossorial clade of lizards, the Crypteira, are recovered as successive sistertaxa to snake lizards; unfortunately for the Gauthier et al. (2012) analysis (see Chapter 6), the concomitant outcome is that anguimorphan lizards are also derived from a burrowing ancestor in that hypothesis, one that is shared with snake lizards. But to return to Caldwell and Lee (1997), because this statement reflects a three-taxon analysis of ophidians (*Pachyrhachis*, scolecophidians, and alethinophidians) in which *Pachyrhachis* is found to be basal, and because the sistergroup to ophidians is not a burrowing lizard or lizard groups (i.e., amphisbaenians), the fossorial hypothesis must be rejected. Even still, the call to reconsider an aquatic ancestry was tentative, but at the same time sensible as it is the condition that is reconstructed by finding that snake lizards share a common ancestor with an aquatic lineage of non-snake lizards. It does not mean it is true, only that it is a logical consequence of the phylogeny. Also, and this is important, Caldwell and Lee (1997:709) never created a scenario story for the evolution of snake lizards from a common

aquatic ancestor shared with mosasaurs, only that the recovered phylogeny indicated that it was a possibility:

> The morphology of *Pachyrhachis* therefore provides surprising and compelling new evidence for the hypothesis that, among lizards, the marine mosasauroids are the nearest relatives of snakes.

And finally, not much needs to be said about the last sentence in Caldwell and Lee (1997) except that it is an accurate portrayal of the cladistics analysis that was conducted in that study. Personally, I am still waiting for the first installment of a fossorial sistergroup hypothesis (but I suspect it will be long in coming).

That none of the studies on snake lizard evolution since Tchernov et al. (2000) have cited anything other than Caldwell and Lee (1997) leaves me shaking my head, to be blunt. This is because I and my coauthors have spent a great deal of time in these last 20 years exploring alternative scenarios to both the burrowing and marine origins scenarios. I earlier cited the discussion from Scanlon et al. (1999), and will cite and quote here a statement from Lee et al. (1999b:518) in our redescription and recharacterization of *Pachyophis woodwardi*, originally described by Nopcsa (1923):

> However, at the moment, whether or not all snakes went through a marine phase in their evolution remains equivocal. Under one scenario consistent with the cladogram in Fig. 4, marine habits are primitive for pythonomorphs…with modern snakes being secondarily terrestrial. This scenario requires two steps.

And:

> However, another equally parsimonious scenario is that the marine habits in mosasauroids and pachyophiids are (convergent) specializations that evolved within each taxon, and that the modern snakes never went through a marine phase. This scenario also requires only two steps.

Further to the point, and from another study oft misquoted and miscited, and certainly the focus of another critical and misquoting analyses (Rieppel and Zaher, 2000a,b), Lee and Caldwell (2000:915) wrote nearly 1000 words on the uncertainty of a marine versus non-burrowing terrestrial origin in their conclusions to their redescription of *Adriosaurus suessi*, concluding, not unexpectedly that there two equally parsimonious hypotheses, consistent with Scanlon et al. (1999). First, that:

> snakes might have terrestrial origins (pachyophiids and mosasauroids each independently invading aquatic habitats)

And second, that:

> ...or they might have undergone a marine phase (marine habits being acquired in the ancestor of pythonomorphs, retained in mosasauroids and pachyophiids, and lost in more derived snakes).

Concluding that:

> This bias against the preservation of any terrestrial members of the snake stem lineage could thus lead to the incorrect inference that all members were marine and that snakes passed through a marine phase in their early evolution.

From the moment that Zaher and Rieppel (1999a) and Tchernov et al. (2000) began to misquote and mischaracterize Caldwell and Lee (1997), there were numerous other studies by myself and my colleagues that were exploring alternatives to the marine and burrowing scenarios, with the most explicit being the one led by Scanlon et al. (1999) who highlighted at least two alternative readings of origins from existing phylogenies. Caldwell (2007a) reviewed and highlighted these errors, as did Caldwell et al. (2015), but as yet to little or no avail, as even the most recent studies concluding burrowing origins and habits for ancient snake lizards have ignored those studies, sticking instead to the misquotations of everyone from Cope (1869) forward (see Hsiang et al., 2015; Yi and Norell, 2015; Da Silva et al., 2018). It is the conclusions on origins and ancient ecologies as highlighted by Caldwell et al. (2015) that serve here as the "bridge over troubled waters" to the final section of this chapter on origins.

## THESIS STATEMENT THREE: LAND HO!

As I have oft stated throughout this book, the first snake lizard "evolutionary relationships" statement ever published belongs to Cope (1869). The ironic aspect of Cope's (1869) birthing of the snake lizard phylogeny and origins debate is that he wasn't even interested in snake lizards per se but was trying to understand the relationships of mosasaurs to other lizards, inclusive, of course, of snake lizards (see Caldwell, 2012). He created the Order Pythonomorpha for the then known mosasaurs (clidastids and mosasaurids) as an equivalent of the Order Ophidia and Order Lacertilia within Squamata. As noted in Chapter 1, and quoted above, Cope (1869:254–255) wrote that mosasaurs, whom he considered to be the closest squamate relative to snake lizards, had "terrestrial representatives unknown to us." I read his statement as indicating he recognized mosasaurs likely had terrestrial cousins and that as secondarily aquatic lizards they also had descended from terrestrial ancestors. This being the case, and the logic is flawless, as are the actual quotes of Cope (1869), then the closest relatives of mosasaurs for Cope, the snake lizards, were first, not descended from mosasaurs (Pythonomorpha), and second, most certainly, like mosasaurs, had at some point in their early evolution a definite phase of terrestrial ancestry as well.

That there has been a consistent phylogenetic hypothesis for the pythonmorph sistergroup hypothesis linking Cope (1869) to the present combined molecular-morphological studies (e.g., Reeder et al., 2015), and nothing of the kind for a burrowing lizard sistergroup hypothesis (not even Vidal and Hedges [2004] despite their claims to having falsified the virtually non-existent marine origins hypothesis), really means very little except that the science and scholarship have been robust with respect to the pythonomorph-snake lizard phylogeny argument. Imperfections of data and observation aside, and they of course are there, the pythonomorph hypothesis is grounded in method and can thus be scrutinized and criticized where necessary. Thus, as Caldwell and Lee (1997) and a host of their collaborative projects have argued, there has never been a one-size-fits-all "marine origins" hypothesis—ever. We argued then, as I argue now here, and have argued for a very long time, it is likely that snake lizards passed through a large-headed, four-legged, short-bodied, terrestrial phase before they entered the marine realm and perhaps simultaneously also radiated into a wide variety of terrestrial habitats. The modern assemblage presents clear evidence of fossorial adaptations and ecologies with commensurate morphological specializations—scolecophidians and atractaspids are but two examples, but these are neither primitive snake lizards nor relict members of basal lineages. They are in fact highly derived in almost every feature they display, with demonstrable emphasis on pedomorphic patterns for their simplified but not primitive states. With respect to ancient snake lizards, there is no doubt in my mind whatsoever that there have been multiple fossorial invasions that we will likely never know anything about at all, as these animals never entered the fossil record (though I am suspicious of the ecology of the still-undescribed snake lizard from the Early Cretaceous of South Africa). But, again, as with marine radiations, early experiments in fossoriality are not primitive for all snake lizards, but derived. Still, none of these seemingly chaotic experiments in terrestrial, marine, or fossorial adaptations and radiations negate the fact that the closest sistergroup among all lizards remains the clade of pythonomorph and ophidiomorph lizards (see Chapter 6). That the probable best origins and ancestral scenario remains a terrestrial one, with subsequent marine radiations by both sisterclades, with extreme aquatic specializations in one lineage and limblessness and aquatic, terrestrial, and fossorial specializations in the other, should come as no surprise whatsoever. Such a "story" is far more complex than the simplistic duality of a polarized marine versus burrowing origins debate

would imply. But this complexity is consistent with a guiding theme of this book, and that is that the history of life is a complex tapestry of patterns, processes, and events, particularly when you factor in the depth of geologic time.

As Caldwell et al. (2015) noted in their discussion of Late Jurassic–Early Cretaceous snake lizard paleobiogeography:

> It is possible that these early snakes were isolated on various islands and continents during the break-up of Pangaea in the mid-Mesozoic…It is also possible that snakes forming these island assemblages arrived as secondarily aquatic invaders.

If snake lizards are indeed to be sourced from an explosive radiation of lizards stem lizards, lepidosaurs, or what have you, in the Late Permian or Early Triassic (see Simões et al., 2018a), then it should come as no surprise whatsoever that by the Late Jurassic to Early Cretaceous (nearly 80 million years later), they were diverse, spread around the world, specialized in their ecologies to both terrestrial and aquatic environments, and just beginning their long "crawl" up through the next 140 million years of earth history to become the snake lizard assemblage we know from today and from the Mesozoic and Cenozoic fossil record.

# CHAPTER SIX

# Ancient Snake Lizard Phylogeny

## WHERE DO MODERN SNAKE LIZARDS BELONG?

I am enough of an artist to draw freely upon my imagination. Imagination
is more important than knowledge. For knowledge is limited,
whereas imagination encircles the world.

ALBERT EINSTEIN as quoted in "What Life Means to
Einstein: An Interview by George Sylvester Vierek,"
*Saturday Evening Post* (October 26, 1929)

We need a dream-world in order to discover the features of the
real world we think we inhabit.

PAUL FEYERABEND
*Against Method* (1975:32)

The title of this chapter reflects my approach to snake lizard phylogeny and to the purpose of this chapter, and that is to reverse the question from, "Where does a fossil taxon fit in snake lizard phylogeny?" to, "Where do modern snake lizards belong in a phylogeny rooted in the ancient past?" I have also drawn from a quote by Albert Einstein on the value of imagination and its relationship to what we think we know, to our epistemic claims. I have done so because imagination is critical to what we do as scholars in demanding at the end of an investigation that our hypotheses make "sense" of the world, but also in the way we "see" old and new data at the beginning of a process of empirical observation. If knowledge is static, then it has become dogma, and the only way to break down

that barrier of absolutes is with newly seen data and newly constructed hypotheses, neither of which can arise in the absence of imagination. You must know something in order to see it differently, and all of this happens before you flip the switch of the algorithm or turn up the heat on the experiment. If imagination is lacking, even the newest fossil snake lizard is simply slotted most awkwardly into a corner of the old, received wisdom.

To address both the need for a bottom–up view of snake lizard phylogeny, and to incorporate the value of imagination, this chapter begins with an exploration of long-standing and recent phylogenies of snake lizard ingroup and sistergroup relationships, driven by what I see as the probabilities of snake lizard and overall lizard phylogeny based on the data set I know the best—morphology and the fossil record. I also present a critical examination of the much more recent use of molecular data of all kinds, for example, nucleotides, proteins, and so on, and the impact and relationship this has had on our understanding of snake lizard phylogeny. In both cases, that is, morphology-based and molecular-based phylogenies (or the more recent simultaneous use of both data sets), the goal of this review and critique is to highlight novelty in this accumulated set of phylogenetic hypotheses and not to present an exhaustive review of every hypothesis in the literature. I therefore highlight a number of studies for the purpose of presenting a critical review of the proposed hypotheses, especially where those hypotheses have been novel and have generated controversy or impacted thought on snake lizard phylogeny. After nearly 160 years of scholarly investigation of these questions, there are some relationships on which there is reasonable consensus, some that remain somewhat contentious, and others where there remains nothing but mystery and vociferous debate. A central theme of this chapter, but most certainly this book, is the critical importance of the fossil record to any and all epistemic and ontic claims regarding snake lizard evolution—thus the reversal of importance in the title. At the same time, it is only recently that the data of the fossil record have played a recognized role in the construction of phylogenies or in the positing of origins scenarios. At the end of this chapter, I fall back on the inspiration of imagination, which in the process of exploring what I see in my mind's eye, my imagination, has led to what I see as a way forward on the problem of snake lizard phylogeny in the guise of formalized method for what I call Fossil Backbone Analysis, the results of which serve as a constraint on phylogenetic topologies similar to the philosophy, method, and application of a Molecular Backbone. But first, I open this chapter with some brief discourse on critical points of philosophy and method and my vision of time, morphology and molecules, and my imagination.

## DEEP TIME, THE FOSSIL BACKBONE, AND PHYLOGENY

As I wrote toward the end of Chapter 4, romanticized as it was, there are a number of scientific disciplines where the students of that discipline have set themselves the task of peering questioningly into the vastness of time to find data that will illuminate both the past and the present—cosmologists, astrophysicists, paleontologists, and geologists immediately come to mind. They all attempt to collect data that are considered to provide insights on processes that led to the creation of a wide variety of phenomena, the fragmentary remains of which still lingering in the universe today can be found, identified for what they are, and used to interpret processes of the past that have brought the universe and its constituents forward to the present. Cosmologists and astrophysicists study fossil light, red-shifting galaxies, the existence and mass of dark matter and what it means, the chemistry and physics of stars, and so on, as they try to unravel the origin and expansion of the universe and its constituent parts. I need not review the data sets used and questions posed by paleontologists and geologists as they work to unravel earth's history of both its organisms and rocks, but in common to all of these scholars and the varied questions, theories, and hypotheses is time, great and vast quantities of it, in fact. So vast and so deep for that matter that the "it" of time is literally unfathomable. As a dimension of what we observe to be real, time is also expressed in this universe in such a way that it displays itself as having direction—it passes, so to speak, at least to our observation of it, in a single direction. While an object may be observed in three dimensions, we only observe that object in relation to the fourth dimension, time, as moving in an unrepeatable direction—things get older, not younger. You can run a stop-motion movie backward to watch time in reverse, but you cannot go further back in time than the moment in which you began filming time going forward. There is no rewind button on time as we experience it.

When I think on the hundreds of millions of years during which snake lizards have been evolving, I try to imagine both the great picture of it in its entirety (which is not possible, of course, but is nevertheless a necessary mental and empirical exercise), and I also seriously work at imagining horizontal slices in ancient time, say, the biotic assemblage of 95 million years ago in all its complexity. I try to imagine those ancient world time-slices in the same way I can "see" the world of today outside my window—a complex linkage of biotic and abiotic processes, fauna and flora and microbes all going about their business, populating ancient seas, subjected to changing climates, and seemingly isolated on continents moving about on dynamic crustal plates. Based on what I "know," nothing is different in terms of processes in those ancient worlds, but neither is anything very much the same when considering the animals and plants. For example, in my mind's eye in the 95 million-year-old deserts of Patagonia, a snake lizard with large back legs, *Najash rionegrina*, whips across the windswept dunes of the Kokorkom Erg in hot pursuit of a terrified saber-toothed, mouse-sized mammal, *Cronopio dentiacutus*. Across the proto-Atlantic ocean from Gondwana Argentina, following the tepid SupraTethys currents, I see the hot surface waters of the greatest carbonate platform in earth history where *Pachyrhachis problematicus* is basking on an emerged rudist reef island, its belly full of reef worms and small fishes, as the still-limbed *Haasiophis terrasanctus* trolls by without blinking, hunting for its own carbonate beach to bask on and lays its belly full of eggs. At the same time, a baby *Xiaophis myanmarensis*, lost on its Noah's Ark of Austral Gondwanan islands, drifting along in plate tectonic fashion toward a far distant future as part of a massive modern Asian continent, has become trapped in a sticky blob of tree resin and is slowly starving to death. In the rocks of ancient Gondwanan South Africa, the skull and skeletal remains of an even more ancient snake lizard, still unnamed as of today, are already 35 million years old and still buried deep in the sediment, as are the even older remains of *Parviraptor*, *Portugalophis*, *Diablophis*, and the grandmother of them all, *Eophis woodwardii*, which had passed out of memory for 70 million years before *Najash* or *Pachyrhachis* were even speciation events in the history of snake lizard evolution. This is what I see in my mind's eye, an ancient world of ancient snake lizards spread around the world long before the modern assemblage were even a mote in the eye of Darwin's mechanisms of evolution.

Why does my mind go here, where is the value in shifting backward in time to feel the sand grains of the ancient Kokorkom Erg sting my face and try to see in my mind's eye ancient snake lizards roaming that ancient landscape? Is it because I am a paleoromantic and feel the sci-fi urge to "time travel?" Well, sure, sometimes yes, but not in this case. Here, when my mind wanders to phylogeny and relationships, to ancestors and descendents, to fundamental sistergroups and ancient evolutionary patterns defined by ancient shared characteristics, I find insight by "seeing"

only snake lizards from the ancient past free of the muddle of the present. In doing so, I can unburden my data set of the noise in the modern assemblage that is the result of the subsequent 95 million years of specializations, adaptations, wrong turns, deadends, convergences, innovations, extinctions and supposed "re-elaborations," "redevelopments," and "re-evolutions" for snake lizards to "return to the surface" (*sensu*, e.g., Walls [1942]; Zaher [1998]). And, to be blunt, my data set is also free of the outright mistakes that have been made (see Chapter 3) in the "evolutionary transforming" and characterizing of modern snake anatomy, phylogeny, and evolution around archetypes and their essential characteristics built around the modern assemblage. I much prefer to define the essentials of snake lizardness around ancient snake lizards and let the chips fall where they may for the modern fauna in the foundational pattern as thus defined by fossils. Molecular systematists talk about a molecular backbone as though molecular data holds epistemic primacy over morphological data because it reflects the material of inheritance…well, maybe and maybe not (I actually do not think so, but more on that later). But if we accept in some fashion the idea of a constraining backbone, then I propose the use of time to build such a backbone. And time in this case manifests itself in the phenotype left over in the fossil record; the morphology of ancient snake lizards serves as the fossil backbone to a phylogeny. Here I find an intellectually satisfying and well-reasoned constraint on any and all phylogenetic hypotheses—as an example, 95 million years ago, today did not exist. Therefore, build a phylogenetic backbone of fossil taxa and then build a complete phylogeny with progressive addition of geologically younger taxa. While I do not have molecules from my fossils, I can reliably present their ancient phenotypes from a slice of ancient time. These data speak to the end point of "genotypes + epigenetics = phenotypes" that existed long before the modern genomic assemblage and its endpoint phenotypes. If molecular systematists cannot resist the use of time calibration points from fossils for their molecular clocks, by logic, then they cannot resist my "fossil backbone" either—and this fossil backbone actually has a backbone.

To elaborate further on the advantages of the "fossil backbone," despite its incompleteness, that ancient snake lizard fossil record of 95 million years ago is missing most of the messy details we must include today as an accommodation for all of the putative data present in today's fauna. Could you really stand on the beaches of ancient Gondwana, you would see those seas and shores to be as biologically diverse and chaotic as is the world of today and you too would get bogged down in the details of the numerous and varied species of snake lizards and their genes, behaviors, anatomies, and ecologies. It is a hard thing to teach, and a harder thing to learn, but not all observations, and not all details, are data (see the debate between Simões et al. [2018b] and Laing et al. [2017] as driven by the original publication on giant data matrices as written by Simões et al. [2017]). Data, as much as we might will it to not be so, are mute and speak to nothing at all without interpretation and questions to constrain their value and make them make sense. Now, if you doubt me and are already preparing some return salvo on the value of all observations and all data, arm cocked with the great left hook and five-fingered knuckle sandwich of "total evidence" ready to make contact with the point of my jaw, then let me remind you shortly before you swing of the wisdom of Willi Hennig (1966)—only the data of shared derived features are used to recover pattern in a cladogram. The remainder of the shared primitive features, the symplesiomorphies, are considered uninformative and excluded from the analysis. This does not mean that five fingers on all primates are not data for another question, only that they are not data for phylogenetic analysis following Hennig (1966). This is a prime example of what I am talking about—and so I repeat, not all observations are data relevant to the question being asked. The only "decider" on that question, and thus on the relevance of data to the question, is the researcher. Change the question, and the valuable observations and thus secondarily the data for the analysis change with the question. As is clear in cladistic analysis, if you ask the question more broadly, then your shared primitive features, the symplesiomorphies, may become relevant synapomorphies due to a new and bigger picture question. Decisions around data are critical decisions to be made by the observer and they must be made wisely and with reference to prior detailed knowledge of comparative anatomy and your study group. Fortunately, Hennig (1966) made it easier for future generations of cladists with his perspective on what are and are not

informative data, thus also paving the way for understanding that a great deal of observable "data" are many times nothing more than noise. Losing the noise is of great value.

Back to my point, for evolutionary biologists and paleontologists, the modern is a mess because it is not clear what data are of value unless you clearly understand the question you are asking. Observation is not objective, it is subjective, and data are not objectively obtained—they are metastatements derived from empirical observation where a question manipulates the "data" extracted from the observation and entered into the "analysis." So it should not sound odd after all that, but it will still be to some, if not many, but, because most of what I will call noise is gone from rock and fossil record, what you are left with is a workable data set that preserves the broad brushstrokes of ancient anatomies and ecologies (I see the potential patterns they might form as the core branching pattern of internal nodes in a cladogram). What remains in the rock record is a distillation with the chaos and noise removed, if you like, making it workable in its incompleteness. It is still necessary to have a reasonably complete data set, make no mistake. An N = 1 is insufficient, as it was for snake lizards when only *Dinilysia* was known from a single specimen in 1901 (Smith-Woodward, 1901). However, the data set has grown today to be a rather remarkable one, and seems to keep growing year by year. You can therefore examine that growing but refined data set and ask a rather straightforward question: "What does this simpler set of observations reveal about key ancient anatomies and the implications these data have for the basic architecture of snake lizard phylogenetic relationships?" In reality, for many paleontologists, this is all they can ever hope to do with their study group because they went extinct a very long time ago and do not have a modern messy data set to contend with, for example, trilobites, sauropods, creodonts, Edicaran whatever-they-ares, and so on, and so on.

At the end of this chapter, I will present my approach to creating a "fossil backbone" by proceeding as though snake lizards went extinct at the end-Cretaceous Event, whatever it was, and that the only data set we have of their existence is the fossil record. What would snake lizard phylogeny look like if that were the case? It is also important to remember that phylogenies are not predictive of what might happen in the future, but rather are reconstructive using anatomy of the phenotype or genotype as "data" informing us of shared, derived similarity due to closest common ancestry. Therefore, the "data" of 95 million years ago present the fossil backbone for the future—a future that will fit onto that backbone as successive branching events. However, the reverse is not true—the past cannot be fitted into the noisy mess that is the present and be made sense of in that manner. If there is a "bauplan" in nature, it is not in an organism, but rather in its connections to its past, and if there is truth in the universe, it is that the networks of past relationships leading to the present forms the framework from which the present has arisen.

## HOMOLOG CONCEPTS: MORPHOLOGY

As an example, numbers of characters are meaningless if the homologue concepts and character constructions fail the criterion, or test, of similarity (Remane, 1952; Patterson, 1982).

FROM SIMÕES ET AL. (2017:199)

To repeat what I wrote in the introduction and Chapter 1, for me as a cladist, homology is first hypothesized (primary homology *sensu* de Pinna, 1991) by comparison of two or more organisms as "similarity" (the Test of Similarity, *sensu* Patterson, 1982) and recovered as synapomorphies via the cladistics Test of Congruence (*sensu* Patterson, 1982). It is congruence that is the final arbiter of what can now be interpreted from synapomorphy distributions as homology. As a morphologist working within a long-standing debate on homology and morphology, this is how I understand homology as a concept and how anatomical comparisons between taxa are illuminated as homologies. Extrapolating from a cladogram to a phylogeny permits the assessment of hypothetical statements of evolutionary relationship derived from a suite of most parsimonius cladograms built upon homologies as read from synapomorphies. Cladograms are built by discovering groupings or clades from clusters of character states/synapomorphies shared between the taxa at the termini of the branches that all first passed the Test of Similarity before being submitted to the Test of Congruence. The selection of cladograms from which to build a phylogeny is guided by the principle of parsimony—evolution need not be parsimonius

at all, but the researcher needs a selection philosophy to guide them in selecting from among the vast number of resultant cladograms, and parsimony is the one that I prefer. In its simplest terms, parsimony dictates the selection of cladograms to be those derived from the fewest number of steps or state transformations. Between Hennig's requirement for informative characters only, that is, synapomorphies, and parsimony as the cladogram selection criterion, cladistics as a method of creating branching diagrams of synapomorphy sets is complete—inferring phylogenetic relationships among the terminal taxa in the shortest cladograms is a simple process requiring only the reasoning that evolution is the causal process that explains the tree topology or pattern of nested groups in cladograms supported by one or more synapomorphies/homologies.

Now, as methodologically simple as cladistics is, the process for constructing homolog concepts from the anatomies of the terminal taxa still requires the observer to think—these are metadata statements framed around primary homolog concepts sensu de Pinna (1991) and Patterson (1982). The data used to discover synapomorphies do not just land in your lap because you can measure a tooth or concoct a ratio of humeral length to length of dentary bone. Instead, characters and their states are created via a process involving extensive background knowledge on the anatomy (molecules to morphology) of the ingroup and outgroup taxa, coupled with creativity and a sense of "what you are looking for"; for example, primary homology concepts, that will be informative on sistergroup relationships, avoid symplesiomorphies, autapomorphies, and so on, and so on (for an opening salvo on rules of character construction, see Simões et al. [2017]). The process of constructing good characters is difficult and time consuming, and far from objective and quantitative, and deeply loaded with subjectivity and qualitative assessments of the data, but all of that is necessary and correct. Character constructions for a cladistics analysis are, in their best form, a thorough, unapologetically interpretive empirical work built on the finest phenomenological examination of the organisms identified as terminal taxa in the analysis. The fossil taxa serving as terminal taxa in a data matrix often have a sample size of one, or if more than one, it is not many, and, they are often incomplete—frustrating? Yes. Problematic, no! This is because the extant taxa serving as terminals are potentially represented by millions of individuals, but such data sets are impractical in their mass, and if you agree with and understand the philosophy behind the method, such volumes of data are pointlessly uninformative, as they all represent one taxon, not many, and so an exemplar of one individual will suffice. Somewhere in that process of eliminating the bulk of the biomass of an extant taxon, the researcher has made a decision to limit the individuals assessed for each terminal to usually the same number as exists for a fossil taxon—one or two. Where does this leave us as experimenters assessing data displayed by our selected terminal

## TABLE 6.1
*Problematic types of character construction in morphological data sets*

| Type/Name | Description |
|---|---|
| **I.** Discrete characters not following basic principles of character formulation | |
| **A**. Character coding leading to logical dependency between characters | |
| 1. Non-additive binary coding | Each state becomes a different character creating redundancy |
| 2. Absence coding | Extra weight for the "absent" condition creating redundancy |
| 3. Compound statement coding | Multiple non-dependent properties as distinct states under the same character (exception: gain/loss of the structure) |
| 4. Compound characters | Multiple non-dependent properties implicit in character description |
| 5. Compound character state coding | Alternative distinct non-dependent properties under the same state |
| 6. Multiple character state variables coding | Two or more non-dependent properties together under the same state |
| 7. Unjustified composite locator coding | Multiple anatomical elements with no evidence to be treated as having evolutionary integration under the same character |
| **B**. Character splitting | Single transformation series broken into two or more |
| **C**. State accretion | One state includes the other states |
| **D**. Conjunction | Two or more structures occur in the same individuals and are treated as homologous (exception: serial homologues) |
| **II.** Continuous Variables | |
| **A**. Data treated as continuous with arbitrary state delimitations | Data treated as continuous, coded either quantitatively or qualitatively, in which states are delimited arbitrarily or not representing the frequency range for each terminal unit (e.g., segment coding) |
| **B**. Continuous data unjustifiably treated as discrete | Continuous data treated as discrete but with no disjoint distribution of its character states |
| **III.** Biogeographical characters | |
| **IV.** Behavioral characters | |
| **V.** Character statements with no, little, or vague explanations | Character descriptions or character states that are ambiguous and likely to result in distinct scoring of taxa depending on the author |
| **VI.** Problems with the interpretation of the morphology during character construction | Misinterpretation of the identity of the structure (locator) being coded |
| **VII.** Taphonomy-biased characters | Characters that are easily altered during transportation or diagenesis of organic remains during the fossilization process, or may have different states when comparing a fresh specimen to a dry or fossil specimen |

SOURCE: Modified from Simões et al., 2017.

taxa? We are left with an abstraction of reality, which is good and necessary because we are not absolute truth seekers, but rather we are arbiters of relative truths who can now proceed forward to derive metastatements on the data (character concepts, character state descriptions, and primary homolog assignments) so to construct testable hypotheses of relative relationships—sistergroup statements—between our selected taxa. That is my approach to morphology as a data set observed in my search to construct homolog concepts using organismal anatomy. As noted, this is a detailed and highly constrained process with what I consider a set of essential rules and guidelines for most critically scrutinizing anatomy and its morphology for well-considered homolog concepts at the root of each primary homolog. Table 6.1 presents an outline of those criteria as was identified and discussed in great detail in Simões et al. (2017) and will not be reviewed in more detail here.

## HOMOLOG CONCEPTS: MOLECULAR DATA

The concept of homology for molecules and molecular data is by no means as straightforward as it is for anatomical data and morphology. As Patterson (1988:603) wrote:

> In morphology the congruence test is decisive in separating homology and nonhomology, whereas with molecular sequence data similarity is the decisive test. Consequences of this difference are that the boundary between homology and nonhomology is not the same in molecular biology as in morphology...

I would also add, following Patterson's observations on species trees versus gene trees, that I am also troubled conceptually by that difference, as I am still allied to axiom of selection that, "Individuals are selected for, but species evolve." As molecules and genes are components of an individual, that they might have gene trees indicative of "evolution" not relevant to the species tree suggests a level of homology not relevant to the species. Separating this out is problematic when I am left to consider the novelty and epistemic claims of molecular phylogenies (e.g., Streicher and Wiens, 2017). This dilemma exists for me because I am looking to find a common homolog concept within a single analysis, yet in truth, it

may not be there. I refer to Patterson (1988:610) yet again, as he discussed Fitch (1970):

> Orthology is homology reflecting the descent of species, and paralogy is homology reflecting the descent of genes; the theoretical distinction concerns whether speciation precedes gene duplication.

These are nontrivial distinctions, which in my view make assessing the compatibility of the root homolog concepts severely problematic. While it is true that molecular data as nucleotide sequences follow a positional pairwise comparison method based on the alignment of nucleotide sequences from a variety of species (Species 1 has a sequence AGCTAGCTAGCT, while Species 2 has sequence ACGTACGTACGT, and Species 3 has sequence ATGCATGCATGC), the homologs are not each base difference across alignments—those are the states and state changes at that position—but rather the homolog concept is in the overall gene to which those state changes are related. The gene homolog concept is derived from the functional product of that gene's transcription, followed by translation, that results in the ribosomal construction of protein. If I continue to apply my concept of homology as a morphologist to the molecular homology problem, then a molecular study of lizard phylogeny, for example, Vidal and Hedges (2004), has used sequence fragments from two primary homologs (genes). Thus, in the case of the study of Vidal and Hedges (2004), all the individuals sampled in the nineteen families of "lizards and amphisbaenians" and seventeen families of "snakes" possessed both RAG1 and C-mos proteins and thus the two primary homologs. Subdividing any one homolog into several thousand character states (states are DNA nucleotides) does not negate the fact that the analysis has still only been conducted on the molecules that code information that ultimately leads to just two proteins. Even if the entire gene was sequenced, the root homolog as I understand homologs would still only represent two proteins/two structural homologs.

Though not consciously intended to address the small number of homologs problem, the trend in lizard phylogenetic studies using molecular data since Vidal and Hedges (2004) has been to add more genes to the data sets along with more nucleotides.

For example, Pyron et al. (2013) used 12 genes, with 12896 basepairs per species and sampled 4161 lizard species, noting that this sample was five times larger than that conducted for anyother study of lizard phylogeny. Two years later, Reeder et al. (2015:2), who blended morphological data with their molecular data, used "46 protein-coding loci and 35,673 characters for each of 161 taxa," or data from 46 genes/homolog concepts. By my count, as with Vidal and Hedges (2004), even these larger numbers of homologs remain an insufficient number of homolog concepts from which to conduct a phylogenetic analysis of more than forty-five taxa. In fact, by comparison to most morphological data sets, this is an embarrassingly small number of characters even if they have been grossly subdivided and atomized into thousands of putative states/homologs. If a complete protein is considered as morphology, which it can be, then morphologically, the thousands of states are merely variance within one homolog. Were mammalian morphologists to present a study based on three hundred "characters" derived only from the lower third molar, the paper would most certainly be rejected with the principal criticism being the absence of other homologs and the weighted analysis constructed around a single tooth.

And, most problematically, whether it is two or forty-six characters, we have not even addressed the root concern I have with such studies, and that is whether any of these genes are anything more than symplesiomorphies; that is, they all produce the same proteins with exactly the same function and are thus cladistically uninformative. The challenge, of course, is that phylogeny can be reconstructed without using maximum parsimony. My rejoinder, though, in the context of symplesiomorphy, is that this is overall similarity as opposed to shared derived similarity, and I therefore ask, "But what do we learn from phylogenetic patterns derived from symplesiomorphic data?" Make no mistake, sensible results are arising from groupings of these variations in nucleotide sequences, and by this I mean the groupings/clades arising from molecule-only analyses are not finding *Boa constrictor* to be a derived hydrophiid snake lizard, nor viperids to be the sister to *Cylindrophis* (without question, though, there are also non-sensible results). But I suspect this is phenomenon arising not from some relation to the homolog concept underpinning molecular data sets, but rather

from approximations of overall similarity (amino acid relevant codon sequences) and the overall similarity of differences of the nucleotide in the amplified sequences. Similar approximations arose out of decades of phenetic analyses of morphology where all observations of any and all kinds, from discrete characters to continuous variables, and inclusive of shared primitive characters, were all blended into a single statistical analysis of overall similarity. The result? Sensible hypotheses of phylogeny were found, but these were recovered in the absence of any sense of either homology or homoplasy in a studied suite of characters.

Recently, and following this unconventional approach to homology where the homolog is recognized before the recovery of the phylogeny, Streicher and Wiens (2017) built a data set from 4178 nuclear loci using ultraconserved elements (these are intergene, non-coding sequences of nucleotides greater than 200 bases long that are identical between three or more species) that resulted in a total nucleotide base data set that was approximately 100 times larger than that of Reeder et al. (2015)—they noted an average of 2738.7 UCEs per taxon. These intergene sequences can vary, though there should be a 200-base-long sequence that is invariant. Presumably Streicher and Wiens (2017) aligned these 4178 nuclear fragments using the invariant UCE sequences and then coded variation in the non-coding remainder of the UCE fragment. Again, I ask the question—where is the homolog concept at the root of these hundreds of thousands of base pairs? Patterson (1988) stated quite clearly that homology and similarity are the same thing under the molecular concept of homology, and I believe Streicher and Wiens (2017) follow that method and philosophy. The variance in non-UCE fragments is used as data for phylogeny reconstruction, not the primitive-similarity-homolog of the UCE itself. The primary homolog possibility I discussed above, linked to a protein as the end product of DNA transcription and translation, does not apply to UCEs, as, by dint of what they are, they do not code for proteins. If there are species or clade differences in the sequences bounding UCEs, they might well reflect nothing more than passive accumulation of mutations displaying cluster effects within and between populations surrounding deletions, substitutions, insertions, and so on, in non-coding regions of the DNA. In the end, the inability to detect a primary homolog and to ask a homolog

THE ORIGIN OF SNAKES

concept question of the data being collected mean that the data are not well understood in relation to the goal of the analysis—again, not all observations are data. They must mean something in relationship to the question being asked (see Patterson, 1982; de Pinna, 1991; Simões et al., 2017).

I believe I grasp thoroughly the notion of homology, real or not, as an operational tool for the recovery of sistergroups, closest relatives, and so on—call them what you want—in recovering phylogeny in order to understand the pattern and process of evolution. While I am a cladist and morphologist, I also see the phylogenetic value of non-parsimony, non-synapomorphy modeling approaches to phylogenetic analysis such as Maximum Likelihood methods or Bayesian analysis (for the latter, see Simões et al., 2018a). However, I must admit that I fail to grasp the homolog concept at the root of the characters as generally approached in studies of molecular data for the reconstruction of phylogeny. I do accept, of course, that many groups of organisms do not lend themselves to morphological assessments of their anatomy in the least—single-celled organisms, for example, both eukaryotes and prokaryotes, and so molecular data is essential. However, for other organisms such as snake lizards, both kinds of data and their attendant epistemic and ontic conundra are available, even if they do not compare well right now. Thus, I see the successes of the Lee (2005a) or Reeder et al. (2015) efforts in using all available data. Again, and to be fair to morphological concepts of homology, not all characters and states are comparable within any one single analysis, as they approach morphology quite differently as well (see Table 6.1 as adapted from Simões et al. [2017], which highlights the variety of character constructions that should not and should be present as character constructions). Therefore, my approach with current data sets composed of observations of anatomy and nucleotide sequences is to use all these data with full awareness of incompatible homolog concepts and search for novelty where it arises.

## STUMBLING IN THE PRESENT: A BRIEF HISTORY OF SNAKE LIZARD PHYLOGENY

Before returning to the importance and impact of the Fossil Backbone to hypotheses of snake lizard phylogeny, it is critical to understand the intellectual stumbling blocks that have been inherited through more than 150 years of discourse on snake lizard ingroup and outgroup phylogeny. Through most of the history of study of lizard phylogeny, until Estes et al. (1988), none of these varied analyses were undertaken using Hennig's (1966) philosophy and method of cladistics analysis. For the balance of those pre-Hennigian studies, this is not a criticism, as they were undertaken using the best tools and approaches of their day. And today, most modern molecular phylogenies are explicitly not cladistic analyses (see above), but still, when I think of those studies and their hypothesized phylogenies, I think critically in terms of sistergroups supported by synapomorphic character distributions. For me, this is an important distinction worth communicating because a phylogeny given by Nopcsa (1923), by Reeder et al. (2015), Gauthier et al. (2012), or Streicher and Weins (2017) must make *sense* (epistemically and ontically) to me in terms of what I understand about the morphology of fossil and living lizards. I therefore, quite consciously, inflict my "knowledge" onto the topologies of alternative phylogenies as a weak test of those ideas. If I find conflict, I have an opportunity to ask new questions and am set on the path of proper falsification tests.

And, again, it should be abundantly clear by now, but, without apology, I am a morphologist and use that data set to address questions of sistergroup relationships and subsequently phylogeny. Naturally, because the fossil record is 100% morphological data, my predilection is toward phylogeny statements that include fossil taxa and their character data. It cannot be understated, so I will restate it here— phylogenetic relationships derived from morphological character states rely 100% on the quality of the character state codings, character constructions, and proposed primary homologies (Simões et al., 2017). What does this mean? It means that if garbage goes into the assessment of primary homologies (*sensu* de Pinna, 1991), then garbage comes out. If authors do not know snake lizard and non-snake lizard anatomy, then they will construct nonsense primary homologies; if they have inherited and uncritically continue to use characters for which there is no homolog concept at the root of the construction, then the character, its states, and its line in the matrix will be empirically empty (e.g., length of skull to length of four dorsal vertebrae means nothing as a homolog concept and neither will the 0s and 1s representing it in the matrix); and, if characters are poorly thought out except to create simplistic and incorrect primary homologs that support desired *a priori* groupings within a cladogram, then nothing is learned in the least. My exhortations are simple—(1) authors should be experts on the biodiversity, comparative anatomy, and paleontology of the group they are systematizing, and (2) their work must be thorough, focused, scholarly, and deeply concerned about doing the empirical heavy lifting when it comes to the conceptualization, construction, observation, and scoring of each and every terminal taxon for each and every character (see Simões et al., 2017, 2018b; Laing et al., 2017).

Here I will review, and review briefly, the long history since Cope (1869) of conflicting phylogenies of snake lizard sistergroup relationships with various lizard clades, followed by an equally brief review of snake lizard interrelationships. I will admit upfront that this is not an exhaustive survey of a rather lengthy literature, but rather stands as my selection of highlights and meaningful points of controversy. I will profile phylogenetic hypotheses generated using morphological data sets, some of which are based only on fossils, some without fossil taxa included, and some where the morphological data set includes both fossil and modern taxa. I will also highlight phylogenies derived from only molecular data, and those that have used both to discover their resultant tree topologies. A variety of parsimony and non-parsimony approaches to data analysis, as well as some precladistic notions of squamate and snake evolutionary relationships, underpin all of these hypotheses. Not too surprisingly, these theoretical underpinnings on method are also blended with deep intellectual and structural biases (burdens) that influenced the construction of characters in those studies. Again, this cannot be an exhaustive review, as there are simply too many phylogenetic studies in the primary literature (i.e., to those authors whose results are excluded, this is not because you were ignored, only triaged based on what I felt I needed to say and review). Thus, I have elected to present phylogenetic schemes that display the range of relationships that have been used as the basis for constructing third-, fourth-, and fifth-order statements on origins, adaptive radiations, functional morphology, and so on,

that underpin the discussion and debate on snake lizard evolution.

## SNAKE LIZARD PHYLOGENY: PROBLEMS OF SISTERGROUP AND INGROUP RELATIONS

*The First Big Picture:* Cope (1869) (see Figure 1.2)—Cope started it all, so he is first, of course, in the brief survey. Though I reviewed his phylogenetic scheme in Chapter 1, it merits brief recounting here as the start point for this section of this chapter. Cope (1869:258) wrote, when speaking about the fifteen characters defining his Pythonomorpha (the extinct group, including all mosasaurs and their kin), that "On account of the ophidian part of their affinities, I have called this order the Pythonomorpha." The "ophidian part of their affinities," as he put it, was in reference to the seven anatomical features he found were shared between snake lizards and mosasaurs to the exclusion of all other lizards. Two of his fifteen characters were unique to pythonomorphs, while the other six were more similar to typical lizards than they were to snake lizards. Cope was explicitly hypothesizing affinities based on shared anatomical features and continued to argue for a closer relationship between snakes and mosasaurs to the end of his life, but never once argued that snake lizards had evolved from mosasaurs (Cope, 1895a,b, 1896a,b).

*Origins Scenarios and Scolecophidians as derived snake lizards:* Janensch (1906)—Janensch's (1906) study presented a redescription and characterization of the Eocene aquatic snake *Archaeophis proavus* Massalongo, 1859. In addition to his description, which was very good, he also presented a scenario for the terrestrial origin of snakes and absolutely no hint of a phylogenetic diagram of any kind. From a translation of Janensch (1906:27) given by John Scanlon, "they probably developed from an unknown terrestrial lizard, not adapted to aquatic life." He was certain that Cope (1869), Kornhuber (1873, 1893, 1901), and Nopcsa (1903) were wrong in the anatomies they suggested pointed to a close relationship between snake lizards and the Mesozooic aquatic lizards, dolichosaurs, aigialosaurs, and mosasaurs. Still, Janensch (1906), in the absence of a phylogenetic hypothesis, was unable to suggest even what kind of "unknown lizard" he envisioned. Making matters worse, he continued to employ the top–down archetype approach, noting that (Janensch, 1906:27): "If we thus see that the living snake-like lizards are exclusively land animals, we must assume that their predecessors gained their body form on the land, but not in water." Moving from his modern-archetype-provides-insight-on-the-ancient-past-approach, he then moved to an adaptationist form of argumentation (Janensch, 1906:27–28):

> Of the forerunners of snakes, it may hence with great probability be assumed that the acquisition of the snake type is to be attributed to adaptation to residence on ground covered with thick vegetation, perhaps also a burrowing way of life, or both conditions at once.

Importantly, though, most of Janensch's (1906) musings favor a terrestrial origin for snake lizards, but he does explore the notion of scolecophidians, his typhlopids, and glauconids, as highly differentiated as fossorial forms, and thus unlikely to be the most primitive snake lizards. Indeed, he goes even further and considers them not to be developed secondarily from "snake living above ground." In my opinion, this is the key and largely overlooked portion of Janensch's (1906) consideration of snake phylogeny. In fact, it is prophetic and consistent with the previous several hundred pages of this treatise where I have argued extensively that there is nothing primitive about scolecophidians in the least.

*Fossils, Sisters, and Cousins:* Nopcsa (1908, 1923) (see Figures 1.3 and 1.4)—Franz Nopcsa followed Cope with two hypotheses, the first of which (Nopcsa, 1908) was an "origins" of snake lizards hypothesis (see Figure 1.3), and the second of which, published in Nopcsa (1923), professed both a sistergroup hypothesis as well as ingroup relationships of what Nopcsa saw as the major groupings of both fossil and living snakes. Nopcsa's (1908) hypothesis proposed that snake lizards were derived from within the fossil dolichosaurs (Owen, 1850; Caldwell, 1999b, 2000b) and that that group shared an unrecognized common ancestor with a lineage known as Varanidae and a second lineage that included Aigialosauridae and Pythonomoprha, within what Nopcsa (1908) called the "Platynota Mesozoica." Nopcsa (1923) revised his earlier idea of deriving snake lizards from dolichosaurs, and instead (see Figure 1.4) hypothesized that "Mesoleptinae and Dolichosaurinae" were a group of sorts that shared a common ancestor with his snake lizard group (the Cholophidia) and the Mosasauridae; that common ancestor was

to be found at a three-taxon node referred to as Aigialosaurinae. Nopcsa's (1923) fossil snake lizard grouping, the Cholophidia, was composed of the snake lizards he was aware of (*Archaeophis provaus*, *Pachyophis woodwardi*, *Palaeophis toliapicus*, and *Simoliophis rochebrunei*) and itself gave rise to the Angiostomata (scolecophidians) and Alethinophidia. Nopcsa (1923) clearly moves the concept of phylogeny forward with his presentation of the phylogenetic tree and explicitly includes both fossil and modern forms in his consideration of phylogenetic relationships. Unfortunately, this is the last time for a long time that any scholar of snake lizard phylogeny elects to do so.

*"Paleotelic Values/Weights" and Early Tree Building:* Camp (1923) (Figure 6.1)—The nearly 200 published pages that Camp (1923) devotes to his "Classification of the Lizards" is a monumental work of early twentieth-century classification. But, like all works of his time, the treatise is deeply constrained in its analysis of phylogeny by the primacy of the classification over a phylogeny, thus supporting the use of paraphyletic groups and shared, primitive characters. Like all studies of its time where the goal was a ranked classification, the philosophical and methodological effort was on binning species into higher taxa based on qualitative summaries of "uniqueness," as though uniqueness and sum total differences amounted to some process or goal of evolution. From his classification scheme, Camp identified two equivalent suborders, Serpentes (snake lizards) and Sauria (non-snake lizards). Phylogenetic hypotheses were not ends unto themselves, but rather served up as individual statements presented inconsistently throughout the text, culminating in Camp's (1923:333) "Chart Illustrating the Phylogeny of and Classification of the Lizards" (Figure 6.1), which was principally used to illustrate his newly developed concept of "paleotelic weights" (Camp, 1923:332). I will discuss Camp's paleotelic concept shortly, but want to first note that, consistent with one of the major themes of this book, Camp's Suborder Serpentes are most certainly lizards, as they are nested quite firmly in the middle of his well-laid-out branching phylogeny as the sistertaxon of his Varanoidea + Mosasauroidea/ Pythonomorpha. I note an important anatomical understanding from Camp (1923) that shows itself in his phylogeny, though it is not so well laid out in his discussion of classification, and that is that he found no anatomical similarities suggestive

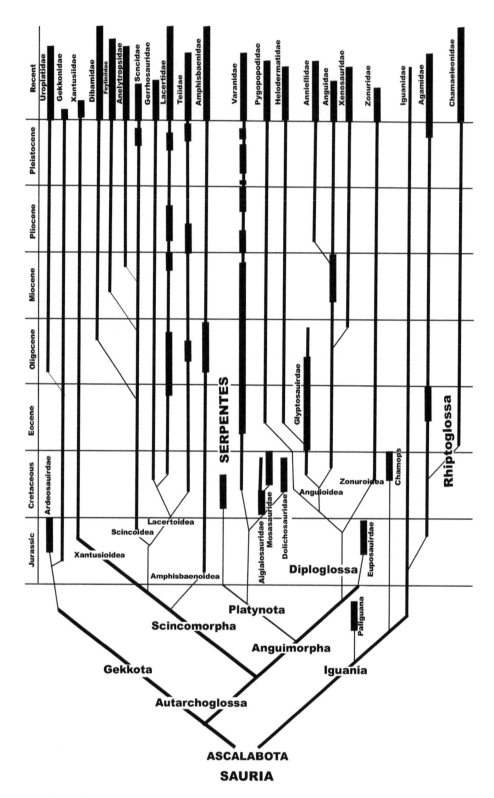

Figure 6.1. "Chart of Phylogeny" redrawn from Camp (1923).

THE ORIGIN OF SNAKES

of a strong phylogenetic signal between snake, amphisbaenian, and dibamid lizards other than limblessness that overwhelmed his estimation of shared features with anguimorphans. As such, Camp (1923:359) held that snake lizards are derived anguimorphan lizards of some kind: "I cannot regard the Serpentes as anything but highly modified anguimorphine lizards near the platynotid stock." While it is impossible to extract that phylogenetic hypothesis from his proposed classification scheme (Camp, 1923:297), he does present very briefly, remembering of course that snake lizards were not detailed in his treatise, as they were not "lizards," an origins scenario for snake lizards (Camp, 1923:313):

> The snakes appear to have arisen from angui-morphid, grass-living lizards...burrowing snakes (Typhlopida and allied families) parallel the burrowing lizards in many profound ways and would seem to be derived from autarchoglossid stock for that reason and because of the paleotelic characters they hold in common with the Anguimorpha.

Despite Camp's (1923:302) claim that snake lizards are lizards derived from within anguimorphans, in the remainder of the text he only wrote 338 words under the heading "Suborder Serpentes" but did provide fourteen examples of anatomical features shared between snake lizards and the various groups of lizards he examined. However, there is no explicit or detailed examination of snake lizard classification or anatomy even though he considered the Suborder to be derived from within his Saurian Section Anguimorpha, Subsection Playnota.

Though origins scenarios for various lizards, including snake lizards, abound in Camp (1923:332), he did build his chart of phylogeny around what he termed characters of "paleotelic value":

> (1) by the distribution of such characters in paleontological sequences, (2) by the time of appearance in the ontogeny, (3) by the possible vestigial or non-functional nature of the structure in the adult...(4) by the degree of complexity of the structure, indicating with reasonable assurance whether or not reversion may have taken place.

If I understand Camp's (1923) approach to synthesizing stratigraphic position (i.e., ancestors had to be found in older rocks than their descendents), the ontogenetic timing of a feature (i.e., primitive features would appear in ontogeny first—a Haeckelian perspective), number of retained vestigial structures (which I interpret to indicate time evolving away from an ancestor and accumulating vestigia along the way), and complexity (as an index of stability against reversals), he was attempting to calculate evolutionary rates by calculating "primitiveness" versus "specialized" in order to understand, phylogenetically, "ancientness" or "antiquity" of a lineage or taxon. He was ranking features and traits to assess antiquity of a taxon, not their primitiveness, and thus understand their importance either to the bottom or top of his phylogeny (Camp, 1923). Moody (1985) argued that the essence of Camp's (1923) approach to phylogeny reconstruction was bordering on cladistics, but I am not so convinced. My interpretation is that he was looking to find a measure of ancientness of a taxon (species, family, order, etc.) by peeling away the noisy overprint of specializations such that a seemingly specialized lizard could be revealed to be a member of an ancient lineage. He listed thirty-four characters, in order of their paleotelic weight (most value for detecting antiquity at number one [three complete branchial arches], least at thirty-four [no endolymphatic sacs]). It is from this calculation of paleotelic weights/values, even though Serpentes was not a lizard, that Camp (1923) was able to conclude that Serpentes is a kind of highly modified anguimorph. Such weights on his most ancient characters (the "paleo" in "paleotelic"), and less to zero weight on more recent features sounds not all like either symplesiomorphies or synapomorphies, and certainly there was nothing to suggest that he could calculate a cluster of cladograms or his "charts of phylogeny" that then required the application of parsimony as a selection criterion. I do agree with Moody (1985), though, that Camp (1923) was the only place where he ever discussed his paleotelic values, weights, and characters, and can only conclude, as did Moody, that this was the end of his methodological musings and he went forward from there to more practical and empirical contributions.

As a curious footnote to Camp (1923), I want to add that his "Chart" is referenced in Estes et al. (1988:141) in the later authors' caption for their figure 7 as a "skiogram." Camp (1923:352) never refers to it as such and only uses the word "skiogram," which should be "skiagram," once in the text in reference to a radiogram image of the pelvis of a skink. His chart of phylogeny, while built on his assessment of primitive and derived characters, was nevertheless constructed with regard to his overall assessment of characters and produced a phylogeny that has not been changed substantially by modern cladistic analyses of morphology and molecules. I only regret he did not treat Serpentes as lizards and explore his paleotelic weights with regard to the ingroup phylogeny of snake lizards—that would have been interesting.

*Origins Stories*: Bellairs and Underwood (1951) and Underwood (1967)—Bellairs and Underwood (1951), in their classic work entitled *The Origin of Snakes*, choose to follow Camp's (1923) classification scheme regarding the Suborder Serpentes and the Suborder Sauria, which also means they also did not deviate from the concept that snake lizards are distinct enough from lizards to be considered an equivalent suborder based on "uniqueness." They created a grade-driven phylogram of sorts in which all Sauria and their Serpentes were derived in a massive cluster from an ancestral "Eosuchian" grade of reptilian organization. In fact, based on their diagram (Bellairs and Underwood, 1951: fig. 8), it is not even clear that they regarded their Squamata/Lacertilia as monophyletic, but rather that the numerous lineages grading into their "Eosuchia" could well have evolved independently of each other from various "eosuchians." I do not believe from what they wrote that this is a necessary hypothesis to be drawn from their work, but the specifics of phylogenetic hypotheses required of modern systematists were of no consequence at the time of this work and so their figure 8 is open to interpretation as it is largely uninformative (the value of Hennig's [1966] work to making systematics as empirical and deductive as possible cannot be overstated).

Bellairs and Underwood (1951:223) felt that the problem of snake lizard sistergroup/closest relative relationships was problematic for understanding origins:

> Turning from the rather unprofitable study of the relationship between the snakes an individual taxonomic groups of lizards, we may now consider the question of the probable habitus of the snake ancestors.

As a result, they chose to propose an adaptive origins scenario for snake lizards in the absence of hypothesizing a closest non-snake lizard relative (which is, as I have stated over and over again, impossible). To support their origins hypothesis with some form of phylogenetic hypothesis, they followed Mahendra (1938: fig. 3) and presented a branching diagram phylogeny of the interrelationships of living forms with leptotyphlopids and typhlopids as successive lineages at the base of the tree (Bellairs and Underwood, 1951: fig. 9). From the summary of their paper (Bellairs and Underwood, 1951:232):

> 7. The affinities of the more primitive families of living snakes are briefly considered; it is suggested that the primitive burrowing snakes are to some extent representative of the early stages in the history of the group.

As this origins scenario and its inductive reasoning in the absence of a non-snake lizard sistergroup hypothesis has already been discussed in Chapter 5, I will not revisit that matter again here, except to say that Bellairs and Underwood (1951) did not explore the origins of snake lizards from foundations of phylogeny, but rather from the framework of adaptationist scenarios. Leptotyphlopids and typhlopids were considered primitive and thus the origin of snake lizards was from a burrowing ancestor. Even though Bellairs and Underwood (1951:194) could not identify a suitable snake lizard ancestor among lizards, they did regard snake lizards as likely being true lizards:

> Although the saurian origin of snakes has occasionally been questioned...it seems clear that the snakes show a much closer relationship with the lizards than with any other group of reptiles.

And:

> Indeed, Camp (1923) has pointed out that the characters separating the lizards from the snakes do not greatly exceed the range of differences found among snake-like lizards.

While Bellairs and Underwood (1951: fig. 9) expressly avoided positing their origins scenario for snake lizards via reference to a phylogenetic hypothesis, with leptotyphlopids at the base of their tree, followed by typhlopids, they concluded that (Bellairs and Underwood, 1951:229):

> On the phylogenetic hypothesis outlined, the primitive burrowing snakes may be regarded as surviving representatives of the earlier stages

in the evolution of the whole group which have failed to acquire those essential modifications necessary for a successful return to epigean life.

For Bellairs and Underwood, the modern assemblage of scolecophidian snake lizards were living fossils that had failed to re-evolve or re-elaborate the necessary secondarily terrestrial anatomies that all the other surface-living modern snake forms had managed, somehow, to accomplish (consistent with Mahendra [1938]). They considered these scolecophidians independently evolved from different lineages of ancient blind, burrowing ophidian lineages (for a recent challenge to their monophyly, see the molecular phylogeny of Miralles et al. [2018]). Though the implications of such a hypothesis were never explicitly examined, the logical outcome of such a statement is that none of the modern lineages shared a common ancestor that was itself a surface-living, re-elaborated form, but that, like their blind snakes, may all have evolved independently from ancient blind, below-ground ancestors. Further support for my supposition comes from their hypothesis statement on the Anilidae (Bellairs and Underwood, 1951:229): "The Anilidae appears, especially on grounds of cranial anatomy, to be the most primitive ophidian family now existing." The logic of this statement is immediately problematic considering that these authors reconstructed their phylogeny (fig. 9) with Anilidae sharing a node with Uropeltidae, and, of course, the blind snakes at the base of the tree. Clearly, Bellairs and Underwood (1951) did not consider the paradox they had just created—blind snake lizards represent the ancestral state and are likely derived from independent ancient blind snake lizards of the most basal kind, but show numerous derived features; comparing anilid snake lizards to modern blind snake lizards, Bellairs and Underwood (1951) considered them the most primitive modern snake lizards. How? And, if so, why not reconstruct the phylogeny in their figure 9 to express that hypothesis? I ask these questions rhetorically, of course, because I have no answers for them. Bellairs and Underwood (1951: fig 9) finish their phylogeny with three more families, in the following order: Xenopeltidae, Boidae, and the pythons. The caenophidian radiation, which is not discussed in the text, is illustrated as unresolved, where the Hydrophiidae, Elapidae, Colubridae, and Viperidae are arise

from a grade somewhere within the crown of xenopeltids, boids, and pythons.

To summarize, Bellairs and Underwood (1951) established the adaptationist/modified Lamarckian scenario for the origin of snakes in the relatively complete absence of a phylogenetic hypothesis that does away with Cope (1869) and Nopcsa (1923), while supporting the classification of Camp (1923). Bellairs and Underwood (1951) did not support Janesch's (1906) and Camp's (1923) grass-living anguimorphan ancestor and origin for snake lizards, but rather sided with Walls (1942) on the burrowing ancestry and origin of snake lizards and on the ingroup phylogenetic topology of Mahendra (1938) (see Chapter 5 on the origins issue). Between Walls (1942) and Bellairs and Underwood (1951), in the stated absence of a phylogenetic hypothesis, the effort of which they found "unprofitable," they set firmly into stone the historical burden of a top–down view of snake lizard anatomy where an origins scenario informed them of phylogeny, not the other way around. Somehow they established, against a very confusing backdrop where they stated that *Cylindrophis*-like snake lizards are the most primitive modern snake lizards, that living scolecophidians are "surviving representatives of the earlier stages" that never returned to the surface to live and thus were at the base of their phylogenetic tree. This origins-myth approach to phylogeny has dominated scholarly research on snake lizards, and lizards in general, ever since, and appears in its same powerful form in Underwood (1967, 1970) and Bellairs (1972), which picks up on snake lizard ingroup "classification" and origins where Bellairs and Underwood (1951) finished off. As noted in Chapter 5, both sets of authors were decidedly against the burrowing origins hypothesis becoming dogma, but it was unavoidable, as history has indicated. The adaptationist approach to such internalized myth stories is powerful, and in the absence of a constraining phylogeny was so powerful as a *causa explanans* for the origins of snake lizards, and thus obviously their phylogeny and anatomical features, that even in the face of contrary evidence, future scholars were trapped by their preconceptions of the correct answer (e.g., Haas [1980a]).

As was noted by Caldwell (2007a), the anatomy of *Pachyrhachis* was not that of a scolecophidian, and as scolecophidians were the primitive snake lizards for Haas, as they had been for Bellairs and Underwood (1951), and because scolecophidians were primitive and burrowing, and snake lizards evolved from burrowing lizard ancestors, then a 95 million-year-old fossil snake lizard could not be marine and thus *Pachyrhachis* could not be a snake lizard. With the appropriate predicates in place, the conclusion was unavoidable. Thus the problem with "burden" as I have explored it in this book—it is powerful, pervasive, dogmatic, and destructive, as it limits what we see and conclude because we think we already know that what we know is true in every absolute sense of the word.

While the problem of historical and intellectual burden does not diminish with the advent and application of cladistics analyses, the use of such a tool ultimately requires the serious reconsideration of data, characters, states, homolog concepts, philosophy, and method. All of this is good. Most important to my mind, though, is the expectation that arises during the era of the cladists that a phylogenetic hypothesis is the foundational requirement and justification for any and all subsequent third-, fourth-, and fifth-order hypotheses constructions, that is, origins scenarios, adaptive frameworks and radiations, and so on.

*Neither Sistergroups nor Ingroups:* Estes et al. (1988) (Figure 6.2a)—The collaborative work by Richard Estes, Jacques Gauthier and Kevin de Queiroz is the break point for hypotheses of lizard phylogeny—it is the first study to use discreet characters and states in a cladistics analysis of their interrelationships. However, despite the progressive use of cladistics, not a single fossil taxon was included in the analysis, and more strange still is that their "conservative tree" (Estes et al., 1988: fig. 6) illustrated snake, amphisbaenid, and dibamid lizards as the termini of branches in phylo-space, that is, not connected to the stem and thus no hypothesis at all. Estes et al. (1988) also did not conduct an ingroup analysis of snake lizards, nor any of their other lizard clade terminals for that matter, which is fine (two of them were at the species level, *Varanus* and *Lanthanotus*, but the others were all "family"-level terminals). It simply means that for the purposes of this discussion, there is not much of value to glean from this work regarding either the sistergroup or ingroup relationships of snake lizards. I mention it here because it is the first cladistics analysis of lizards, inclusive of amphisbaenians, dibamids, and snakes, and in general replicates the pattern hypothesized by Camp (1923).

*Ingroup Analysis and the Primitive Snake Lizard:* Rieppel (1988a,b) (Figure 6.2b)—At the same time as

Estes et al. (1988) avoided the problems created by including fossil taxa and snake, amphisbaenian, and dibamid lizards in their taxon-character matrix, Rieppel (1988a,b) published his thorough and classically constructed *Review of the Origin of Snakes*. Rieppel (1988a,b: fig. 3) is, I believe, the first published cladogram of snake lizard ingroup relationships based on morphology (though admittedly his study as presented did not include a taxon-character matrix, nor metrics or collections of shortest trees, etc.), regardless of taxonomic level, that also included a fossil snake lizard taxon, in this case *Dinilysia patagonica*. I refer to this work as classic in its construction, as it follows very much the "bauplan" so to speak of Camp (1923) and Bellairs and Underwood (1951) in examining in detail a number of soft and hard tissue anatomies and their morphological differences within snake lizards and in some cases through comparison to various lizard groups. Rieppel (1988a,b:89–93) also makes direct comparisons to the anatomy of *Dinilysia patagonica*, devoting five pages to a discussion of its importance to understanding snake lizard phylogeny and evolution and concluding that it be "…inserted as a plesion between the Scolecophidia and the Alethinophidia in the hierarchy of snake classification: it represents the plesiomorphous sister-group of the Alethinophdia."

He also mentions *Pachyrhachis* and the now referred genus, *Estesius*, in the context of a consideration of fossils, but follows Haas (1980a,b) in continuing to treat both taxa as aigialosaurs, not snake lizards (Rieppel, 1988a,b:97–98). Not too surprisingly, Rieppel (1988a,b) follows his predecessors, principally Haas (1962, 1973), but certainly Bellairs and Underwood (1951) and Underwood (1967), and reconstructs Scolecophidia at the base of his tree, followed by *Dinilysia*, then Anilioidea, Booidea, Achrochordoidea, and the Colubroidea.

*Sistergroups, Pachyrhachis and Cretaceous Anguimorphs:* Caldwell and Lee (1997), Zaher (1998), and Caldwell (2000a,b) (Figure 6.3a,b)—Lee (1997a,b) was the first published cladistic analysis to place snake lizards in a close sistergroup position to mosasaurid lizards; admittedly though, this analysis was not a sistergroup-level analysis including all snake and non-snake lizards. Rather, Lee's (1997a,b) hypothesis was restricted to just playtnotans, snake lizards, and mosasauroids and their kin—but it got the ball rolling, and set the stage for the collaborative work undertaken by Lee and Caldwell in 1996 that

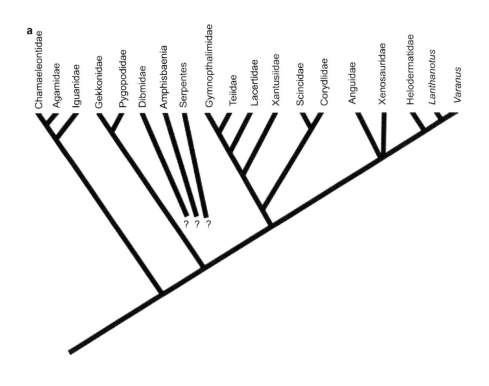

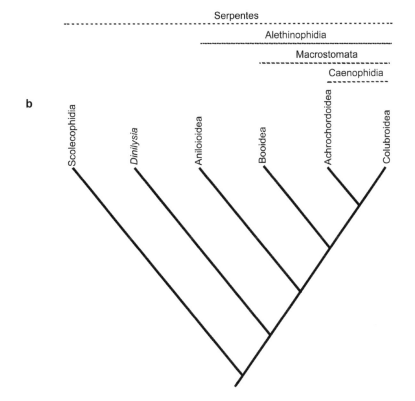

Figure 6.2. Global lizard phylogenetic relationships as hypothesized by (a) Estes et al. (1988); snake lizard ingroup relationships as presented by (b) Rieppel (1988a,b).

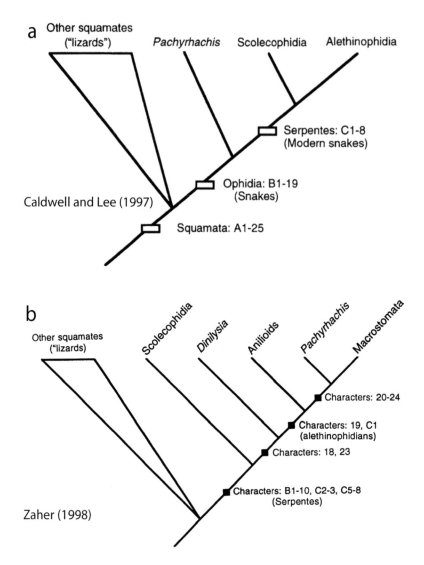

**Figure 6.3.** Global lizard phylogenetic relationships as hypothesized by (a) Caldwell and Lee (1997) with a focus on snake lizard ingroup phylogeny; Snake lizard ingroup relationships as presented by (b) Zaher (1998).

resulted in Caldwell and Lee (1997) on *Pachyrhachis problematicus*. Caldwell and Lee (1997) did not use Lee's (1997a,b) matrix, but rather obtained their phylogenetic hypothesis using a highly modified matrix version of Estes et al. (1988) that also used newly introduced characters from Scanlon (1996) and Lee (1997a,b). As it included representatives of all families and the fossil taxon *Pachyrhachis*, it was the first full cladistic test (taxon-character matrix, parsimony algorithm, presentation of suite of most parsimonius cladograms, etc.) of all lizards including at least one fossil taxon. The core hypothesis arising from their analysis was that scolecophidian snake lizards were not, and I

repeat not, the most basal/primitive members of Serpentes/Ophidia. For the very first time since Cope (1869), or even the ad hoc hypothesis of Janensch (1906), scolecophidians were forced from the bottom of the tree to an uncertain spot among crown snake lizards. The general read of Caldwell and Lee (1997) was that *Pachyrhachis* (Figure 6.3a) was unequivocally the most primitive snake lizard ever, but this was not the case. Instead, the core takeaway message should have been that scolecophidians were far more derived—that the received wisdom was a mischaracterization of their anatomy. That message was unfortunately overlooked in favor of finding every way possible to reinsert them at their proper position at the base of the tree. And, of course, the overarching hypothesis of the sistergroup relationships of snake lizards among all lizards was mischaracterized as an ancestor descendent story, an origins scenario, and for the next few years was generally ignored in favor of arguing to keep scolecophidians at the base of the tree and not hypothesize an origins scenario based on sistergroup relationships, but rather on having reaffirmed that scolecophidians were primitive snake lizards.

The first attempt to reassert the position of scolecophidians at the base of the tree was not long in coming; in fact, it happened almost immediately in the form of a brief reply by Zaher (1998) (Figure 6.3b), who took a fragment of the taxon-character matrix of Caldwell and Lee (1997)—a mere eighteen characters, and created seven of his own. Zaher (1998) did not analyze snake lizard relationships in a global analysis of all lizards, which means he did not retest a single aspect of Caldwell and Lee's (1997) study, nor any component of Estes et al. (1988). The published phylogeny appears similar to that of Caldwell and Lee (1997) (compare Figure 6.3a,b), but this is an artifact of how Zaher chose to represent his cladogram as a phylogeny. To his entirely new and small data set, he coded his select characters based on his interpretations of morphology for scolecophidians, Dinilysia patagonica, Pachyrhachis, and a divided Alethinophidia composed of "Macrostomata" and "Anilioids." However, Zaher (1998) did not bother to address the "pythonomorph" hypothesis, and of course could not with his data matrix, as his sole goal was the placement of scolecophidians at the base of the tree at all costs. While Zaher's (1998) hypothesis is in fact a legitimate matrix in its own right, it was not a retest of Caldwell and Lee (1997), it was a new hypothesis derived from very different data and did not posit a key position—a sistergroup hypothesis for Serpentes within lizards.

Caldwell (2000a,b) rebutted Zaher (1998) at the level of Zaher's (1998) homolog concepts and character constructions, which is the only level of criticism of a phylogenetic hypothesis that actually provides a falsification test of each individual character (see Simões et al., 2017). The result of Caldwell's (2000a,b) reinterpretations of Zaher's (1998) characters and states resulted in three most parsimonius cladograms all of which placed Pachyrhachis at the base of the tree, but, most importantly, consistently placed scolecophidians well within the crown of the modern assemblage, the most important and prescient of which is Caldwell (2000a: fig. 1c), where Scolecophidia and Macrostomata form a clade; the sistergroup is a clade composed of Dinilysia and Anilioids, and Pachyrhachis is reconstructed as the basal sister to those two clades. As I will shortly examine in a bit more detail, the struggle to maintain scolecophidians as the basalmost snake lizards is ironic considering that only a few years later Apesteguía and Zaher (2006) and Zaher et al. (2009) published phylogenetic hypotheses finding the basalmost snake lizard to be Najash rionegrina, a very Dinilysia-like snake lizard and so begins the slow march of scolecophidians up the tree of snake lizards (see Wilson et al., 2010; Longrich et al., 2012; Caldwell et al., 2015) in long argument circling back to the hypothesis of Caldwell and Lee (1997).

*A Story Retold From Ingroup Relationships:* Tchernov et al. (2000), Rieppel and Zaher (2000a), and Rieppel et al. (2003a)—The phylogenetic narrative connecting these three studies and their varied data sets and intentions, is consistent with that initiated by Zaher (1998)—scolecophidians are the most primitive snake lizards and Pachyrhachis and all other fossil snake lizards are some kind of macrostomatan or at least alethinophidian snake lizard. In none of these published phylogenies is there a single attempt made to place Serpentes within a global analysis of all other lizards. The methodological approach, and thus philosophical and empirical assumption, is that snake origins can be divined from nothing more than the identification of the anointed basalmost, most primitive snake lizard from an ingroup analysis. Presumed falsifications of the global lizard sistergroup studies of Lee (1997a,b), Caldwell and Lee (1997), Lee and Caldwell (1998), Lee (1998), Caldwell (1999a,b), and Lee and Caldwell (2000) were consistently announced through the results of ingroup analyses as rendered by Zaher (1998), Zaher and Rieppel (1999a), Tchernov et al. (2000), Rieppel and Zaher (2000a,b), Zaher and Rieppel (2002), and Rieppel et al. (2003a,b). The language of testing and superior hypotheses was given as though the character critiques given were superior Tests of Similarity to those presented in the critiqued works, regardless of the fundamental incompatibility of the claimed falsification test to the data and hypotheses being falsified (Patterson,

1982; Rieppel and Kearney, 2002). My response to those criticisms, and the core of my two-decade-long research program, is what forms the core of Chapter 3 of this treatise and so I will not review it again here. However, the purpose of discussing these studies in this chapter is that they reaffirm the basal position of scolecophidians—ferociously so in my opinion, suggesting most centrally that ancient snake lizards such as *Pachyrhachis* and *Dinilysia*, and then eventually *Haasiophis terrasanctus* (Tchernov et al., 2000) and *Wonambi naarcortensis* (Rieppel et al., 2003b), are derived macrostomatan snake lizards firmly nested within the modern assemblage. While such hypotheses might indeed be accurate, what it also suggests is that the crown of Serpentes is in fact extremely old and that 95 million-year-old snake lizard fossils can quite simply be slotted into the branches and stems of modern diversity, particularly if "well enough is left alone" and scolecophidians remain at the base of the tree.

When Tchernov et al. (2000) described *Haasiophis*, they presented a phylogenetic hypothesis based on two most parsimonius cladograms where *Haasiophis* and *Pachyrhachis* formed a clade at the base of the Macrostomata and where scolecophidians were reconstructed as the most basal snake lizards. Based on this cladogram, they claimed to have falsified the hypothesis of Caldwell and Lee (1997:2012): "As macrostomatan snakes, Haasiophis and Pachyrhachis have no particular bearing on snake-mosasauroid relationships or snake origins." This statement highlights the reason studies like that of Rieppel and Zaher (2000a) require critical assessment. The hypothesis presented by Caldwell and Lee (1997) was a sistergroup hypothesis of the position of snake lizards—all snake lizards, Serpentes, Ophidia, or whatever term is applied to that clade, both fossil and modern—not an ingroup hypothesis of snake lizard phylogeny. The result presented by Caldwell and Lee (1997) did not require *Pachyrhachis* to be present in that data set in order to result in the shortest cladograms they presented. All that was required was a taxon line, that is, Serpentes and its characters as shared with mosasauroids and all other lizards in the Caldwell and Lee (1997) data matrix in order to achieve that result. The only valid falsification test of Caldwell and Lee (1997) would have been to use the same taxon suite they presented but with rescored characters. All studies claiming falsification of the Caldwell and Lee (1997) hypothesis using ingroup tests to find a different basalmost snake lizard

have failed to meet the necessary expectations of falsification tests and can be ignored as falsifications (e.g., Zaher, 1998; Tchernov et al., 2000; Rieppel and Zaher, 2000a). I have made this statement numerous times (Caldwell, 2000a,b, 2007a) and will reiterate it again here. There is no difference between such claims and the simple example I give here: I publish an experimental protocol in which I claim to have observed cold fusion; another laboratory claims to have falsified my result in producing evidence for low-temperature cold fusion using an entirely different experimental protocol. Obviously, the experimental protocols must be the same, particularly for the claimed falsification test which must in fact be more precise and show greater rigor then even the original claim had been able to present.

The Tchernov et al. (2000) hypothesis is a perfectly valid statement of ingroup relationships based on the data employed by those authors, but the macrostomatan sistergroup position of *Pachyrhachis* and *Haasiophis* has nothing to do with the sistergroup relationships of all snake lizards among all lizards. This misconstrued notion of most-primitive known member of a clade as indicative of clade origins is non-science, and, as I discussed it in Chapter 5, will not examine it further here.

Another example, perpetuating the same error, was the study by Rieppel and Zaher (2000a) where they claimed to have revised and falsified Lee (1998), who created a taxon character matrix of 230 characters for 22 terminal taxa inclusive of all lizard families. However, Rieppel and Zaher's (2000a) revision once again did not merely recode Lee's (1997a,b) characters, but also built their matrix from a select terminal taxon list (ten, not twenty-two) that by intention included only *Heloderma*, *Lanthanotus*, *Varanus*, Mosasauroidea, *Sineoamphisbaena*, Amphisbaenia, Dibamidae, Scolecophidia, *Dinilysia*, *Pachyrhachis*, Anilioidea, and Macrostomata. They also added 10 characters to Lee's (1998) 230, thus creating a new matrix of 240 characters scored for ten taxa and a hypothetical ancestor. Snake lizards were found to present the same ingroup topology as that argued for by Zaher (1998) with a clade formed by Dibamids and amphisbaenids as their sistergroup and where *Sineoamphisbaena* was found as sister to this entire clade (this taxon is now recognized to be a lacertoid/teioid of some kind and has nothing to do with amphisbaenids, dibamids,

THE ORIGIN OF SNAKES

or snake lizards (see Kearney [2003] versus Wu et al. [1996]). The *Heloderma–Varanus* clade was reconstructed as the sister to the *Sineoamphisbaena–*Serpentes clade. The outgroup of choice in this case, which varied wildly from that of Zaher (1998) who used twenty-two family level lizards as his outgroup, was a hypothetical ancestor created by following Lee's (1998) lepidosaur model.

In sum total, it is difficult to understand how this phylogenetic scheme was conceived as a repudiation of Lee (1998). A phylogenetic tree, resulting from a cladistic analysis, is based first and foremost on character distributions to form synapomorphy clusters (e.g., clades), creating internal nodes within a tree-form. The second most critical aspect informing synapopmorphy clusters are the taxa possessing the characters that are reconstructed at internal nodes and shared synapomorphically with other terminals at various levels within the cladogram—if you have five terminals, you will have a very different internal tree form than if you have ten terminals. Rieppel and Zaher (2000a) went even further in moving away from an empirical test of Lee (1998) when they used a data set of primary homologs conceived of, as Lee (1998) did, for all lizards including snake lizards, but reduced the terminal taxon sample to reflect primary homolog assignments for only three families from Lee's (1998) sample of twenty-two families. Like Tchernov et al. (2000), Rieppel and Zaher's (2000a) professed falsification of Lee (1998) was based on incompatible tests. Making it worse in my opinion is that it was not clear as to why they assembled the ingroup as they did; that is, why are *Varanus* and Dibamids in an ingroup exclusive of all other scleroglossan lizards? Where are the skinks, lacertoids, anguids, and so on? And why include amphisbaenians and not just simply use the complete data set of Lee (1998) with your own recodings of problem characters, show the results of that test, and then proceed to add ten new characters? Those would be appropriate to the methodology and to claims of falsification or verification. Once again, comparable methodologies produce comparable results, non-comparable ones do not give answers that permit falsification of the hypothesis at the root of the criticism. You can change characters, that is, make character scoring corrections, and then justifiably claim to have falsified hypotheses of phylogeny by a primary falsification of character state scoring

(see Simões et al., 2017). But if you change taxa, you cannot claim to have falsified the results of the study in question.

As a final example of an ingroup analysis geared toward idealizing yet again the ancestral snake lizard archetype, and finding scolecophidians to be at the base of the snake lizard tree, I will highlight the Rieppel et al. (2003b) recharacterization of *Wonambi naarcortensis*. While this study and its anatomical descriptions and interpretations of *Wonambi* was firmly rebutted by Scanlon (2005a,b), Rieppel et al. (2003b) presented a phylogenetic analysis where they concluded that Wonambi was not an Australian madtsoiid, but rather was a misidentified member of the Macrostomata. To find this result, Rieppel et al. (2003b: fig. 14) constructed a small data matrix of forty-five characters from a mixed bag of sixteen fossil and modern snake lizards, some at the species level and others at the level of family, and then used five non-snake lizards (amphisbaenids, dibamids, mosasauroids, *Lanthanotus*, and *Varanus*) to serve as outgroups to their sixteen ingroup taxa. To say the least, this ingroup analysis resulted in forty-five characters that largely had no vector of transformation, as the five outgroup taxa were all state "0" for all characters as scored. Scolecophidians were scored for only nine characters, where a "1" or "2" state was assigned, meaning they at least shared some snake lizard synapomorphies, but overall would be dragged to the base of the tree by the lizard features of no less than five outgroups covering the range of aquatic and burrowing morphologies; it is important to note that Rieppel et al. (2003b: fig. 14) appear not to have required the outgroup to be monophyletic compared to the ingroup and thus amphisbaenids and dibamids are successive branches on a Hennigian Comb leading to scolecophidians.

The value of using these five taxa as outgroups (recall they were all ingroup taxa in Rieppel and Zaher [2000a]) makes no methodological sense as there cannot be any clear character polarities for the sixteen ingroup taxa. That there was any resolution whatsoever in "Serpentes" is surprising, but that macrostomatans were reconstructed as a nearly complete polytomy is no surprise at all considering that resolution of their relationships was not the goal. Other than providing further assurance that scolecophidians were, yet again, basal snakes, a result that was impossible to avoid in relationship to their experimental design, the value and purpose of this phylogenetic hypothesis are unclear.

*Derivatives of Lee and Scanlon (2002a,b)/Scanlon (2006):* Apesteguía and Zaher, 2006; Wilson et al., 2010 (Figure 6.4a,b)—Interest in the paleontology and evolution of snake lizards was seriously stimulated by the "blizzard" (Cundall and Irish, 2008:358) of the debate highlighted previously, with the end result that a number of new specimens from a wide a varied number of localities around the world (e.g., India, Argentina, Australia) quickly made their way into the literature. Equally powerful, and thus very relevant to the science, was the socially polarizing effect of such debates on different research groups. Each sought to use taxon-character matrices that reflected their shared perspective on properly constructed Tests of Similarity *sensu* Rieppel and Kearney (2002). A surprisingly neutral data set that came to sit at the core of a number of studies and hypotheses was that of Scanlon (2006), which interestingly enough was derived from Lee and Scanlon (2002b).

From this data set, two major hypotheses were presented for two very important Mesozoic snake lizards, *Najash rionegrina* Apesteguía and Zaher, 2006 (Figure 6.4a) and *Sanajeh indicus* Wilson et al., 2010 (Figure 6.4b), as well as later revisions and redescriptions of both *Najash* (Zaher et al., 2009) and *Dinilyisa* (Zaher and Scanferla, 2012). Using the core data matrix of Lee and Scanlon (2002b) as built upon in Scanlon (2006), both Apesteguía and Zaher (2006) and Wilson et al. (2010) found scolecophidians to be either more derived than *Najash*, or in a basal polytomy with *Najash* (which means of course in the Wilson et al. [2010] cladograms, *Najash* was more basal than scolecophidians). While the remainder of the hypothesis presented by Apesteguía and Zaher (2006) is the normal alethinophidian to macrostomatan placement of taxa, the Wilson et al. (2010) hypothesis is quite the contrary and presents instead a much overlooked and rather controversial suite of relationships that parallel the clade implied by my Type I *Dinilysia-Najash* and Type I Anilioid skulls. Wilson et al. (2010) find *Xenopeltis* and *Loxocemus* to be outside of Macrostomata, with which I agree, and that madstoiids, modern anilioids, and *Dinilysia* and *Najash* are all nested in and around those other fossil and modern taxa. Scolecophidians are still at the base of their tree in a polytomy with *Najash*,

but things are beginning to move in a positive direction. The Wilson et al. (2010) findings are consistent with observations made by Caldwell and Calvo (2008:361), where we described a new skull of *Dinilysia patagonica* and compared it element for element to *Najash rionegrina*, and then to *Xenopeltis*, *Anilius*, and *Cylindrophis*:

> We expect, contra Apesteguía and Zaher (2006), that future studies comparing *Najash* and *Dinilysia* will find these two Argentinian Cretaceous snakes…to form a clade of large-bodied, big-headed Gondwanan snakes with affinities to the well-known South American…and Australian Madtsoiidae.

And:

> In general terms, the condition observed for *Dinilysia* and *Najash* is strikingly similar to the partial crest system and relatively large stapedial footplate of *Cylindrophis*, *Anilius* and *Xenopeltis*

CALDWELL, PERS. OB.

*Big Data Sets and Sistergroup Analyses:* Conrad (2008) and Gauthier et al. (2012) (Figure 6.5a,b)—As new data from new fossils were impacting the anatomical data sets of snake lizard systematists, so too were the giant data sets of molecular systematists. By the early part of the twenty-first century, these giant molecular data sets of supposedly greater and greater resolution moving toward ultimate truth were producing data sets with 4–5000 nucleotides from two or more gene loci. Two hundred morphological characters could simply not keep up—or so it was thought despite the evidence to the contrary as presented by Lee (2005a) in his critique of Vidal and Hedges (2004) (for details, see below). But despite Lee's (2005a) effort to show that ~4000 base pairs would not "swamp" a significantly smaller data set, the angst still grew in the morphological community. And what was the result…?

Behold the rise of the giant morphological data set! For mammals, the clear winner remains O'Leary et al. (2013) at 4541 phenomic characters; for dinosaurs, at 1500 characters, the prize is held by Godefroit et al. (2013); and for lizards, including snakes, it was Conrad (2008) first at 363 characters, followed by Gauthier et al. (2012) at 610 characters. If number means anything at all, the cladistic analysis of lizards, modern and fossil, is clearly falling behind the work being

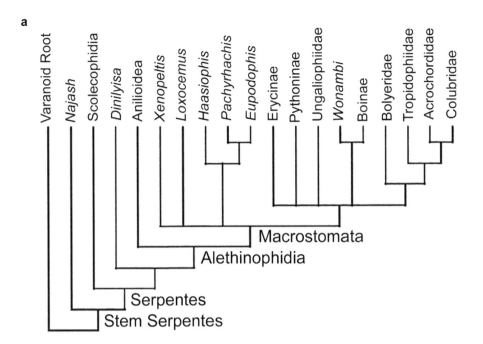

Apesteguia and Zaher (2006)

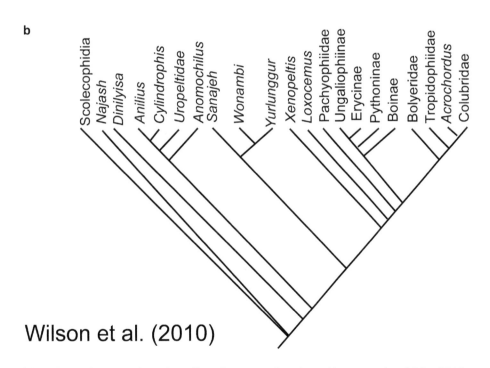

Wilson et al. (2010)

**Figure 6.4.** Snake lizard ingroup relationships. (a) *Najash rionegrina* as hypothesized by Apesteguía and Zaher (2006); (b) *Sanajeh* indicus as hypothesized by Wilson et al. (2010).

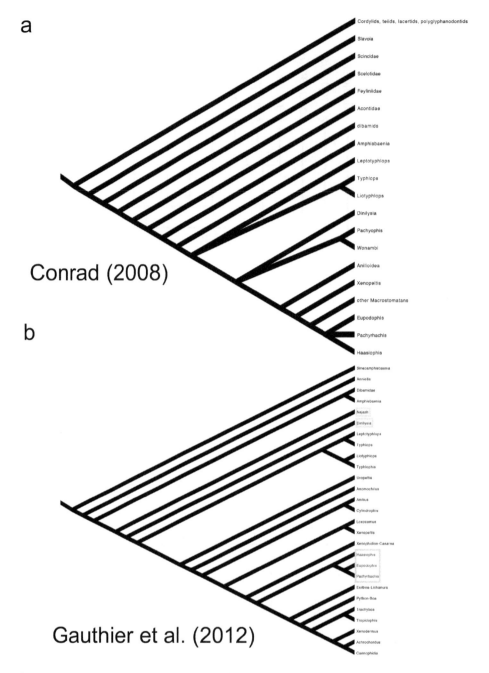

a

Cordylids, teiids, lacertids, polyglyphanodontids
Slavoia
Scincidae
Scelotidae
Feyliniidae
Acontidae
dibamids
Amphisbaenia
Leptotyphlops
Typhlops
Liotyphlops
Dinilysia
Pachyophis
Wonambi
Anilioidea
Xenopeltis
other Macrostomatans
Eupodophis
Pachyrhachis
Haasiophis

Conrad (2008)

b

Sineoamphisbaenia
Anniella
Dibamidae
Amphisbaenia
Najash
Dinilysia
Leptotyphlops
Typhlops
Liotyphlops
Typhlophis
Uropeltis
Anomochilus
Anilius
Cylindrophis
Loxocemus
Xenopeltis
Xenophidion-Casarea
Haasiophis
Eupodophis
Pachyrhachis
Exiliboa-Lichanura
Python-Boa
Trachyboa
Tropidophis
Xenodermus
Achrochordus
Caenophidia

Gauthier et al. (2012)

Figure 6.5. Broad patterns (numerous terminals omitted for clarity) of global lizard phylogenetic relationships as hypothesized by: (a) Conrad (2008); (b) Gauthier et al. (2012).

done on both dinosaurs and mammals. However, number of characters is not everything, a point made most forcefully by Simões et al. (2017:19) in their detailed argument that the quality outweighs quantity regardless of the desire to create gigantic data sets (see also the criticism by Laing et al. [2017] and the rebuttal by Simões et al. [2018b]):

> The notion that morphological data…need to be codified on a "…scale comparable to that for genomic data…" (O'Leary et al., 2013, p. 662)

THE ORIGIN OF SNAKES

is flawed. Reasoned and logically constructed character statements and concepts, regardless of number, will always be superior to large numbers of poorly constructed characters.

Though both Conrad (2008) and Gauthier et al. (2012) presented very large data sets in support of their phylogenies and have been scrutinized for character quality as a result (Simões et al., 2017), they did reconstruct snake lizard relationships within the context of a global analysis of lizards. Conrad's (2008) hypothesis (Figure 6.5a) recovered snake lizards as a derived member of a clade he named the "Scincophidia," nested within his "Scincoidea," which included, not too surprisingly, all scincid, dibamid, amphisbaenid, and snake lizards. The three scolecophidian taxa Conrad (2008) included in his analysis, which are present as distinct genera as terminals, form a clade at the bottom of his "Serpentes"; that is, they are the most basal/primitive snake lizards. He also included six fossil taxa, again at the generic level, and three snake lizard terminals (*Xenopeltis*, Anilioids, and Other Macrostomatans). The sistergroup to the Scincoidea were the Lacertoidea, both of which, along with the Anguimorpha, formed his Autarchoglossa (Conrad, 2008: fig. 61). Problematic character constructions aside (see Simões et al., 2017), Conrad's (2008) selection of terminals from among the modern snake lizard assemblage is curious. I would simply ask: Why break out the three scolecophidian groups at the generic level as a professed test of scolecophidian monophyly, include six fossil taxa, one extant taxon at the generic level, and a second terminal as "Anilioids" when there are only three to code for (*Anilius*, *Cylindrophis*, and *Anomochilus*), and then assume the monophyly of the "Other Macrostomata?" I would guess, based on a close reading of his character state codings, that Conrad (2008) found the polymorphisms within "Other Macrostomatans" intractable and was set on a course to support the received view that the archetypal snake lizard was in fact exemplified by *Typhlops*, *Leptotyphlops*, and *Liotyphlops*, all of which would easily be "pulled," so to speak, to the base of the snake lizard clade by the host of burrowing "Scincophidia" found to be successive sistergroups to Serpentes. However, to be fair to Conrad (2008:137), his conclusions on snake origins were consistent with his phylogenetic results, and, most importantly, remained empirical and thus scientific by not extending beyond his phylogeny

thus miring him in ad hoc scenario constructions based on morphology, paleoenvironment, and so on. I also consider it accurate of Conrad (2008) to be concerned about his finding in the context of convergent anatomy and the complete absence of a supportive suite of anatomies from fossil exemplars of ancient scincophidian snake lizards.

Gauthier et al.'s (2012) character constructions have also been scrutinized by Simões et al. (2017) and need not be reviewed in detail here. However, the pattern of relationships presented in Gauthier et al. (2012) is important (Figure 6.5b), as it reproduces the ingroup results arising from the study of *Najash* as given by Apesteguía and Zaher (2006), also placing *Dinilysia* as slightly more derived than *Najash*. For the first time, a non-Caldwell and Lee-style global analysis of all lizards failed to support the long-held truth that scolecophidians are the most basal snake lizards. It is important to note that this result was recovered even though all snake lizards were nested in what Gauthier et al. (2012) called the Crypteira, a polphyletic assemblage of limbless or limb-reduced lizards including the anguid *Anniella pulchra*, all modern amphisbaenians, modern dibamids, and the fossil macrocephalosaur, *Sineoamphisbaena hexatabularis*. Consistent with Conrad (2008), Gauthier et al. (2012) also avoided ad hoc origins hypotheses for snake lizards even though they did name their limbless clade the Crypteira.

## ENTER THE MOLECULE

While the paleontologists were doing battle over new and old snake lizard fossils, or as Cundall and Irish (2008:358) wrote, exchanging a "a blizzard of literature on a few controversial fossils," the molecular systematists began working away on recovering fragments of genes from an ever-increasing number of extant snake and non-snake lizards. In fact, in the earliest days of molecular data and its use in phylogenetic studies, the molecules used were entire proteins such as albumins (e.g., Cadle, 1988), not nucleotides from nuclear or mitochondrial DNA. There is no doubt that the sheer numbers of claimed data points were intimidating to morphologists, as the literature was riddled with claims along the lines of, "Morphology is dead, long live the molecule" (e.g., Scotland et al., 2003). While bombastic and ridiculous as all data bear on questions of phylogeny, the threat was real, as these ideas were being published in the literature and arriving with confident language couched in terms of the absolutes of logical positivism and with numbers of claimed homologs in the thousands masquerading as homologs, "characters," and "states" all at the same time (this still remains a key problem, in my opinion). While my position on the sample size employed by these studies remains steadfast (i.e., two loci equal two genes, which equals two proteins or two homologs and therefore is decidedly insufficient for phylogenetic analysis, regardless of the number of basepairs used as characters or states), the "fear factor" that morphological data would be swamped by these literally gigantic nucleotide data sets was real. It has changed in recent years (e.g., Jenner, 2004; Lee, 2005a), but usually now bears some mask of special apology to defend it (e.g., MorphoEvoDevo *sensu* Wanninger, 2015), though Reeder et al. (2015) made no such apologies.

Molecules and Morphology I: Vidal and Hedges (2004) (Figure 6.6a,b) versus Lee (2005a) (Figure 6.7a)—Vidal and Hedges's (2004) molecular phylogenetic analysis of lizard interrelationships was a groundbreaking study for several reasons, and thus also a point of remarkable controversy. First, these authors recovered a phylogenetic pattern where iguanian lizards moved from the base of the tree to a position nested high in the

crown as the sistertaxon to snake lizards with anguimorph lizards as sister to that clade (for morphologists, this bordered on heresy). Second, they used only nucleotide data from two genes (C-mos and RAG-1) for a grand total of ~4000 sites and completely ignored morphological characters except where pre-existing morphology-based phylogenies provided suggestions on the position of fossil snake lizards and mosasaurs. Vidal and Hedges (2004) produced their consensus molecular phylogeny (Figure 6.6a) from their posterior Bayesian probability analysis; mapped onto this consensus phylogeny are the bootstrap values obtained from their maximum parsimony (MP), maximum likelihood (ML), and minimum evolution (ME) consensus trees. They also presented a second phylogeny (Figure 6.6b) that included two fossil lineages: pachyophiid snake lizards and mosasaurid lizards. Serpentes was monophyletic but unresolved, and mosasaurs were inferred to be anguimorph lizards whose phylogenetic relationships were perfectly resolved as the sistergroup to living varanid lizards with anguids as basal. Stated support for these relationships was literature based, that is, pachyophiids are snake lizards of uncertain relationship (Caldwell and Lee, 1997 versus Tchernov et al., 2000), and mosasaurs are derived aquatic varanoids (Lee, 1997a,b; Rieppel and Zaher, 2000a). From this phylogenetic sketch (it is hard to call it an empirically supported

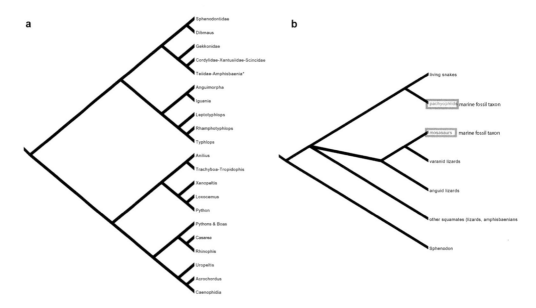

**Figure 6.6.** Global lizard phylogenetic relationships as hypothesized by Vidal and Hedges (2004). (a) Snake lizard ingroup and sistergroup relationships; (b) sketch phylogeny including pachyophiids and mosasaurs of Vidal and Hedges (2004).

THE ORIGIN OF SNAKES

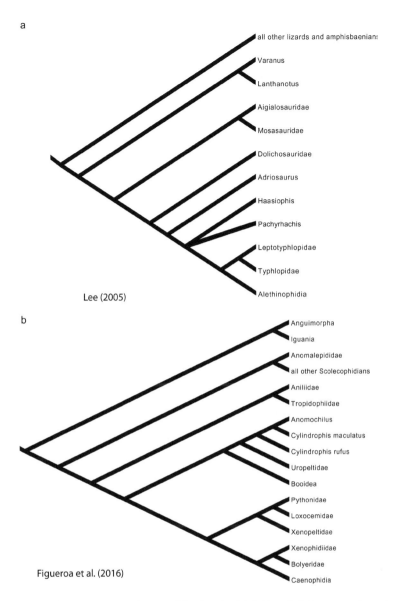

a

all other lizards and amphisbaenians
Varanus
Lanthanotus
Aigialosauridae
Mosasauridae
Dolichosauridae
Adriosaurus
Haasiophis
Pachyrhachis
Leptotyphlopidae
Typhlopidae
Alethinophidia

Lee (2005)

b

Anguimorpha
Iguania
Anomalepididae
all other Scolecophidians
Aniliidae
Tropidophiidae
Anomochilus
Cylindrophis maculatus
Cylindrophis rufus
Uropeltidae
Booidea
Pythonidae
Loxocemidae
Xenopeltidae
Xenophidiidae
Bolyeridae
Caenophidia

Figueroa et al. (2016)

**Figure 6.7.** Broad patterns (numerous terminals omitted for clarity) of global lizard phylogenetic relationships as hypothesized by (a) Lee (2005a); snake lizard ingroup relationships as hypothesized by (b) Figueroa et al. (2016).

hypothesis, as they simply "drew" in mosasaurs and pachyophiids were they saw fit), Vidal and Hedges (2004) proposed a snake lizard origins scenario, which, as stated in the title of the paper, was, "Molecular evidence for a terrestrial origin of snakes." While I would not disagree that at some point in the ancient past snake lizards were terrestrial tetrapods, the empirical quality of Vidal and Hedges's (2004) second phylogeny is virtually non-existent, and this is a key problem if this is their justification for their major conclusion. No data were used to create that phylo-sketch, merely a

pencil and reference to the literature as supporting verification of their choices, making that sketch absolutely unfalsifiable and thus rendering their conclusions on origins non-science.

Responding to a variety of points, Lee (2005a) presented a critique of Vidal and Hedges (2004) using their molecular data set and simultaneously analyzing it with his own morphological data set (followed by successive studies [Lee, 2005b,c]). He first set about to constrain relationships in his morphological analysis around "the molecular backbone" for the modern assemblage as presented in Vidal and Hedges's (2004) first hypothesis (Figure 6.6a) and then ran the analysis with all fossil and modern taxa in his morphological data set (Lee, 2005a: fig. 1b), with the result that snake lizards were recovered within a pythonomorph clade ((Aigialosauridae, Mosasauridae) (Dolichosauridae (Adriosaurus (Pachyrhachis, Haasiophis, Serpentes)))) that was itself the sister to Iguania. Anguimorphs formed an Hennigian comb leading to the Iguanian-Pythonomorph clade. Lee's (2005a) combined evidence analysis included a much more detailed sampling of snake lizards and found a similar tree topology (Figure 6.7a) to that of his molecular backbone constrained morphological data only analysis. Lee's (2005a:227) major conclusions from his analysis were as follows:

> ...if the relationships between living taxa are constrained to the proposed molecular tree, with fossil forms allowed to insert in their optimal positions within this framework, mosasaurs cluster with snakes rather than with varanids...the molecular data do not refute the phylogenetic evidence for a marine origin of snakes.

What is important to understand about the Lee (2005a) study, which also was a focal point of Caldwell (1999a,b), Lee and Caldwell (2000), and Palci and Caldwell (2007), is that the relationships of snake lizards with mosasaurs is not consequential (mosasaurs are not some direct ancestor to snake lizards), but rather that mosasaurs are included within a radiation of Mesozoic lizards that includes a large number of much smaller-bodied, limb-reduced, axially elongate, and pachyostotic forms including dolichosaurs, pontosaurs, and adriosaurs. The hypothesis presented by Lee (2005a) followed Caldwell (1999a) and Lee and Caldwell (2000) in including those fossil lizards as terminal taxa. None of the studies rendering critiques of the Caldwell and Lee's (1997) hypothesis have ever included any of these fossil marine lizards, even though they have been critical components of the data set of Caldwell and Lee in their various studies. They were also missing in Vidal and Hedges (2004). These experimental oversights are illustrative of the problem of appropriate test of falsification and criticism—there must be comparable experiments.

In the case of what Palci and Caldwell (2007: fig. 4; 2010) described as the Ophidiomorpha

(dolichosaurs to Serpentes), as all of these terminal taxa are aquatic, going backward down the tree from Serpentes, an aquatic/marine origin scenario is unavoidable as a logical possibility. As noted by Lee (2005a) (Figure 6.7a), there must still have been a terrestrial phase in the origin of snake lizards, but it would be back at a point in time and below the node shared in common, at the origin of the pythonomorph lineage from within an even more ancient anguimorph radiation.

*Scolecophidians Are Not Monophyletic:* Figueroa et al. (2016) (note: see also the recent study by Miralles et al. [2018]) (Figure 6.7b)—I have highlighted the maximum likelihood analysis of Figueroa et al. (2016) because that study recovered a phylogenetic hypothesis not supportive of the monophyly of the Scolecophidia (Figure 6.7b). However, as a counterpoint, the data as analyzed also resulted in an odd configuration for anilioid snake lizards, where they too were not a monophyletic assemblage with *Anilius* in a clade with Tropidophiidae, which were basal alethinophidians, while *Cylindrophis*, *Anomochilus*, and Uropeltidae formed a clade with *Xenopeltis*, *Loxocemus*, and Pythonidae that itself was sister to "boiids." While the non-monophyly of the scolecophidian snake lizards makes sense to me, this secondary finding does not. However, again, data sampling and homolog concepts mean everything—Figueroa et al. (2016) constructed a data set of 1745 terminal taxa and +9000 nucleotides from ten different gene loci that include C-mos and RAG-1, plus eight other genes representing a mix of mitochondrial, ribosomal, and nuclear loci. Again, while basepairs were numerous, the total number of homolog concepts still only number ten, which remains an infinitesimally small number of characters from which to reconstruct the relationships of 1745 terminal taxa. To make matters worse, some taxa were incompletely represented in terms of gene coverage within the available sequences; that is, *Anomochilus* was represented by only two of the ten gene loci and thus presented an enormous amount of missing data to the analysis.

*Molecules and Morphology II:* Reeder et al. (2015) (Figure 6.8a)—This study, in contrast to almost all recent studies since Vidal and Hedges (2004), followed the principle of all evidence as espoused by Lee (2005a,c), and followed up on by Simões et al. (2018a) by simultaneously analyzing molecular and morphological data. Reeder et al. (2015) used a modified and slightly enlarged

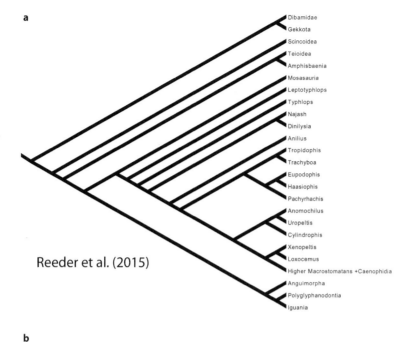

a

Dibamidae
Gekkota
Scincoidea
Teioidea
Amphisbaenia
Mosasauria
Leptotyphlops
Typhlops
Najash
Dinilysia
Anilius
Tropidophis
Trachyboa
Eupodophis
Haasiophis
Pachyrhachis
Anomochilus
Uropeltis
Cylindrophis
Xenopeltis
Loxocemus
Higher Macrostomatans +Caenophidia
Anguimorpha
Polyglyphanodontia
Iguania

Reeder et al. (2015)

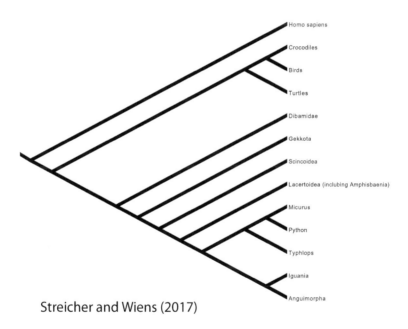

b

Homo sapiens
Crocodiles
Birds
Turtles
Dibamidae
Gekkota
Scincoidea
Lacertoidea (including Amphisbaenia)
Micurus
Python
Typhlops
Iguania
Anguimorpha

Streicher and Wiens (2017)

Figure 6.8. Broad patterns (numerous terminals omitted for clarity) of global lizard phylogenetic relationships as hypothesized by: (a) Reeder et al. (2015); (b) Streicher and Wiens (2017).

version of Gauthier et al. (2012) (691 characters), to which they added a significant molecular data set (nucleotides for 46 genes) for 161 modern lizards and 49 fossil taxa. In contrast to Lee (2005a,c), they found support for Vidal and Hedges (2004) finding that Iguania nested within the crown of lizards, not at its base. Serpentes was found to be the sistergroup position of their "Mosasauria" (Figure 6.8a), while that clade was in a larger clade inclusive of Anguimorpha, Polyglyphanodontia, and Iguania—this superclade of crown lizards was Reeder et al.'s (2015:1) "Toxicofera":

These results further demonstrate the importance of combining fossil and molecular information… Thus, our results caution against estimating fossil relationships without considering relevant molecular data, and against placing fossils into molecular trees (e.g. for dating analyses) without considering the possible impact of molecular data on their placement.

The elegance with which they implore their fellow scholars to make use of all the data is commendable, to say the least. As I think I have made clear with molecular data sets, on their own, there are still too few homolog concepts in current data sets to satisfy my requirement for sufficient empirical data to at least resolve all internal nodes. I recognize that each base pair is treated as a distinct character and so it appears as though there are, for example, 10,000 characters with one of four states, but I continue to disagree, as the homolog concept cannot be identified in such a character, only overall similarity or difference in the protein-coding gene nucleotide sequence, assuming alignments are perfect of course. However, as a kind of balance to my concerns, Reeder et al. (2015) address some of these issues by using all evidence together, fossil and modern, and morphology and molecules, resulting in hypotheses that make sense.

*Molecules and Absolute Answers*: Streicher and Wiens (2017) (Figure 6.8b)—As discussed above, Streicher and Wiens (2017) presented a phylogenetic hypothesis derived from the analysis of a data set of nucleotide sequences obtained only from Ultraconserved Elements found in nuclear DNA—they did not use any other sequence data, nor, of course, did they follow Reeder et al. (2015) and use any morphological data. In contrast to the other studies highlighted in this chapter, Streicher and Wiens (2017) did not claim to have discovered any novel relationships not previously hypothesized by one or more of the molecular or morphological studies highlighted in this chapter. Rather, what Streicher and Wiens (2017:3) did claim was (boldface added):

> Our results **strongly resolve** the phylogenetic placement of snakes among lizard families, as the sister group to anguimorphs and iguanians… they provide **overwhelming evidence to conclusively resolve** the contentious placement of iguanians…by species-tree analyses of thousands of loci…

It is because of the language of absolutes, which I flatly reject as non-science, and of course because of the assumption that bigger is better (see Simões et al., 2017a) that I am presenting a critique here, not so much of their results, which admittedly lack novelty, but because of the non-science of their claims made with high profile in the literature.

As scholars of science, we are responsible for the philosophy and method at the foundation of how we collect, interpret, and analyze data; how we construct hypotheses drawn from our analyses; and the language we use to communicate our conclusions. Language is a key element of our scholarship, as has been argued numerous times in this book, and serves as a philosophical reflection of what we claim to know and what we think is real. In this sense, any empirical claims of the discovery of absolutes cannot be verified to be universally true, though such language is common throughout Streicher and Wiens (2017). From the title: (1) "resolve the origin"; (2) from page 2 (quoted above), "Concomitantly, they provide overwhelming evidence to conclusively resolve the contentious placement of iguanians"; (3) from page 4, "We resolve two contentious aspects of squamate phylogeny using approximately 100 times more loci than in previous studies, and with species tree methods."

Claims of absolute knowledge using the word "resolve" were consistently made by Streicher and Wiens (2017), though, interestingly, the authors used the relative epistemic language of cladistics, that is, "sister groups" and "clades" in discussing groups and their relationships. This language is at odds with their certainty of resolution of relationships and origins, as a sistergroup is not an absolute ancestor, but rather is a relative statement of relationship arising from synapomorphy distributions that "create" "clades" within a cladogram. For example, in a three-taxon analysis including a turtle, dog, and chicken, the turtle would be the sistergroup to the chicken, and the dog the sistergroup to the chicken-turtle clade. Add a lizard to the data set, and the information being analyzed changes such that the lizard and chicken now form a clade where the lizard is now the sistergroup to chickens, and the turtle is the sistergroup to the lizard-chicken clade; should we add a crocodile to our analysis, the crocodiles will now be the sistergroup to chickens. While relationships are resolved in this example, resolution of relationships is relative to the data being analyzed, not absolute regardless of the data.

From the language used by Streicher and Wiens (2017), to "conclusively resolve," it can only be assumed that the original chicken-turtle clade is inviolate and can never be subject to revision. This is not how science works, neither by design nor by intention, but it is the root of dogma.

And, finally, to go back to my opening critique, where is the rest of the data, that is, morphology? Even if homolog concepts can be developed for molecular data of any and all kinds, molecules are only one data stream informing us on phylogeny, not the only one. While certainly trendy, valuable, and new, nucleotide sequence data, both nuclear and mitochondrial, and UCEs, whatever the value (see above), only reflect one level of organization within the hierarchy of dependent properties that is a complete organism. To ignore morphology completely, and presume the absolute answers of phylogeny can be found in isolation within the genome alone, is simply incorrect. All evidence is of consequence to unraveling the evolutionary past, and this is not a statement pledging allegiance to total evidence—I am talking about all evidence at every level in the hierarchy of biological organization. Deep time and ancient cladogenic events, regardless of mechanism, are beyond discovery using molecules alone (after all, >99% of all things to have lived are extinct and long ago gave up their nucleotides to the vastness of time, but not their preservable anatomy, which remains trapped in the rocks), and so, if for no other reason, every answer in deep phylogeny will be relative, not absolute, whether you choose to use morphology alongside molecules or not.

So, what does that mean with respect to the claim Streicher and Wiens (2017) made to have "conclusively resolved" the phylogenetic position of iguanian and snake lizards? It means nothing more than the resolution provided by previous hypotheses: like it or not, Streicher and Wiens's (2017) hypothesis is a relative answer based solely on the data used to analyze the question. It is not an absolute answer for the simple reason that UCEs and hundreds of thousands of their nucleotides are not absolute data. No data are absolute answers to anything, so neither are the hypotheses built from them—such claims are empirically, epistemically, and ontically empty. UCEs are what they are, nucleotide sequences caught up in the nuclear DNA of organisms, and not absolute answers to

any one question, though they might well be useful data when combined with all other relevant data to address exceedingly complex questions such as those constructed to investigate ancient evolutionary events.

*Molecules and Morphology III*: Simões et al. (2018a) (Figure 6.9)—Simões et al. (2018a) adhered to the philosophical and methodological approaches taken by the likes of Lee (2005a,b,c) and Reeder et al. (2015) for global lizard phylogenetic studies—that is, the simultaneous analysis of morphological and molecular data. Simões et al. (2018a) used a wide variety of phylogenetic testing methods (maximum parsimony and Bayesian). The most important step forward taken by this study is not limited to its phylogenetic results, which are intriguing, but rather in its empiricism—literally every single terminal taxon was personally observed by Tiago Simões and/or myself, and Simões personally studied 99% of the terminal taxa. Thus, the terminals were not coded from the literature, and the characters and states and homolog concepts were all created de novo and not from the use of pre-existing data matrix. These new characters and states were also scrutinized against the criteria laid out in Simões et al. (2017). The scale and quality of this study is unique, and while I am admittedly extremely biased toward it as both coauthor and supervisor of Dr. Simões, I feel it is, at the very least, philosophically and methodologically consistent with the issues raised in this book and its various chapters and criticisms.

With respect to the findings reported by Simões et al. (2018a), the results bear on the phylogenetic relationships of all sauropsid reptiles, as the focal taxon in the study was *Megachirella wachtleri*, the oldest known lizard, but we also examined the interrelationships of all lizards in that same global analysis (Figure 6.9). In a nutshell, in the combined evidence relaxed-clock Bayesian inference analysis (Figure 6.9), the Mosasauria were found to be the closest lizard sistergroup to snake lizards, and as a clade, was itself in the sistergroup position to a large clade composed of the Anguimorpha and Iguania. This result is contrary to all previous morphological studies from Camp (1923) to Gauthier et al. (2012), but is consistent with Streicher and Wiens (2017), Vidal and Hedges (2004), Reeder et al. (2015), and so on. Most importantly, though, Simões et al. (2018a) in the "dated" phylogeny (Figure 6.9) most clearly

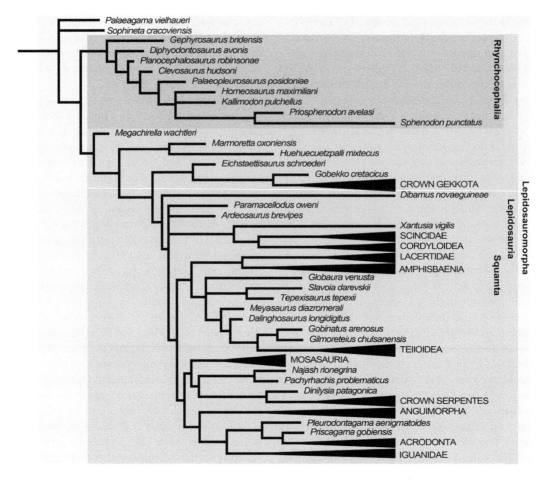

**Figure 6.9.** Broad patterns (numerous terminals omitted for clarity) of lepdiosaur phylogeny (majority-rule consensus tree) of Simões et al. (2018a) derived from a combined evidence relaxed-clock Bayesian inference analysis using both molecular and morphological data. Tip and node dating was done using the fossilized birth–death tree model. As noted by Simões et al. (2018a), the numbers at the nodes indicate posterior probabilities and the dashed line is the PermoTriassic mass extinction event. (Modified from Simões et al. 2018a.)

recover a much older time of origin for lizards in the Upper Permian, at least 260 million years ago, and note that the Triassic was not the period of origin for lizards, but was a time of radiation (see also Jones et al. [2013] following the same theme, but more broadly on lepidosaurian reptiles). As you can imagine, this has profound impact on the likely timing of appearance of the first, and most incredibly ancient, snake lizard. Studies looking to discuss the origin of snake lizards in the Lower Cretaceous, at roughly 110–130 million years ago (Hsiang et al., 2015), are most likely more than 100 million years too late to that party. Even the oldest known snake lizards reported by Caldwell et al. (2015), at ~160–170 million years old, are quite likely 70–80 million years younger than the probable appearance of the first snake lizards.

## SUMMARY

So what is it, after nearly 160 years of morphological and now molecular data sets being applied to questions of snake lizard phylogeny, that we can claim to know? Or perhaps, put another way, where do the points of consensus lie between varying factions of polarized morphological systematists and between morphological and molecular data sets? To be honest, there are not many, and none of them include the base of the tree and how the fossil taxa form clades with themselves or with the modern assemblage, nor how they influence early branching patterns within Serpentes. The position of scolecophidians to fossil and modern snake lizards is extremely unclear, and, of course, there is still no agreed-upon hypothesis for the closest sistergroup of snake lizards among all lizards (fossil and modern). The big questions and broad patterns seem not to be very close to any kind of resolution or agreement. If there are points of agreement they, lie among the modern assemblage way up in the crown and mostly center on the caenophidians. That is to say that *Acrochordus* and the colubroid radiation of viperids, colubrids, and elapids do seem to form a robust clade and a rather recent radiation—even here, though, there is disagreement on the details of that clade structure; there are a number of studies that present data in support of the hypothesis that colubrids are not a monophyletic assemblage and their position within Colubroidea is not well understood in the least; as Pyron et al. (2011:340) noted:

> The higher-level classification of Colubroidea has been in flux as new molecular results contradict traditional taxonomy, and new phylogenies and taxonomies contradict each other.

This is as it should be, new data in the form of taxa, fossil or modern, morphology and/ or molecules, should most certainly scramble existing hypotheses, or at least test them severely. The proof that the system of inquiry is working is the constant assault on existing ideas and their supporting data. The list of uncertainties also extends to groups such as scolecophidians; as reconstructed by Figueroa et al. (2016), they are not a monophyletic group even if they do belong at the base of the phylogeny (note: see also, Miralles et al. [2018]). It is abundantly clear from this treatise that I find them to be a highly derived grade of snake lizards that likely nest within the colubroid radiation and also are not a monophyletic group. As for the rest of the tree below Caenophidia, well, that is certainly poorly supported, including the concepts of boiine and pythonine (e.g., Figueroa et al., 2016), and of course the "anilioid" radiation as a series of successive branches on a Hennigian Comb, rather than being reconstructed as clade. Siegal et al. (2011), using data from the reproductive organs, hypothesized a close relationship between *Anilius* (Pipe Snake; Type I Skull) and the two genera *Trachyboa* and *Tropidophis* (Dwarf Boa; Type I skull), coherent with the Vidal et al. (2007b) Amerophidia.

If this is as far as we have gotten in the 150 years since Cope (1869), is there anything else that can we do to move forward on our understanding of major clade points in ancient snake lizard phylogeny? I think so, and the remainder of this chapter explains why and how and takes us back to the beginning of this chapter and my philosophy and method of what I call the Fossil Backbone.

## "AND NOW FOR SOMETHING COMPLETELY DIFFERENT..."

### End-Mesozoic Snake Lizard Phylogeny—The Fossil Backbone

At the beginning of this chapter, I promised to present a new method that might assist in understanding the structure of problematic internal nodes in both snake lizard ingroup and sistergroup phylogenetic analyses (Figures 6.10 and 6.11). I have identified the problem creating this chaos as arising from the infliction of the essential characters of modern snake lizards onto the fossil record and forcing ancient snake lizards into a topology constrained by the modern assemblage, and not the other way around. I have referred to my possible solution, or at very worst a test of alternative topologies, as "Fossil Backbone Analysis." After stumbling around through nearly 150 years of wrangling, haggling, battling, and debating the phylogenetic relationships of snake lizards, it is time for something new. We can of course wait for new data in the form of fossils— and, trust me, these rewrite the story every time we find one—or more complete genomes, molecular data sets, and so on, or we can approach an old story with new eyes. Here are the new eyes of the Fossil Backbone, presented in two phases: (1) my "Imagination Experiment," which is not part of the formal protocol at all, but was part of my own intellectual process in arriving at a more formal version; (2) the more empirical protocol, what I refer to as "The Empirical Test" of Fossil Backbone Analysis.

### The Imagination Experiment

Imagine if you can a modern world without snake lizards. It is hard to imagine our mythologies and iconographies without "snakes" to represent such things as infinity (the snake eating its tail) or flying feather dragon gods (the Aztec god Quezlcoatl) or evil (the devil as a four-legged snake tempting Eve in the Garden). What if snake lizards had gone extinct at the end of the Mesozoic Era along with the big non-avian dinosaurs, many marine reptiles, and a host of other terrestrial

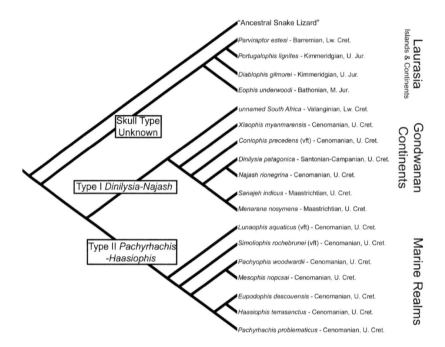

**Figure 6.10.** The Fossil Backbone and Snake Lizard Phylogeny at the end of the Cretaceous. Phylogenetic hypothesis derived from the Mesozoic-aged fossil snake lizard taxa. Had snake lizards gone extinct at the Cretaceous–Palaeogene Boundary, this is what snake lizard phylogeny would look like based on the available taxa and their preserved data.

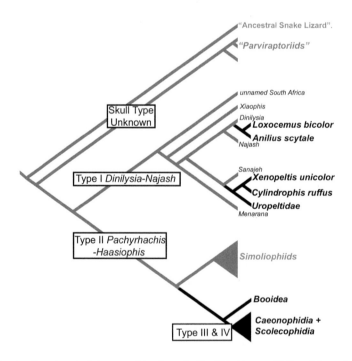

**Figure 6.11.** Imagination Experiment phylogeny arising from the Fossil Backbone with Type I through IV skull types added to the fossil clades and terminal taxa with whom they share anatomical similarities as discussed throughout this book.

THE ORIGIN OF SNAKES

creatures? One outcome of this event would have meant Harry Green would have written a book about "normal lizards" in direct competition with Eric Pianka and Laurie Vitt! Apart from Professor Green's sadness at such a possibility, I am left to ask, "How would we reconstruct the phylogenetic relationships of this obscure little group of extinct limbless lizards if that had happened, and all we knew of them was from the fossil record?" Admittedly, it is hard to imagine this iconic group of lizards as little more than an oddity of extinction, but it has happened to countless clades and lineages of organisms, so snake lizards simply got lucky. But for this exercise, let us imagine that Serpentes or Ophidia, or what have you, went extinct at the end of the Cretaceous. Therefore, the "data set" for this phylogenetic exercise revolves around the seventeen genera and species of snake lizards profiled in this book, including the two form taxa *Simoliophis* and *Coniophis*, and includes as an "outgroup" of sorts the "Ancestral Snake Lizard," and the characteristics of these snake lizards as discussed throughout this book. I have not designated a formal outgroup as this first step I am describing is not a cladistic analysis, but rather very much a subjective and qualitative sketch of what I see as relationships based on shared, derived anatomies. I have not included *Tetrapodophis* because it is not a snake lizard (see Chapters 2 and 3 for data).

The results of my thumb up, tongue out, squinting, one-eyed assessment imagination experiment of the fossil backbone of ancient snake phylogeny pretty much match the contents of this book beginning in Chapter 1 where I identified my four skull groupings, Chapter 2 where I surveyed the Mesozoic fossil taxa of relevance to the core theses at the foundation of this treatise, and Chapter 3 where I reviewed, somewhat exhaustively, the anatomy of homolog concepts at the root of the phylogenetic debate on snake lizard sistergroup and ingroup relationships. Figure 6.10 is, as a result, rather predictable. First, we do not know very much about the anatomy of what I have called the "parviraptorids," and thus I cannot confidently assess the skull type as either Type I *Dinilysia–Najash* or Type II *Pachyrhachis–Hassiophis*. The maxilla of *Parviraptor* certainly appears to be more similar to that of modern python than it does to that of *Dinilysia* (see Caldwell et al., 2015). At the same time, a single skull element is not a complete skull, and the qualities of the otic capsule and suspensorium are critical to making a determination of skull types as I have identified them. There is no doubt, however, in my mind, as a qualitative exercise, that the parviraptorids are indeed snake lizards, but they also are certainly of a kind we currently know very little about. I have elected to connect the parviraptorids to the stem shared by all other more crownward Mesozoic snake lizards, but they are not stem snake lizards, and besides (Note: I despise the term "stem," as it is merely a modern code-term for "ancestor" and does nothing to help define snake lizardness in a cladogram—but I digress), the position of the parviraptorids is by no means a confident placement and could also easily be represented as a polytomy, or they might, with more information, be found as basal members of the clade including *Najash* and *Dinilysia*, or perhaps as Type II fossil snake lizards. It is also remotely possible that they were still four-legged snakes, as suggested by Caldwell et al. (2015) and represent an even more ancient radiation of snake lizards than do the Gondwanan "dinilysioids" or the circumglobal marine pachyophiids. I suspect this last scenario is much closer to reality than the others.

The temporal gap between the last currently known "parviraptoriid" and the first "dinilysioid," the unnamed South African taxon, is essentially zero. Of course, here we have uncomparables, jaws versus a braincase, and thus cannot link the two together in any meaningful way. The gap in the fossil record between *Parviraptor estesi* and *Najash rionegrina* and the *Pachyrhachis* assemblage, all of which share comparable elements, is significant (~45–50 million years). What is clear from these shared morphologies and the temporal gap presented (knowing full well that is only missing data, not a cessation of evolution and adaptive radiations) is that in this time frame, with the evolution of pachyophiids and "dinilysioids," we see the rise of the modern snake lizard body form as exemplified by at least two distinct lineages (Figure 6.11). Arising from within my "dinilysioids" are a number of extinct lineages found as fossils in the Cenozoic of Austral Gondwana (e.g., *Yurlungurr* and *Wonambi*) (Scanlon and Lee, 2000; Scanlon, 2005a,b, 2006), South American Gondwana (the madtsoiid radiation) (Albino, 1986; Albino and Brizuela, 2014), and as numerous newly recognized madtsoiidlike aquatic, marine, and giant-bodied snake lizards, such as *Gigantophis* and *Palaeophis* (Rio and Mannion, 2017). Most importantly, though, when my Imagination Experiment adds to this "dinilysioid" clade the modern fauna based on the skull types recognized in this treatise, *Anilius* and *Loxocemus* (and probably the tropidophiids as well) nest within the clade formed by *Dinilysia* and *Najash*, and *Cylindrophis, Anomochilus*, xenopeltids, and uropeltids all nest within the Indian Gondwana radiation including *Sanajeh* (Figure 6.11).

As this is my thought experiment, I will continue—the pachyophiid + "macrostomatan" clade is composed of Type II–IV skull types, inclusive of scolecophidians (which are easily explained as extreme degenerate pedomorphs of highly derived crown snake lizards, not primitive snake lizards from which all other snake lizards arise). This clade typifies the essential, archetypal category of "snake," incorrect as it is, and can be seen to arise sometime in the early part of the Upper Cretaceous, consistent with hypotheses for the origin of all "snakes" by recent studies such as that of Hsiang et al. (2015). However, in my thought experiment, the modern assemblage of snake lizards includes other snake lizards, and is inclusive of anilioids etc. which are not part of this archetype of snake lizards (of course, neither are scolecophidians), which means that the archetypal assemblage of "snakes" are "polyphyletic" in the sense that they do not share a common ancestor represented by a modern snake lizard (the archetype problem arising from the received view of scolecophidians as exemplars of the primitive snake condition). Thus for the usual snake lizard clade Alethinophidia, if a formal test was administered and this same result found, we would define the crown clade as inclusive of "dinilysioids" + pachyophiids and "macrostomate" snake lizards inclusive of scolecophidians. This same outcome arises from all studies finding, for example, *Najash* as the most primitive snake lizard (e.g., Apesteguía and Zaher, 2006) as *Najash* is clearly an alethinophidian in terms of anatomical features, and thus so are scolecophidians as they are crownward of *Najash*.

A fascinating outcome of the Imagination Experiment is that in the context of the geologic time, snake lizard evolution can clearly be seen as a series of adaptive radiations and evolutionary pulses, not one single "origins event" in the early Late Cretaceous *sensu* Hsiang et al. (2015). The "origins event" of Hsiang et al. (2015) defines the origins of the essential modern snake lizard, but certainly not the origins of snake lizards. This

is an important rewrite on the received wisdom of snake lizard origins that can only be derived from the fossil record of these animals. Similar to the data and phylogenetic results presented by Simões et al. (2018a) and used as the foundation for hypothesizing a Permo-Triassic origin and subsequent radiation of lizards, the origin and subsequent radiation of snake lizards is much, much older than has ever been previously considered, and likely is at least middle to late Triassic in age. And now, for the empirical test and proposed Fossil Backbone Analysis.

## The Empirical Tests

The last step is to take the inspiration expressed in Figures 6.10 and 6.11 and ask the same questions, but using a formalized data set of a taxon-character matrix (Figures 6.12 and 6.13). Not too surprisingly, and with absolutely no manipulation of characters and states from published analysis and matrices, the results are extremely similar. My proposed methodology works in two parts: (1) construct an independent data set for the phylogenetic analysis of a "fossils only" data set and run that analysis to produce a Fossil Backbone topology as a constraint on a future analysis of all taxa and characters; (2) conduct post-hoc analyses of existing matrices of all taxa and characters by excluding all members of the modern assemblage from the analysis.

Phase 1, as described above, will form the core of future research publications presenting phylogenetic relationships on snake lizards specifically, and all lizards generally, and will not be explored further here. However, Phase 2 is described in more detail here, as existing matrices can easily be tested and compared to my "Imagination Experiment" and thus relate to the core themes of this book.

Phase 2 of the "Fossil Backbone Analysis" of snake lizard ingroup relationships was conducted using the data matrix presented by Caldwell et al. (2015) in their study on the oldest known snake lizards. I simply removed all geologically younger terminal taxa from the analysis using the end-Maastrichtian as my temporal cut-off—snake lizards went extinct with non-avian dinosaurs and a host of marine reptiles (Figure 6.10). I did not manipulate characters nor change any of the published parameters, but simply deleted taxa. *Najash* and *Dinilysia* are unresolved

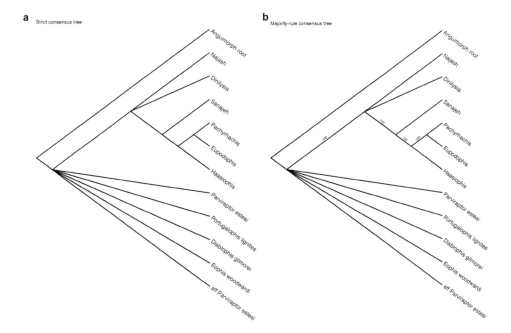

**Figure 6.12.** Fossil Backbone Analysis and snake lizard sistergroup relationships using all fossil snake lizards to the end-Maastrichtian (vertebral form taxa excluded), derived by excluding all younger and modern terminal taxa from the data matrix of Caldwell et al. (2015). (a) Strict consensus; (b) majority rule consensus tree.

in even the Majority-Rule Consensus Tree, but the pachyophiids form a distinct clade; the parviraptoriids, on the other hand, which suffer from an absence of defining data, are polytomy at the bottom of the tree.

The sistergroup analysis, on the other hand, is far more interesting, as it provides both a sistergroup hypothesis for snake lizards and also an ingroup hypothesis. The data matrix manipulated for this Fossil Backbone Analysis was the version of Gauthier et al. (2012) presented in Caldwell et al. (2015: fig. 4a). For this analysis of all lizards, I removed almost all modern taxa from the analysis except for those non-snake lizard groups with a Mesozoic-aged exemplar from the fossil record. My methodological rationale here is to retain some measure of diversity reflecting the characters and homolog concepts underlying the character states, as I did not recode those for this Fossil Backbone Analysis. However, to be frank, in terms of philosophy and method, a Fossil Backbone Analysis should be conducted from an original taxon character matrix that is specific to the included terminal taxa. The analyses presented here, ingroup and sistergroup (Figures 6.12 and 6.13), are all *post-hoc* analyses and will suffer from uninformative primary homolog concepts and

character constructions. As with the application of a molecular backbone to morphological data sets, the data set for a Fossil Backbone Analysis should be independent of the data set to which the Fossil Backbone is then applied as a constraint on tests of phylogenetic topology.

While it is not my intent here to review Figure 6.13 in detail, it is important to note that snake lizards are nested within Palci and Caldwell's (2007) Ophidiomorpha, which itself is nested within Pythonomorpha. Pythonomorpha is in an unresolved scleroglossan polytomy with a wide variety of lizard clades, with Gauthier et al.'s (2012) Polyglyphanodontia in the sisterposition to those unresolved scleroglossans. While that "Mosasauria" sistergroup hypothesis is consistent with Reeder et al. (2015), and even the analysis of Martill et al. (2015) on *Tetrapodophis*, it is the ingroup relationships of snake lizards (Figure 6.13b) from the Majority Rule Consensus Tree that are intriguing, as they align perfectly with the results of my imagination experiment in Figure 6.10 where *Dinilysia* and *Najash* form a distinct clade separate form the monophyletic pachyophiids. The parviraptorids are still an unresolved polytomy, but this is what snake lizard phylogeny would have looked like, give or take a little bit, at the end of the Maastrichtian.

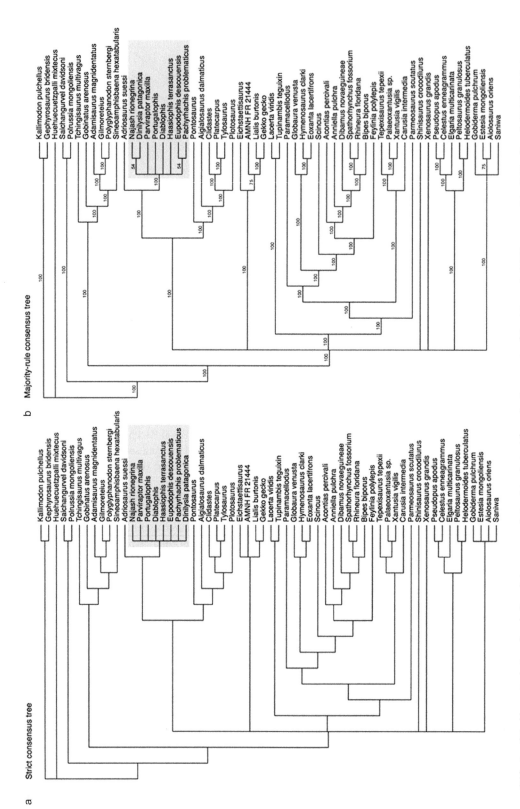

Figure 6.13. Fossil Backbone Analysis and snake lizard sistergroup relationships derived from the data matrix of Gauthier et al. (2012) presented by Caldwell et al. (2015). (a) Strict Consensus; (b) majority rule consensus tree. Gray box indicates snake lizard ingroup relationships in the context of a sistergroup analysis of all lizards. Snake lizards are nested within Ophidiomorpha, which itself is nested within Pythonomorpha. Pythonomorpha is in an unresolved scleroglossan polytomy.

The fossil record of snake lizard vertebral form taxa suggests that the radiation of the modern assemblage was underway in the Late Cretaceous (see Hsiang et al., 2015), and this overprint of fossil snake lizard data would most certainly make this simple three-taxon statement of parviraptorids, dinilysioids, and pachyophiids more complex. But as an experiment in constructing a Fossil Backbone, this posthoc analysis of taking existing taxon-character matrices and excising terminal taxa around a geological timeline provides an important perspective on the essential anatomies of ancient snake lizards, the problem of top–down essentialism in the first place, and a backbone constraint worth using as a test of all evidence analyses, both morphology and molecules, of all lizards generally and snake lizards specifically. I can only hope that I have achieved the goals of this chapter as stated in the title and presented my reasons for electing to place the modern assemblage within the context of the foundation created hundreds of millions of years ago, which is in all truth, actually how it happened, whether my version is even mildly accurate or not.

To close this chapter, I am concluding with an excellent, insightful, and prescient quote from a review paper by Coates and Ruta (2000:505–506), written at the height of Cundall and Irish's (2008:359) "blizzard" of fossil papers:

> Clearly, none of the hypotheses summarized above is likely to be the last word on snake evolution. These trees will continue to be tested and modified by the addition of novel characters and taxa, as well as revised interpretations of existing data…

# CHAPTER SEVEN

# Beginnings

## WHERE DO WE GO FROM HERE?

Each answer only rattles the question harder.

ADAPTED FROM SNAKEBIT BY LESLIE ANTHONY (2008:149)

In late 1996 I read John Horgan's book *The End of Science*, which was as much about the "End" of pretty much everything, from philosophy to progress to theology, as it was about the end of "science." At the same time, I had also just finished reading Paul Feyerabend's *Against Method* (Feyerabend, 1975) and Thomas Kuhn's *The Structure of Scientific Revolutions* (Kuhn, 1962), inspired as I was to do so by my then postdoc mentor Olivier Rieppel. Horgan, as was widely known then and still recognized today, had provocatively concluded that "science" had answered all the big picture stuff and was pretty much over, that science had reached its acme and was now just "filling in the blanks." The contrast for me then, as it remains now, was from what I took away from Kuhn and Feyerabend. To this day, I find them to be arguing that the revolution is never ending because science is not about its techniques, claimed results, temporary hypotheses, or even big picture theory stuff. Rather, for Kuhn and Feyerabend, the revolution or the anarchistic struggle is constant because each new piece of data, hypothesis, or theory must struggle to overcome the inertia and burden of the human side of science, i.e. its politics, history, social structures, and conventions, that build on what we claim is real and what we think we know.

At same time that I was wrestling with Feyerabend, Kuhn, and Horgan, I was well into my fourth read of my little dog-eared copy of Charles Darwin's *On the Origin of Species* (Darwin, 1859) and had started into my first read of the English translation of Karl Popper's *The Logic of Scientific Discovery* (Popper, 1959)—after Kuhn and Feyerabend, I needed to know Popper, and I had and have always enjoyed reading Darwin. Out of that mash-up of contrasting ideas on what science is, that it might have come to its end, and, of course, Darwin's insights on natural selection as a process leading to evolution, it is fair to say I was most profoundly impacted by a marvelous blend of the rule breaking/mind-bending "quantum mechanics" of Feyerabend versus the real world "general mechanics" of Popper, what constituted "proof" of evolution, and whether my chosen scholarly discipline was actually in its twilight. I have continued to think on these things ever since.

In 2018, as I ponder the future of snake lizard science by assessing its past, and have written it down in the pages of this book, it is fair to ask if Horgan (1996) might still be relevant. First, as should be clear from the Preface and Introduction, Horgan, Kuhn, Feyerabend, and Popper all coincide in time and space for me with Jerusalem and the 1996 Snakes with Legs World Tour. The publication of the *End of Science* coincided quite literally with the beginning of science for me. At the time, I simply disagreed with Horgan as yet another prophet outside the walls of the city warning of yet another end of all things by a vengeful god, and I said so in pencil scribbles all over the pages of his book as I read it—the fact that he interviewed Kuhn, Feyerabend, Popper, and a host of my other heroes only meant the book was that much more interesting and provocative to read.

Today, however, as I conclude this book, as I reflect on Horgan's central thesis of science as having answered the key questions of the universe, I continue to see just the opposite; that is, that science generally, and the science of snake lizard paleontology and evolution specifically, have barely gotten out of the starting gates. I see nothing but hubris in thinking otherwise—the same hubris that upheld Ptolemy and the geocentric solar system as the "truth" for nearly 1500 years, long after many people had come up with an alternative hypothesis. For me, science can never be at its end because truth always remains relative, not absolute. In 1996, I turned to Feyerabend going back to 1975, because where Horgan thought everything that was big and worth knowing had been discovered for its realness and truth and that nothing new and big would ever be found again, Feyerabend (1975:14) had this to say about such claims:

> Indeed, one of the most striking features of recent discussions in the history and philosophy of science is the realization that events and developments, such as….the Copernican Revolution…occurred only because some thinkers either decided not to be bound by certain "obvious" methodological rules, or because they unwittingly broke them.

Feyerabend resonated with me in terms of what science is, not as a result of method, but as a point of inquiry that rejects the notion of absolutes and their evolution into dogma—the ideas that arise during these revolutions are neither big nor small, they simply are new. Horgan might be right in arguing that many "big" things have indeed been found, but it does not mean that all of them are absolutely correct, nor even slightly correct for that matter, nor that they cannot be replaced by something else yet unimagined. There

have been numerous "right answers" that survived all conceivable tests for long centuries, though in the end they were eventually found false in some aspect of their data or resulting hypotheses, and revised. If it happened in the past, it will happen in the future. The big stories of 500 years ago might well be "solved," but there will be big stories 500 years from now that will rewrite what we think we know today. This is not a sci-fi defense of the future, but a simple fact—no knowledge is perfect; in fact, no claims to knowledge can be. Thinking otherwise encourages current knowledge to transform into dogma, and that must be avoided at all costs—the burden is unthinkable.

Let me illustrate with my seriously simplified and truncated version of the bare bones of the Copernican Revolution. As I see it, the "shock" of the Copernican Revolution should really not have been much of a shock at all, as Copernicus versus Ptolemy was little more than a switching around of which body revolved around which body (sun around earth versus earth around sun) and not some horrendous mistake of observation and mathematics. The eye-opener for me occurred during my first walk along the Philosenweg in Heidelberg, which was also one of Emmanuel Kant's favorite walks and why I crossed the Neckar River in the first place. As you stroll along the "weg" today, there are a number of Kantian moments portrayed as historical markers, one of which tells the story of a conversation held between Kant and one his students. The story, apocryphal or not, goes something like this:

> Kant and the student were strolling along the Philosenweg one morning, and the student mused of the rising sun, "How could anyone have thought the sun was in orbit round the earth?" Kant is said to have paused for a moment, and then replied, "True, but if we imagine that it actually does orbit the earth, how different would it look from what we see before us right now?"

I was in awe as I stood there, reading that historic marker and soaking in the view of Heidelberg and the sun low in the sky over the Neckar River—absolute awe. I began to see Copernicus through to Galileo as having produced a rather slight revision of Ptolemy, meriting perhaps a *Nature* or *Science* paper today—with the cover, of course—but certainly not a cultural and social revolution. But history tells a different story, as we are aware, and as Feyerabend and numerous others have argued, history records a process of cultural and social revolution linked to the Renaissance and the rise of secularism. Why? Because Ptolemy's model was not the issue at all. The problem was not a scientific one, but rather one of the political and social intransigence for change represented in this case by the Church of Rome right alongside all manner of long-held beliefs and expectations of how the world worked. Horgan's certainty that this revolution revealed big picture science stuff, and thus incontrovertible truth, was not winning a great deal of support in my mind.

Blending Kant, Kuhn, but mostly Feyerabend into my read of Horgan, it was clear to me that Ptolemy's science was actually pretty good—he saw a solar system from his Greco-Roman flat earth that was utterly consistent with his observations and those of millions of other people who had come long before him, and many who came after him. Copernicus did some excellent empirical work and correctly reorganized the geocentric model, make no mistake, but let us not forget that Copernicus and later Galileo, and a lot of other people, also had the advantage of Columbus, and telescopes, and the observation that the earth was indeed a sphere. The problem for Ptolemy was that his geocentric solar system was co-opted by an entirely independent political and social system with an intellectual agenda driven not by the science of the day, but by historical traditions and religious text that placed humans, and god, at the center of everything; that science and religion became essentially the same thing for so long also deeply complicated the problem. Linked together, religion-science knew that humans were on the earth and not on the sun, and as humans were the pinnacle of creation, then the sun could not be at the center of everything—simple and dogmatic and utterly unrelated to Ptolemy's science, but yet it controlled the outcomes of Ptolemy's work far into the future and suppressed all contradictory evidence and ideas. What we had here was ancient science being used to support ancient religious text—there is beauty in such harmony, but it is not science and it easily becomes dogma.

The real Copernican Revolution was not in the falsification of Ptolemy's science, which it did, and good on it, but rather the real revolution was in the non-science battle against intractable agenda driven non-science dogmas. The Copernican Revolution and its protagonists had no choice but to undertake anarchy as a means of defeating dogma. In this sense, I came to understand that Horgan's book would have been best titled *The End of Dogma*, not the *End of Science*, because, as Feyerabend had correctly noted, the anarchy required to make the kind of changes brought about by the Copernican Revolution had very, very little to do with a method of science. The data were already there, they simply needed fresh eyes and minds, a bit of wordsmithing here and there, and the excoriation of dogma in order to let one small rewrite supersede the pretty good first run of Ptolemy. Science from my perspective is not about its answers or discoveries, big or small, it is about a philosophy and then a method that never settles for absolutes. Religion and other dogmatic systems of human organization have final "truth" answers, but science must not. From this point of view, the Horgans of the world can have their "End of whatever…" all they like, but we are not at the end of something just because they say so.

We rest on a tiny bluish planet in a solar system caught up in the gravitational pull of one arm of a spiral galaxy in a universe of endless galaxies all linked together under rules we understand not in the least—considering how little we know, the chances are great we are in for some major paradigm shifts going forward. The hubris of presuming we have found all the big answers, no matter how provocative when it is repeated, time and again, is really rather silly. I would guess someone said the same thing shortly after Ptolmey's big idea broke, "Well, that one is solved. No point in looking further, we have now solved the problem of the cosmos." We have no doubt discovered a few cool things that have allowed us to overturn some dumb ideas and subvert some dogmatic "truths," but that is about as far as we have gotten to date—subverting our own dogmas, that is, not figuring out very many of the "how, what, where, when, and why" questions of the universe and its bits and pieces. I mean, honestly, we do not even have a complete catalog of the all the things in our solar system that likely have answers to questions we have not even thought of yet. It is not a truism to state that we are just getting started on our understanding of quantum particles, strings, star formation, black holes, alternate universes, evolution, the formation of the solar system, the origin of metazoans, the true size of the periodic table, the mystery of genetic inheritance, the black hole of epigenetics, the symbiosis of fungi and algae and how they make lichens, the diversity of fungi, what viruses really are, the social behavior of honey bees, and another 10 billion other things we do not know ... oh, and yes, the origins and evolution of snake lizards. That is why I think Horgan is gloriously wrong—he confused ending dogma with ending knowledge. They are simply not the same thing.

## SNAKE LIZARDS AND THE PHILOSOPHER'S PLAYBOOK

So there I was in early 1996 with Mike Lee in Jerusalem, with *Pachyrhachis problematicus* Haas, 1979, and *Against Method*. Today, I am finishing a book on how I see the state of knowledge on the paleontology and evolution of snake lizards. I can honestly say that I did not purposefully set out to change anything, but rather thought that because I had new data and saw things differently that these "insights" would be well received—naïveté is so charming, and hindsight really is 20:20. I now see the mini-revolt Mike Lee and I set in motion, and how we were time and again accused of violating method and ignoring well-known facts that long ago had absolutely answered the questions we were resurrecting, to be fascinating. Right out of the philosopher's playbook. Now, make no mistake, the recent history of the raging mini-battle around the evolution of snake lizards is not a Copernicus/Galileo-scale event. The Vatican has not issued a writ by the Pope on this one, nor was Mike Lee ever under house arrest in Adelaide. But as Cundall and Irish (2008:358) noted, there was and has continued to be a "blizzard" of papers presenting point and counterpoint on fossil and modern snake lizard anatomy, phylogeny, and origins scenarios. The teacup is still rocked by its tiny tempests even today and there appears to be no end in sight (see Martill et al., [2015] for their version of snake lizard origins and the debate and supposed positions taken by various groups of workers).

Today, as I conclude this book, if I can lay claim to what I think is one incontrovertible fact, it is that the current mélange of fossils, skeletons, data, and hypotheses make it clear that the story of snake lizard evolution is far more wildly complex, intricate, and chaotic than anyone had ever suspected (Figure 7.1). From my point of view, what this also means is that the simple, linear, and deeply teleological story of, "Lizard goes blind, loses legs, spends good part of Mesozoic

**Figure 7.1.** Cover art for this book, of an unknown ancient snake lizard with four legs, hunting in an ancient mangrove swamp in an unknown time and place. (Illustration by Copyright Julius Csotonyi.)

underground, evolves into snake (somehow), and then blind and blinking in the sunshine, comes back to surface and re-evolves just about everything to become a modern snake, while some of its kind stay underground to become the modern blind burrowing snake," is not even mildly close to the truth. But, it is quite the fairy tale, to say the least.

While I remain a philosophical Feyerabendian, a Kuhnian, and a bit of a happy anarchist (though I still like meeting with groups of people to have hearty debates, pint in hand, on all things interesting and important, which means I am not much of an anarchist in the least), I do not feel at all that my work on snake lizard evolution has stepped outside of the rational methodology of science, sensu Karl Popper. It may have been, and may remain as evidenced by the positions and data espoused in this book, contrary to the closely held views of others, but it is hardly anarchistic in its presentation, method, or philosophy. In fact, it is the very opposite. I have employed all of the appropriate tools of empirical observation in order to be deeply constrained by the repeatability of my observations, the clarity of my presentations, and the collection of detailed anatomical observations expressed as the root data for my hypotheses and

THE ORIGIN OF SNAKES

those of my collaborations with my colleagues. My skill as a comparative anatomist and observer has improved, and thus so have my conceptualizations of primary homologs, what homologs are and what homology is, and what makes for a good character construction—but I do not believe I am finished here yet, as there will always be ground for improvement. In fact, if I could not read my own work and find improvements in my character constructions, outright mistakes in my work, and so on, then I would not be much of a scholar at all, as I would truly have learned nothing through these 20-some-odd years. It is absolutely true (for me) that today I know a great deal more about snake lizards than I did when Mike Lee and I opened that drawer in Jerusalem and suddenly had to learn something about snake lizards. A maxim I preach to my graduate students from the office where I sit with my feet on my desk is, "Always be the first to revise your own work," and so I hope I have vastly improved on Caldwell and Lee (1997) these past 20 years. The data set is certainly larger and the ideas expressed in the constantly evolving hypotheses are resonating more deeply, I hope, with the real complexity of nature and not with the simplicity of my human perspective and that of others, though this last part is unavoidable.

## WHAT I THINK WE KNOW: BEGINNINGS

To conclude this long essay on beginnings, and by that I reference this book and my empirical and theoretical works on everything "lizard," we are barely at the beginning of asking questions on the evolution and interrelationships of snake lizards, not the end. The core thesis of this book, and of my empirical and theoretical work, is that the only way forward on the most pressing questions of snake lizard origins and their subsequent evolution, that is, the picture at the base of that ancient phylogenetic tree, can only be answered by the discovery of more and more new snake lizard fossils (Figure 7.1). The modern fauna can be tapped for their complete species-level genome, and I predict, as we have discovered from study of the phenotypes of the modern assemblage, that when we do we will only discover that most of the nucleotide sequences of a complete genome are nothing but noisy data on the modern assemblage and inform us no better of ancient events than do the phenotypes that scale up from genotype in those same modern snake lizards. The complete genome of an extant snake lizard, like that same animal's myology, neurology, and osteology, must certainly bear some evidence of the history that created it, but I doubt that enough of it remains to build a satisfactory pattern to "solve" phylogeny once and for all. At the same time that every living organism reflects some part of its ancient roots, it reflects a great deal more of its recent roots—recall here that I am talking about recent roots that might be 10 or even 30 million years old, so not recent at all in reality. But in consideration of those deep time relationships where we are now looking to understand pattern and process going back beyond 200 million years, the only place we can look is within the crinkled pages of the tome of life that is the fossil record. This is not a romanticized statement extolling the virtues of fossil data, it is simply a factual one. We are left with organisms of the past, and all of their surviving parts and their ancient ecologies, to inform us on the spectacle that has been some 4 billion years of earth history and the evolution of life on this planet. Only the fossil record, for all of its incompleteness, connects the present to its past, or better yet, the past to the riddle of the present.

So back I go to the fossil record and to what I have learned through my several decades of studentship on ancient snake lizards and their modern and fossil kin. In very matter-of-fact terms, some of these new specimens will no doubt come from older units of rock than could possibly be anticipated by current hypotheses derived from molecular clock calibrations. Current best attempts have put the timing of snake lizard origins at approximately 120 million years ago (Hsiang et al., 2015). In contrast, the currently oldest known snake lizards from the Middle Jurassic (Caldwell et al., 2015) clearly indicate that the temporal origins of the clade are even older. The recent report on the oldest known "squamate"/lizard, *Megachirella wachtleri* (Simões et al., 2018a) from the Middle Triassic of northern Italy, makes it very clear that the radiation of lizards was well underway much earlier than ever was suspected, affirming the predictions made by Evans and Borsuk-Bialynicka (2009). My guess? The founding cladogenic events leading to the snake lizard lineage or lineages are likely much older than the Middle Triassic— probably Permo-Triassic and likely driven by the breakup of the supercontinent Pangaea. In classical population genetic terms, populations would have been separated from each other and formed population isolates on the Gondwanan and Laurasian landmasses, which by the Jurassic were again splitting apart, forming even smaller continental isolates (modern continent-sized cratonic masses). As the Mesozoic Era progressed, snake lizards moved around the world, some carried to new worlds by the continents on which they rode as they continued to evolve in isolation, while others most certainly radiated around the world using the seas as their thoroughfares. Their story is spellbindingly complex and not simple at all. We barely understand it in the least, but its major patterns are not unknowable, they are simply ghostly and will take time to unravel, as they must be rebuilt from the phantoms that have been left behind in the rock record.

Where the modern spectacle of our world is breathtakingly beautiful and I am happy here, I also like to wander in my mind's eye to the island jungles of ancient Myanmar, threading my way through a tropical forest dripping with sap, alive with a colorful tapestry of snake lizards, beetles, ancient birds and dinosaurs, and long gone flora that no one will ever see. I like to stand in the ancient deserts of Gondwana sipping mate with *Najash* (Figure 7.2) and watch the sunrise to the east, knowing it has been shining for hours already on a lost and vast reef system spanning the dreamtime Tethys Sea filled with hundreds of thousands of seagoing snake lizards with tiny back legs. But, most critically, I am beginning to see now a host

of ancient snake lizards that we have not found evidence for yet in the fossil record, the anatomies of which are building in my imagination in time, space, and morphology, from the increasingly more ancient relicts we are finding from *Najash* all the way back to *Eophis*. I am seeing four-legged snake lizards with serpent heads and bodies living along coastlines and in the middle of continents, and as I go further back in time, their limbs are longer and their bodies shorter, but yet their heads remain serpentlike in form and evolving function (Figure 7.1). And when I come back to the here and now, I am left with nothing but wonder that anything of these ancient lost worlds and times ever even had a chance to enter the fossil record, let alone survive to the present day.

The concept behind my proposed Fossil Backbone Analysis seeks to understand those ancient roots of the great phylogenetic tree of snake lizard evolution, and so this book has been focused on the fossil record of the Mesozoic as the key data source for such questions. While these data speak to that "root," the snake lizards of the Mesozoic Era created lineages that occupied myriad habitats around the world and clearly survived the end-Cretaceous extinction event to enter the Cenozoic and begin radiating and diversifying into the various clades that now represent the modern assemblage. Scolecophidians are one of those lineages, though I do not think they are monophyletic in the least (e.g., Miralles

**Figure 7.2.** Two *Najash*, Fernando Garberoglio on the left, Michael Caldwell on the right, taking a break wondering about their own existence, to play poker and share maté, the national drink of a "far-off-in-the-future-Argentina" that is yet to evolve. (Illustration by Copyright Fernando Garberoglio.)

et al., 2018) and that like all higher caenophidian/ colubroid snake lizards, they represent the modern radiation, not the base of the tree. In this book and within its expressed data and arguments, I think I have shown why I see them as exceptionally highly derived snake lizards nested high up within macrostomatans if not well within the colubroid radiation. The same is true of the burrowing origins scenario and its constraining Ocular Backbone as driven by the "eye evolution myth"—it is an idea based on circumstantial evidence, as Bellairs (1972) put it, that became dogma. Many aspects of the empirical evidence used to build it over the years remain valid and important, and genes and molecules still may have something to say on the sistergroup relationships of scolecophidians (e.g., Pyron et al., 2011; Reeder et al., 2015), but the burrowing origins scenario needs to abandoned.

While the Cenozoic fossil record and its relevance to snake lizard evolution and the modern assemblage is the subject of a future treatise, it is sufficient to state here that we have barely scratched the surface of those data and the secrets they hold. One of the newest and most exciting discoveries, in my opinion, is that snake lizards entered the marine realm as secondary aquatic adaptations numerous times in the Mesozoic, and from the Cenozoic record it appears they entered it numerous times again − the probing question is whether or not secondary adaptations to marine environments in snake lizards is a constrained one-way street as it has become for most other secondarily adapted aquatic tetrapods (e.g., cetaceans, mosasauroid lizards, etc.). If it is not, and I suspect it is not, the question going forward is how this potential capacity to move between continents by using marine systems as routes of adaptive radiation and migration may have influenced ancient and modern snake lizard biogeography and diversity. The answers can only be guessed at until we undertake a concerted study focused on those data and such questions. To say the least, there is so much to learn from the available data, let alone future and unsuspected fossils and their new data, that it is abundantly clear that the science of snake lizard evolution is only just beginning, and that, as Les Anthony (2008) recalls it, "Each answer only rattles the question harder."

# REFERENCES

Abdeen, A. M., Abo-Taira, A. M., and Zaher, M. M. 1991a. Further studies on the ophidian cranial osteology: The skull of the Egyptian blind snake *Leptotyphlops cairi* (Family Leptotyphlopidae). I. The cranium. A: The median dorsal bones, bones of the upper jaw, circumorbital series and occipital ring. *Journal of the Egyptian German Society of Zoology*, 5: 417–437.

Abdeen, A. M., Abo-Taira, A. M., and Zaher, M. M. 1991b. Further studies on the ophidian cranial osteology: The skull of the Egyptian blind snake *Leptotyphlops cairi* (Family Leptotyphlopidae). I. The cranium. B: The otic capsule, palate and temporal bones. *Journal of the Egyptian German Society of Zoology*, 5: 439–455.

Abdeen, A. M., Abo-Taira, A. M., and Zaher, M. M. 1991c. Further studies on the ophidian cranial osteology: The skull of the Egyptian blind snake *Leptotyphlops cairi* (Family Leptotyphlopidae). II. The lower jaw and the hyoid apparatus. *Journal of the Egyptian German Society of Zoology*, 5: 457–467.

Abdel-Fattah, Z. A., Gingras, M. K., Caldwell, M. W., and Pemberton, S. G. 2009. Sedimentary environments and depositional characters of the Middle to Upper Eocene whale-bearing succession in the Fayum depression, Egypt. *Sedimentology*, 57: 446–476.

Abdel-Fattah, Z. A., Gingras, M. K., Caldwell, M. W., Pemberton, S. G., and MacEachern, J. A. 2016. The *Glossifungites* Ichnofacies and sequence stratigraphic analysis: A case study from Middle to Upper Eocene Successions in Fayum, Egypt. *Ichnos*, 23: 157–179.

Abel, O. 1924. *Die Eroberungsziige der Wirbeltiere in die Meere der Vorzeit*. Gustav Fischer, Jena, Germany.

Albino, A. 1986. Nuevos Boidae Madtsoiinae en el Cretácico tardío de Patagonia (Formación Los Alamitos, Río Negro, Argentina). In Bonaparte, J. F. (ed.) *Actas IV Congreso Argentino de Paleontología y Bioestratigrafía*. Inca Editorial, Mendoza, Argentina, pp. 15–21.

Albino, A., and Brizuela, S. 2014. An overview of the South American fossil squamates. *The Anatomical Record*, 297: 349–368.

Albino, A., Carrillo-Briceño, J. D., and Neenan, J. M. 2016. An enigmatic aquatic snake from the Cenomanian of Northern South America. *PeerJ*, 4: e2027.

Albino, A. A., and Caldwell, M. W. 2003. Hábitos de vida de la serpiente cretácica *Dinilysia patagonica* Woodward, 1901. [English Translation: Life habits of the Cretaceous snake *Dinilysia patagonica* Woodward, 1901]. *Ameghiniana*, 40: 407–414.

Allen, R. E. 2006. *Plato: The Republic*. Yale University Press New Haven, U.S.A., pp. 1–400.

Anderson, H. T. 1936. The jaw musculature of the phytosaur, *Machaeroprosopus*. *Journal of Morphology*, 59(3): 549–587.

Anderton, R. 1985. Clastic facies models and facies analysis. *Geological Society, London, Special Publications*, 18: 31–47.

Andrews, C. W. 1901. Preliminary notes on some recently discovered extinct vertebrates from Egypt. Part II. *Geological Magazine*, 8: 436–444.

Anthony, L. 2008. *Snakebit: Confessions of a Herpetologist*. Greystone Books, Vancouver, Canada, pp. 1–304.

Apesteguía, S., Agnolin, F. L., and Claeson, K. 2007. Review of Cretaceous dipnoans from Argentina

(Sarcopterygii: Dipnoi) with descriptions of new species. *Revista del Museo Argentino de Ciencias Naturales*, nueva serie, 9: 27–40.

Apesteguía, S., Agnolin, F. L., and Lio, G. L. 2005. An early Late Cretaceous Lizard from Patagonia, Argentina. *Comptes Rendu Palevol*, 4: 311–315.

Apesteguía, S., de Valais, S., Gonzalez, J. A., Gallna, P. A., and Agnolin, F. L. 2001. The tetrapod fauna of "La Buitrera", new locality from the basal late Cretaceous of North Patagonia, Argentina. *Journal of Vertebrate Paleontology, Supplemental*, 21: 29A.

Apesteguía, S., and Garberoglio, F. 2013. The return of *Najash*: New, better preserved specimens change the face of the basalmost snake. *Journal of Vertebrate Paleontology Program and Abstracts*, 2013: 79.

Apesteguía, S., and Novas, F. E. 2003. Large Cretaceous sphenodontian from Patagonia provides insight into lepidosaur evolution in Gondwana. *Nature*, 425: 609–612.

Apesteguía, S., Veiga, G. D., Sánchez, M. L., Argüello Scotti, A., and Candia Halupczok, D. J. 2016. Kokorkom, el desierto de los huesos: Grandes dunas eólicas en la Formación Candeleros (Cretácico Superior), Patagonia Argentina. *Ameghiniana*, 54(Suppl.): 7.

Apesteguía, S., and Zaher, H. 2006. A Cretaceous terrestrial snake with robust hindlimbs and a sacrum. *Nature*, 440: 1037–1040.

Aristotle. The History of Animals. Translated by D'Arcy Wentworth Thompson and Provided by The Internet Classics Archive, http://classics.mit.edu//Aristotle/history_anim.html

Ashmead, W. H. 1896. Rhopalosomidae. A new family of fossorial wasps. *Proceedings of the Entomological Society of Washington*, 3-4: 303–310.

Auen, E. L., and Langebartel, D. A. 1977. The cranial nerves of the colubrid snakes *Elaphe* and *Thamnophis*. *Journal of Morphology*, 154: 205–222.

Auge, M., and Rage, J.-C. 2006. Herpetofaunas from the upper Paleocene and Lower Eocene of Morocco. *Annales de Paleontologie*, 92: 235–253.

von Baer, K. E. 1828. *Über Entwickelungsgeschichte der Thiere. Beobachtung und reflexion. Erster Thiel. [On the Developmental History of the Animals. Observations and Reflections, First Part]*. Gebrüdern Borntrager, Königsberg. pp. 1–271.

Bahl, K. N. 1937. Skull of *Varanus monitor*. *Records of the Indian Museum*, 39: 133–174.

Baird, I. L. 1960. A survey of the periotic labyrinth in some representative recent reptiles. *University of Kansas, Science Bulletin*, 41: 891–981.

Baird, I. L. 1970. The anatomy of the reptilian ear. In Gans, C., and Parsons, T. S. (eds.) *Biology of the Reptilia 2*. Academic Press, London, pp. 193–275.

Baszio, S. 2004. *Messelophis variatus* n. gen. n.sp. from the Eocene of Messel: A tropidophine snake with affinities to Erycinae (Boidae). *Courier Forschungs Institute, Senckenberg*, 252: 47–66.

Baumeister, L. 1908. Beitrage zur Anatomie und Physiologie der Rhinophiden. *Zoologische Jahrbuch, Abteilung Anatomie*, 26: 423–526.

Baur, G. 1892. On the morphology of the skull in the Mosasauridae. *Journal of Morphology*, 7: 1–22.

Baur, G. 1895. Cope on the temporal part of the skull, and on the systematic position of the Mosasauridae—A reply. *American Naturalist*, 29: 998–1002.

Baur, G. 1896. The paroccipital of the Squamata and the affinities of the Mosasauridae once more. A rejoinder to Professor E. D. Cope. *American Naturalist*, 30: 143–152.

de Beer, G. S. 1926. Studies on the Vertebrate Head. II. The Orbito-temporal Region of the Skull. *Quarterly Journal of the Microscopical Sciences, London*, 70: 263–370.

de Beer, G. S. 1937. *The Development of the Vertebrate Skull*. Clarendon Press, Oxford, pp. 1–554.

Bell, G. L., Jr. 1997. Chapter 11. A phylogenetic revision of North American and Adriatic Mosasauroidea. In Callaway, J. M., and Nicholls, E. L. (eds.) *Ancient Marine Reptiles*. Academic Press, San Diego, pp. 293–332.

Bellairs, A. d'A. 1972. Comments on the evolution and affinities of snakes. In Joysey, K., and Kemp, T. (eds.) *Studies in Vertebrate Evolution*. Oliver and Boyd, Edinburgh, UK, pp. 157–172.

Bellairs, A. d'A., and Boyd, J. D. 1947. The lachrymal apparatus in lizards and snakes. I. The brille, the orbital glands, lachrymal canaliculi and origin of the lachrymal duct. *Proceedings of the Zoological Society of London*, 117: 81–108.

Bellairs, A. d'A., and Boyd, J. D. 1950. The lachrymal apparatus in lizards and snakes. II. The anterior part of the lachrymal duct and its relationship with the palate and with the nasal and vomeronasal organs. *Proceedings of the Zoological Society of London*, 120: 269–309.

Bellairs, A. d'A., and Kamal, A. M. 1981. The chondrocranium and the development of the skull in recent reptiles. In Gans, C., and Parsons, T. S. (eds.) *Biology of the Reptilia 11*. Academic Press, London, pp. 1–263.

Bellairs, A. d'A., and Underwood, G. 1951. The origin of snakes. *Biological Reviews*, 26: 193–237.

Bertin, T. J., Thivichon-Prince, B., LeBlanc, A. R., Caldwell, M. W., and Viriot, L. 2018. Current perspectives on tooth implantation, attachment, and replacement in Amniota. *Frontiers in Physiology*, 9, doi:10.3389/fphys.2018.01630

THE ORIGIN OF SNAKES

Bolkay, S. J. 1925. *Mesophis nopcsai* n. g., n. sp., ein neues, schlangenahnliches Reptil aus der unteren Kreide (Neocom) von Bilek-Selista (Ost-Hercegovina). *Glasnika Zemaljskog Mujeza u Bosni i Hercegovini*, 27: 125–126.

Boltt, R. E., and Ewer, R. F. 1964. The functional anatomy of the head of the puff adder, *Bitis arietans* (Merr.). *Journal of Morphology*, 114: 83–106.

Boulenger, G. 1891. Notes on the osteology of *Helodema horridum* and *H. suspectum* with remarks on the systematic position of the Helodermatidae and on the vertebrae of the Lacertilia. *Proceedings of the zoological Society of London*, 1891: 109–118.

Boulenger, G. A. 1893. *Catalogue of the Snakes in the British Museum (Natural History). Volume I, Containing the Families Typhlopidae, Glauconidae, Boidae, Ilysiidae, Uropeltidae, Xenopeltidae, and Colubridae aglyphae.* Printed by order of the Trustees, London, 448 pp.

Boulenger, G. A. 1894. *Catalogue of the Snakes in the British Museum (Natural History). Volume II, Containing the Conclusion of the Colubridae Agliphae.* Printed by order of the Trustees, London, 382 pp.

Boulenger, G. A. 1896. *Catalogue of the Snakes in the British Museum (Natural History). Volume II, Containing the Conclusion of the Colubridae (Opisthoglyphae and Proteroglyphae), Amblycephalidae, and Viperidae.* Printed by order of the Trustees, London, 727 pp.

Bradford, E. W. 1954. *An histological investigation of the dental tissues of vertebrates with special reference to the phylogeny of dentine and its bearing on the structure of the tissue in man.* Unpublished PhD thesis, St. Andrews University, St. Andrews, Scotland.

Brazeau, M. D., and Friedman, M. 2015. The origin and early phylogenetic history of jawed vertebrates. *Nature*, 520: 490–497.

Brochinski, A. 2000. "Lizards from the Guimarota Coal Mine". In Martin, T., and Krebs, B. (eds.) *Guimarota: A Jurassic Ecosystem.* Verlag Dr. Friedrich Pfeil, Berlin, pp. 59–68.

Brock, G. T. 1941. The skull of *Acontias meleagris*, with a study of the affinities between lizards and snakes. *Journal of the Linnean Society–Zoology*, 41: 71–88.

Brongniart, A. 1800a. Essai d'une classification naturelle des reptiles. I^ere partie, establissement des ordres. *Bulletin des Sciences, par la Société Philomatique*, 11: 81–82.

Brongniart, A. 1800b. Essai d'une classification naturelle des reptiles. II^e partie, formation et disposition des genres. *Bulletin des Sciences, par la Société Philomatique*, 11: 89–91.

Buatois, L. A., Mángano, M. G., and Aceñolaza, F. G. 2002. *Trazas fósiles.* 1a ed. Museo Paleontológico Egidio Feruglio, Trelew, Argentina, 382 p.

Budney, L. A. 2004. *A Survey of Tooth Attachment Histology in Squamata: The Evaluation of Tooth Attachment Classifications and Characters.* Unpublished Master's Thesis, Univeristy of Alberta, pp. 1–229.

Budney, L. A., Caldwell, M. W., and Albino, A. 2006. Unexpected tooth socket histology in the Cretaceous snake *Dinilysia*, with a review of amniote dental attachment tissues. *Journal of Vertebrate Paleontology*, 26: 138–145.

Buffrénil, V. d., and Rage, J.-C. 1993. La "pachyostose" vertébrale de *Simoliophis* (Reptilia, Squamata): données comparatives et considérations fonctionnelles. *Annales de Paléontologie (Vertébrés)*, 79: 315–335.

de Burlet, H. M. 1934. Vergleichende anatomie des stato-akustischen organs. A, die innere ohrsphäre. B, die mittlere ohrsphäre. In Bolk, L., Göppert, E., Kallius, E., and Lubosch, W. (eds.) *Handbuch der Vergleichenden Anatomie der Wirbeltiere.* Urban und Schwarzenberg, Berlin, Germany, 2, pp. 1293–1432.

Cadle, J. E. 1988. Phylogenetic relationships among advanced snakes: A molecular perspective. *University of California Publications in Zoology*, 119: 1–77.

Caldwell, M. W. 1997. Modified perichondral ossification and the evolution of paddle-like limbs in ichthyosaurs and plesiosaurs. *Journal of Vertebrate Paleontology*, 17: 534–547.

Caldwell, M. W. 1999a. Squamate phylogeny and the relationships of snakes and mosasauroids. *Zoological Journal of the Linnean Society*, 125: 115–147.

Caldwell, M. W. 1999b. Description and phylogenetic relationships of a new species of *Coniasaurus* Owen, 1850 (Squamata). *Journal of Vertebrate Paleontology*, 19: 438–455.

Caldwell, M. W. 2000a. On the phylogenetic relationships of *Pachyrhachis* within snakes: A response to Zaher (1998). *Journal of Vertebrate Paleontology*, 20: 181–184.

Caldwell, M. W. 2000b. An aquatic squamate reptile from the English Chalk: *Dolichosaurus longicollis* Owen, 1850. *Journal of Vertebrate Paleontology*, 20: 720–735.

Caldwell, M. W. 2003. "Without a leg to stand on": Evolution and development of axial elongation and limblessness in tetrapods. *Canadian Journal of Earth Sciences*, 40: 573–588.

Caldwell, M. W. 2006. A new species of *Pontosaurus* (Squamata, Pythonomorpha) from the Upper Cretaceous of Lebanon and a phylogenetic analysis of Pythonomorpha. *Memorie della Società Italiana di Scienze Naturali e del Museo Civico di Storia Naturale di Milano*, 34: 1–43.

Caldwell, M. W. 2007a. Snake phylogeny, origins, and evolution: The role, impact, and importance of fossils (1869–2006). In Anderson, J. S., and

Sues, H.-D. (eds.) *Major Transitions in Vertebrate Evolution.* Bloomington and Indianapolis. Indiana University Press, Indiana, pp. 253–302.

Caldwell, M. W. 2007b. Ontogeny, anatomy and attachment of the dentition in mosasaurs (Mosasauridae: Squamata). *Zoological Journal of the Linnean Society,* 149: 687–700.

Caldwell, M. W. 2012. A challenge to categories: "What if anything, is mosasaur". *Bulletin de la Societé National de Geologique de France,* 183: 7–33.

Caldwell, M. W., and Albino, A. 2001. Palaeoenvironment and palaeoecology of three Cretaceous snakes: *Pachyophis, Pachyrhachis,* and *Dinilysia. Acta Palaeontologica Polonica,* 46: 203–218.

Caldwell, M. W., and Albino, A. 2002. Exceptionally preserved skeletons of the Cretaceous snake *Dinilysia patagonica* Woodward, 1901. *Journal of Vertebrate Paleontology,* 22: 861–866.

Caldwell, M. W., Budney, L. A., and Lamoureux, D. 2003. Histology of thecodont-like tooth attachment tissues in mosasaurian squamates. *Journal of Vertebrate Paleontology,* 23: 622–630.

Caldwell, M. W., and Calvo, J. 2008. Details of a new skull and articulated cervical column of *Dinilysia patagonica* Woodward. *Journal of Vertebrate Paleontology,* 28: 349–362.

Caldwell, M. W., and Cooper, J. 1999. Redescription, palaebiogeography, and palaeoecology of *Coniasaurus crassidens* Owen, 1850 (Squamata) from the English Chalk (Cretaceous; Cenomanian). *Zoological Journal of the Linnean Society,* 127: 423–452.

Caldwell, M. W., and Lee, M. S. Y. 1997. A snake with legs from the marine Cretaceous of the Middle East. *Nature,* 386: 705–709.

Caldwell, M. W., Nydam, R. L., Palci, A., and Apesteguía, S. 2015. The oldest known snakes from the Middle Jurassic-Lower Cretaceous provide insights on snake evolution. *Nature Communications,* 6: 5996.

Caldwell, M. W. and Palci, A. 2007. A new basal mosasauroid from the Cenomanian (Upper Cretaceous) of Slovenia with a review of mosasauroid phylogeny and evolution. *Journal of Vertebrate Paleontology,* 27: 863–880.

Caldwell, M. W., and Palci, A. 2010. A new species of marine ophidiomorph lizard, *Adriosaurus skrbinensis,* from the Upper Cretaceous of Slovenia. *Journal of Vertebrate Paleontology,* 30: 747–755.

Caldwell, M. W., Reisz, R., Nydam, R., Palci, A., and Simoes, T. R. 2016. *Tetrapodophis amplectus* (Crato Formation, Lower Cretaceous, Brazil) is not a snake.

Abstracts of Papers, 76th Annual Meeting, Society of Vertebrate Paleontology. *Supplement to the online Journal of Vertebrate Paleontology,* 108.

Callison, G. 1967. Intracranial mobility in mosasaurs. *The University of Kansas Paleontological Contributions,* 26: 1–15.

Camp, C. L. 1923. Classification of the Lizards. *Bulletin of the American Museum of Natural History,* 48: 289–481.

Camp, C. L. 1942. California mosasaurs. *Memoirs of the University of California,* 13: 1–68.

Candia Halupczok, D. J., Sánchez, M. L., Veiga, G., and Apesteguía, S. 2018. Dinosaur tracks in the Kokorkom Desert, Candeleros Formation (Cenomanian, Upper Cretaceous), Patagonia Argentina: Implications for deformation structures in dune fields. *Cretaceous Research,* 83: 194–206.

Caprette, C. I., Lee, M. S. Y., Shine, R., Mokany, A., and Downhower, J. F. 2004. The origin of snakes (Serpentes) as seen through eye anatomy. *Biological Journal of the Linnean Society,* 81: 469–482.

Carroll, R. L. 1988. *Vertebrate Paleontology and Evolution.* Freeman & Sons, New York, 698 pp.

Chalifa, Y. 1985. *Saurorhamphus judeaensis* (Salmoniformes: Enchodontidae), a new longirostrine fish from the Cretaceous (Cenomanian) of Ein Jabrud, near Jerusalem. *Journal of Vertebrate Paleontology,* 5: 181–193.

Chalifa, Y., and Tchernov, E. 1982. *Pachyamia latimaxillaris,* new genus and species (Actinopterygii: Amiidae), from the Cenomanian of Jerusalem. *Journal of Vertebrate Paleontology,* 2: 269–285.

Charas, M. 1669. *Nouvelles expériences sur la vipere, ou l'on verra une description exacte de toutes ses parties, la source de son venin, ses divers effects, et les remedes exquis que les artistes peuvent tirer de la vipere, tant pour la guerison de ses morsures, que pour celle de plusieurs autres maladies.* Paris.

Clements, R. G. 1993. Type-section of the Purbeck Limestone Group, Durlston Bay, Swanage, Dorset. *Proceedings of the Dorset Natural History and Archaeological Society,* 114: 181–206.

Close, M., and Cundall, D. 2014. Snake lower jaw skin: Extension and recovery of a hyperextensible keratinized integument. *Journal of Experimental Zoology,* 321: 78–97.

Close, M., Perni, S., Franzini-Armstrong, C., and Cundall, D. 2014. Highly extensible skeletal muscle in snakes. *Journal of Experimental Biology,* 217: 2445–2448.

Coates, M., and Ruta, M. 2000. Nice snake, shame about the legs. *Trends in Evolution and Ecology,* 15: 503–507.

Cohn, M., and Tickle, C. 1999. Developmental basis of limblessness and axial patterning in snakes. *Nature,* 399: 474–479.

Conrad, J. L. 2008. Phylogeny and systematics of Squamata (Reptilia) based on morphology. *Bulletin of the American Museum of Natural History*, 310: 1–182.

Cope, E. D. 1864. On the characters of the higher groups of Reptilia, Squamata, and especially of the Diploglossa. *Proceedings of the Academy of Natural Sciences of Philadelphia*, 16: 224–231.

Cope, E. D. 1869. On the reptilian orders Pythonomorpha and Streptosauria. *Proceedings of the Boston Society of Natural History*, 12: 250–261.

Cope, E. D. 1875. The Vertebrata of the Cretaceous Formations of the West. In Hayden, F. V. (ed.) *United States Geological Survey of the Territories*. Government Printing Office, Washington, DC, Vol. II, pp. 1–302.

Cope, E. D. 1878. Professor Owen on the Pythonomorpha. *Bulletin of the United States Geological and Geographical Survey of the Territories*, 4: 299–311.

Cope, E. D. 1895a. Baur on the temporal part of the skull, and on the morphology of the skull in the Mosasauridae. *American Naturalist*, 29: 855–859.

Cope, E. D. 1895b. Reply to Dr. Bauer's critique of my paper on the Parooccipital bone of the scaled reptiles and the systematic position of the Pythonomorpha. *American Naturalist*, 29: 1003–1005.

Cope, E. D. 1896a. Criticism of Dr. Bauer's rejoinder on the homologies of the paroccipital bone, etc. *American Naturalist*, 30: 147–149.

Cope, E. D. 1896b. Boulenger on the difference between Lacertilia and Ophidia; And on the Apoda. *American Naturalist*, 30: 149–152.

Cope, E. D. 1900. *The Crocodilians, Lizards and Snakes of North America*. Government Printing Office, Washington, DC, 1294 pp.

Corbella, H., Novas, F. E., Apesteguía, S., and Leanza, H. A. 2004. First fission-track age for the dinosaur-bearing Neuquen Group (Upper Cretaceous), Neuquen basin, Argentina. *Revista del Museo Argentino de Ciencias Naturales*, nueva serie, 6: 227–232.

Cruickshank, R. D., and Ko, K. 2003. Geology of an amber locality in the Hukawng Valley, Northern Myanmar. *Journal of Asian Earth Sciences*, 21: 441–455.

Cundall, D. 1983. Activity of head muscles during feeding by snakes: A comparative study. *American Zoologist*, 23: 383–396.

Cundall, D. 1995. Feeding behaviour in *Cylindrophis* and its bearing on the evolution of alethinophidian snakes. *Journal of Zoology, London*, 237: 353–376.

Cundall, D., and Gans, C. 1979. Feeding in water snakes: An electromyographic study. *Journal of Experimental Zoology*, 209: 189–208.

Cundall, D., and Greene, H. W. 2000. Feeding in snakes. In Schwenk, K. (ed.) *Feeding*. Academic Press, San Diego, pp. 293–333.

Cundall, D., and Irish, F. 2008. The snake skull. In Gans, C., Gaunt, A. S., and Adler, K. (eds.) *Biology of the Reptilia, the Skull of Lepidosauria, Vol. 20, Morphology H*. Society for the Study of Amphibian and Reptiles, University Heights, Ohio, pp. 349–692.

Cundall, D., and Rossman, D. A. 1993. Cepahlic anatomy of the rare Indonesian snake *Anomochilus weberi*. *Zoological Journal of the Linnean Society*, 109: 235–273.

Cundall, D., Tuttman, C., and Close, M. 2014. A model of the anterior esophagus in snakes, with functional and developmental implications. *The Anatomical Record*, 297: 586–598.

Cundall, D., Wallach, V., and Rossman, D. A. 1993. The systematic relationships of the snake genus *Anomochilus*. *Zoological Journal of the Linnean Society*, 109: 275–299.

Cuny, G., Jaeger, J. J., Mahboubi, M., and Rage, J.-C. 1990. Le plus ancien serpents (Reptilia, Squamata) connus. Mise au point sur l'age geologique des serpents de la partie moyenne du Crétacé. *Comptes rendu des śances de l'Academie des Sciences, Paris, II*, 311: 1267–1272.

Cuvier, G. 1798. Tableau élémentaire de l'Histoire naturelle des animaux, Paris, 770 pp.

Cuvier, G. 1805. *Leçons d'anatomie compare*. Baudouin, Imprimeur de L'Institute, Paris, pp. 1–805.

Cuvier, G. 1824. *Reserches sur les Ossemens Fossiles*. Novelle ed. G. Dufour and Ed. D'Ocagne, Paris, Vol. 5, Pt. II.

Da Silva, F. O., Fabre, A.-C., Savriama, Y., Ollonen, J., Mahlow, K., Herrel, A., Muller, J., and Di-Poi, N. 2018. The ecological origins of the snakes as revealed by skull evolution. *Nature Communications*, 9: 376, doi:10.1038/s41467-017-02788-3

Darwin, C. 1859. *On the Origin of Species by Means of Natural Selection*. John Murray, London, 516 pp.

Darwin, C., and Wallace, A. R. 1858. On the tendency of species to form varieties; and on the perpetuation of varieties and species by natural means of selection. [Read 1 July] *Journal of the Proceedings of the Linnean Society of London. Zoology*, 3: 45–50.

Deufel, A., and Cundall, D. 2010. Functional morphology of the palato-maxillary apparatus in "Palatine dragging" snakes (Serpentes: Elapidae: *Acanthophis, Oxyuranus*). *Journal of Morphology*, 271: 73–85.

Di-Poi, N., Montoya-Burgos, J. I., Miller, H., Pourquie, O., Milinkovitch, M. C., and Duboule, D. 2010. Changes in Hox genes' structure and function during the evolution of the squamate body plan. *Nature*, 464: 99–103.

Diedrich, C. J., Caldwell, M. W., and Gingras, M. 2011. Stratigraphy, sedimentology, palaeoecology and palaeoenvironment of the sabkha and tidal flat to lagoons of the Cenomanian (Upper Cretaceous) of Hvar Island, Croatia, on the Adriatic Carbonate Platform. *Carbonates and Evaporites*, 26: 381–399.

Dollo, L. 1903. Les ancestres des mosasauriens. *Bulletin de la science de France et Belgique*, 19: 1–11.

Douglas, D.A., and Gower, D. J. 2010. Snake mitochondrial genomes: Phylogenetic relationships and implications of extended taxon sampling for interpretations of mitogenomic evolution. *Genomics*, 11: 14.

Duméril, A. 1853. Prodrome de la classification des reptiles ophidiens. *Memoires de l'Académie scientifique de Paris*, 23: 399–535.

Duméril, A., and Bibron, G. 1844. *Erpétologie Générale, ou Histoire Naturelle Complète des Reptiles.* Librairie Encyclopédique de Roret, Paris, VI.

Eberth, D. A. 1993. Depositional environments and facies transitions of dinosaur-bearing Upper Cretaceous redbeds at Bayan Mandahu (Inner Mongolia, People's Republic of China). *Canadian Journal of Earth Sciences*, 30: 2196–2213.

Edmund, A. G. 1960. Tooth replacement phenomena in the lower vertebrates. *Royal Ontario Museum, Life Sciences Contributions*, 52: 1–190.

Estes, R., de Quieroz, K., and Gauthier, J. 1988. Phylogenetic relationships within Squamata. In Estes, R., and Pregill, G. (eds.) *Phylogenetic Relationships of the Lizard Families.* Stanford University Press, Stanford, CA, pp. 119–281.

Estes, R., Frazzetta, T. H., and Williams, E. E. 1970. Studies on the fossil snake *Dinilysia patagonica* Woodward: Part 1. Cranial morphology. *Bulletin of the Museum of Comparative Zoology*, 140: 25–74.

Evans, S. E. 1991. A new lizard-like reptile (Diapsida:Lepidosauromorpha) from the Middle Triassic of England. *Zoological Journal of the Linnean Society*, 103: 391–412.

Evans, S. E. 1994. A new anguimorph lizard from the Jurassic and Lower Cretaceous of England. *Palaeontology*, 37: 33–49.

Evans, S. E. 1996. Parviraptor (Squamata: Anguimorpha) and other lizards from the Morrison Formation at Fruita, Colorado. In Morales, M. (ed.) *The Continental Jurassic.* Museum of Northern Arizona Bulletin, Flagstaff, Arizona, Vol. 60, pp. 243–248.

Evans, S. E. 2008. The skull of lizards and Tuatara. In Gans, C., Gaunt, A. S., and Adler, K. (eds.) *Biology of the Reptilia, the Skull of Lepidosauria, Vol. 20, Morphology H.* Society for the Study of Amphibian and Reptiles, University Heights, Ohio, pp. 1–347.

Evans, S. E. 2016. Chapter 9. The Lepidosaurian ear: Variations on a theme. In Clack, J. A., Fay, R. R., and Popper, A. N. (eds.) *Evolution of the Vertebrate Ear: Evidence from the Fossil Record.* ASA Press, Cham, Switzerland, pp. 245–284.

Evans, S. E., and Borsuk-Bialynicka, M. 2009. A small lepidosauromorph reptile from the Early Triassic of Poland. *Palaeontologia Polonica*, 65: 179–202.

Evans, S. E., Manabe, M., Noro, M., Isaji, S., and Yamaguchi, M. 2006. A long bodied lizard from the Lower Cretacous of Japan. *Palaeontology*, 49: 1143–1165.

Evans, S. E., and Milner, A. R. 1994. Middle Jurassic microvertebrate assemblages from the British Isles. In Fraser, N. C., and Sues, H.-D. (eds.) *In the Shadow of the Dinosaurs: Early Mesozoic Tetrapods.* Cambridge University Press, Cambridge, U.K., pp. 303–321.

Fabre, P.-H., Hautier, L., Dimitrov, D., and Douzery, E. J. P. 2012. A glimpse on the pattern of rodent diversification: A phylogenetic approach. BMC *Evolutionary Biology*, 12: 1–19.

Fastovsky, D. E., and Bercovici, A. 2016. The Hell Creek Formation and its contribution to the Cretaceous–Paleogene extinction: A short primer. *Cretaceous Research*, 57: 368–390.

Fernandes, A. C. S., and Carvalho, I. D. S. 2006. Invertebrate ichnofossils from the Admantina Formation (Bauru Basi, Late Cretaceous), Brazil. *Revistas Brasiliera Paleontologia*, 9: 211–220.

Feyerabend, P. 1975. *Against Method.* 3rd ed. Verso Books, New York, pp. 1–279.

Figueroa, A., McKelvy, A. D., Grismer, L. L., Bell, C. D., and Lailvaux, S. P. 2016. A Species-Level Phylogeny of Extant Snakes with Description of a New Colubrid Subfamily and Genus. *PLOS ONE*, 11(9): e0161070.

Fitch, W. M. 1970. Distinguishing homologous from analogous proteins. *Systematic Zoology*, 19: 99–113.

Folie, A., and Codrea, V. 2005. New lissamphibians and squamates from the Maastrichtian of Hațeg Basin, Romania. *Acta Palaeontologica Polonica*, 50: 57–71.

Follmi, K. B. 1989. Evolution of the Mid-Cretaceous Triad. *Lecture Notes in the Earth Sciences*, 23: 1–153.

Fox, R. C. 1975. Fossil snakes from the upper Milk River Formation (Upper Cretaceous), Alberta. *Canadian Journal of Earth Sciences*, 12: 1557–1563.

Frazzetta, T. H. 1966. Studies on the morphology and function of the skull in the Boidae (Serpentes). Part II. Morphology and function of the jaw apparatus in *Python sebae* and *Python molurus. Journal of Morphology*, 118: 217–296.

Frazzetta, T. H. 1970. Studies on the fossil snake *Dinilysia patagonica* Woodward. Part II. Jaw machinery in the earliest snakes. *Forma et Functio*, 3: 205–221.

Frazzetta, T. H. 1999. Adaptations and significance of the cranial feeding apparatus of the Sunbeam snake (*Xenopeltis unicolor*): Part 1. Anatomy of the skull. *Journal of Morphology*, 239: 27–43.

Freeman, E., Ten Cate, A. R., and Mills, H. R. 1975. Development of a gomphosis by tooth germ implants in the parietal bone of the mouse. *Archives of Oral Biology*, 20: 139–140.

Gadow, H. F. 1888. On the modifications of the first and second visceral arches with special reference to the homologies of the auditory ossicles. *Philosophical Transactions of the Royal Society*, 179: 451–485.

Gaengler, P. 2000. Evolution of tooth attachment in lower vertebrates to tetrapods. In Teaford, M. F., Smith, M. M., and Ferguson, M. W. J. (eds.) *Development, Function and Evolution of Teeth*. Cambridge University Press, Cambridge, pp. 173–185.

Galli, K. G. 2014. Fluvial architecture element analysis of the Brushy Basin Member, Morrison Formation, western Colorado, USA. *Volumina Jurassica*, 12: 69–106.

Gans, C. 1961. The feeding mechanism of snakes and its possible evolution. *American Zoologist*, 1: 217–227.

Gans, C., and Montero, R. 2008. An Atlas of amphisbaenian skull anatomy. In Gans, C., Gaunt, A. S., and Adler, K. (eds.) *Biology of the Reptilia, the Skull of Lepidosauria, Vol. 20, Morphology H*. Society for the Study of Amphibian and Reptiles, University Heights, Ohio, pp. 621–738.

Garberoglio, F., Gómez, R. O., Simões, T., Caldwell, M. W., and Apesteguía, S. 2019a. The evolution of the axial skeleton intercentrum system in snakes revealed by new data from the Cretaceous snakes *Dinilysia* and *Najash*. *Scientific Reports*, 9(1276): 1–10.

Garberoglio, F., Gómez, R. O., Apesteguía, S., Caldwell, M. W., Sánchez, M. L., and Veiga, G. 2019b. A new specimen with skull and vertebrae of *Najash rionegrina* (Lepidosauria: Ophidia) from the early Late Cretaceous of Patagonia. *Journal of Systematic Palaeontology*.

Gardner, J., and Cifelli, R. 1999. A primitive snake from the Cretaceous of Utah. *Special Papers in Palaeontology*, 60: 87–100.

Gartner, G. E. A., and Greene, H. W. 2008. Adaptation in the African egg-eating snake: A comparative approach to a classic study in evolutionary functional morphology. *Journal of Zoology*, 275: 368–374.

Gaupp, E. 1900. Das Chondrocranium von *Lacerta agilis*. *Anatomischer Hefte*, 15: 433–595.

Gauthier, J. A. 1982. Fossil xenosaurid and anguid lizards from the Early Eocene Wasatch Formation, southeast Wyoming, and a revision of Anguoidea. *Contributions to Geology, University of Wyoming*, 21: 7–54.

Gauthier, J. A., Estes, R., and de Queiroz, K. 1988. A phylogenetic analysis of Lepidosauromorpha. In Estes, R., and Pregill, G. (eds.) *Phylogenetic Relationships of the Lizard Families*. Stanford University Press, Stanford, CA, pp. 14–98.

Gauthier, J. A., Kearney, M., Maisano, J. A., Rieppel, O., and Behlke, A. D. B. 2012. Assembling the squamate tree of life: Perspectives from the phenotype and the fossil record. *Bulletin of the Peabody Museum of Natural History*, 53: 3–308.

Gauthier, J. A., Kluge, A. G., and Rowe, T. 1988. Amniote phylogeny and the importance of fossils. *Cladistics*, 4: 105–209.

Gayet, M. 1980. Recheres sur l'ichthyofaune Cenomanienne des monts de Judee: "Les Acanthopterygiens". *Annales de Palebntologie (Vertebraes)*, 66: 75–128.

Gianechini, F. A., Makovicky, P. J., and Apesteguía, S. 2017. The cranial osteology of *Buitreraptor gonzalezorum* Makovicky, Apesteguía, and Agnolín, 2005 (Theropoda, Dromaeosauridae), from the Late Cretaceous of Patagonia, Argentina. *Journal of Vertebrate Paleontology*, 37: e1255639, doi:10.1080/02724634.2017.1255639

Godefroit, P., Cau, A., Dong-Yu, H., Escuillié, F., Wenhao, W., and Dyke, G. 2013. A Jurassic avialan dinosaur from China resolves the early phylogenetic history of birds. *Nature*, 498: 359–362.

Gray, J. E. 1825. A synopsis of the genera of reptiles and amphibians, with a description of some new species. *Annals of Philosophy, New Series*, 10: 193–217.

Greene, H. 1997. *Snakes: The Evolution of Mystery in Nature*. University of California Press, Berkeley, CA.

Greene, H. W. 1983. Dietary correlates of the origin and radiation of snakes. *American Zoologist*, 23: 431–441.

Greene, H. W. 1984. Feeding behavior and diet of the eastern coral snake, *Micrurus fulvius*. In Seigel, R. A., Hunt, L. E., Knight, J. L., Malaret, L., and Zuschlag, N. L. (eds.) *Vertebrate Ecology and Systematics: A Tribute to Henry S. Fitch*. The University of Kansas Museum of Natural History Special Publication, Vol. 10, Lawrence, KS, pp. 147–162.

Greer, A. E. 1987. Limb reduction in the lizard genus *Lerista*. 1. Variation in the number of phalanges and presacral vertebrae. *Journal of Herpetology*, 21: 267–276.

Greer, A. E. 1990. Limb reduction in the scincid lizard genus *Lerista*. 2. Variation in the bone complements of the front and rear limbs and the number of postsacral vertebrae. *Journal of Herpetology*, 24: 142–150.

Greer, A. E. 1991. Limb reduction in squamates: Identification of the lineages and discussion of the trends. *Journal of Herpetology*, 25: 166–173.

Gregory, J. T. 1951. Convergent evolution: The jaws of Hesperornis and the mosasaurs. *Evolution*, 5: 345–354.

Gregory, W. K., and Noble, G. K. 1924. The origin of the mammalian alisphenoid bone. *Journal of Morphology*, 39: 435–463.

Groombridge, B. C. 1979. Variations in morphology of the superficial palate of henophidian snakes and some possible systematic implications. *Journal of Natural History, London*, 13: 447–475.

Gómez, R. O. 2011. A snake dentary from the Upper Cretaceous of Patagonia. *Journal of Herpetology*, 45: 230–233.

Gómez, R. O., Báez, A. M., and Rougier, G. W. 2008. An anilioid snake from the Upper Cretaceous of northern Patagonia. *Cretaceous Research*, 29: 481–488.

Haas, G. 1930. Über das Kopfskelett und die Kaumuskulatur der Typhlopiden und Glauconiiden. *Zool. Jahrb. Abt. Anat. Ontog. Tiere*, 52: 1–94.

Haas, G. 1955. The systematic position of Loxocemus bicolor Cope (Ophidia). *American Museum Novitates*, 1748: 1–8.

Haas, G. 1959. Bemerkungen über die Anatomie des Kopfes und des Schädels der Leptotyphlopidae (Ophidia), speziell von *L. macrorhynchus* Jan. *Vierteljahresschr Naturforschung Gesellschaft Zürich*, 104: 90–104.

Haas, G. 1962. Remarques concernant les relations phylogéniques des diverses familles d'ophidiens fondées sur la différénciation de la musculature mandibulaire. *Colloqium Internationale Centre Nationale Recherece des Science*, 104: 215–241.

Haas, G. 1964. Anatomical observations on the head of Liotyphlops albirostris (Typhlopidae, Ophidia). *Acta Zoologica*, 45: 1–62.

Haas, G. 1968. Anatomical observations on the head of Anomalepis aspinosus (Typhlopidae, Ophidia). *Acta Zoologica*, 48: 63–139.

Haas, G. 1973. Muscles of the jaws and associated structures of the Rhyncocephalia and Squamata. In Gans, C., Bellairs, A. D., and Parsons, T. S. (eds.) *Biology of the Reptilia*. Academic Press, New York, Vol. 4, pp. 285–490.

Haas, G. 1979. On a new snakelike reptile from the Lower Cenomanian of Ein Jabrud, near Jerusalem. *Bulletin du Muséum national d'Histoire naturelle, Paris, Series 4*, 1: 51–64.

Haas, G. 1980a. Pachyrhachis problematicus Haas, snakelike reptile from the Lower Cenomanian: Ventral view of the skull. *Bulletin du Muséum national d'Histoire naturelle, Paris, Series IV*, 2: 87–104.

Haas, G. 1980b. Remarks on a new ophiomorph reptile from the Lower Cenomanian of Ein Jabrud, Israel. In Jacobs, L. L. (ed.) *Aspects of Vertebrate History, in Honor of E.H. Colbert*. Museum of Northern Arizona Press, Flagstaff, AZ, pp. 177–102.

Hall, B. K. 2005. *Bones and Cartilage*. 2nd ed. Academic Press, San Diego, pp. 1–920.

Haluska, F., and Alberch, P. 1989. The cranial development of Elaphe obsoleta (Ophidia, Colubridae). *Journal of Morphology*, 178: 37–55.

Hampton, P. M., and Moon, B. R. 2013. Gape size, its morphological basis, and the validity of gape indices in western diamond-backed rattlesnakes (Crotalus atrox). *Journal of Morphology*, 274: 194–202.

Hanken, J., and Hall, B. K. (eds.) 1993a. *The Skull. Volume 1, Development*. University of Chicago Press, Chicago, pp. 1–587.

Hanken, J., and Hall, B. K. (eds.) 1993b. *The Skull. Volume 2, Patterns of Structural and Systematic Diversity*. University of Chicago Press, Chicago, pp. 1–566.

Hanken, J., and Hall, B. K. (ed.) 1993c. *The Skull. Volume 3, Functional and Evolutionary Mechanisms*. The University of Chicago Press, USA, pp. 1–460.

Harrington, S. M., and Reeder, T. W. 2017. Phylogenetic inference and divergence dating of snakes using molecules, morphology and fossils: New insights into convergent evolution of feeding morphology and limb reduction. *Biological Journal of the Linnean Society*, 121: 379–394.

Hasse, C. 1873. Zur Morphologie des Labyrinthes der Vögel. In Hasse, C. (ed.) *Anatomische Studien*. Wilhelm Englemann, Leipzig, Germany, Vol. 1, pp. 189–224.

Head, J. J., Bloch, J. I., Hastings, A. K., Bourque, J. R., Cadena, E. A., Herrera, F. A., Polly, P. D., and Jaramillo, C. A. 2009. Giant boid snake from the Paleocene neotropics reveals hotter past equatorial temperatures. *Nature*, 457: 715–718.

Head, J. J., and Polly, P. D. 2015. Evolution of the snake body form reveals homoplasy in amniote Hox gene function. *Nature*, 520: 86–89.

Hecht, M. K. 1982. The vertebral morphology of the Cretaceous snake Dinilysia patagonica Woodward. *Neues Jarhbuch für Geologie und Paläontologie, Monatschafte*, 117: 130–146.

Heise, P. J., Maxson, L. R., Dowling, H. G., and Hedges, S. B. 1995. Higher-level snake phylogeny inferred from mitochondrial DNA sequences of 12S rRNA and 16S rRNA genes. *Molecular Biology and Evolution*, 12: 259–265.

Hennig, W. 1966. *Phylogenetic Systematics, Translated by D. Davis and R. Zangerl*. University of Illinois Press, Urbana, IL, pp. 1–263.

Herrel, A., Vincent, S. E., Alfaro, M. E., Van Wassenbergh, S., Vanhooydonck, B., and Irschick, D. J. 2008. Morphological convergence as a consequence of extreme functional demands: Examples from the feeding system of natricine snakes. *Journal of Evolutionary Biology*, 21: 1438–1448.

Hoffstetter, R. 1939. Contribution à l'étude des Elapidae actuels et fossiles et de l'ostéologie des ophidiens. *Archives du Museum d'Histoire Naturelle de Lyon*, 15: 1–78.

Hoffstetter, R. 1959. Un serpent terrestre dans le Crétacé du Sahara. *Bulletin de la Société géologique de France, Paris*, 7: 897–902.

Hoffstetter, R. 1961. Nouveaux restes d'un serpent boidé (*Madtsoia madagascarensis* nov. sp.) dans le Crétacé supériure de Madagascar. *Bulletin du Muséum national d'Histoire naturelle, Paris, Series II*, 33: 152–160.

Hoffstetter, R. 1967. Remarques sur les dates d'implantations de differents groupes de serpents terrestres en Amerique du Sud. *Comptes Rendue de Sommaire Sceances de la Societe Geologique du France*, 1967: 93–94.

Hoffstetter, R. 1968. Review of "A Contribution to the Classification of Snakes" by G. Underwood. *Copeia*, 1968: 2-1-2103.

Hoffstetter, R., and Gasc, J.-P. 1969. Vertebrae and ribs of modern reptiles. In Gans, C., Bellairs, A. d'A., and Parsons, T. S. (eds.) *Biology of the Reptilia*. Academic Press, London, Vol. I, pp. 201–310.

Holder, L. A. 1960. The comparative morphology of the axial skeleton in the Australian Gekkonidae. *Journal of the Linnean Society, Zoology*, 44: 300–335.

Holliday, C. M., and Witmer, L. M. 2007. Archosaur adductor chamber evolution: Integration of musculoskeletal and topological criteria in jaw muscle homology. *Journal of Morphology*, 268: 457–484.

Holman, J. A. 2000. *Fossil Snakes of North America. Origin, Evolution, Distribution, Palaeoecology*. Indiana University Press, Bloomington, Indiana, pp. 1–357.

Holmes, R. 1989. The skull and axial skeleton of the Lower Permian anthracosaurid amphibian *Archeria crassidisca* Cope. *Palaeontolographica*, 207: 161–206.

Horgan, J. 1996. *The End of Science: Facing the Limits of Science in the Twilight of the Scientific Age*. Broadway Books, New York, pp. 1–336.

Hsiang, A. Y., Field, D. J., Webster, T. H., Behlke, A. D. B., Davis, M. B., Racicot, R. A., and Gauthier, J. A. 2015. The origin of snakes: Revealing the ecology, behavior, and evolutionary history of early snakes using genomics, phenomics, and the fossil record. *BMC Evolutionary Biology*, 15: 87.

Hsiou, A. S., Albino, A. M., Medeiros, M. A., and Santos, R. A. B. 2014. The oldest Brazilian Snakes from the Cenomanian (Early Late Cretaceous). *Acta Palaeontologica Polonica*, 59: 635–642.

Hückel, U. 1970. Die Fischschiefer von Haqel und Hjoula in der Oberkreide des Libanon. *N. Jb. Geol. Palaont. Abh., Stuttgart*, 135: 113–149.

Hückel, U. 1974a. Vergleich des Mineralbestandes der Plattenkalke Solnhofens und des Libanon mit anderen Kalken. *N. Jb. Geol. Palaont. Abh., Stuttgart*, 145: 153–182.

Hückel, U. 1974b. Geochemischer Vergleich der Plattenkalke Solnhofens und des Libanon mit anderen Kalken. *N. Jb. Geol. Palaont. Abh., Stuttgart*, 145: 279–305.

Hull, D. 1988. *Science as a Process*. University of Chicago Press, Chicago, pp. 1–586.

Huxley, T. H. 1858. Theory of the vertebrate skull. The Croonian Lecture. *Proceedings of the Royal Society, London*, 9: 381–433.

Huxley, T. H. 1864. *Lectures on the Elements of Comparative Anatomy*. John Churchill & Sons, London, pp. 1–303.

Huxley, T. H. 1868a. Remarks upon *Archaeopteryx lithographica*. *Proceedings of the Royal Society of London*, 6: 243–248.

Huxley, T. H. 1868b. On the animals which are most nearly intermediate between birds and reptiles. *Annals of the Magazine of Natural History*, 2: 66–75.

Huxley, T. H. 1870. Further evidence of the affinity between the dinosaurian reptiles and birds. *Proceedings of the Geological Society of London*, 26: 12–31.

Infante, C. R., Mihala, A. G., Park, S., Wang, J. S., Johnson, K. K., Lauderdale, J. D., and Menke, D. B. 2015. Shared enhancer activity in the limbs and phallus and functional divergence of a limb-genital cis-regulatory element in snakes. *Developmental Cell*, 35: 107–119.

Janensch, W. 1906. Uber *Archaeophis proavus* Mass., eine schlange aus dem Eocan des Monte Bolca. *Beitrage zur Palaontologie und Geologie Oestreich-Ungarns und des Orients*, 19: 1–33.

Jenner, R. 2004. Accepting partnership by submission? Morphological phylogenetics in a molecular millennium. *Systematic Biology*, 53: 333–342.

Johnston, P. 2014. Homology of the jaw muscles in lizards and snakes–A solution from a comparative gnathostome approach. *The Anatomical Record*, 297:5 74–585.

Jones, M. E., Anderson, C. L., Hipsley, C. A., Müller, J., Evans, S. E., and Schoch, R. R. 2013. Integration of molecules and new fossils supports a Triassic origin for Lepidosauria (lizards, snakes, and tuatara). *BMC Evolutionary Biology*, 13: 208.

Jones, M. E. H., Curtis, N., O'Higgins, P., Fagan, M., and Evans, S. E. 2009. The head and neck muscles associated with feeding in *Sphenodon* (Reptilia: Lepidosauria: Rhynchocephalia). *Palaeontologia Electronica*, 12(2): 7A, 56p, http://palaeo-electronica.org/2009_2/179/index.html

Jurkovsek, B., Toman, M., Ogorelec, B., Sribar, L., Drobne, K., and Poljak, M. 1996. *Geological Map of the Southern Part of the Trieste-Komen Plateau*. Institut za Geologijo, Geotechniko in Geofiska, Ljubljana, Slovenia.

Kaltcheva, M. M., and Lewandoski, M. 2015. Evolution: Enhanced footing for snake limb development. *Current Biology*, 26: R1226–R1245.

Kamal, A. M. 1965. On the cranio-vertebral joint and the relation between the notochord and occipital condyle in Squamata. *Proceedings of the Egyptian Academy of Sciences*, 19: 1–12.

Kamal, A. M. 1966. On the process of rotation of the quadrate cartilage in Ophidia. *Anatomischer Anzieger*, 118: 87–90.

Kamal, A. M., and Hammouda, H. G. 1965a. The chondrocranium of the snake, *Eryx jaculus*. *Acta Zoologica*, 46: 167–208.

Kamal, A. M., and Hammouda, H. G. 1965b. The columella auris of the snake *Psammophis sibilans*. *Anatomischer Anzeiger*, 116: 124–138.

Kardong, K. V. 1986. Kinematics of swallowing in the yellow rat snake, *Elaphe obsoleta quadrivittata*: A reappraisal. *Japanese Journal of Herpetology*, 11: 96–109.

Kardong, K. V., and Berkhoudt, H. 1998. Intraoral transport of prey in the reticulated *Python*; Tests of a general tetrapod feeding model. *Zoology*, 101: 7–23.

Kardong, K. V., Dullemeijer, P., and Fransen, J. A. M. 1986. Feeding mechanism in the rattlesnake *Crotalus durissus*. *Amphibia-Reptilia*, 7: 271–302.

Kearney, M. 2003. The phylogenetic position of *Sineoamphisbaena hexatabularis* reexamined. *Journal of Vertebrate Paleontology*, 23: 394–403.

Kearney, M., and Rieppel, O. 2006. An investigation into the occurrence of plicidentine in the teeth of squamate reptiles. *Copeia*, 2006: 337–350.

Keighley, D. G., and Pickerill, R. K. 1995. The ichnotaxa *Palaeophycus* and *Planolites*: Historical perspectives and recommendations. *Ichnos*, 3: 301–309.

Kellner, A. W. A., and Campos, D. A. 1999. Vertebrate paleontology in Brazil—A review. *Episodes*, 22: 238–251.

Kindler, C., Chèvre, M., Ursenbacher, S., Böhme, W., Hille, A., Jablonski, D., Vamberger, M., and Fritz, U. 2017. Hybridization patterns in two contact zones of grass snakes reveal a new Central European snake species. *Scientific Reports*, 7: 7378, doi:10.1038/s41598-017-07847-9

de Klerk, W. J., Forster, C. A., Sampson, S. D., Chinsamy, A., and Ross, C. F. 2000. A new coelurosaurian dinosaur from the Early Cretaceous of South Africa. *Journal of Vertebrate Paleontology*, 20: 324–332.

Kley, N. J. 2001. Prey transport mechanisms in blindsnakes and the evolution of unilateral feeding systems in snakes. *American Zoologist*, 41: 1321–1337.

Kley, N. J. 2006. Morphology of the lower jaw and suspensorium in the Texas blindsnake, *Leptotyphlops dulcis* (Scolecophidia: Leptotyphlopidae). *Journal of Morphology*, 267: 494–515.

Kley, N. J., and Brainerd, E. L. 1996. Internal concertina swallowing: A critical component of alethinophidian feeding systems. *American Zoologist*, 36: 81A.

Kley, N. J., and Brainerd, E. L. 1999. Feeding by mandibular raking in a snake. *Nature*, 402: 369–370.

Kley, N. J., and Kearney, M. 2008. Chapter 17. Adaptations for digging and burrowing. In Hall, B. K. (ed.) *Fins into Limbs: Evolution, Development, and Transformation*. University of Chicago Press, Chicago, pp. 284–309.

Kluge, A. 1993a. *Aspidites* and the phylogeny of pythonine snakes. *Records of the Australian Museum, Supplement*, 19: 1–77.

Kluge, A. 1993b. *Calabaria* and the phylogeny of erycine snakes. *Zoological Journal of the Linnean Society*, 107: 293–351.

Kocurek, G. 1999. The aeolian rock record. In Goudie, A. S., Livingstone, I., and Stokes, S. (eds.) *Aeolian Environments, Sediments and Landforms*. Wiley, Chichester, UK, pp. 239–259.

Kornhuber, A. 1873. Über einen neuen fossilen saurier aus Lesina. Herausgegeben von der k. k. geologischen Reichsanstalt. *Wien*, 5(4): 75–90.

Kornhuber, A. 1893. *Carsosaurus marchesetti*, ein neuer fossiler Lacertilier aus den Kreideschichten des Karstes bei Komen. *Abhandlungen der geologischen Reichsanstalt Wien*, 17: 1–15.

Kornhuber, A. 1901. *Opetiosaurus bucchichi*, eine neue fossile Eidechse aus der unteren Kreide von Lesina in Dalmatien. *Abhandlungen der geologischen Reichsanstalt Wien*, 17: 1–24.

Kuhn, T. S. 1962. *The Structure of Scientific Revolutions* Chicago. University of Chicago Press, Chicago, pp. 1–212.

Kvon, E. Z., Kamneva, O. K., Melo, U. S., Barozzi, I., Osterwalder, M., Mannion, B. J., Tissiéres, V. et al. 2016. Progressive loss of function in a limb enhancer during snake evolution. *Cell*, 167: 633–642.

Laduke, T. C., Krause, D. W., Scanlon, J. D., and Kley, N. J. 2010. A Late Cretaceous (Maastrichtian) snake assemblage from the Maevarano Formation, Mahajanga Basin, Madagascar. *Journal of Vertebrate Paleontology*, 30: 109–113.

Laing, A. M., Doyle, S., Gold, M. E. L., Nesbitt, S. J., O'Leary, M. A., Turner, A. H., Wilberg, E. W., and Poole, K. E. 2017. Giant taxon-character matrices: The future of morphological systematics. *Cladistics*, 34: 333–335.

Langer, W. 1961. Über das Alter der Fischschiefer von Hvar-Lesina (Dalmatien). *Neues Jarhbuch für Geologie und Päleontologie, Monatschafte*, 1961: 329–331.

Lawson, R., Slowinski, J. B., and Burbrink, F. T. 2004. A molecular approach to discerning the phylogenetic placement of the enigmatic snake *Xenophidion schaeferi* among the Alethinophidia. *Journal of Zoology, London*, 263: 285–294.

Leal, F., and Cohn, M. J. 2016. Loss and re-emergence of legs in snakes by modular evolution of Sonic hedgehog and HOXD enhancers. *Current Biology*, 26: 2966–2973.

Leal, F., and Cohn, M. J. 2017. Developmental, genetic, and genomic insights into the evolutionary loss of limbs in snakes. *Genesis*, 56: 1–12.

Leanza, H. A., Apesteguía, S., Novas, F. E., and de la Fuente, M. S. 2004. Cretaceous terrestrial beds from the Neuquén Basin (Argentina) and their tetrapod assemblages. *Cretaceous Research*, 25: 61–87.

LeBlanc, A., Caldwell, M. W., and Lindgren, J. 2013. Aquatic adaptation, cranial kinesis and the skull of the mosasaurine *Plotosaurus bennisoni*. *Journal of Vertebrate Paleontology*, 33: 349–362.

LeBlanc, A. R. H., Brink, K. S., Cullen, T., and Reisz, R. R. 2017a. Evolutionary implications of tooth attachment versus tooth implantation: A case study using dinosaur, crocodilian, and mammal teeth. *Journal of Vertebrate Paleontology*, 37: e1354006.

LeBlanc, A. R. H., Lamoureux, D. O., and Caldwell, M. W. 2017b. Mosasaurs and snakes have a periodontal ligament: Timing of calcification, not tissue complexity, determines tooth attachment mode in reptiles. *Journal of Anatomy*, 231: 869–885.

LeBlanc, A. R. H., and Reisz, R. R. 2013. Periodontal ligament, cementum, and alveolar bone in the oldest herbivorous tetrapods, and their evolutionary significance. *PLOS ONE*, 8: e74697.

LeBlanc, A. R. H., Reisz, R. R., Brink, K. S., and Abdala, F. 2016a. Mineralized periodontia in extinct relatives of mammals shed light on the evolutionary history of mineral homeostasis in periodontal tissue maintenance. *Journal of Clinical Periodontology*, 43: 323–332.

LeBlanc, A. R. H., Reisz, R. R., Evans, D. C., and Bailleul, A. M. 2016b. Ontogeny reveals function and evolution of the hadrosaurid dinosaur dental battery. *BMC Evolutionary Biology*, 16: 152, doi:10.1186/s12862-016-0721-1

Lee, M. S. Y. 1997a. The phylogeny of varanoid lizards and the affinities of snakes. *Philosophical Transactions of the Royal Society, Series B*, 352: 53–91.

Lee, M. S. Y. 1997b. On snake-like dentition in mosasaurian lizards. *Journal of Natural History*, 31: 303–314.

Lee, M. S. Y. 1998. Convergent evolution and character correlation in burrowing reptiles: Towards a resolution of squamate phylogeny. *Biological Journal of the Linnean Society*, 65: 369–453.

Lee, M. S. Y. 2005a. Molecular evidence and marine snake origins. *Biology Letters*, 1: 227–230.

Lee, M. S. Y. 2005b. Squamate phylogeny, taxon sampling, and data congruence. *Organisms, Diversity and Evolution*, 5: 25–45.

Lee, M. S. Y. 2005c. Phylogeny of snakes (Serpentes): Combining morphological and molecular data in likelihood, Bayesian and parsimony analyses. *Systematics and Biodiversity*, 5: 371–389.

Lee, M. S. Y., Bell, G. L., and Caldwell, M. W. 1999a. The origins of snake feeding. *Nature*, 400: 655–659.

Lee, M. S. Y., and Caldwell, M. W. 1998. Anatomy and relationships of *Pachyrhachis*, a primitive snake with hindlimbs. *Philosophical Transactions of the Royal Society, Series B*, 353: 1521–1552.

Lee, M. S. Y., and Caldwell, M. W. 2000. *Adriosaurus* and the affinities of mosasaurs, dolichosaurs, and snakes. *Journal of Paleontology*, 74: 915–937.

Lee, M. S. Y., Caldwell, M. W., and Scanlon, J. D. 1999b. A second primitive marine snake: *Pachyophis woodwardi* Nopcsa. *Journal of Zoology*, 248: 509–520.

Lee, M. S. Y., Palci, A., Jones, M. E. H., Caldwell, M. W., Holmes, J. D., and Reisz, R. R. 2016. Aquatic adaptations in the four limbs of the snake-like reptile *Tetrapodophis* from the Lower Cretaceous of Brazil. *Cretaceous Research*, 66: 194–199.

Lee, M. S. Y., and Scanlon, J. D. 2001. On the lower jaw and intramandibular septum in snakes and anguimorph lizards. *Copeia*, 2001: 531–535.

Lee, M. S. Y., and Scanlon, J. D. 2002a. The Cretaceous marine squamate *Mesoleptos* and the origin of snakes. *Bulletin of the Natural History Museum, London*, 68: 131–142.

Lee, M. S. Y., and Scanlon, J. D. 2002b. Snake phylogeny based on osteology, soft anatomy and ecology. *Biological Journal of the Linnean Society*, 77: 333–401.

Leidy, J. 1865. Cretaceous reptiles of the United States. *Smithsonian Contributions to Knowledge*, 192: 1–135.

Lessmann, M. H. 1952. Zur labialen Pleurodontie an Lacertilier Gebissen. *Anatomischer Anzeiger*, 99: 35–67.

Linnaeus, C. 1735. *Systema Naturae, Sive Regna Tria Naturae Sistematice Proposita per Classes, Ordines, Genera et Species.* Lugduni Batavorum, Leiden, Netherlands, 12 pp.

Linnaeus, C. 1740. *Systema naturæ in quo naturae regna tria, secundum classes, ordines, genera, species, systematice proponuntur.* Editio secunda, auctior, Stockholm, 80 pp.

Linnaeus, C. 1756. *Systema naturae sistens regna tria naturæ in classes et ordines, genera et species redacta, tabulisque æneis illustrata.* Lugduni Batavorum, Leiden, Netherlands, 227 pp.

Linnaeus, C. 1758. *Systema Naturae per regna tria naturae, secundum classes, ordines, genera, species, cum characteribus, differentiis, synonymis, locis. Editio decima, reformata.* Laurentius Salvius, Stockholm, 824 pp.

List, J. C. 1966. Comparative Osteology of the snake families Typhlopidae and Leptotyphlopidae. *Illinois Biological Monographs*, 36: 1–112.

Longrich, N. R., Bhullar, B.-A. S., and Gauthier, J. A. 2012. A transitional snake from the Late Cretaceous period of North America. *Nature*, 488: 205–208.

Luan, X., Walker, C., Dangaria, S., Ito, Y., Druzinsky, R., Jarosius, K., Lesot, H., and Rieppel, O. 2009. The mosasaur tooth attachment apparatus as paradigm for the evolution of the gnathostome periodontium. *Evolution and Development*, 11: 247–259.

Lucas, E. A. 1896. A new snake from the Eocene of Alabama. *Proceedings of the United States National Museum*, 21: 637–638.

Lydekker, R. 1893. The dinosaurs of Patagonia. Anales del Museo de la Plata. *Seccion de Palaeontologia*, 2: 1–14.

Lyson, T. R., Rubidge, B. S., Scheyer, T.S., de Queiroz, K., Schachner, E. R., Smith, R. M. H., Botha-Brink, J., and Bever, G. S. 2016. Fossorial Origin of the Turtle Shell. *Current Biology*, 26: 1887–1894.

Mahendra, B. C. 1936. Contributions to the osteology of the Ophidia. I. The Endoskeleton of the So-called "Blind-Snake", *Typhlops braminus* Daud. *Proceedings of the Indian Academy of Sciences*, 3: 128–142.

Mahendra, B. C. 1938. Some remarks on the phylogeny of the Ophidia. *Anatomischer Anzeiger*, 86: 347–356.

Mahler, D. H., and Kearney, M. 2006. The palatal dentition in squamate reptiles: Morphology, development, attachment, and replacement. *Fieldiana, Zoology*, NS 108, 1540: 1–61.

Maisey, J. G. 1993. Tectonics, the Santana lagerstätten, and the implications for late Gondwanan biogeography. In Goldblatt, P. (ed.) *Biological Relationships between Africa and South America.* Yale University Press, New Haven, CT, pp. 435–454.

Makovicky, P. J., Apesteguía, S., and Agnolín, F. L. 2005. The earliest dromaeosaurid theropod from South America. *Nature*, 437: 1007–1011.

Makovicky, P. J., Apesteguía, S., and Gianechini, F. A. 2012. A new coelurosaurian theropod from the La Buitrera fossil locality of Río Negro. Argentina. *Fieldiana Life and Earth Sciences*, 5: 90–98.

Malnate, E. V. 1960. Systematic division and evolution of the colubrid snake genus *Natrix*, with comments on the subfamily Natricinae. *Proceedings of the Academy of Natural Sciences, Philadelphia*, 112: 41–71.

Marsh, O. C. 1871. Notice of some new fossil reptiles from the Cretaceous and Tertiary formations. *American Journal of Science, Series 3*, 1: 447–459.

Marsh, O. C. 1880. New characters of mosasauroid reptiles. *American Journal of Science, Series 3*, 19: 83–87.

Marsh, O. C. 1892. Notice of new reptiles from the Laramie Formation. *American Journal of Science*, 43: 449–453.

Martill, D. M., Bechly, G., and Loveridge, R. F. (eds.) 2007. *The Crato Fossil Beds of Brazil: Window into an Ancient World.* Cambridge University Press, Cambridge, pp. 577–581.

Martill, D. M., Tischlinger, H., and Longrich, N. R. 2015. A four-legged snake from the Early Cretaceous of Gondwana. *Science*, 349: 416–419.

Martin, T., and Krebs, B. 2000. *Guimarota. A Jurassic Ecosystem.* Verlag Dr. Friedrich Pfeil, München, Germany, pp. 1–155.

Martins, A., Passos, P., and Pinto, R. 2018. Unveiling diversity under the skin: Comparative morphology study of the cephalic glands in threadsnakes (Serpentes: Leptotyphlopidae: Epictinae). *Zoomorphology*, 137: 433–443.

Massalongo, D. A. B. 1859. *Specimen photographicum animalium quorundam platarumque fossilium agri Veronensis.* Vicentini-Franchini, Verona, Italy, pp. 1–101.

Mattison, C. 1995. *The Encyclopedia of Snakes.* Checkmark Books Ltd., New York, pp. 1–257.

Maxwell, E. E., Caldwell, M. W., and Lamoureux, D. O. 2010. Tooth histology in the Cretaceous ichthyosaur *Platypterygius australis*, and its significance for the conservation and divergence of mineralized tooth tissues in amniotes. *Journal of Morphology*, 272(2): 129–135.

Maxwell, E. E., Caldwell, M. W., and Lamoureux, D. 2011a. Evolution of tooth histology, attachment, and replacement in the Ichthyopterygia. *Palaeontologische Zeitschrift*, doi:10.1007/s12542-011-0115-z

Maxwell, E. E., Caldwell, M. W., and Lamoureux, D. 2011c. The structure and phylogenetic distribution of amniote plicidentine. *Journal of Vertebrate Paleontology*, 31: 553–561.

Maxwell, E. E., Caldwell, M. W., Lamoureux, D., and Budney, L. 2011b. Histology of tooth attachment tissues and plicidentine in *Varanus* (Reptilia: Squamata), and a discussion of the evolution of amniote tooth attachment. *Journal of Morphology*, 272: 1170–1181.

McDowell, S. B. 1961. Review of "Systematic division and evolution of the colubrid snake genus *Natrix*, with comments on the subfamily Natricinae" by Edmond V. Malnate. *Copeia*, 1961: 502–506.

McDowell, S. B. 1967. The extracolumella and tympanic cavity of the "earless" monitor lizard, *Lanthanotus borneensis*. *Copeia*, 1967: 154–159.

McDowell, S. B. 1974. A catalogue of the snakes of New Guinea and the Solomons, with special reference to those in the Bernice P. Bishop Museum. Part I. Scolecophidia. *Journal of Herpetology*, 8: 1–57.

McDowell, S. B. 1975. A catalogue of the snakes of New Guinea and the Solomons, with special reference to those in the Bernice P. Bishop Museum. Part II. Anilioidea and Pythoninae. *Journal of Herpetology*, 9: 1–79.

McDowell, S. B. 1979. A catalogue of the snakes of New Guinea and the Solomons, with special reference to those in the Bernice P. Bishop Museum. Part III. Boinae and Acrochordoidea (Reptilia, Serpentes). *Journal of Herpetology*, 13: 1–92.

McDowell, S. B. 1987. Systematics. In Siegel, R. A., Collins, J. T., and Novak, S. S. (eds.) *Snakes: Ecology and Evolutionary Biology*. Macmillan, New York, pp. 3–50.

McDowell, S. B. 2008. The Skull of Serpentes. In Gans, C., Gaunt, A. S., and Adler, K. (eds.) *Biology of the Reptilia. Vol. 21. Morphology I. The Skull and Appendicular Locomotor Apparatus of Lepidosauria*. Society for the Study of Amphibians and Reptiles, Ithaca, NY, Contributions in Herpetology. 24: vii + 784 pp.

McDowell, S. B., and Bogert, C. M. 1954. The systematic position of *Lanthanotus* and the affinities of the anguimorph lizards. *Bulletin of the American Museum of Natural History*, 105: 1–142.

McKerrow, W. S., Johnson, R. T., and Jakobson, M. E. 1969. Palaeoecological studies in the Great Oolite at Kirtlington. *Oxfordshire*, 12: 56–83.

McMillan, I. 2003. *The Foraminifera of the Late Valanginian to Hauterivian (Early Cretaceous) Sundays River Formation of the Algoa Basin, Eastern Cape Province, South Africa*. South African Museum, Cape Town, South Africa, pp. 1–274.

Mead, J. I. 2013. Scolecophidia (Serpentes) of the Late Oligocene and Early Miocene, North America, and a fossil history overview. *Geobios*, 46: 225–231.

Merrem, B. 1820. *Versuch eines Systems der Amphibia*. Johann Christian Krieger, Marburg, Germany, 192 pp.

Middleton, G. V. 1973. Johannes Walther's Law of the Correlation of Facies. *Geological Society of America Bulletin*, 84: 979–988.

Miller, W. 2011. *Trace Fossils: Concepts, Problems, Prospects*. 1st ed. Elsevier Science, Amsterdam, pp. 1–632.

Miralles, A., Marin, J., Markus, D., Herrel, A., Hedges, S. B., and Vidal, N. 2018. Molecular evidence for the paraphyly of Scolecophdia and its evolutionary implications. *Journal of Evolutionary Biology*, 31: 1782–1793.

Mohabey, D. M., Head, J. J., and Wilson, J. A. 2011. A new species of the snake *Madtsoia* from the Upper Cretaceous of India and its paleobiogeographic implications. *Journal of Vertebrate Paleontology*, 31: 588–595.

Moody, S. M. 1985. Charles L. Camp and his 1923 Classification of Lizards: An Early Cladist? *Systematic Biology*, 34: 216–222.

Muir, R. A., Bordy, E. M., and Prevec, R. 2015. Lower Cretaceous deposit reveals evidence of a post-wildfire debris flow in the Kirkwood Formation, Algoa Basin, Eastern Cape South Africa. *Cretaceous Research*, 56: 161–179.

Muir, R. A., Bordy, E. M., and Prevec, R. 2017. Lithostratigraphy of the Kirkwood Formation (Uitenhage Group), including the Bethelsdorp, Colchester and Swartkops Members, South Africa. *South African Journal of Geology*, 120: 281–293.

de Muizon, C., Gayet, M., Lavenu, A., Marshall, L. G., Sige, B., and Villaroel, C. 1983. Late Cretaceous vertebrates, including mammals, from Tiumpampa, southcentral Bolivia. *Geobios*, 16: 747–753.

Myers, T. S., Tabor, N. J., Jacobs, L. L., and Mateus, O. 2012. Palaeoclimate of the Late Jurassic of Portugal: Comparison with the Western United States. *Sedimentology*, 59: 1695–1717.

Müller, J. 1831. Beiträge zur Anatomie und Naturgeschichte der Amphibien. *Zeitschrift für Physiologie*, 4: 90–275.

Müller, J., Hipsley, C. A., Head, J. J., Kardjilov, N., Hilger, A., Wuttke, M., and Reisz, R. R. 2011. Eocene lizard from Germany reveals amphisbaenian origins. *Nature*, 473: 364–367.

Neuman, V. H., and Cabrera, L. 1999. *Una nueva propuesta estratigrafica para la tectonosecuencia post-rifte de la cuenca de Araripe, noreste de Brasil*. In Boletim Do 5, simposio sobre o Cretaceo do Brasil, São Paulo, pp. 279–285.

Nopcsa, F. B. 1903. Über die varanusartigen lacerten Istriens. *Beiträge zur Paläontologie und Geologie Österreich-Ungarns und des Orients*, 15: 31–42.

Nopcsa, F. B. 1908. Zur Kenntnis der fossilen Eidechsen. *Beiträge zur Paläontologie und Geologie Österreich-Ungarns und des Orients*, 21: 33–62.

Nopcsa, F. B. 1923. *Eidolosaurus* und *Pachyophis*. Zwei neue Neocom-Reptilien. *Palaeontographica*, 65: 99–154.

Nopcsa, F. B. 1925. Ergebnisse der Forschungsreisen Prof. E. Stromers in den Wüsten Ägypten. II. Wirbeltier-Reste der Baharîje-Stufe (unterstes Cenoman). 5. Die *Symoliophis*-Reste. Abhandlungen Bayerischstaatsamlung Akademische Wissschaften, Math.-naturwisschaften. Abt., 30: 1–27.

Norell, M. A. 1992. Taxic origin and temporal diversity: The effect of phylogeny. In Novacek, M. J., and Wheeler, Q. D. (eds.) *Extinction and phylogeny*. Columbia University Press, New York, pp. 89–118.

Néraudeau, D., Perrichot, V., Dejax, J., Masure, E., Nel, A., Philippe, M., Pierre Moreau, P., François Guillocheau, F., and Guyot, T. 2002. Un nouveau gisement à ambre insectifère et à végétaux (Albien terminal probable): Archingeay (Charente-Maritime, France). *Geobios*, 35: 233–240.

Néraudeau, D., Vullo, R., Gomez, B., Girard, V., Lak, M., Videt, B., Dépré, É., and Perrichot, V. 2009. Amber, plant and vertebrate fossils from the Lower Cenomanian paralic facies of Aix Island (Charente-Maritime, SW France). *Geodiversitas*, 31: 13–27.

O'Leary, M. A., Bloch, J. I., Flynn, J. J., Gaudin, T. J., Giallombardo, A., Giannini, N. P., Goldberg, S. L. et al. 2013. The placental mammal ancestor and the post–K-Pg radiation of placentals. *Science*, 339: 662–667.

Oliveira, L., Jared, C., da Costa Prudente, A. L., Zaher, H., and Antonazzi, M. M. 2008. Oral glands in dipsadine "goo-eater" snakes: Morphology and histochemistry of the infralabial glands in *Atractus reticulatus, Dipsas indica,* and *Sibynomorphus mikanii*. *Toxicon*, 58: 898–913.

Olori, J. C., and Bell, C. J. 2012. Comparative skull morphology of uropeltid snakes (Alethinophidia: Uropeltidae) with special reference to disarticulated elements and variation. *PLOS ONE*, 7: e32450, doi:10.1371/journal.pone.0032450

Oppel, M. 1811. *Die Ordnungen, Familien, und Gattungen der Reptilien als Prodrom einer Naturgeschichte derselben.* J. Lindauer, München, Germany, 86 pp.

Osborn, J. W. 1984. From reptile to mammal: Evolutionary considerations of the dentition with emphasis on tooth attachment. *Symposia of the Zoological Society of London*, 52: 549–574.

Ostrom, J. 1962. On the constrictor dorsalis muscles of *Sphenodon. Copeia*, 1962: 732–735.

Owen, R. 1840. *Odontography; or, a Treatise on the Comparative Anatomy of the Teeth; Their Physiological Relations, Mode of Development, and Microscopic Structure, in the Vertebrate Animals.* Hippolyte Bailliere, London.

Owen, R. 1841. XXI. Description of some Ophidiolites (Palaeophis toliapicus) from the London Clay at Sheppey, indicative of an extinct species of Serpent. *Transactions of the Geological Society of London,* 6: 209–210.

Owen, R. 1849. On the Archetype and Homologies of the Vertebrate Skeleton (London, 1848), p. 177.

Owen, R. 1850. Description of the fossil reptiles of the Chalk Formation. In Dixon, F. (ed.) *The Geology and Fossils of the Tertiary and Cretaceous Formations of Sussex.* Longman, Brown, Green, and Longman, London, pp. 378–404.

Owen, R. 1863. On the *Archaeopteryx* of Von Meyer, with a description of the fossil remains of a long-tailed species from the lithographic stone of Solnhofen. *Philosophical Transactions of the Royal Society,* 153: 33–47.

Owen, R. 1877. On the rank and affinities of the reptilian class of Mosasauridae, Gervais. *Quarterly Journal of the Geological Society of London,* 33: 682–715.

Palci, A., and Caldwell, M. W. 2007. Vestigial forelimbs and axial elongation in a 95-million-year-old non-snake squamate. *Journal of Vertebrate Paleontology*, 27: 1–7.

Palci, A., and Caldwell, M. W. 2010. Redescription of *Acteosaurus tommasinii* Von Meyer, 1860, and a discussion of evolutionary trends within the Clade Ophidiomorpha. *Journal of Vertebrate Paleontology*, 30: 94–108.

Palci, A., and Caldwell, M. W. 2013. Primary homologies of the circumorbital bones of snakes. *Journal of Morphology,* 274: 973–986.

Palci, A., and Caldwell, M. W. 2014. The Upper Cretaceous snake *Dinilysia patagonica* Smith-Woodward, 1901, and the Crista Circumfenestralis of snakes. *Journal of Morphology,* 275: 1187–1200.

Palci, A., Caldwell, M. W., and Nydam, R. 2013. Reevaluation of the anatomy of the Cenomanian (Upper Cretaceous) hind-limbed marine fossil snakes Pachyrhachis, Haasiophis and Eupodophis. *Journal of Vertebrate Paleontology*, 33: 1328–1342.

Palci, A., Lee, M. S. Y., and Hutchinson, M. N. 2016. Patterns of postnatal ontogeny of the skull and lower jaw of snakes as revealed by micro-CT scan data and three-dimensional geometric morphometrics. *Journal of Anatomy*, 229: 723–754.

Palci, A., Hutchinson, M. N., Caldwell, M. W., and Lee, M. S. Y. 2017. The morphology of the inner ear of squamate reptiles and its bearing on the origin of snakes. *Royal Society Open Science,* 4: 170685.

Palci, A., Hutchinson, M. N., Caldwell, M. W., Scanlon, J. D., and Lee, M. S. Y. 2018. Palaeoecological inferences for the fossil Australian snakes Yurlunggur and *Wonambi* (Serpentes, Madtsoiidae). *Royal Society Open Science*, 5: 172012.

Palmer, T. J., and Jenkyns, H. C. 1975. A carbonate island barrier from the Great Oolite (Middle Jurassic) of central England. *Sedimentology*, 22: 125–135.

Paparella, I., Palci, A., Nicosia, U., and Caldwell, M. W. 2018. A new fossil marine lizard with soft tissues from the Late Cretaceous of Southern Italy. *Royal Society Open Science*, 5(6): 172411, doi:10.6084/m9

Parker, H. W., and Grandison, A. G. C. 1977. *Snakes – A Natural History*. British Museum of Natural History, London, pp. 1–112.

Parker, W. K. 1878a. On the structure and development of the skull in the Lacertilia. Part I. On the skull of the common lizards (*Lacerta agilis, L. viridis*, and *Zootoca vivipara*). *Proceedings of the Royal Society, London*, 28: 214–218.

Parker, W. K. 1878b. XII. On the structure and development the skull in the common snake (tropidonotus natrix). *Philosophical Transactions of the Royal Society*, 169: 385–425.

Patchell, F. C., and Shine, R. 1986. Hinged teeth for hard-bodied prey: A case of convergent evolution between snakes and legless lizards. *Journal of Zoology*, 208: 269–275.

Patterson, C. 1982. Morphological characters and homology. In Joysey, K. A., and Friday, A. E. (eds.) *Problems of Phylogenetic Reconstruction*. Academic Press, London, pp. 21–74.

Patterson, C. 1988. Homology in classical and molecular biology. *Molecular Biology and Evolution*, 5: 603–625.

Peyer, B. 1912. Die Entwicklung des Schadelskelettes von *Vipera aspis*. *Morphologishcer Jahrbuch*, 44: 563–621.

Peyer, B. 1968. *Comparative Odontology*. The University of Chicago Press, Chicago.

Pianka, E. R., and Vitt, L. J. 2003. *Lizards: Windows to the Evolution of Diversity*. University of California Press, Berkeley, CA, 346 pp.

Pierce, S. E., and Caldwell, M. W. 2004. Redescription and phylogenetic position of the Adriatic (Upper Cretaceous; Cenomanian) dolichosaur, *Pontosaurus lesinensis* (Kornhuber, 1873). *Journal of Vertebrate Paleontology*, 24: 373–386.

de Pinna, M. 1991. Concepts and tests of homology in the cladistic paradigm. *Cladistics*, 7: 367–394.

Pinto, R. R., Martins, A. R., Curcio, F., and Ramos, L. d. O. 2015. Osteology and Cartilaginous Elements of *Trilepida salgueiroi* (Amaral, 1954) (Scolecophidia: Leptotyphlopidae). *The Anatomical Record*, 298: 1722–1747.

Pol, D., and Apesteguía, S. 2005. New Araripesuchus remains from the Early Late Cretaceous (CenomanianeTuronian) of Patagonia. *American Museum Novitates*, 3490: 1–38.

Polcyn, M. J., Jacobs, L. L., and Haber, A. 2005. A morphological model and CT assessment of the skull of *Pachyrhachis problematicus* (Squamata, Serpentes), a 98 million year old snake with legs from the Middle East. *Palaeontologia Electronica*, 8(1): 26A, 24p, http://palaeo-electronica.org/paleo/2005_1/polcyn26/issue1_0.5.htm

Poole, D. F. G. 1967. Phylogeny of tooth tissues: Enameloid and enamel in recent vertebrates, with a note on the history of cementum. In Miles, A. E. W. (ed.) *Structural and Chemical Organization of the Teeth*. Academic Press, New York, pp. 111–149.

Popper, K. R. 1959. *The Logic of Scientific Discovery*. Basic Books, New York, pp. 1–479.

Presley, R. 1993. Preconception of adult structural pattern in the analysis of the developing skull. In Hanken, J., and Hall, B.K. (eds.) *The Skull Volume 1*. University of Chicago Press, Chicago, pp. 347–377.

Pyron, R. A. 2010. A likelihood method for assessing molecular divergence time estimates and the placement of fossil calibrations. *Systematic Biology*, 59: 185–194.

Pyron, R. A., Burbrink, F. T., Colli, G. R., de Oca, A. N. M., Vitt, L. J., Kuczynski, C. A., and Wien, J. J. 2011. The phylogeny of advanced snakes (Colubroidea), with discovery of a new subfamily and comparison of support methods for likelihood trees. *Molecular Phylogenetics and Evolution*, 58: 329–342.

Pyron, R. A., Burbrink, F. T., and Wiens, J. J. 2013. A phylogeny and updated classification of Squamata, including 4161 species of lizards and snakes. *BMC Evolutionary Biology*, 13: 93, doi:10.1186/1471-2148-13-93

Pyron, R. A., Hendry, C. R., Chou, V. M., Lemmon, E. M., Lemmon, A. R., and Burbrink, F. T. 2014. Effectiveness of phylogenomic data and coalescent species-tree methods for resolving difficult nodes in the phylogeny of advanced snakes (Serpentes: Caenophidia). *Molecular Phylogenetics and Evolution*, 81: 221–231.

Queral-Regil, A., and King, R. B. 1998. Evidence for phenotypic plasticity in snake body size and relative head dimensions in response to amount and size of prey. *Copeia*, 1998: 423–429.

Rage, J.-C. 1977. La position phyletique de Dinilysia patagonica, serpent du Cretace superieur. *Comptes Rendus de l'Academie des Sciences*, 284: 1765–1768.

Rage, J.-C. 1984. *Serpentes*. Handbuch der Paläoherpetologie Teil 11. Gustav Fischer Verlag, Stuttgart, Germany, p. 80.

Rage, J.-C., and Albino, A. 1989. *Dinilysia patagonica* (Reptilia, Serpentes); materiel vertebral additionnel du

Cretacé superieur d'Argentine. Etude complementaire des vertebres, variations intraspecifiques et intracolumnaires. *Neues Jarhbuch für Geologie und Paläontologie, Monatschefte*, 1989: 433–447.

Rage, J.-C., and Escuillié, F. 2000. Un nouveau serpent bipede du Cénomanien (Crétacé). Implications phyletiques. *Comptes Rendu de la Academie des Sciences, Paris, Serie Ia*, 330: 1–8.

Rage, J.-C., and Escuillié, F. 2003. The Cenomanian: Stage of hindlimbed snakes. *Carnets de Géologie/Notebooks on Geology*, 1: 1–11.

Rage, J.-C., Prasad, G. V., and Bajpai, S. 2004. Additional snakes from the uppermost Cretaceous (Maastrichtian) of India. *Cretaceous Research*, 25: 425–434.

Rage, J.-C., and Richter, A. 1994. A snake from the Lower Cretaceous (Barremian) of Spain: The oldest known snake. *Neues Jarhbuch für Geologie und Pälaontologie, Monatschefte*, 1994: 561–565.

Rage, J.-C., Vullo, R., and Néraudeau, D. 2016. The mid-Cretaceous snake *Simoliophis rochebrunei* Sauvage, 1880 (Squamata: Ophidia) from its type area (Charentes, southwestern France): Redescription, distribution, and palaeoecology. *Cretaceous Research*, 58: 234–253.

Rage, J.-C., and Werner, C. 1992. Mid-Cretaceous (Cenomanian) snakes from Wadi Abu Hashim, Sudan: The earliest snake assemblage. *Palaeontologia Africana*, 35: 85–110.

Redi, F. 1668. *Esperienze intorno alla generezione degli' inseti.* Florence, Italy.

Reeder, T. W., Townsend, T. M., Mulcahy, D. G., Noonan, B. P., Wood, P. L., Jr., Sites, J. W., Jr. and Wiens, J. W. 2015. Integrated analyses resolve conflicts over squamate reptile phylogeny and reveal unexpected placements for fossil taxa. *PLOS ONE*, 10(3): e0118199, doi:10.1371/journal.pone.0118199

Riboulleau, A., Schnyder, J., Riquier, L., and Decninck, J.-F. 2007. Environmental change during the Early Cretaceous in the Purbeck-type Durlston Bay section (Dorset, Southern England): A biomarker approach. *Organic Geochemistry*, 11: 1804–1823.

Rieppel, O. 1976. The homology of the laterosphenoid bone in snakes. *Herpetologica*, 32: 426–428.

Rieppel, O. 1977. Studies on the skull of the Henophidia (Reptilia: Serpentes). *Journal of Zoology*, 181: 145–173.

Rieppel, O. 1978. The evolution of the naso-frontal joint in snakes and its bearing on snake origins. *Zeitschrift für zoologische Systematik und Evolutionsforschung*, 16: 14–27.

Rieppel, O. 1979a. A cladistic classification of primitive snakes based on skull structure. *Journal of Zoological Systematics and Evolutionary Research*, 17: 140–150.

Rieppel, O. 1979b. The braincase of *Typhlops* and *Leptotyphlops* (Reptilia: Serpentes). *Zoological Journal of the Linnaean Society*, 65: 161–176.

Rieppel, O. 1980a. *The Phylogeny of Anguinomorph Lizards.* Birkhauser Verlag, Basel, Switzerland, pp. 86.

Rieppel, O. 1980b. The sound-transmitting apparatus in primitive snakes and its phylogenetic significance. *Zoomorphology*, 96: 45–62.

Rieppel, O. 1980c. The evolution of the ophidian feeding system. *Zoological Journal*, 103: 551–564.

Rieppel, O. 1981. The hyobranchial skeleton in some little known lizards and snakes. *Journal of Herpetology*, 15: 433–440.

Rieppel, O. 1983. A comparison of the skull of *Lanthanotus borneensis* (Reptilia: Varanoidea) with the skull of primitive snakes. *Zeitschrift für zoologische Systematik und Evolutionsforschung*, 21: 142–153.

Rieppel, O. 1984. The cranial morphology of the fossorial lizard genus *Dibamus* with a consideration of its phylogenetic relationships. *Journal of Zoology, London*, 204: 289–327.

Rieppel, O. 1985. The recessus scalae tympani and its bearing on the classification of reptiles. *Journal of Herpetology*, 19: 373–384.

Rieppel, O. 1988a. A review of the origin of snakes. In Hecht, M. K., Wallace, B., and Prance, G. T. (eds.) *Evolutionary Biology.* Plenum Press, USA, Vol. 22, pp. 37–130.

Rieppel, O. 1988b. *Fundamentals of Comparative Biology.* Birkäuser Verlag, Germany, pp. 202.

Rieppel, O. 1992. The skull in a hatchling of *Sphenodon punctatus. Journal of Herpetology*, 26: 80–84.

Rieppel, O. 1996. Miniaturization in tetrapods: Consequences for skull morphology. In Miller, PJ (ed.) *Miniature Vertebrates: The Implications of Small Body Size, Vol. 69. Symposia of the Zoological Society of London.* Claredon Press, Oxford, pp. 47–61.

Rieppel, O. 2012. "Regressed" macrostomatan snakes. *Fieldiana (Life and Earth Scince)*, 2012: 99–103.

Rieppel, O., and Grande, L. 2007. The anatomy of the fossil varanoid lizard *Saniwa ensidens* Leidy, 1870, based on a newly discovered complete skeleton. *Journal of Paleontology*, 81: 643–665.

Rieppel, O., and Head, J. 2004. New specimens of the fossil snake genus *Eupodophis* Rage and Escuillié, from Cenomanian (Late Cretaceous) of Lebanon. *Memorie della Societá Italiana di Scienze Naturali e del Museo Civico di Storia Naturale di Milano*, 32: 1–26.

Rieppel, O., and Kearney, M. 2001. The origin of snakes: Limits of a scientific debate. *Biologist*, 48: 110–114.

Rieppel, O., and Kearney, M. 2002. Similarity. *Biological Journal of the Linnean Society*, 75: 59–82.

Rieppel, O., and Kearney, M. 2005. Tooth replacement in the Late Cretaceous mosasaur *Clidastes*. *Journal of Herpetology*, 39: 168–172.

Rieppel, O., Kley, N. J., and Maisano, J. A. 2009. Morphology of the skull of the white-nosed blindsnake, *Liotyphlops albirostris* (Scolecophidia: Anomalepididae). *Journal of Morphology*, 270: 536–557.

Rieppel, O., Kluge, A. G., and Zaher, H. 2003b. Testing the phylogenetic relationships of the Pleistocene snake *Wonambi naracoortensis* Smith. *Journal of Vertebrate Paleontology*, 22: 812–829.

Rieppel, O., and Maisano, J. A. 2007. The skull of the rare Malaysian snake *Anomochilus leonardi* Smith, based on high-resolution X-ray computed tomography. *Zoological Journal of the Linnaean Society*, 149: 671–685.

Rieppel, O., and Zaher, H. 2000a. The intramandibular joint in squamates, and the phylogenetic relationships of the fossil snake *Pachyrhachis problematicus* Haas. *Fieldiana, Geology, New Series*, 43: 1–69.

Rieppel, O., and Zaher, H. 2000b. The braincases of mosasaurs and *Varanus*, and the relationships of snakes. *Zoological Journal of the Linnean Society*, 129: 489–514.

Rieppel, O., and Zaher, H. 2001. The development of the skull in *Acrochordus granulatus* (Schneider) (Reptilia: Serpentes), with special consideration of the Otico-occipital complex. *Journal of Morphology*, 249: 252–266.

Rieppel, O., Zaher, H., Tchernov, E., Jacobs, L. L., and Polcyn, M. J. 2003a. The anatomy and relationships of *Haasiophis terrasanctus*, a fossil snake with well-developed hind limbs from the mid-Cretaceous of the Middle East. *Journal of Paleontology*, 77: 336–358.

Rio, J. P., and Mannion, P. D. 2017. The osteology of the giant snake *Gigantophis garstini* from the upper Eocene of North Africa and its bearing on the phylogenetic relationships and biogeography of Madtsoiidae. *Journal of Vertebrate Paleontology*, 37: e1347179.

Rochebrune, A. T. 1881. Memoire sur les vertebres des ophidiens. *Journal de l'anatomie et de la physiologie normales et pathologiques de l'homme et des animaux, Paris*, 17: 185–229.

Rodríguez-López, J. P., Clemmensen, L. B., Lancaster, N., Mountney, N. P., and Veiga, G. D. 2014. Archaen to Recent aeolian sand systems and their sedimentary record: Current understanding and future prospects. *Sedimentology*, 61: 1487–1534.

Rogers, R. R., Hartman, J. H., and Krause, D. W. 2000. Stratigraphic analysis of Upper Cretaceous Rocks in the Mahajanga Basin, northwestern Madagascar: Implications for ancient and modern faunas. *Journal of Geology*, 108: 275–301.

Romer, A. S. 1956. *Osteology of the Reptiles: A Comparative Summary of the Reptile Skeleton, Living and Fossil, with a Classification of the Reptile Family*. The University of Chicago Press, Chicago, pp. 1–772.

Ross, C., Sues, H.-D., and de Klerk, W. J. 1999. Lepidosaurian remains from the Lower Cretaceous Kirkwood Formation of South Africa. *Journal of Vertebrate Paleontology*, 19: 21–27.

Rossman, D. A., and Eberle, W. G. 1977. Partition of the genus *Natrix*, with preliminary observations on evolutionary trends in natricine snakes. *Herpetologica*, 33: 34–43.

Rougier, G. W., Apesteguía, S., and Gaetano, L. C. 2011. Highly specialized mammalian skulls from the Late Cretaceous of South America. *Nature*, 479: 98–102.

Rupik, W. 2002. Early development of the adrenal glands in the grass snake *Natrix natrix* L. (Lepidosauria, Serpentes). *Advances in Anatomy, Embryology, and Cell Biology*, 161: 1–102.

Rupke, N. 1993. Richard Owen's vertebrate archetype. *History of Science Society*, 84: 231–251.

Russell, D. 1967. Systematics and morphology of American mosasaurs. *Peabody Museum of Natural History, Yale University Bulletin*, 23: 1–241.

Samankassou, E., Tresch, J., and Strasser, A. 2005. Origin of peloids in Early Cretaceous deposits, Dorset, South England. *Facies*, 51: 264–273.

Samuelson, P. A. 1975. Alvin H. Hansen, 1887–1975. *Newsweek*, June 16: 72.

Sánchez, M. L., Heredia, S., and Calvo, J. O. 2006. Paleoambientes sedimentarios del Cretácico Superior de la Formación Plottier (Grupo Neuquén), Departamento Confluencia, Neuquén. *Revista de la Asociación Geológica Argentina*, 61: 3–18.

Sauvage H. E. 1880. Sur l'existence d'un reptile du type ophidien dans les couches à Ostrea columba des Charentes. In *Comptes rendus hebdomadaires des séances de l'Académie des Sciences*, Paris, t. 91, pp. 671–672.

Savitsky, A. H. 1981. Hinged teeth in snakes: An adaptation for swallowing hard-bodied prey. *Science*, 212: 346–349.

Scanferla, A. 2016. Postnatal ontogeny and the evolution of macrostomy in snakes. *Royal Society Open Science*, 3: 160612.

Scanferla, A., and Bhullar, B.-A. S. 2014. Postnatal development of the skull of *Dinilysia patagonica* (Squamata-Stem Serpentes). *Anatomical Record*, 297: 560–573.

Scanferla, A., Zaher, H., Novas, F. E., de Muizon, C., and Cespedes, R. 2013. A new snake skull from the Paleocene of Bolivia Sheds light on the evolution of

Macrostomatans. *PLOS ONE*, 8: e57583, doi:10.1371/journal.pone.0057583

Scanferla, C. A., and Canale, J. I. 2007. The youngest record of the Cretaceous snake genus *Dinilysia* (Squamate, Serpentes). *South American Journal of Herpetology*, 2: 76–81.

Scanferla, C. A., Smith, K. R., and Schaal, S. F. K. 2016. Revision of the cranial anatomy and phylogenetic relationships of the Eocene minute boas *Messelophis variatus* and *Messelophis ermannorum* (Serpentes, Booidea). *Zoological Journal of the Linnean Society*, 176: 182–206.

Scanlon, J. D. 1996. *Studies in the palaeontology and systematics of Australian snakes.* Unpublished PhD thesis, University of New South Wales, Sydney, pp. 648.

Scanlon, J. D. 2004. First known axis vertebra of a madtsoiid snake (*Yurlungurr camfieldensis*) and remarks on the neck of snakes. *The Beagle, Records of the northern Territory Museum of Arts and Sciences*, 20: 207–215.

Scanlon, J. D. 2005a. Cranial morphology of the Plio-Pleistocene giant madtsoiid snake *Wonambi naracoortensis*. *Acta Palaeontologica Polonica*, 50: 139–180.

Scanlon, J. D. 2005b. Australia's oldest known snakes: *Patagoniophis*, *Alamitophis*, and cf. *Madtsoia* (Squamata: Madtsoiidae) from the Eocene of Queensland. *Memoirs of the Queensland Museum (Proceedings of the Conference of Australasian Vertebrate Evolution, Palaeontology and Systematics)*, 51: 215–223.

Scanlon, J. D. 2006. Skull of the large non-macrostomatan snake *Yurlungurr* from the Australian Oligo-Miocene. *Nature*, 439: 839–842.

Scanlon, J. D., and Lee, M. S. Y. 2000. The Pleistocene serpent *Wonambi* and the early evolution of snakes. *Nature*, 403: 416–420.

Scanlon, J. D., Lee, M. S. Y., Caldwell, M. W., and Shine, R. 1999. The palaeoecology of the primitive snake *Pachyrhachis*. *Historical Biology*, 13: 127–152.

Schaal, S., and Baszio, S. 2004. *Messelophis ermannorum* n.sp. eine neue Zwergboa (Serpentes: Boidae: Tropidopheinae) aus dem Mittal-Eozän von Messel. *Courier Forschungs Institute, Senckenberg*, 252: 67–77.

Schaal, S., and Ziegler, W. 1992. *Messel: An Insight into the History of Life and of the Earth.* Oxford University Press, Oxford, pp. 1–322.

Scholz, H., and Hartman, J. H. 2007. Paleoenvironmental reconstruction of the Upper Cretaceous Hell Creek Formation of the Williston Basin, Montana, USA: Implications from the quantitative analysis of unionoid bivalve taxonomic diversity and morphologic disparity. *Palaios*, 22: 24–34.

Schudack, M. 2000. Geological setting and dating of the Guimarota beds. In Martin, T., and Krebs, B. (eds.) *Guimarota. A Jurassic Ecosystem.* Verlag Dr. Friedrich Pfeil, München, Germany, pp. 21–26.

Schulze, E.-D., Beck, E., and Müller-Hohenstein, K. 2005. *Plant Ecology.* Springer, Berlin, pp. 702.

Scotland, R. W., Olmstead, R. G., and Bennett, J. R. 2003. Phylogeny reconstruction: The role of morphology. *Systematic Biology*, 52: 539–548.

Segall, M., Cornette, R., Fabre, A.-C., Godoy-Diana, R., and Herrel, A. 2016. Does aquatic foraging impact head shape evolution in snakes? *Proceedings of the Royal Society, B*, 283: 20161645.

Sherman, P. M. 2003. Effects of land crabs on leaf litter distributions and accumulations in a mainland tropical rain forest. *Biotropica*, 35: 365–374.

Shimer, H. W. 1903. Adaptations to aquatic, arboreal, fossorial and cursorial habits in mammals. *The American Naturalist*, 444: 819–825.

Shine, R., and Wall, M. 2008. Interactions between locomotion, feeding and bodily elongation during the evolution of snakes. *Biological Journal Linnean Society*, 95: 293–304.

Shinohara, A., Campbell, K. L., and Suzuki, H. 2003. Molecular phylogenetic relationships of moles, shrew moles, and desmans from the new and old worlds. *Molecular Phylogenetics and Evolution*, 27: 247–258.

Siegal, D. S., Miralles, A., and Aldridge, R. D. 2011. Controversial snake relationships supported by reproductive anatomy. *Journal of Anatomy*, 218: 342–348.

Simoés, B. F., Sampaio, F. L., Jared, C., Antoniazzi, M. M., Loew, E. R., Bowmake, J. K., Rodriguez, A et al. 2015. Visual system evolution and the nature of the ancestral snake. *Journal of Evolutionary Biology*, 28: 1309–1320.

Simões, T. R. 2012. Redescription of *Tijubina pontei*, an Early Cretaceous lizard (Reptilia; Squamata) from the Crato Formation of Brazil. *Anais da Academia Brasileira de Ciências (2012)*, 84: 1–15.

Simões, T. R., Caldwell, M. W., Bernardi, M., Talanda, M., Palci, A., Vernygora, O., Bernardini, F., Mancini, L., and Nydam, R. L. 2018a. The origin of squamates revealed by a Middle Triassic lizard from the Italian Alps. *Nature*, 557: 706–709.

Simões, T. R., Caldwell, M. W., and Kellner, A. W. A. 2014. A new Early Cretaceous lizard species from Brazil, and the phylogenetic position of the oldest known South American squamates. *Journal of Systematic Palaeontology*, doi:10.1080/14772019.2014.947342

Simões, T. R., Caldwell, M. W., Palci, A., and Nydam, R. L. 2017. Squamate phylogeny and giant taxon-character matrices: Quality of character constructions remains critical regardless of size. *Cladistics*, 33: 198–219.

Simões, T. R., Caldwell, M. W., Palci, A., and Nydam, R. L. 2018b. Giant taxon-character matrices II: A response to Laing et al. (2017). *Cladistics*, doi:10.1111/cla.12231

Sliskovic, T. 1970. Die stratigraphische lage der schichten mit Pachyophiidae aus Seliste bei Bileca (Ostherzegowina). *Bulletin Scientifique, Yougoslavie, Zagreb*, 15: 389–390.

Slowinski, J. B., and Lawson, R. 2002. Snake phylogeny: Evidence from nuclear and mitochondrial genes. *Molecular Phylogenetics and Evolution*, 24: 194–202.

Smit, A. L. 1949. Skedelmorfologie en -kinese van *Typhlops delalandii* (Schlegel). *South African Journal of Science*, 45: 117–140.

Smith-Woodward, A. 1896. On some extinct reptiles from Patagonia of the genera *Miolania*, *Dinilysia*, and *Genyodectes*. *Proceedings of the Zoological Society, London*, 1901: 169–184.

Smith-Woodward, A. 1901. On some extinct reptiles from Patagonia of the genera *Miolania*, *Dinilysia*, and *Genyodectes*. *Proceedings of the Zoological Society, London*, 1901: 169–184.

Sood, M. S. 1946. The anatomy of the vertebral column in Serpentes. *Proceedings of the Indian Academy of Sciences*, B, 28: 1–26.

Streicher, J. W., and Wiens, J. J. 2017. Phylogenomic analyses of more than 4000 nuclear loci resolve the origin of snakes among lizard families. *Biology Letters*, 13: 20170393.

Tchernov, E., Rieppel, O., Zaher, H., Polcyn, M. J., and Jacobs, L. L. 2000. A fossil snake with limbs. *Science*, 287: 2010–2012.

Ten Cate, A. R. 1976. Development of the periodontal membrane and collagen turnover. In Pode, D. F. G., and Stack, M. V. (eds.) *The Eruption and Occlusion of Teeth*. Butterworths, London, pp. 281–289.

Ten Cate, A. R. 1989. *Oral Histology: Development, Structure, and Function*. 3rd ed. The C.V. Mosby Company, St. Louis, 497 pp.

Ten Cate, A. R., and Mills, H. R. 1972. The development of the periodontium: The origin of alveolar bone. *Anatomical Records*, 173: 69–77.

Tomes, C. S. 1875a. On the structure and development of the teeth of Ophidia. *Philosophical Transactions of the Royal Society of London*, 165: 297–302.

Tomes, C. S. 1875b. On the development of the teeth of the newt, frog, slowworm, and green lizard. *Philosophical Transactions of the Royal Society of London*, 165: 285–296.

Tomes, C. S. 1882. *A Manual of Dental Anatomy: Human and Comparative*. Presley Blakiston, Philadelphia.

Tomes, C. S. 1898. *A Manual of Dental Anatomy: Human and Comparative*. 5th ed. J. and A. Churchill, London, 596 pp.

Townsend, T. M., Larson, A., Louis, E., and Macey, J. R. 2004. Molecular phylogenetics of Squamata: The position of snakes, amphisbaenians and dibamids, and the root of the squamate tree. *Systematic Biology*, 53: 735–757.

Triviño, L. N., Albino, A. M., Dozo, M. T., and Williams, J. D. 2018. First natural endocranial cast of a fossil snake (cretaceous of Patagonia, Argentina). *Anatomical Record*, 301: 9–20.

Tsuihiji, T., Kearney, M., and Rieppel, O. 2006. First report of a pectoral girdle muscle in snakes, with comments on the snake neck-trunk boundary. *Copeia*, 2006: 206–215.

Tsuihiji, T., Kearney, M., and Rieppel, O. 2012. Finding the neck–trunk boundary in snakes: Anteroposterior dissociation of myological characteristics in snakes and its implications for their neck and trunk body regionalization. *Journal of Morphology*, 273: 992–1009.

Turner, C., and Peterson, F. 2004. Reconstruction of the Upper Jurassic Morrison Formation extinct ecosystem—A synthesis. *Sedimentary Geology*, 167: 309–355.

Tyson, E. 1682–3. *Vipera caudisoma Americana*, or the anatomy of a rattlesnake, dissected at the repository of the Royal Society in January 1682–3. *Philosophical Transactions of the Royal Society of London*, 13: 561–576.

Uetz, P., Freed, P. & Hošek, J. (eds.) 2018. The Reptile Database, http://www.reptile-database.org

Underwood, G. 1957. *Lanthanotus* and the anguinomorph lizards: A critical review. *Copeia*, 1957: 20–30.

Underwood, G. 1967. *A Contribution to the Classification of Snakes*. British Museum of Natural History, London, pp. 179.

Underwood, G. 1970. The eye. In Gans, C., and Parsons, T. S. (eds.) *Biology of the Reptilia*. Academic Press, London, 2, pp. 1–97.

Underwood, G. 1976. Simplification and degeneration in the course of evolution of squamate reptiles. *Colloques Internationaux C.N.R.S.*, 266: 341–352.

Valença, L. M. M., Neuman, V. H., and Mabesoone, J. M. 2003. An overview on Calloviane Cenomanian intracratonic basins of northeast Brazil: Onshore stratigraphic record of the opening of the southern Atlantic. *Geologica Acta*, 1: 261–275.

Van Erve, A., and Mohr, B. A. R. 1988. Palynological investigations of the Late Jurassic microflora from the vertebrate locality Guimarota coalmine (Leiria, Central Portugal). *Neues Jahrbuch für Geologie und Paläontologie, Monatshefte,* 1988: 246–262.

Vasile, Ş, Csiki-Sava, Z., and Venczel, M. 2013. A new madtsoiid snake from the Upper Cretaceous of the Hateg Basin, western Romania. *Journal of Vertebrate Paleontology,* 33: 1100–1119.

Versluys, J. 1898. Die mittlere und /iussere Ohrsphfire der Lacertilia und Rhynchocephalia. *Zoologischer Jahrbuch (Anat),* 12: 161–406.

Viana, M. S. S., and Neuman, V. H. L. 2002. Membro Crato da Formação Santana, Chapada do Araripe, CE. Riquíssimo registro de fauna e flora do Cretáceo. In Schobbenhas, C., Campos, D. A., Queiroz, E. T., Winge, M., and Born, M. L. C. B. (eds.) *Sítios Geológicos e Paleontológicos do Brasil.* DNPM/CPRM/ SIGEP, Brasília, Brazil, pp. 113–120.

Vidal, N., B, S., and Hedges, S. B. 2002. Higher-level relationships of snakes inferred from four nuclear and mitochondrial genes. *Compte Rendu de Biologie,* 325: 977–985.

Vidal, N., Delmas, A., David, P., Cruaud, C., Couloux, A., and Hedges, S. B. 2007a. The phylogeny and classification of caenophidian snakes inferred from seven nuclear protein-coding genes. *Compte Rendu de Biologie,* 330: 182–187.

Vidal, N., Delmas, A.-S., and Hedges, S. B. 2007b. The higher-level relationships of alethinophidian snakes inferred from seven nuclear and mitochondrial genes. In Henderson, R. W., and Powell, R. (eds.) *Biology of the Boas and Pythons.* Eagle Mountain Publishing, Eagle Mountain, UT, pp. 27–33.

Vidal, N., and Hedges, S. B. 2004. Molecular evidence for a terrestrial origin of snakes. *Proceedings of the Royal Society of London B,* 271: S226–S229.

Vidal, N., and Hedges, S. B. 2009. The molecular evolutionary tree of lizards, snakes, and amphisbaenians. *Comptes Rendu de Biologie,* 332: 129–139.

Villar, D., and Odom, D. T. 2015. When the snake lost its limbs, what did the mouse and lizard say? *Developmental Cell,* 35: 3–4.

Vincent, S. E., Brandley, M. C., Herrel, A., and Alfaro, M. E. 2009. Convergence in trophic morphology and feeding performance among piscivorous natricine snakes. *Journal of Evolutionary Biology,* 22: 1203–1211.

Vincent, S. E., Moon, B. R., Herrel, A., and Kley, N. J. 2007. Are ontogenetic shifts in diet linked to shifts in feeding mechanics? Scaling of the feeding apparatus in the banded watersnake *Nerodia fasciata. The Journal of Experimental Biology,* 210: 2057–2069.

Vullo, R., Neraudeau, D., and Videt, B. 2002. Un facies de type falun dans la Cenomanien basal de Charente-Maritime (France). *Annales de Paléontologie,* 89: 171–189.

Wagler, J. G. 1830. *Naturaliches System der Amphibien, mit vorangehender Classification der Säugethiere un Vögel: ein Beitrag zur vergleichenden Zoologie.* J.G. Cotta'scchen Buchhandlung, Munchen, Germany, Vol. 8, 354 pp.

Wallach, V. 1984. A new name for *Ophiomorphus colberti* Haas, 1980. *Journal of Herpetology,* 18: 329.

Walls, G. L. 1940. Ophthalmological implications for the early history of snakes. *Copeia,* 1940: 1–8.

Walls, G. L. 1942. The vertebrate eye and its adaptive radiation. *Bulletin of the Cranbrook Institute of Science,* 19: 1–785.

Wanninger, A. 2015. Morphology is dead—Long live morphology? Integrating MorphoEvoDevo into molecular EvoDevo and phylogenomics. *Frontiers in Ecology and Evolution,* 3: 54, doi:10.3389/fevo.2015.00054

Wegener, A. von. 1912. Die Entstehung der Kontinente. *Geologische Rundschau,* 3: 276–292.

West, I. 1975. Evaporites and associated sediments of the basal Purbeck Formation (Upper Jurassic) of Dorset. *Proceedings of the Geologists' Association,* 86: 205–225.

Westwood, J. O. 1843. *Arcana Entomologica, or Illustrations of Rare, New or Interesting Insects.* William Smith, London, Vol. II, pp. 1–192.

Wever, E. G. 1978. *The Reptile Ear, Its Structure and Function.* Princeton University Press, Princeton, NJ, pp. 1024.

Wiens, J. J., Hutter, C. R., Mulcahy, D. G., Noonan, B. P., Townsend, T. M., Sites, J. W., and Reeder, T. W. 2012. Resolving the phylogeny of lizards and snakes (Squamata) with extensive sampling of genes and species. *Biology Letters,* 8: 1043–1046.

Wiens, J. J., Kuczinski, C. A., Smith, S. A., Mulcahy, D. G., Sites, J. W., Townsend, T. M., and Reeder, T. W. 2008. Branch lengths, support, and congruence: Testing the phylogenomic approach with 20 nuclear loci in snakes. *Systematic Biology,* 57: 420–431.

Wiens, J. J., Kuczynski, C. A., Townsend, T., Reeder, T. W., Mulcahy, D. G., and Sites, J. W. 2010. Combining phylogenomics and fossils in higher-level squamate reptile phylogeny: Molecular data change the placement of fossil taxa. *Systematic Biology,* 59: 674–688.

Wilcox, T. P., Zwickl, D. J., Heath, T. A., and Hillis, D. M. 2002. Phylogenetic relationships of the dwarf boas and a comparison of Bayesian and bootstrap measures of phylogenetic support. *Molecular Phylogenetics and Evolution,* 25: 361–371.

THE ORIGIN OF SNAKES

Williston, S. W. 1898. Mosasaurs. *Kansas University, Geological Survey*, 4: 81–221.

Williston, S. W. 1904. The relationships and habits of the Mosasaurs. *Journal of Geology*, 12: 43–51.

Wilson, J. A., Mohabey, D. M., Peters, S. E., and Head, J. J. 2010. Predation upon hatchling dinosaurs by a new snake from the Late Cretaceous of India. *PLoS Biology*, 8: e1000322.

Winchester, L., and d'Bellairs, A. 1977. Aspects of vertebral development in lizards and snakes. *Journal of Zoology*, 181: 495–525.

Wollenberg, K. C., and Measey, J. 2009. Why colour in subterranean vertebrates? Exploring the evolution of colour patterns in caecilian amphibians. *Journal of Evolutionary Biology*, 22: 1046–1056.

Woltering, J. M. 2012. From lizard to snake; behind the evolution of an extreme body plan. *Current Genomics*, 13: 289–299.

Woltering, J. M., Vonk, F. J., Müller, H., Bardine, N., Tuduce, I. L., de Bakker, M. A. G., Knöchel, W., Sirbu, I. O., Durston, A. J., and Richardson, M. K. 2009. Axial patterning in snakes and caecilians: Evidence for an alternative interpretation of the Hox code. *Developmental Biology*, 332: 82–89.

Wu, X.-C., Brinkman, D. B., and Russell, A. P. 1996. *Sineoamphisbaena hexatabularis*, an amphisbaenian (Diapsida: Squamata) from the Upper Cretaceous redbeds at Bayan Mandahu (Inner Mongolia, People's Republic of China), and comments on the phylogenetic relationships of the Amphisbaenia. *Canadian Journal of Earth Sciences*, 33: 541–577.

Wyatt, R. J. 1996. A correlation of the Bathonian (Middle Jurassic) succession between Bath and Burford, and its relation to that near Oxford. *Proceedings of the Geologists' Association*, 107: 299–322.

Xing, L., Caldwell, M. W., Chen, R., Nydam, R. L., Palci, A., Simões, T. R., McKellar, R. C. et al. 2018. A Mid-Cretaceous embryonic-to-neonate snake in amber: Ancient snake ontogeny and ecological diversity. *Science Advances*, 4: eaat5402.

Yi, H., and Norell, M. A. 2015. The burrowing origin of modern snakes. *Science Advances*, 1: e1500743.

Young, B. A. 1998. The comparative morphology of the intermandibular connective tissue in snakes (Reptilia: Squamata). *Zoologischer Anzieger*, 237: 59–84.

Zaher, H. 1998. The phylogenetic position of *Pachyrhachis* within snakes (Squamata, Serpentes). *Journal of Vertebrate Paleontology*, 18: 1–3.

Zaher, H., Apesteguía, S., and Scanferla, C. A. 2009. The anatomy of the Upper Cretaceous snake *Najash rionegrina* Apesteguía and Zaher, 2006, and the evolution of limblessness in snakes. *Zoological Journal of the Linnean Society*, 156: 801–826.

Zaher, H., and Rieppel, O. 1999a. The phylogenetic relationships of *Pachyrhachis problematicus*, and the evolution of limblessness in snakes (Lepidosauria, Squamata). *Comptes Rendu de l' Academie des Sciences, Paris*, 329: 831–837.

Zaher, H., and Rieppel, O. 1999b. Tooth implantation and replacement in Squamates, with special reference to mosasaur lizards and snakes. *American Museum Novitates*, 3271: 1–19.

Zaher, H., and Rieppel, O. 2000. A brief history of snakes. *Herpetological Reivew*, 31: 73–76.

Zaher, H., and Rieppel, O. 2002. On the phylogenetic relationships of the Cretaceous snakes with legs, with special reference to *Pachyrhachis problematicus* (Squamata, Serpentes). *Journal of Vertebrate Paleontology*, 22: 104–109.

Zaher, H., and Scanferla, C. A. 2012. The skull of the Upper Cretaceous snake *Dinilysia patagonica* Smith-Woodward, 1901, and its phylogenetic position revisited. *Zoological Journal of the Linnean Society*, 164: 194–238.

Zorn, M., Caldwell, M. W., and Wilson, M. V. H. 2005. Sedimentology of the Devonian MOTH locality: A reformulated taphonomic perspective of a fish lagerstätten. *Canadian Journal of Earth Sciences*, 42: 1–13.

# INDEX

Mesozoic
    aquatic lizards, 196, 228–229
    era, 251, 264–265
    fossil record, 37
    lizards, 246
    Mesozoic-aged exemplar, 255
    record, 39
    snake lizard, 55
Metazoans origin, 261–262
Miadana, 174
Microstomata, 12
Microstomy, 143–146
Middle ear, 83, 108
    osteology of plioplatecarpine mosasaurids, 109
Mid to Late Jurassic, 40–41
Minimum evolution (ME), 244–245
Mitochondrial DNA, 243–244
ML, *see* Maximum likelihood
MNCN, *see* Museo de Ciencias Naturales de Caracas
Modern assemblage, 18, 199–200
Modern iguanas, 10
Modern snake, 2, 9–10, 16–17
Modern snake lizards, 35, 159
    columella and additional elements of, 113–114
Modern snake lizard skull, 18
    bare-bones, 29–34
    bewildering disparity of modern phenotypes, 19–20
    cranial unit, 21–26
    hyobranichal unit, 28–29
    skull basics, 20–21
    skull units, 21
    snout unit, 26–27
    suspensorial–mandibular unit, 27–28
Molecular-based phylogenies, 220
Molecular data, 225–227
Molecular phylogenetic analysis, 244
Molecules, 225, 243–250
*Morelia*, 160
Morphoclinal character evolution, 72
"Mosasauria" sistergroup hypothesis, 255
Mosasaurian, 200
Mosasauridae, 246
Mosasaur middle ear, 112–113
Mosasauroidea, 238–239
Mosasauroids, 15
Mosasauroid sistergroup hypothesis, 212
Mosasaurs, 93–94, 158, 228–229
    ancestry and origins, 211–212
*Mosasaurus hoffmannii*, 117
MP, *see* Maximum parsimony
MPCA, *see* Museo Paleontológico "Carlos Ameghino"
MSNM, *see* Museo di Storia Naturale di Milano
MUCP, *see* Museo Museo de la Universidad del Comahue
*Musculus longissimus capitis*, 159
*Musculus rectus capitis*, 159
Museo de Ciencias Naturales de Caracas (MNCN), 50
Museo di Storia Naturale di Milano (MSNM), 47
Museo Museo de la Universidad del Comahue (MUCP), 55
Museo Paleontológico "Carlos Ameghino" (MPCA), 56

## N

*Najash*, 133, 150, 153, 162, 174, 181–182, 221, 254, 264
    facies analysis and association, 185–188
    fieldwork, 188–191

geological overview, 183
    Kokorkom Desert, 183–184
*Najash rionegrina*, 38, 50–52, 56, 77, 83, 99, 101–103, 106, 128, 148, 161–162, 172, 175, 182–183, 221, 237, 240, 253–254
*Natrix natrix* (Colubrid snake lizard), 23, 109
Neo-Darwinian mechanisms, 194
Neontology, phoenix of, 14–15
Neuquén Group, 54
NHMUK R48388, 41–42
NHMUK R8551, 43
Non-arbitrary boundary points, 194
Non-atlas-axis intercentra, 158
Non-coding sequences, 226
Non-evolutionary approach, 3
Non-macrostamatans, 28, 150
Non-macrostomatan snakes, 140
    lizards, 83
Non-macrostomate, 146–148
Non-parsimony approach, 227
Nonscolecophidians, 20
Non-snake lizards, 74, 177–178, 215; *see also* Snake lizards
    columella and extracolumella of, 111–112
    visual cell system, 207
Non-supernatural phenomena, 10
Non-synapomorphy modeling approach, 227
North American fossil snake, *see Boavus occidentalis*
North America, Upper Cretaceous of, 179–180
*Nqwebasaurus thwazi*, 172
Nucleotide sequences, 225, 248

## O

Obsession, 194
Ocular Backbone, 204, 265
Oligocene, 39
Ontogeny, 132–133
*Ophidia*, 12, 194, 211
Ophidia, 236–237
*Ophidia macrostomata*, 129, 136–137
Ophidii, 11, 136
Ophidiomorpha, 246, 255
*Ophiomorphus*, *see Pachyrhachis*
*Ophiomorphus colberti*, 45, 105
Ophisaurine anguids, 152
*Ophisaurus*, 16, 202–203
*Ophisaurus apodus*, 69, 153
*Ophisaurus gracilis* (legless anguid lizard), 7
Origin myths as opposed to scientific hypotheses
    corrections to twenty years of misrepresentation, 212–216
    discarding burrowing, 200
    evolutionary relationships, 216
    "eye" on birth of myth, 200–205
    "fossorial," 207–210
    inverting phylogeny and revisiting eye, 205–207
    marine, 210–211
    mosasaurs and snake lizard ancestry and origins, 211–212
    observation, 194
    problem with, 197–200
    statements, 195–197
*Origin of Species* (Darwin), 3
Orthology, 225
Osteocementum, 126

# AUTHOR

**Michael Wayne Caldwell** was born and raised in Alberta, where as a small child he developed his love for fossils. He strayed for a time from his passion for science and paleontology, but made his way back after a "Devonian Fossil Epiphany" on Parker's Ridge, Sunwapta Pass, on the Saskatchewan Glacier, with his then two small boys, Garrett and Landon, in the summer of 1987. He was left with no choice but to pursue his second undergraduate degree, a BSc Honors Paleontology, that was completed at the University of Alberta in 1991. He became a doctoral student of Dr. Robert L. Carroll's at McGill University, graduating in November 1995, held a brief postdoctoral position at George Washington University, with Dr. James Clark (January–March 1996), went on the "Snake with Legs World Tour" with Dr. Michael S. Y. Lee (April–May 1996), and returned to the Field Museum, Chicago, (June 1996), where he finished out his postdoctoral fellowship with Dr. Olivier Rieppel. He was employed as a research scientist at the Canadian Museum of Nature in Ottawa, Canada (April 1998 to July 2000) until he accepted a position as an assistant professor at the University of Alberta. He is now full professor and has served as chair of the Department of Biological Sciences for ten years; he remains cross-appointed to both Biological Sciences and the Department of Earth and Atmospheric Sciences, and most recently served as founding president of the Canadian Society of Vertebrate Paleontology (2013–2017).

T - #0891 - 101024 - C328 - 254/178/15 - PB - 9781032177694 - Gloss Lamination